The Ecology of Phycotoxins

Preface

Human hands have changed the seas, yet life finds ways to adapt.

After completing the text for the following chapters on the ecology of phycotoxins, the desire to engage in any sports in a non-pristine aquatic environment—or to consume wild-caught fish or shellfish—is reduced to almost zero. The risk of poisoning seems to be everywhere. One cannot help but wonder: How do marine dinoflagellates, for example, know that we are harmless swimmers and not predators trying to end their existence? They release their highly potent toxins into the water regardless of their intentions. Don't they?

Of course, this is a very subjective—and somewhat fearful—perspective. It is not truly representative of nature and largely overlooks the fact that humans profoundly contribute to the production of these toxins. Humans have altered aquatic ecosystems, overexploited both marine and freshwater resources, and polluted them with liquid and solid waste. And while these ecosystems are no longer as pristine as we would like, they remain vibrant living worlds—home to billions of organisms. Across the oceans and in countless freshwater lakes, a form of large-scale chemical warfare has unfolded between primary producers themselves as well as between these producers and their predators. And humans continue to wonder: What is the original function of these energetically costly toxins?

In the first chapters, we present overviews of cyanotoxins and marine toxins "in action," illustrating the ways these compounds affect organisms, ecosystems, and even human health. We explore their modes of action, ecological consequences, and the often dramatic events associated with harmful algal blooms. In the following chapter, we turn to the origins of phycotoxins, gathering scattered evidence about their ecological functions long before marine algae or cyanobacteria faced significant predation on Earth. We examine hypotheses about their roles in competition, defense, and resource acquisition, aiming to understand why these organisms invest so much energy in producing such potent compounds. Finally, in the last chapter, we present evidence that the predators of toxic algae and cyanobacteria are not completely defenseless; several of them have developed and continue to develop means of dealing with the toxins and even exploiting the energy and nutrient content their prey has invested in these "defense" mechanisms.

By highlighting these interactions, we not only uncover the subtle balance of aquatic ecosystems but also find a measure of optimism: Despite widespread human-induced disturbances, nature retains remarkable resilience, adaptation, and ingenuity.

České Budějovice, Czech Republic
Berlin, Germany
December 2025

Thora Lieke
Christian E. W. Steinberg

Notes from the Authors

- Throughout the books, names and abbreviations of genes are written in lower-case italics and the abbreviations of the corresponding proteins in capital letters.
- Chemical structures were drawn with ACD/ChemSketch, unless otherwise noted.
- Taxonomy of fishes follows "fishbase" and "Eschmeyer's Catalog of Fishes," of invertebrates "Encyclopedia of Life," "World Register of Marine Species," of algae "AlgaeBase," of macrophytes "Plants of the World Online," and in general "NCBI Taxonomy Browser." Additional recent revisions have been used where appropriate—for example, in the giant shrimp genus *Penaeus*. However, there is no guarantee that an obsolete scientific name has not crept into this treatise.
- Papers published in predatory journals have not found their way into this book, if these journals have been identified as such. However, there are also journals still listed in "The Directory of Open Access Journals" that practice incredibly short intervals between submission and acceptance of papers (often less than 10 days) and still pretend to subject submissions to rigorous review. Recent reviews in these critical journals show that the skepticism is justified, as the reviews are often quite incomplete, shallow, and superficial. To address this concern, some journals have begun to publish reviewer comments and author responses. As long as critical journals that do not yet disclose comments and responses are listed in this directory, these papers have not been removed from this book.

Competing Interests The authors have no competing interests to declare that are relevant to the content of this manuscript.

Contents

The World of Phycotoxins: Ecology, Human Impact, and Global Drivers

1

1.1 The Chemical Arsenal of Algae

From the tiniest microscopic algae emerge some of the most powerful natural toxins known to science, capable of influencing entire ecosystems and even human health. Despite their microscopic origins, these compounds wield outsized effects, shaping food webs, altering species interactions, and challenging our ability to manage aquatic environments. Phycotoxins rank among nature's most intriguing and potent molecules. Produced by a variety of photosynthetic microorganisms—including cyanobacteria (colloquially: "blue-green algae"), dinoflagellates, diatoms, and other phytoplankton—they appear in marine, freshwater, and brackish environments worldwide. These compounds exhibit remarkable diversity in their chemical structures, physical properties, and modes of biological action, a testament to their intricate biosynthesis through secondary metabolic pathways Carmichael (1992). Table 1.1 provides a comprehensive overview of some of the most significant phycotoxins, highlighting their sources, toxicity, and ecological impact. These toxins represent key examples of compounds that can profoundly affect aquatic ecosystems, human health, and aquaculture. In the following chapters, we will explore each of these toxins in detail, examining their mechanisms of action, environmental prevalence, and real-world consequences.

The discovery of phycotoxins dates back several decades, initially triggered by observations of massive fish kills and human illnesses linked to contaminated drinking water or seafood. The earliest documented report of phycotoxin poisoning occurred in 1793 in Canada during Captain George Vancouver's expedition along the British Columbia coastline (Kao 1993). Several crew members consumed contaminated mussels harvested near Poison Cove in Nootka Sound, leading to numbness, dizziness, and severe illness, with at least one fatality reported. Indigenous coastal tribes in the region reportedly recognized and avoided shellfish during harmful algal bloom events, suggesting even earlier unrecorded cases. Phycotoxins disrupt critical physiological processes by blocking sodium channels, inhibiting protein phosphatases, or interfering with neurotransmission, resulting in neurotoxic,

T. Lieke, C. E. W. Steinberg, *The Ecology of Phycotoxins*,
https://doi.org/10.1007/978-3-032-18380-4_1

Table 1.1 Overview of most significant phycotoxins, their major modes of action and environmental impact

Toxin	Produced by	Poisoning type	Mode of action	Environmental severity/ecological impact
Microcystins (MCs)	*Microcystis*, *Planktothrix*, *Dolichospermum* spp.	Hepatotoxic (microcystin poisoning)	Inhibit protein phosphatases PP1/PP2A → cytoskeletal collapse, oxidative stress, liver failure	Most globally prevalent freshwater toxin; causes mass fish kills, livestock deaths, and chronic human exposure
Saxitoxins (STX group)	*Alexandrium*, *Gymnodinium*, *Cylindrospermopsis*, *Aphanizomenon* spp.	Paralytic shellfish poisoning (PSP)	Blocks voltage-gated Na^+ channels → paralysis, respiratory arrest	Extremely potent neurotoxins, cause global PSP outbreaks; impacts fisheries and human health
Domoic acid (DA)	*Pseudo-nitzschia* spp. (diatoms)	Amnesic shellfish poisoning (ASP)	Overstimulates glutamate (AMPA/kainate) receptors → neuronal damage	Major marine issue; linked to mass mortality in sea lions, birds, and fish; recurring coastal blooms
Brevetoxins & Ciguatoxins	*Karenia brevis*, *Gambierdiscus* spp.	Neurotoxic shellfish/ciguatera fish poisoning (NSP, CFP)	Bind to Na^+ channels → persistent depolarization	Highly potent and widespread; major cause of marine mammal deaths and human intoxication
Cylindrospermopsin (CYN)	*Cylindrospermopsis*, *Raphidiopsis* spp.	Cytotoxic/hepatotoxic	Inhibits protein synthesis; causes oxidative DNA damage	Expanding in range with climate change; long persistence in water and sediments
Okadaic acid (OA)	*Dinophysis*, *Prorocentrum* spp.	Diarrhetic shellfish poisoning (DSP)	Inhibits PP1/PP2A → hyperphosphorylation of proteins	Common coastal HAB toxin; major economic losses in aquaculture
Spirolides	*Alexandrium*, *Karenia* spp.	Fast-acting toxins (FATs)	Block nicotinic acetylcholine receptors	Highly potent neurotoxins; cause rapid shellfish and fish mortality
β-N-methylamino-L-alanine (BMAA)	Various cyanobacteria	Neurodegenerative (ALS-like)	Misincorporates into proteins → misfolding, oxidative stress	Chronic exposure risk; biomagnifies; debated link to human neurodegeneration
Karlotoxins	*Karlodinium* spp.	Cytotoxic / Ichthyotoxic	Disrupt membrane integrity and ion balance	Localized but severe fish kills in estuaries and lagoons

Aetokthonotoxin (AETX)	*Aetokthonos hydrillicola*	Avian vacuolar Myelinopathy (AVM)	Causes demyelination in brain and spinal cord	Emerging freshwater toxin; devastating for waterfowl and raptors
Anatoxin-a (ATX-a)	*Dolichospermum*, *Phormidium* spp.	Neurotoxic (fast-acting)	Nicotinic acetylcholine receptor agonist → persistent depolarization, paralysis	Extremely fast-acting; lethal to wildlife and livestock, but short-lived in the environment; limited human cases
Nodularin (NOD)	*Nodularia* spp.	Hepatotoxic	Inhibit protein phosphatases PP1/PP2A → cytoskeletal collapse, liver damage	Localized freshwater blooms; potent like microcystin but less widespread; risk to fish and livestock
Prymnesins	*Prymnesium parvum*	Ichthyotoxic	Membrane disruption, hemolysis	Massive freshwater and brackish fish kills; increasing global prevalence in aquaculture and natural ecosystems
Oxylipins	Various marine diatoms	Developmental/sublethal	Interfere with embryonic development in invertebrates	Mostly ecological/sublethal effects; can reduce reproductive success of zooplankton and benthic species; low human relevance

hepatotoxic, or cytotoxic effects. Consequently, they pose significant risks not only to aquatic organisms but also to wildlife and humans exposed to contaminated water or food.

Quantifying the hazard posed by different phycotoxins is inherently difficult and somewhat subjective, as toxicity can vary across species, their physiological state, gender, environmental conditions, bloom years, exposure pathways, and resulting toxic effects. To provide a more systematic assessment, we calculated a Hazard Index (HI) for the major phycotoxins, integrating three key dimensions. Each toxin was scored on: (i) acute and chronic toxicity (based on LD_{50}/LC_{50} values, biochemical potency, and neuro- or hepatotoxic effects), (ii) environmental impact and prevalence (frequency of blooms, persistence in water or sediments, and involvement in fish kills or shellfish poisonings), and (iii) human and aquaculture relevance (reported cases of poisoning, economic losses, or ecological disruption). These three components were weighted as 40% for toxicity, 35% for environmental impact, and 25% for human/aquaculture relevance, and combined into a single index ranging from 1 (low hazard) to 10 (high hazard) (Fig. 1.1).

Yet, these natural compounds are more than threats. Over time, advances in analytical chemistry, molecular biology, and ecotoxicology have revealed a broad spectrum of these toxins and uncovered their intricate interactions within aquatic ecosystems. Beyond direct toxicity, these compounds can have profound indirect effects on aquatic ecosystems by altering food web dynamics, influencing biodiversity, reducing oxygen levels, and shifting species composition—suppressing sensitive organisms while favoring those that are toxin-tolerant. They also impact grazers and predators, triggering cascading effects across multiple trophic levels. Understanding these toxins is essential not only for environmental monitoring,

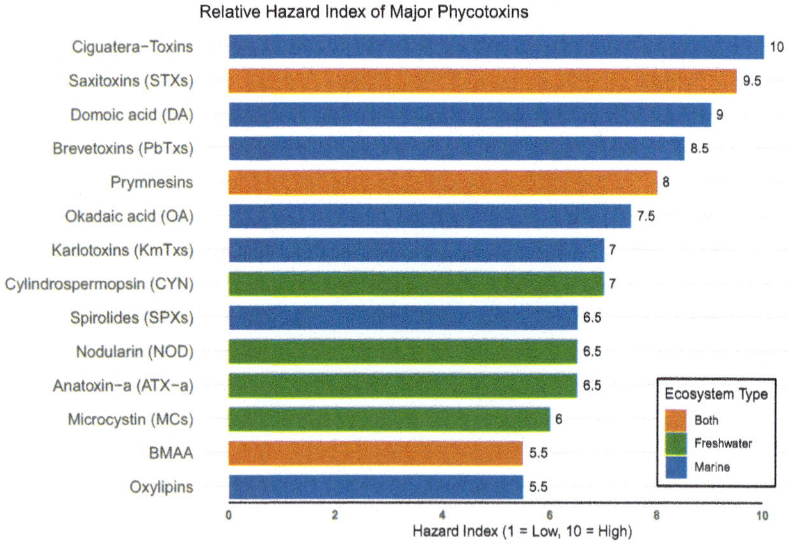

Fig. 1.1 Relative hazard index of major phycotoxins. © T.Lieke

public health, and sustainable aquaculture but also for appreciating the delicate balance of aquatic ecosystems.

1.2 Global Trends Driving Phycotoxin Proliferation

The frequency and intensity of harmful algal blooms (HABs)—excessive algae growth that produces phycotoxins—are rising worldwide, driven by several interconnected global trends. There are five important global factors that drive this increase: (i) climate change with global warming, (ii) eutrophication and nutrient loading, (iii) hydrological changes and water management, (iv) globalization and species translocation, and (v) pollution and chemical stressors.

(i) **Climate change, with global warming**, is undoubtedly among the major factors contributing to the increase in HABs. Rising global temperatures directly influence the growth and distribution of phycotoxin-producing algae by enhancing water column stratification and extending the duration of favorable growing conditions (Gobler 2020; Tester et al. 2020). Warmer temperatures have been shown to favor toxin-producing phytoplankton over non-toxic species, contributing to the global rise in harmful algal blooms (Brandenburg et al. 2019; Griffith and Gobler 2020). Elevated water temperatures accelerate the metabolic and reproductive rates of harmful algal species, providing them with a competitive advantage that promotes bloom formation and persistence. This thermal boost not only increases the frequency and intensity of harmful algal blooms (HABs) but also enables these species to expand into previously unsuitable temperate regions, where they can survive. Harmful algal blooms have been increasingly reported, even in remote regions such as the Arctic Circle, where toxins like STX and DA are emerging as serious threats to coastal communities. The strong dependence of Arctic residents on shellfish as a dietary staple heightens the risk of exposure, while the transfer of these toxins through multiple trophic levels further amplifies their ecological and health impacts (Anderson et al. 2021a, 2022; Lefebvre et al. 2025). Consequently, climate change acts as a key driver, creating broader spatial and temporal windows for toxin production, thereby amplifying the risks of phycotoxin contamination in diverse aquatic ecosystems (Committee 2021).

Another important factor is (ii) **eutrophication and nutrient loading** due to anthropogenic activities, such as agricultural runoff, wastewater discharge, and industrial effluents. These inputs increase the availability of nitrogen and phosphorus in aquatic systems, promoting excessive primary productivity and creating favorable conditions for the growth of algae that produce phycotoxins (Zhang et al. 2022a). Under these nutrient-enriched conditions, harmful species often outcompete non-toxic phytoplankton species, forming dense blooms that can produce high levels of toxins.

(iii) **Hydrological changes and water management** further amplify the conditions favorable for harmful algal blooms. Modifications to hydrological regimes—including reduced river flow and increased water retention in reservoirs—create stagnant and warm aquatic environments that promote HAB formation.

Slower-moving waters concentrate nutrients and maintain stable thermal layers, providing ideal conditions for the growth and persistence of phycotoxin-producing algae.

(iv) **Globalization and Species Translocation**: Human activities such as shipping, aquaculture, and trade have accelerated the movement of harmful algal species into previously uncolonized regions. Ballast water discharge, hull fouling, and deliberate introductions allow phycotoxin-producing algae to establish in new habitats, often encountering favorable thermal and nutrient conditions created by climate change and eutrophication (Bolch and de Salas 2007; Anderson et al. 2012; Hu et al. 2024). This translocation not only expands the geographic reach of toxic blooms but also increases genetic diversity among invasive populations, potentially enhancing bloom resilience and toxicity, and linking human globalization directly to ecological risk.

Lastly, (v) **Pollution and Chemical Stressors**: Aquatic ecosystems exposed to pollutants like heavy metals, pesticides, and organic contaminants, including antibiotics, experience heightened stress that can influence phycotoxin production (Alarcan et al. 2018; Agathokleous et al. 2022; Celano et al. 2025). These chemical stressors can synergistically interact with HABs, increasing toxin synthesis and amplifying ecological and public health impacts and risks. Pollution-driven stress can also weaken competing non-toxic species, further tipping the balance in favor of phycotoxin-producing algae (Anderson et al. 2012). These synergistic interactions may exacerbate harmful effects even when individual contaminants are present at low concentrations, thereby compounding risks to aquatic organisms and ecosystems. When combined with warming, eutrophication, and hydrological changes, chemical pollutants create a multifactorial environment that intensifies both the frequency and severity of toxic algal events.

And as if the drivers are not causing enough trouble on their own, they are highly interconnected, often amplifying the effects of one another. Rising temperatures and altered precipitation patterns from climate change create warmer, more stratified waters, which enhance the growth rates and geographic range of phycotoxin--producing algae. Nutrient enrichment from eutrophication feeds this growth, allowing toxic species to outcompete non-toxic phytoplankton and form dense blooms. Hydrological modifications, such as reduced river flows and stagnant reservoirs, concentrate nutrients and heat, thereby further promoting the persistence of blooms. In addition, human-mediated translocation spreads harmful species to new regions, where they encounter these favorable environmental conditions, increasing bloom frequency and resilience. At the same time, chemical pollutants act synergistically with these factors, stressing ecosystems, enhancing toxin production, and weakening competitors. Together, these drivers create a feedback loop: each factor reinforces the others, resulting in more frequent, widespread, and toxic algal blooms with severe ecological and public health consequences (Fig. 1.2).

Although global patterns in HAB occurrence, toxicity, and ecological or economic risk are still poorly quantified, there is widespread evidence suggesting that these events are becoming more frequent and impactful (Fig. 1.3). Together,

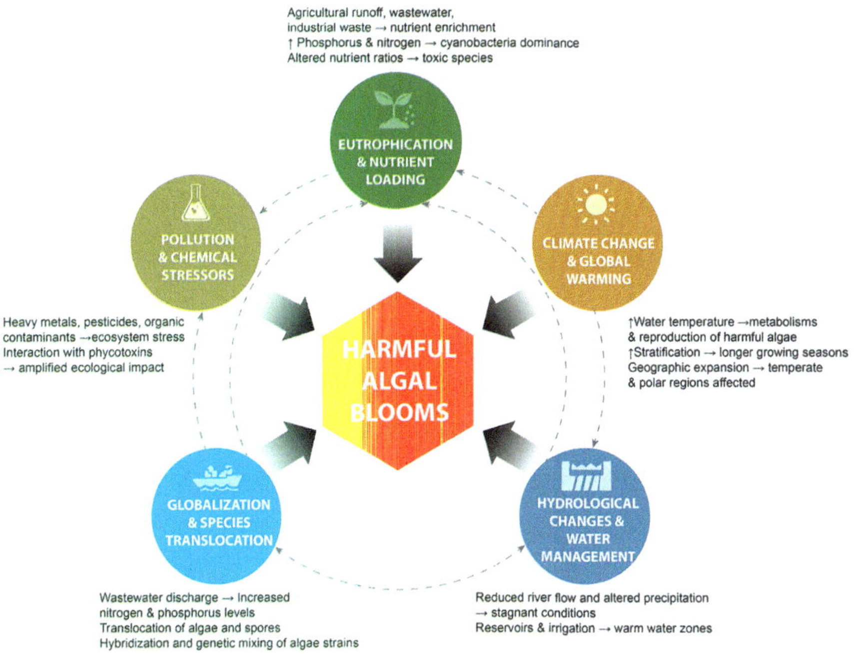

Fig. 1.2 Overview of environmental factors supporting HABs © T.Lieke

eutrophication and global warming are expanding the geographic range of HABs into new regions, threatening freshwater reservoirs, coastal zones, and open oceans.

Based on a statistical analysis of a global dataset extracted from the Harmful Algae Event Database and the Ocean Biodiversity Information System (OBIS) for the period 1985–2018, Hallegraeff et al. (2021a) suggested that the apparent global increase in harmful algal events may partly reflect intensified monitoring efforts and the expansion of aquaculture rather than a uniform rise in bloom occurrence. They emphasized that trends should be interpreted within regional and species-specific contexts (Pitcher and Louw 2021). Experts broadly agree that, as global climate change advances through both natural and human-driven forces, HABs will likely expand their range and intensify in many regions worldwide (Griffith and Gobler 2020; Trainer et al. 2020). Recent evidence confirms that coastal phytoplankton blooms are indeed expanding and intensifying in the twenty-first century (Dai et al. 2023). Global mapping has revealed an increase in marine and lacustrine algal blooms over the past decades (Anderson et al. 2012; Hou et al. 2022). Figure 1.4 illustrates recorded algal toxin–poisoning events from 1970–1980 (blue) and 1981–1991 (yellow) in panel A, as well as the documented events from 2015–2025 in panel B (data from the HAIS database), clearly demonstrating the sharp increase in reported occurrences over time.

Selected HAB case studies illustrate increases in amnesic shellfish toxins and paralytic shellfish toxins (PST) in marine mammals in the Arctic Pacific (Lefebvre

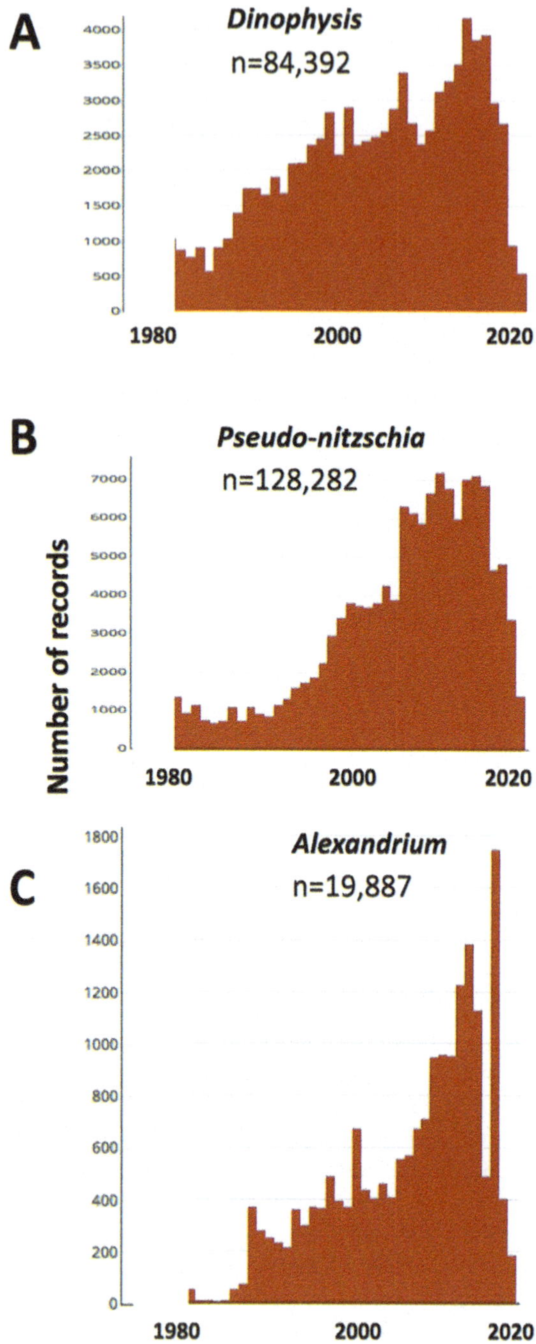

Fig. 1.3 Increases between 1985 and 2018 of global observations of the causative organisms: Total number of global observations from Ocean Biodiversity Information System (OBIS) of causative microalgal organisms of *Dinophysis* spp., *Pseudo-nitzschia* spp., and *Alexandrium* spp. (from Hallegraeff et al. (2021a), credit to Springer Nature)

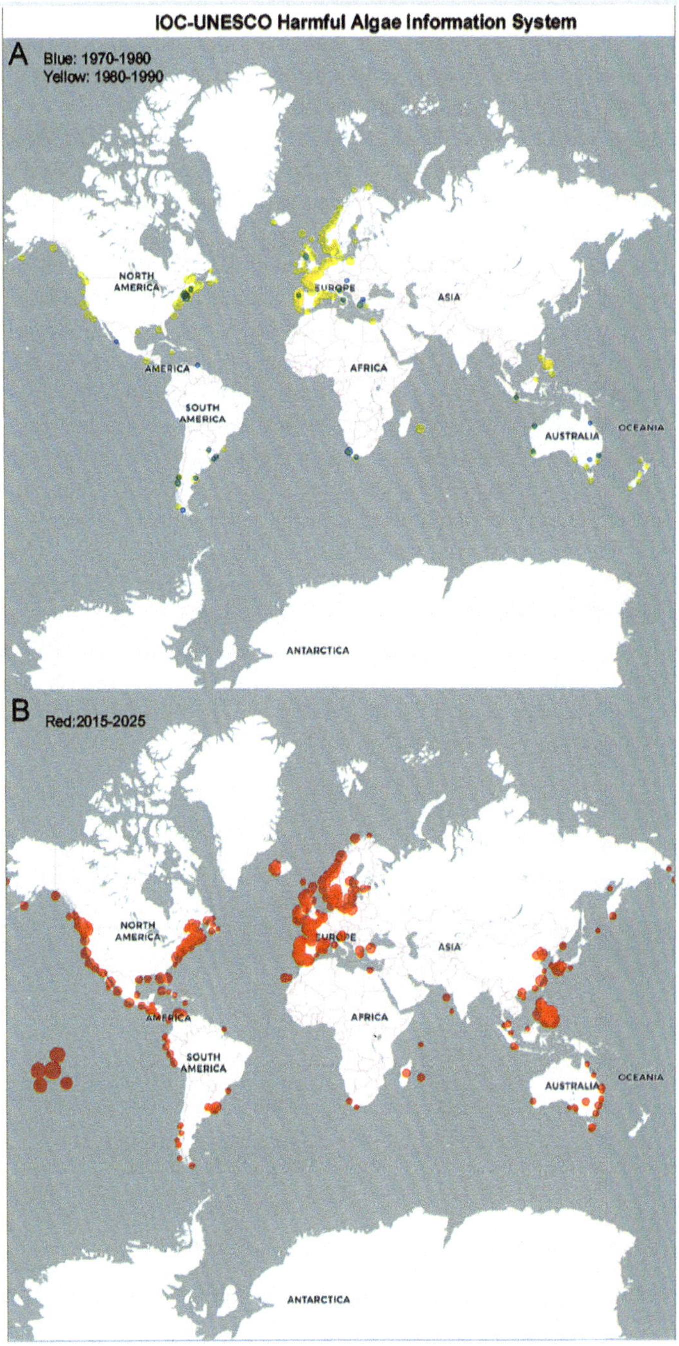

Fig. 1.4 Geographical distribution of HABs, along with the increase in intensity and awareness from 1970 to 2025. (Source: IOC-UNESCO https://data.hais.ioc-unesco.org/ ©T. Lieke)

et al. 2016), ciguatera poisoning (CP) in the Canary Islands (Chinain et al. 2021), and green (do Rosário Gomes et al. 2014) and red *Noctiluca* surface water discoloration (Hallegraeff et al. 2021b) in the Indian Ocean and southeast Australia, respectively; benthic *Ostreopsis* mats and aerosol poisoning since the 2000s in the Mediterranean (Zingone et al. 2021), amnesic shellfish toxins on the northeast and west coasts of the USA (McKibben et al. 2017; Clark et al. 2019; Anderson et al. 2021b), and *Pseudochattonella* blooms causing fish mortality in northern Europe and Chile (Trainer et al. 2020; Karlson et al. 2021).[1]

With respect to harmful cyanobacterial blooms (cyanoHABs) in freshwaters, global warming over the past six decades (IPCC 2021) has increased their occurrence and exposure to their toxic effects (Paerl and Otten 2013a, b; Trombetta et al. 2019). One example: Ancient DNA in the shallow, hyper-eutrophic Lake Taihu, China, reveals the increase of potentially toxic cyanobacteria with climate change (Zhang et al. 2023). A steady increase in *Microcystis* abundance over the past 30 years was correlated with increasing temperatures and decreasing wind speeds, but not with temporal trends in lake water nutrient concentrations, highlighting recent climate impacts on potentially increasing toxin-producing taxa.

However, the perception of inland HABs, as shown in Fig. 1.5 and described by Ho et al. (2019), also raises concerns. There is an imbalance in the geographical coverage of HAB studies. Most studies have focused on lakes in high-income countries, with approximately 65% of the literature focusing on just 20 inland lakes. Remotely sensed data have provided insights into HAB events in over 20,000 lakes worldwide, indicating a clear need for widespread in situ studies of HABs. HAB

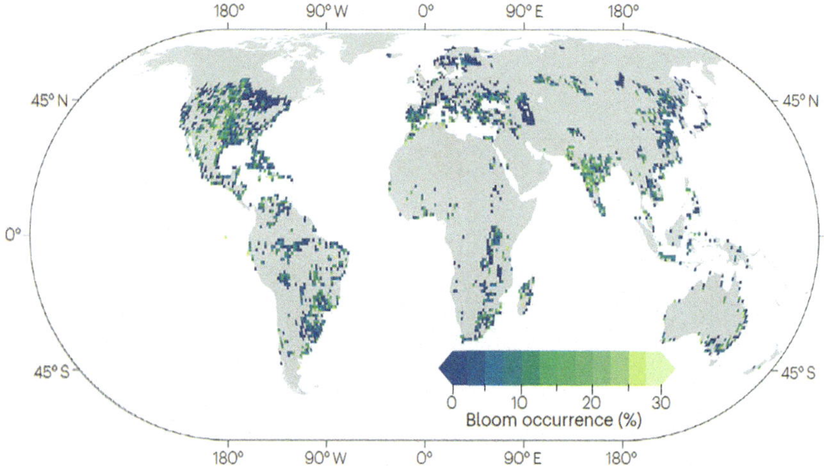

Fig. 1.5 Global occurrence patterns of harmful lacustrine algal blooms between 1982 and 2019 aggregated into 1° × 1° grid cells and expressed as a percentage of total observational bloom number over the period (from Feng et al. (2024), with permission from Springer Nature)

[1] From Hallegraeff et al. (2021a), credit to Springer Nature, appropriately referenced

events are known to be prevalent in a number of African countries and several countries in Asia (including Indonesia, Pakistan, and the Philippines) and South America (such as Chile and Colombia). However, there is limited research on HABs in these regions (Feng et al. 2024).

1.3 HABs and Their Impacts on Nature, Society, and the Economy

In addition to increased observational efforts, the widespread perception that HABs are increasing is driven by the expanding scale and nature of their societal impacts. Socially and economically, HABs pose major challenges to fisheries and aquaculture—through fish losses, stock contamination, and reduced employment opportunities—as well as to recreation and tourism, where access to affected coastal areas is often restricted. Additional expenses arise from the need for continuous monitoring, management, and prevention measures, as well as from the removal of dead fish and accumulated algal biomass. Public health costs further add to the burden, driven by the need for medical treatments—both physical and mental—and the potential long-term effects of toxin exposure. According to the U.S. National Office for Harmful Algal Blooms (https://hab.whoi.edu/), the average annual economic cost of HABs nationwide is conservatively estimated at approximately $50 million, with public health impacts accounting for nearly $20 million (42%), followed by commercial fisheries ($18 million), recreation and tourism ($seven million), and monitoring and management ($two million). These figures are highly uncertain due to incomplete data and the difficulty of assigning monetary values to many HAB impacts. Individual events, however, can cause severe local and regional losses: for example, a paralytic shellfish poisoning (PSP) outbreak in New England resulted in $12–20 million in losses in Massachusetts alone, ongoing PSP intoxication has limited Alaskan commercial shellfish development ($six million annually), brown tide blooms devastated Long Island's bay scallop industry ($two million), and Pfiesteria--like outbreaks in Chesapeake Bay in 1997 caused approximately $43 million in losses to seafood-related businesses. Especially under warm conditions, when HABs can reach very high densities, they can even affect the taste and odor of drinking water and pose health risks to humans if cyanotoxins are not fully removed during water treatment. These examples highlight the substantial and sometimes dramatic economic consequences of HABs.

Harmful algal blooms have severe impacts on ecosystems and the environment. In recent years, rising environmental awareness and improved monitoring have drawn greater attention to these events, highlighting their frequency and severity and underscoring the urgent need to protect the health of aquatic ecosystems. HABs disrupt aquatic ecosystems by altering food web dynamics, reducing biodiversity, and causing oxygen depletion, which can lead to widespread fish kills. Dense blooms reduce light penetration and limit the growth of submerged vegetation, thereby altering primary productivity and the quality of habitats. These ecological disturbances are further intensified by climate-driven stressors such as warming waters, altered stratification, and changes in hydrology, creating feedback loops that can increase the frequency, persistence, and severity of blooms. Phycotoxins produced during blooms accumulate across multiple trophic levels, affecting both pelagic and benthic organisms. Recent large-scale ecological disasters illustrate the severity of these impacts: massive fish mortality events in the River Oder in 2022 (Mora et al. 2024; Escher et al. 2024) and in the Red Sea in 2023 (Tietbohl et al. 2025); unprecedented seabird die-offs along the west coast of North America during 2015–2016 (Piatt et al. 2020); and extensive mass mortality of California sea lions and dolphins in 2024 along the California coast and in the southeast Bering Sea (NOAA-Fisheries 2024). Cyanobacterial blooms have triggered some of the most shocking and large-scale wildlife die-offs in recent history. In 1999, the mysterious deaths of approximately 30,000 lesser flamingos (*Phoenicopterus minor*) were linked to cyanobacterial blooms in East African soda lakes (Krienitz et al. 2003; Krienitz and Schagerl 2016). More recently, bald eagle mortality in Arkansas, USA, was attributed to vesicular myelin disease caused by the epiphytic cyanobacterium *Aetokthonos hydrillicola* (Breinlinger et al. 2021). The death of 356 African savannah elephants in Botswana in 2020 was also linked to the ingestion of microcystins (MCs) at watering holes (Wang et al. 2021). Clearly, MCs pose a widespread threat with various modes of action, as outlined in Fig. 1.6, with a useful, strong human protection aspect.

These events highlight the far-reaching and often cascading effects of HABs on aquatic and coastal ecosystems. Given the breadth and significance of this issue, we explore the ecological consequences of HABs in detail in Chap. 4.

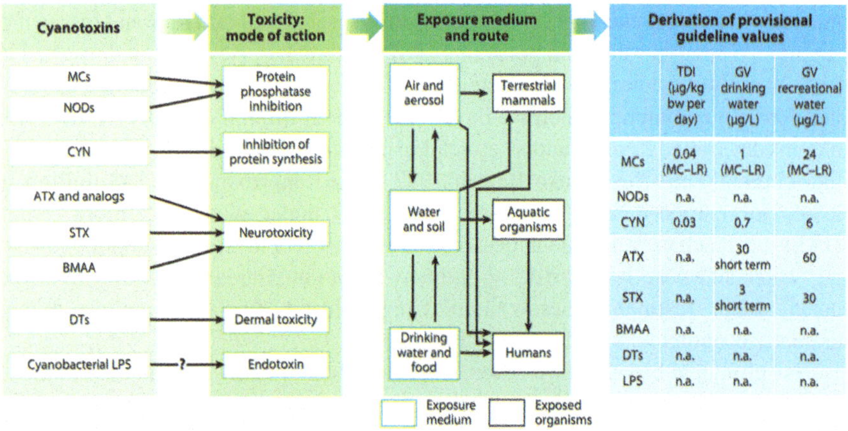

Fig. 1.6 The aggregated exposure pathways of cyanotoxins. The provisional guideline values (GVs) for MCs, NODs, CYN, ATX, and STX are presented in accordance with the current WHO GVs. Note that these GVs are provisional due to limitations in the toxicological data, including the availability of data for only one congener. *ATX* anatoxins, *BMAA* β-methylamino-L-alanine, *CYN* cylindrospermopsin, *DT* dermatoxin, *LPS* lipopolysaccharide, *MC* microcystin, *MC-LR* microcystin-leucine-arginine, *NOD* nodularin, *STX* saxitoxin, *T&O* taste and odor, *TDI* tolerable daily intake, *WHO* World Health Organization (from Song et al. (2023), credit to Annual Reviews)

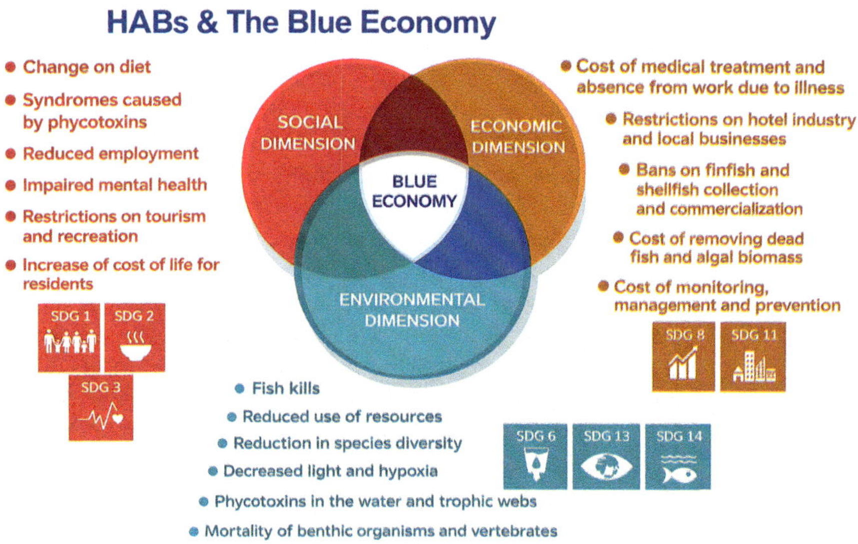

Fig. 1.7 Impacts of Harmful Algal Blooms (HABs) on the Blue Economy Across Social, Economic, and Environmental Dimensions (Alves de Souza et al. (2022), with permission from Springer Nature)

The successful management and mitigation of HAB threats—and, where possible, the prevention of their occurrence—are fundamental to achieving several indicators of the Sustainable Development Goals (SDGs) outlined in the UN Agenda 2030. Closely linked to the objectives of the 2021–2030 United Nations Decade of Ocean Science for Sustainable Development and the principles of the blue economy (Lee et al. 2020), these include food security (SDG 2), human health (SDG 3), clean water and sanitation (SDG 6), sustainable communities (SDG 11), climate action (SDG 13), and the conservation of life below water (SDG 14). For a more detailed discussion, readers are referred to Chap. 10, *Harmful Algae in the Blue Economy* (Alves de Souza et al. 2022), which gives a global and local overview of the disruptive impacts of HABs on the blue economy (Fig. 1.7).

Taken together, the environmental, social, and economic consequences of HABs underscore the need for integrated, cross-sectoral strategies that support both ecosystem resilience and sustainable human use of aquatic resources.

1.4 Human Activities as Drivers of HABs

While HABs clearly impose significant economic, social, and ecological costs, it is essential to acknowledge that human economic activities also impact the frequency, intensity, and distribution of these blooms. As mentioned above, climate change and eutrophication are among the most significant anthropogenic drivers influencing the frequency, intensity, and distribution of HABs, and both are severely driven by human activities. The burning of fossil fuels, industrial emissions, deforestation, thawing of permafrost soils, and intensive agriculture have dramatically increased atmospheric concentrations of greenhouse gases, particularly carbon dioxide (CO_2), methane (CH_4), and nitrous oxide (N_2O). These gases trap heat in the atmosphere, leading to global warming, ocean heating, and widespread alterations in weather and hydrological cycles and support growth and toxin production of HAB (Hattenrath-Lehmann et al. 2015).

Rising water temperatures accelerate algal metabolism and reproduction, extending the growing season and geographical range of many bloom-forming and toxin--producing species (Brandenburg et al. 2019; Gobler 2020; Anderson et al. 2021b). Additionally, warmer conditions favor stratification—the layering of water columns that limits vertical mixing—creating stable, nutrient-rich surface waters where harmful algae can thrive (Steinberg and Hartmann 1988). Xu et al. (2023) demonstrated that warming by +4 °C increased DA concentrations in the diatom *Pseudo--nitzschia multiseries* up to 11-fold. The results indicate that *P. multiseries* can rapidly acclimate to temperature fluctuations and may gradually adapt to sustained warming, suggesting enhanced bloom potential and toxicity in future ocean conditions.

The story of climate change does not end in the atmosphere—the excess CO_2 we release also dissolves into our waters, slowly shifting their chemistry and transforming marine and freshwater worlds. Approximately one-quarter of anthropogenic CO_2 emissions are absorbed by the oceans (De Vries et al. 2023; Gruber et al. 2019, 2023), where CO_2 reacts with seawater to form carbonic acid, thereby lowering the pH and altering the carbonate chemistry. This acidification can have multiple, complex effects on HABs. Changes in pH influence nutrient availability, trace metal solubility, and the physiological performance of phytoplankton. Lower water pH can promote the proliferation and toxicity of various HAB species, including red tide algae like *Alexandrium fundyense* and other toxic algae such as *Vicicitus globosus* (Hattenrath-Lehmann et al. 2015; Riebesell et al. 2018). In a field experiment near the Canary Islands, elevated CO_2 levels above 600 ppm resulted in large blooms of *V. globosus*, which adversely impacted plankton communities and marine food webs. In contrast, acidification can stress competitors, such as calcifying organisms, thereby reducing grazing pressure on HAB species.

Climate change also alters precipitation and storm patterns, intensifying both droughts and extreme rainfall events. Prolonged droughts concentrate nutrients in water bodies, while heavy rains and flooding increase terrestrial runoff, transporting fertilizers, (dissolved) organic matter (DOM), and pollutants from agricultural lands and urban areas into rivers, lakes, and coastal zones, leading to eutrophication (Lieke et al. 2025; Zhao et al. 2025). Certain phytoplankton, including those forming HABs and mixotrophic species, can utilize dissolved organic nitrogen (DON) and dissolved organic phosphorus (DOP) as nutrient sources when inorganic nutrients are limited, aiding their persistence and proliferation under nutrient scarcity (Reinl et al. 2022). For example, the dinoflagellate *Prorocentrum shikokuense* (previously *P. donghaiense)* exhibits similar growth rates on both inorganic phosphate and organic phosphorus sources, such as ATP, illustrating its ability to utilize DOP under phosphorus limitation (Lu et al. 2022). Studies in estuarine and freshwater systems have shown that certain phytoplankton genera, such as *Cyclotella*—which are not typically considered harmful but can cause unpleasant odors and affect drinking water quality—are capable of utilizing organic phosphorus compounds as a nutrient source. This ability enables them to thrive in ecosystems where the recycling of organic nutrients is high, highlighting the crucial role of dissolved organic matter in supporting phytoplankton growth, even under low inorganic nutrient conditions (Liu et al. 2023). Additionally, many HAB species prefer reduced forms of nitrogen (e.g., ammonium) and can subsist on organic nitrogen sources such as urea and amino acids, which often dominate nutrient pools in eutrophic, stratified waters (Papacek et al. 2024). This ability effectively gives them a competitive advantage over other species in low-inorganic-nutrient environments, allowing them to bloom even after pulses of inorganic nutrients are depleted (Moschonas et al. 2017).

The increased use of coastal waters for aquaculture has been a key driver of occasionally disastrous, long-lasting economic impacts from HABs, leading to an increased awareness of new harmful algal species and toxin types. Between 2010 and 2020, aquaculture production increased 16-fold, with the most significant increases in East Asia, Latin America, and the Caribbean, and stabilization in North America and Europe (FAO 2020). This situation is illustrated in Fig. 1.8. Aquaculture sites and eutrophication hotspots exhibit a high degree of overlap, suggesting that the relationship is likely causal: aquaculture is one of the leading causes of seawater eutrophication.

Industrial activities contribute to HAB development by simultaneously altering water temperature and nutrient availability. Many industries, such as power plants and manufacturing facilities, discharge heated water from their cooling processes into nearby rivers, lakes, or coastal zones, raising local temperatures and promoting

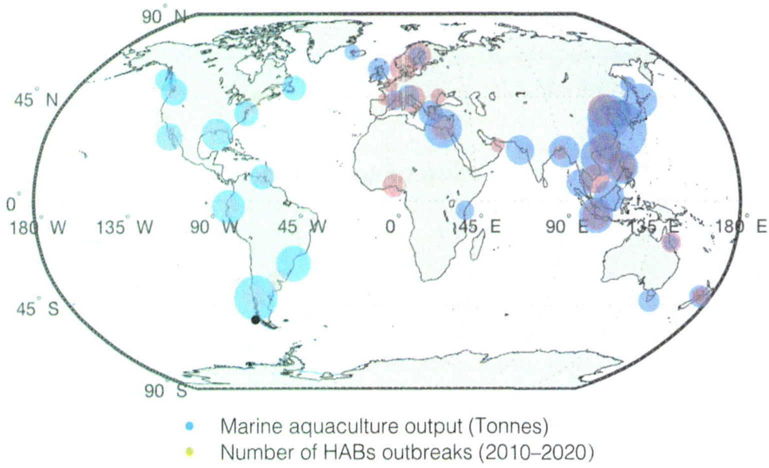

Fig. 1.8 The distribution of major global aquaculture sites and eutrophication hotspots in the last decade (from Lan et al. (2024), credit to MDPI, Basel)

thermal stratification that favors the persistence of blooms. At the same time, industrial effluents from food processing, fertilizer production, and chemical manufacturing often contain nitrogen, phosphorus, and organic compounds that enrich aquatic systems and stimulate algal growth. When combined, these thermal and nutrient inputs create warm, nutrient-rich, and stratified environments ideally suited for HAB proliferation, particularly in industrialized estuaries and coastal regions where cumulative discharges amplify ecological stress.

The topic of hormesis,[2] which will be explored in greater detail below with regard to animals and plants, is also relevant to harmful algae themselves, such as cyanobacteria. For instance, Agathokleous et al. (2022) reported that low levels of contaminants stimulate the growth of harmful algae and accumulate their toxins, as demonstrated with *Microcystis* sp., the most notorious cyanobacteria in eutrophic freshwater systems. Although non-target organisms, cyanobacteria are highly sensitive to antibiotics due to their bacterial-like structure (Chen and Guo 2012; Zhang et al. 2022b). Work by Jiang et al. (2020a, b, c, 2021) concludes that pollution is associated with the intensification of algal blooms. Contaminant concentrations well below the toxicity threshold promote harmful colonization, increase cyanotoxin production and release into the environment (Fig. 1.9), and reduce the

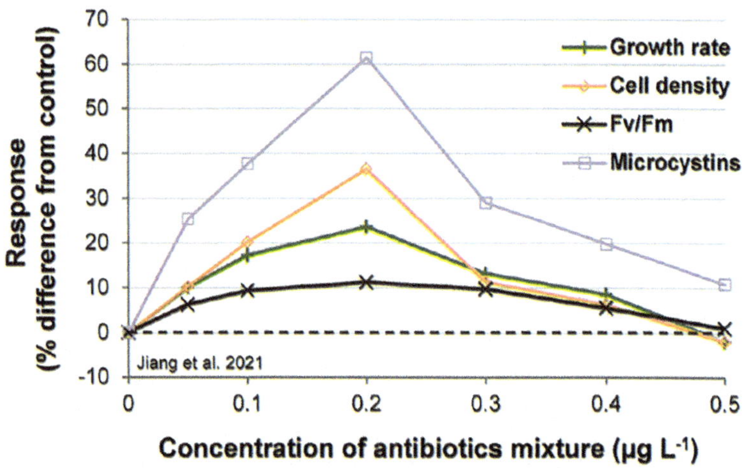

Fig. 1.9 Response of *Microcystis aeruginosa* to low, environmentally realistic concentrations of antibiotic mixtures. At these trace concentrations included in the study, there were no adverse effects on these studied endpoints; however, the concentration-response relationship suggests that adverse effects would be expected at antibiotic concentrations greater than 0.5 μg L^{-1}. The antibiotics were amoxicillin, ciprofloxacin, spiramycin, sulfamethoxazole, and tetracycline, and the exposure lasted 14 days under aseptic conditions in a constant-temperature illuminating incubator. F_v/F_m: maximum photochemical quantum yield of photosystem II (PSII). Microcystins refer to the total concentrations (μg mL^{-1}). The data are based on Jiang et al. (2021); data extraction and calculation by Agathokleous et al. (2022), figure taken with permission from the American Chemical Society

[2] From ancient Greek *hormáein* 'to excite' (Calabrese 2014)

effectiveness of algicides in controlling algal blooms. Given that colonization and phycotoxin production and release are enhanced by contaminant concentrations below the toxicological threshold and are widespread in the environment, the impact of contaminants on harmful algal blooms is more widespread than previously thought. More details on this issue can be found in Steinberg (2025)'s review and in a recent paper by Celano et al. (2025). Climate change and nutrient enrichment, known mechanisms underlying algal blooms, are thus joined by low-level contaminants as an additional causal mechanism.

These are only a few examples of how human activities play a central role in driving HABs by simultaneously altering the physical, chemical, and biological conditions of aquatic ecosystems. Human activities and HABs are linked in a complex, bidirectional feedback loop. Anthropogenic CO_2 emissions drive ocean warming and acidification, while nutrient enrichment from agriculture, aquaculture, wastewater, and urban runoff generates eutrophication hotspots, providing the biological fuel for blooms. Aquaculture, in particular, can both contribute to and suffer from blooms: nutrient-rich effluents from fish and shellfish farms stimulate HAB growth, while the blooms themselves compromise farmed species through toxin accumulation, oxygen depletion, or even anoxic "dead zones," stressing aquatic organisms and increasing mortality. These pressures often lead to intensified aquaculture efforts—higher stocking densities, additional feed, and expansion into new areas—which can further increase nutrient loading and create conditions for even more frequent or severe blooms. Similarly, coastal infrastructure, shipping, and other industrial activities modify hydrodynamics and water quality, amplifying bloom development. The ecological, social, and economic consequences are profound: fisheries and aquaculture face harvest closures, production losses, and product contamination, while monitoring and mitigation measures create substantial financial and management burdens (Fig. 1.10). This self-reinforcing cycle illustrates that HABs are not just a symptom of human activity but are actively shaped by it. The challenge is compounded by the complex ecology of blooms, the variability of toxin production, and the difficulties in predicting the timing and intensity of events. Consequently, effective monitoring and mitigation strategies are essential but remain difficult to implement consistently. Understanding the links between anthropogenic pressures, bloom proliferation, and toxin impacts is therefore critical not only for safeguarding human and ecosystem health but also for sustaining the economic viability of coastal fisheries and aquaculture operations.

1.5 How Animals Deal with Algae-Born Xenobiotics

Marine and freshwater herbivores seem to have lost the chemical war against phycotoxins—for now; not least because humans have often unwittingly contributed to it. Aquatic animals are only partially prepared to cope with the toxins in their environment and especially those found in their food. But are they completely defenseless? Certainly not! They possess diverse innate and trained or adapted defensive techniques, more or less adapted to the particular situation. However, when the

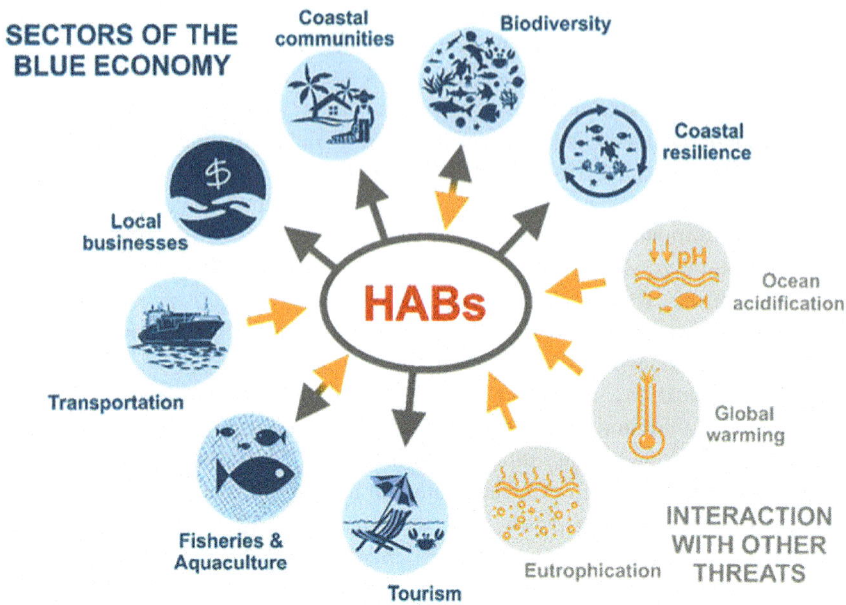

Fig. 1.10 Harmful algal blooms (HABs) interact with various sectors of the blue economy as well as other environmental stressors. Grey arrows represent sectors that are negatively impacted by HABs, while orange arrows indicate factors that contribute to or promote bloom formation (from Alves de Souza et al. (2022) with permission from Springer Nature)

natural environment changes too rapidly, leading to inappropriate and even toxic exposures, these animals have a serious problem and lose, at least temporarily, the chemical warfare race. We are going to witness this race and try to draw some conclusions for applied aquatic sciences, such as aquatic animal farming.

Despite the wildlife and human casualties, HABs, whether in oceans or freshwaters, are giant chemo-ecological experiments between chemically armed primary producers and their competitors and predators of different trophic levels. Sure, most of the poor creatures are currently losing the chemical weapons race, but are they completely defenseless? If they are not completely defenseless, can we learn anything from the animals' response? And if so, what? Studies on this subject are not very frequent because many researchers only find what they are looking for, and when they hear the term "toxin," they apply such high exposure doses that the animals have no choice but to die.

Based on considerations made a quarter of a century ago, Schreck (2010) confirmed that stressors, including toxic exposures, of different severity, duration, and frequency, can lead to different response patterns at the organism level. The physical fitness of fish can be enhanced by low levels of stress, whereas it can be reduced by high levels or prolonged durations of stress. This biphasic dose-response phenomenon is called "hormesis." It has gradually become a key concept in the biological sciences (Calabrese 2014) and plays a crucial role in environmental risk assessment

(Agathokleous et al. 2018). However, it is very reluctantly used when combined with phycotoxins. To our knowledge, only a few papers have described this dose-response phenomenon for cyanotoxins and none for marine phycotoxins, often without using the term "hormesis" itself. Most studies do not adequately cover the low dose range, if at all. This means that these studies intentionally overlook or ignore a significant part of the physiological and biomolecular response of exposed organisms. The authors behind many of these studies are not seeking to uncover as much as possible about their subject. A few striking examples involving cyanotoxins and animals, as well as plants (to underscore the universality of these mechanisms), reveal the wealth of knowledge that is often overlooked when exposure doses are set too high, focusing solely on toxicity rather than the full spectrum of biological responses.

Example 1.1 In rainbow trout, Rymuszka et al. (2007) found that lymphocyte proliferation increased significantly when exposed to low concentrations of MC-LR, but decreased significantly when exposed to high concentrations. Similarly, low doses of lyophilized cyanobacteria powder increased lysozyme activity in crucian carp, while high doses led to results with opposite trends (Qiao et al. 2013). Very few studies comprehensively explain such biphasic consequences; thus, future studies should explore the underlying mechanisms.

Example 1.2 Lin et al. (2017) observed a hormetic response in male zebrafish exposed to MC-LR. Specifically, subchronic exposure to environmentally relevant concentrations of MC-LR has dualistic effects on the innate immune system of fish, with inflammatory activation at low concentrations (0.3, 1.0, and 3.0 $\mu g\ L^{-1}$), but immune inhibition at high concentrations (10 and 30 $\mu g\ L^{-1}$). Details of the underlying mechanism are shown in Fig. 1.11.

Example 1.3 Zhao et al. (2024) observed a similar hormetic response to the major toxic component of *M. aeruginosa* exudates, phytosphingosine. In their study on the endemic golden-line barbel (*Sinocyclocheilus grahami*), low concentrations of the toxin stimulated immune activity, while higher concentrations led to immunosuppression, with the greatest number of genes expressed at lower exposure levels.

Our fourth example demonstrates the astonishment of the authors when a toxin, at low doses, does not induce toxic effects: Guzmán-Guillén et al. (2017) investigated cylindrospermopsin effects on carrots (*Daucus carota*) grown in soil and irrigated with a *Chrysosporum* (*Aphanizomenon*) *ovalisporum* extract containing environmentally relevant concentrations of cylindrospermopsin (10 and 50 $\mu g\ L^{-1}$). "The results show that exposure to cylindrospermopsin had no clear negative effect on the growth or photosynthesis of *D. carota*, and even led to an increase in both parameters."

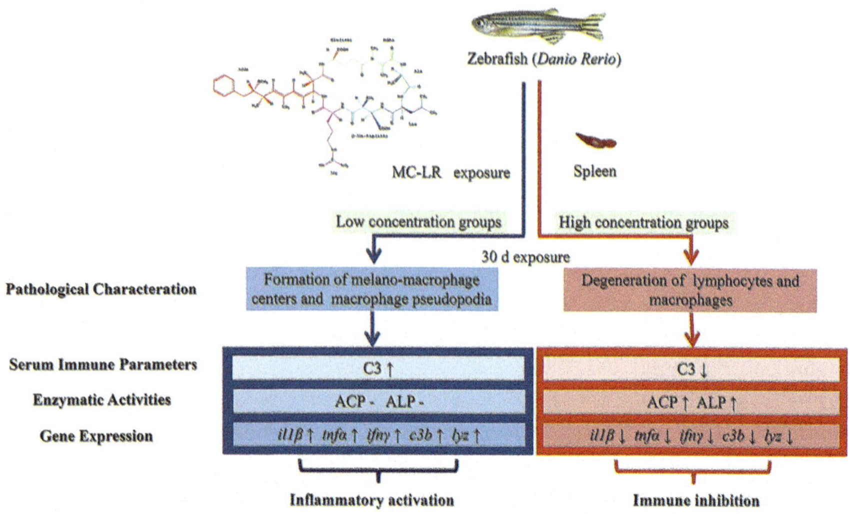

Fig. 1.11 Hormetic immunomodulation of sub-chronic microcystin-LR exposure on the innate--immune defense system in male zebrafish (from Lin, Hou et al. (2017), with permission from Elsevier). *C3* serum complement C3, *ACP* acid phosphatase, *ALP* alkaline phosphatase, *il1β interleukin 1β*, a pro-inflammatory cytokine, *tnfα tumor necrosis factor α*, a pro-inflammatory cytokine, *ifnγ interferon-γ*, *c3b complement c3*, *lyz lysozym*

Example 1.2 raises some general, future-oriented questions. How do the beneficial effects of low-dose MC exposure translate to other life-history traits? If there are other improved traits, are they passed on to subsequent generations? If so, what are the underlying mechanisms? Are they epigenetic mechanisms and the basis of adaptive responses? What is the function of the gut microbiota? And: Can these mechanisms, alone or in combination, explain the fact that, for example, marine copepods from regions where HABs are more common are more tolerant to algal xenobiotics than those from regions where HABs are less common or intense?

In the following chapters, we will explore:

- *Chapter 2: Cyanotoxins in Action*—Unveiling the impacts of freshwater toxins on ecosystems, food webs, and human health.
- *Chapter 3: Marine Phycotoxins in Action*—A deep dive into the ocean's most notorious chemical weapons and their ecological consequences.
- *Chapter 4: Ecological Significance*—Understanding how harmful algal blooms shape aquatic communities and ecosystem function.
- *Chapter 5: Tolerance Modes*—How aquatic animals defend themselves in the ongoing chemical arms race against phycotoxins. We will place a special emphasis on epigenetic mechanisms, such as DNA methylation and histone modification.

Join us on a fascinating journey to uncover how aquatic herbivores are closing the gap in the chemical arms race against phycotoxin producers—revealing remarkable and unexpected mechanisms of adaptive response.

References

Agathokleous E, Kitao M, Calabrese EJ (2018) Environmental hormesis and its fundamental biological basis: rewriting the history of toxicology. Environ Res 165:274–278. https://doi.org/10.1016/j.envres.2018.04.034

Agathokleous E, Peñuelas J, Azevedo RA, Rillig MC, Sun H, Calabrese EJ (2022) Low levels of contaminants stimulate harmful algal organisms and enrich their toxins. Environ Sci Technol 56(17):11991–12002. https://doi.org/10.1021/acs.est.2c02763

Alarcan J, Biré R, Le Hégarat L, Fessard V (2018) Mixtures of lipophilic phycotoxins: exposure data and toxicological assessment. Mar Drugs 16(2). https://doi.org/10.3390/md16020046

Alves de Souza C, Mardones JI, Yñiguez AT, Le Bihan V, Guillotreau P, Gatti CMI, Richlen ML, Larsen J, Berdalet E (2022) Harmful Algae. In: Urban ER Jr, Ittekkot V (eds) Blue economy: an ocean science perspective. Springer Nature Singapore, Singapore, pp 287–317. https://doi.org/10.1007/978-981-19-5065-0_10

Anderson DM, Cembella AD, Hallegraeff GM (2012) Progress in understanding harmful algal blooms: paradigm shifts and new technologies for research, monitoring, and management. Annu Rev Mar Sci 4:143–176. https://doi.org/10.1146/annurev-marine-120308-081121

Anderson DM, Fachon E, Pickart RS, Lin P, Fischer AD, Richlen ML, Uva V, Brosnahan ML, McRaven L, Bahr F, Lefebvre K, Grebmeier JM, Danielson SL, Lyu Y, Fukai Y (2021a) Evidence for massive and recurrent toxic blooms of *Alexandrium catenella* in the Alaskan Arctic. Proc Natl Acad Sci USA 118(41):e2107387118. https://doi.org/10.1073/pnas.2107387118

Anderson DM, Fensin E, Gobler CJ, Hoeglund AE, Hubbard KA, Kulis DM, Landsberg JH, Lefebvre KA, Provoost P, Richlen ML, Smith JL, Solow AR, Trainer VL (2021b) Marine harmful algal blooms (HABs) in the United States: history, current status and future trends. Harmful Algae 102. https://doi.org/10.1016/j.hal.2021.101975

Anderson DM, Fachon E, Hubbard K, Lefebvre KA, Lin P, Pickart R, Richlen M, Sheffield G, van Hemert C (2022) Harmful algal blooms in the Alaskan Arctic: an emerging threat as the ocean warms. Oceanography 35(3–4):130–139. https://doi.org/10.5670/oceanog.2022.121

Bolch CJS, de Salas MF (2007) A review of the molecular evidence for ballast water introduction of the toxic dinoflagellates *Gymnodinium catenatum* and the *Alexandrium* "tamarensis complex" to Australasia. Harmful Algae 6(4):465–485. https://doi.org/10.1016/j.hal.2006.12.008

Brandenburg KM, Velthuis M, Van de Waal DB (2019) Meta-analysis reveals enhanced growth of marine harmful algae from temperate regions with warming and elevated CO_2 levels. Glob Change Biol 25(8):2607–2618. https://doi.org/10.1111/gcb.14678

Breinlinger S, Phillips TJ, Haram BN, Mareš J, Martínez Yerena JA, Hrouzek P, Sobotka R, Henderson WM, Schmieder P, Williams SM, Lauderdale JD, Wilde HD, Gerrin W, Kust A, Washington JW, Wagner C, Geier B, Liebeke M, Enke H, Niedermeyer THJ, Wilde SB (2021) Hunting the eagle killer: a cyanobacterial neurotoxin causes vacuolar myelinopathy. Science 371(6536):eaax9050. https://doi.org/10.1126/science.aax9050

Calabrese EJ (2014) Hormesis: a fundamental concept in biology. Microbial Cell (Graz, Austria) 1(5):145–149. https://doi.org/10.15698/mic2014.05.145

Carmichael WW (1992) Cyanobacteria secondary metabolites-the cyanotoxins. J Appl Bacteriol 72(6):445–459. https://doi.org/10.1111/j.1365-2672.1992.tb01858.x

Celano MR, Vilar MCP, Pacheco ABF, Azevedo SMFDOE (2025) Azithromycin at environmentally relevant concentrations increases the growth and microcystin content in the cyanobacterium *Microcystis aeruginosa*. Water Air Soil Pollut 237(2):98. https://doi.org/10.1007/s11270-025-08662-z

Chen JQ, Guo RX (2012) Access the toxic effect of the antibiotic cefradine and its UV light degradation products on two freshwater algae. J Hazard Mater 209-210:520–523. https://doi.org/10.1016/j.jhazmat.2012.01.041

Chinain M, Gatti CMI, Darius HT, Quod JP, Tester PA (2021) Ciguatera poisonings: a global review of occurrences and trends. Harmful Algae 102:101873. https://doi.org/10.1016/j.hal.2020.101873

Clark S, Hubbard KA, Anderson DM, McGillicuddy DJ, Ralston DK, Townsend DW (2019) *Pseudo-nitzschia* bloom dynamics in the Gulf of Maine: 2012–2016. Harmful Algae 88:101656. https://doi.org/10.1016/j.hal.2019.101656

Committee GS (2021) Guidelines for the study of climate change effects on HABs. IOC manuals guides

Dai Y, Yang S, Zhao D, Hu C, Xu W, Anderson DM, Li Y, Song X-P, Boyce DG, Gibson L, Zheng C, Feng L (2023) Coastal phytoplankton blooms expand and intensify in the 21st century. Nature 615(7951):280–284. https://doi.org/10.1038/s41586-023-05760-y

De Vries T, Yamamoto K, Wanninkhof R, Gruber N, Hauck J, Müller JD, Bopp L, Carroll D, Carter B, Chau TTT, Doney SC, Gehlen M, Gloege L, Gregor L, Henson S, Kim JH, Iida Y, Ilyina T, Landschützer P, Le Quéré C, Munro D, Nissen C, Patara L, Pérez FF, Resplandy L, Rodgers KB, Schwinger J, Séférian R, Sicardi V, Terhaar J, Triñanes J, Tsujino H, Watson A, Yasunaka S, Zeng J (2023) Magnitude, trends, and variability of the global ocean carbon sink from 1985 to 2018. Global Biogeochem Cy 37(10):e2023GB007780. https://doi.org/10.1029/2023GB007780

do Rosário Gomes H, Goes JI, SGP M, Buskey EJ, Basu S, Parab S, Thoppil P (2014) Massive outbreaks of *Noctiluca scintillans* blooms in the Arabian Sea due to spread of hypoxia. Nat Commun 5(1):4862. https://doi.org/10.1038/ncomms5862

Escher BI, Ahlheim J, Böhme A, Borchardt D, Brack W, Braun G, Colbourne JK, Dann JP, Gessner J, Jahnke A, König M, Klüver N, Krauss M, Lee J, Li X, Lips S, Orsini L, Rinke K, Schmitt-Jansen M, Scholz S, Schulze T, Spahr S, Ulrich N, Weitere M, Varga E (2024) Mixtures of organic micropollutants exacerbated in vitro neurotoxicity of prymnesins and contributed to aquatic toxicity during a toxic algal bloom. Nature Water 2(9):889–898. https://doi.org/10.1038/s44221-024-00297-4

FAO (2020) The state of world fisheries and aquaculture 2020. In: Sustainability in action. Abbreviation 2. Food and agriculture Organization of the United Nations, Rome. https://doi.org/10.4060/ca9229en

Feng L, Wang Y, Hou X, Qin B, Kuster T, Qu F, Chen N, Paerl HW, Zheng C (2024) Harmful algal blooms in inland waters. Nat Rev Earth Environ 5(9):631–644. https://doi.org/10.1038/s43017-024-00578-2

Gobler CJ (2020) Climate change and harmful algal blooms: insights and perspective. Harmful Algae 91:101731. https://doi.org/10.1016/j.hal.2019.101731

Griffith AW, Gobler CJ (2020) Harmful algal blooms: a climate change co-stressor in marine and freshwater ecosystems. Harmful Algae 91:101590. https://doi.org/10.1016/j.hal.2019.03.008

Gruber N, Clement D, Carter BR, Feely RA, van Heuven S, Hoppema M, Ishii M, Key RM, Kozyr A, Lauvset SK, Lo Monaco C, Mathis JT, Murata A, Olsen A, Perez FF, Sabine CL, Tanhua T, Wanninkhof R (2019) The oceanic sink for anthropogenic CO_2 from 1994 to 2007. Science 363(6432):1193–1199. https://doi.org/10.1126/science.aau5153

Gruber N, Bakker DCE, DeVries T, Gregor L, Hauck J, Landschützer P, McKinley GA, Müller JD (2023) Trends and variability in the ocean carbon sink. Nat Rev Earth Environ 4(2):119–134. https://doi.org/10.1038/s43017-022-00381-x

Guzmán-Guillén R, Campos A, Machado J, Freitas M, Azevedo J, Pinto E, Almeida A, Cameán AM, Vasconcelos V (2017) Effects of *Chrysosporum* (*Aphanizomenon*) *ovalisporum* extracts containing cylindrospermopsin on growth, photosynthetic capacity, and mineral content of carrots (*Daucus carota*). Ecotoxicology 26(1):22–31. https://doi.org/10.1007/s10646-016-1737-4

Hallegraeff GM, Anderson DM, Belin C, Bottein M-YD, Bresnan E, Chinain M, Enevoldsen H, Iwataki M, Karlson B, McKenzie CH, Sunesen I, Pitcher GC, Provoost P, Richardson

A, Schweibold L, Tester PA, Trainer VL, Yñiguez AT, Zingone A (2021a) Perceived global increase in algal blooms is attributable to intensified monitoring and emerging bloom impacts. Commun Earth Environ 2(1):117. https://doi.org/10.1038/s43247-021-00178-8

Hallegraeff GM, Schweibold L, Jaffrezic E, Rhodes L, MacKenzie L, Hay B, Farrell H (2021b) Overview of Australian and New Zealand harmful algal species occurrences and their societal impacts in the period 1985 to 2018, including a compilation of historic records. Harmful Algae 102:101848. https://doi.org/10.1016/j.hal.2020.101848

Hattenrath-Lehmann TK, Smith JL, Wallace RB, Merlo L, Koch F, Mittelsdorf H, Goleski JA, Anderson DM, Gobler CJ (2015) The effects of elevated CO_2 on the growth and toxicity of field populations and cultures of the saxitoxin-producing dinoflagellate, *Alexandrium fundyense*. Limnol Oceanogr 60(1):198–214. https://doi.org/10.1002/lno.10012

Ho JC, Michalak AM, Pahlevan N (2019) Widespread global increase in intense lake phytoplankton blooms since the 1980s. Nature 574(7780):667–670. https://doi.org/10.1038/s41586-019-1648-7

Hou X, Feng L, Dai Y, Hu C, Gibson L, Tang J, Lee Z, Wang Y, Cai X, Liu J, Zheng Y, Zheng C (2022) Global mapping reveals increase in lacustrine algal blooms over the past decade. Nat Geosci 15(2):130–134. https://doi.org/10.1038/s41561-021-00887-x

Hu W, Su S, Mohamed HF, Xiao J, Kang J, Krock B, Xie B, Luo Z, Chen B (2024) Assessing the global distribution and risk of harmful microalgae: a focus on three toxic *Alexandrium* dinoflagellates. Sci Total Environ 948:174767. https://doi.org/10.1016/j.scitotenv.2024.174767

IPCC (2021) Contribution of working group I to the sixth assessment report of the intergovernmental panel on climate change. Abbreviation 2

Jiang Y, Chen Y, Yang S, Li R (2020a) Phylogenetic relationships and genetic divergence of paralytic shellfish toxin- and cylindrospermopsin-producing *Cylindrospermopsis* and *Raphidiopsis*. Harmful Algae 93:101792. https://doi.org/10.1016/j.hal.2020.101792

Jiang Y, Liu Y, Zhang J (2020b) Antibiotic contaminants reduced the treatment efficiency of UV-C on *Microcystis aeruginosa* through hormesis. Environ Pollut 261:114193. https://doi.org/10.1016/j.envpol.2020.114193

Jiang Y, Liu Y, Zhang J (2020c) Antibiotics induced alterations in cell density, photosynthesis, microcystin synthesis and proteomic expression of *Microcystis aeruginosa* during $CuSO_4$ treatment. Aquat Toxicol 222:105473. https://doi.org/10.1016/j.aquatox.2020.105473

Jiang Y, Liu Y, Zhang J (2021) Mechanisms for the stimulatory effects of a five-component mixture of antibiotics in *Microcystis aeruginosa* at transcriptomic and proteomic levels. J Hazard Mater 406:124722. https://doi.org/10.1016/j.jhazmat.2020.124722

Kao CY (1993) Paralytic shellfish poisoning. In: Falconer IR (ed) Algal toxins in seafood and drinking water. Academic Press, San Diego, pp 75–86. https://doi.org/10.1016/B978-0-08-091811-2.50009-5

Karlson B, Andersen P, Arneborg L, Cembella A, Eikrem W, John U, West JJ, Klemm K, Kobos J, Lehtinen S, Lundholm N, Mazur-Marzec H, Naustvoll L, Poelman M, Provoost P, De Rijcke M, Suikkanen S (2021) Harmful algal blooms and their effects in coastal seas of northern Europe. Harmful Algae 102:101989. https://doi.org/10.1016/j.hal.2021.101989

Krienitz L, Schagerl M (2016) Tiny and tough: Microphytes of east African Soda Lakes. Soda Lakes of East Africa, In, pp 149–177. https://doi.org/10.1007/978-3-319-28622-8_6

Krienitz L, Ballot A, Kotut K, Wiegand C, Pütz S, Metcalf JS, Codd GA, Pflugmacher S (2003) Contribution of hot spring cyanobacteria to the mysterious deaths of lesser flamingos at Lake Bogoria. Kenya FEMS Microbiol Ecol 43(2):141–148. https://doi.org/10.1016/S0168-6496(02)00387-2

Lan J, Liu P, Hu X, Zhu S (2024) Harmful algal blooms in eutrophic marine environments: causes, monitoring, and treatment. Water 16(17). https://doi.org/10.3390/w16172525

Lee K-H, Noh J, Khim JS (2020) The blue economy and the United Nations' sustainable development goals: challenges and opportunities. Environ Int 137:105528. https://doi.org/10.1016/j.envint.2020.105528

Lefebvre KA, Quakenbush L, Frame E, Huntington KB, Sheffield G, Stimmelmayr R, Bryan A, Kendrick P, Ziel H, Goldstein T, Snyder JA, Gelatt T, Gulland F, Dickerson B, Gill V (2016) Prevalence of algal toxins in Alaskan marine mammals foraging in a changing arctic and subarctic environment. Harmful Algae 55:13–24. https://doi.org/10.1016/j.hal.2016.01.007

Lefebvre KA, Charapata P, Stimmelmayr R, Lin P, Pickart RS, Hubbard KA, Bill BD, Sheffield G, Bowers EK, Anderson DM, Fachon E, Thoman R (2025) Bowhead whale faeces link increasing algal toxins in the Arctic to ocean warming. Nature 644(8077):693–698. https://doi.org/10.1038/s41586-025-09230-5

Lieke T, Jung-Schroers V, Rebl A (2025) Freshwater browning as a hidden threat—transcriptomic responses in fish gills exposed to fulvic acid. J Hazard Mater 488:137260. https://doi.org/10.1016/j.jhazmat.2025.137260

Lin W, Hou J, Guo H, Qiu Y, Li L, Li D, Tang R (2017) Dualistic immunomodulation of subchronic microcystin-LR exposure on the innate-immune defense system in male zebrafish. Chemosphere 183:315–322. https://doi.org/10.1016/j.chemosphere.2017.05.079

Liu J, Ding X, Xia X, Zhou L, Liu W, Lai Y, Ke Z, Tan Y (2023) Dissolved organic phosphorus promotes *Cyclotella* growth and adaptability in eutrophic tropical estuaries. Appl Environ Microbiol 90(1):e01637–e01623. https://doi.org/10.1128/aem.01637-23

Lu S, Ou L, Dai X, Cui L, Dong Y, Wang P, Li D, Lu D (2022) An overview of *Prorocentrum donghaiense* blooms in China: species identification, occurrences, ecological consequences, and factors regulating prevalence. Harmful Algae 114:102207. https://doi.org/10.1016/j.hal.2022.102207

McKibben SM, Peterson W, Wood AM, Trainer VL, Hunter M, White AE (2017) Climatic regulation of the neurotoxin domoic acid. Proc Natl Acad Sci USA 114(2):239–244. https://doi.org/10.1073/pnas.1606798114

Mora D, Schlüsener M, Fischer H, Kleinteich J, Schulz M, Ternes T, Thiel J, Wick A, Krenek S (2024) From genes to toxins: profiling *Prymnesium parvum* during a riverine harmful algal bloom. Harmful algae 102644. https://doi.org/10.1016/j.hal.2024.102644

Moschonas G, Gowen RJ, Paterson RF, Mitchell E, Stewart BM, McNeill S, Glibert PM, Davidson K (2017) Nitrogen dynamics and phytoplankton community structure: the role of organic nutrients. Biogeochemistry 134(1):125–145. https://doi.org/10.1007/s10533-017-0351-8

NOAA-Fisheries (2024) Toxic algal bloom affecting California Sea lions and dolphins. Abbreviation 2

Paerl HW, Otten TG (2013a) Blooms bite the hand that feeds them. Science 342(6157):433–434. https://doi.org/10.1126/science.1245276

Paerl HW, Otten TG (2013b) Harmful cyanobacterial blooms: causes, consequences, and controls. Microb Ecol 65(4):995–1010. https://doi.org/10.1007/s00248-012-0159-y

Papacek JR, Inglett PW, Phlips EJ, Lasi MA (2024) Nitrogen and phosphorus uptake kinetics in cultures of two novel picoplankton groups responsible for a recent bloom event in a subtropical estuary (Indian River lagoon, Florida). Front Mar Sci 11:1256901. https://doi.org/10.3389/fmars.2024.1256901

Piatt JF, Parrish JK, Renner HM, Schoen SK, Jones TT, Arimitsu ML, Kuletz KJ, Bodenstein B, García-Reyes M, Duerr RS, Corcoran RM, Kaler RSA, McChesney GJ, Golightly RT, Coletti HA, Suryan RM, Burgess HK, Lindsey J, Lindquist K, Warzybok PM, Jahncke J, Roletto J, Sydeman WJ (2020) Extreme mortality and reproductive failure of common murres resulting from the Northeast Pacific marine heatwave of 2014-2016. PLoS One 15(1):e0226087. https://doi.org/10.1371/journal.pone.0226087

Pitcher GC, Louw DC (2021) Harmful algal blooms of the Benguela eastern boundary upwelling system. Harmful Algae 102:101898. https://doi.org/10.1016/j.hal.2020.101898

Qiao Q, Liang H, Zhang X (2013) Effect of cyanobacteria on immune function of crucian carp (*Carassius auratus*) via chronic exposure in diet. Chemosphere 90(3):1167–1176. https://doi.org/10.1016/j.chemosphere.2012.09.025

Reinl KL, Harris TD, Elfferich I, Coker A, Zhan Q, De Senerpont Domis LN, Morales-Williams AM, Bhattacharya R, Grossart H-P, North RL, Sweetman JN (2022) The role of organic nutrients in structuring freshwater phytoplankton communities in a rapidly changing world. Water Res 219:118573. https://doi.org/10.1016/j.watres.2022.118573

Riebesell U, Aberle-Malzahn N, Achterberg EP, Algueró-Muñiz M, Alvarez-Fernandez S, Arístegui J, Bach LT, Boersma M, Boxhammer T, Guan W, Haunost M, Horn HG, Löscher CR, Ludwig A, Spisla C, Sswat M, Stange P, Taucher J (2018) Toxic algal bloom induced by ocean acidification disrupts the pelagic food web. Nat Clim Chang 8(12):1082–1086. https://doi.org/10.1038/s41558-018-0344-1

Rymuszka A, Sierosławska A, Bownik A, Skowroński T (2007) *In vitro* effects of pure microcystin-LR on the lymphocyte proliferation in rainbow trout (*Oncorhynchus mykiss*). Fish Shellfish Immunol 22(3):289–292. https://doi.org/10.1016/j.fsi.2006.06.002

Schreck CB (2010) Stress and fish reproduction: the roles of allostasis and hormesis. Gen Comp Endocrinol 165(3):549–556. https://doi.org/10.1016/j.ygcen.2009.07.004

Song L, Jia Y, Qin B, Li R, Carmichael WW, Gan N, Xu H, Shan K, Sukenik A (2023) Harmful cyanobacterial blooms: biological traits, mechanisms, risks, and control strategies. Annu Rev Environ Res 48(48):123–147. https://doi.org/10.1146/annurev-environ-112320-081653

Steinberg CEW (2025) Cyanobacteria and the food chain: "cyanobacteria are also not eaten with pleasure". In: Aquatic animal nutrition: plant compounds and dietary obstacles. Springer Nature, Cham, pp 459–554. https://doi.org/10.1007/978-3-031-97988-0_8

Steinberg CEW, Hartmann HM (1988) Planktonic bloom-forming cyanobacteria and the eutrophication of lakes and rivers. Freshw Biol 20(2):279–287. https://doi.org/10.1111/j.1365-2427.1988.tb00452.x

Tester PA, Litaker RW, Berdalet E (2020) Climate change and harmful benthic microalgae. Harmful Algae 91:101655. https://doi.org/10.1016/j.hal.2019.101655

Tietbohl MD, Genevier LGC, Krieger EC, Kattan A, Wang Y, Gokul EA, Rodriguez Bravo LM, Palm L, Mele G, Hoteit I, Johnson MD (2025) Extreme marine heatwave linked to mass fish kill in the Red Sea. Sci Total Environ 975:179073. https://doi.org/10.1016/j.scitotenv.2025.179073

Trainer VL, Moore SK, Hallegraeff G, Kudela RM, Clement A, Mardones JI, Cochlan WP (2020) Pelagic harmful algal blooms and climate change: lessons from nature's experiments with extremes. Harmful Algae 91:101591. https://doi.org/10.1016/j.hal.2019.03.009

Trombetta T, Vidussi F, Mas S, Parin D, Simier M, Mostajir B (2019) Water temperature drives phytoplankton blooms in coastal waters. PLoS One 14(4):e0214933. https://doi.org/10.1371/journal.pone.0214933

Wang H, Xu C, Liu Y, Jeppesen E, Svenning J-C, Wu J, Zhang W, Zhou T, Wang P, Nangombe S, Ma J, Duan H, Fang J, Xie P (2021) From unusual suspect to serial killer: cyanotoxins boosted by climate change may jeopardize megafauna. Innovation 2(2). https://doi.org/10.1016/j.xinn.2021.100092

Xu D, Zheng G, Brennan G, Wang Z, Jiang T, Sun K, Fan X, Bowler C, Zhang X, Zhang Y, Wang W, Wang Y, Li Y, Wu H, Li Y, Fu FX, Hutchins DA, Tan Z, Ye N (2023) Plastic responses lead to increased neurotoxin production in the diatom *pseudo-nitzschia* under ocean warming and acidification. ISME J 17(4):525–536. https://doi.org/10.1038/s41396-023-01370-8

Zhang P, Peng C, Zhang J, Zhang J, Chen J, Zhao H (2022a) Long-term harmful algal blooms and nutrients patterns affected by climate change and anthropogenic pressures in the Zhanjiang Bay. China Front Mar Sci 9:849819. https://doi.org/10.3389/fmars.2022.849819

Zhang Y, He D, Bu Z, Li Y, Guo J, Li Q (2022b) The transcriptomic analysis revealed sulfamethoxazole stress at environmentally relevant concentration on the mechanisms of toxicity of cyanobacteria *Synechococcus* sp. J Environ Chem Eng 10(3):107637. https://doi.org/10.1016/j.jece.2022.107637

Zhang J, Shi K, Paerl HW, Rühland KM, Yuan Y, Wang R, Chen J, Ge M, Zheng L, Zhang Z, Qin B, Liu J, Smol JP (2023) Ancient DNA reveals potentially toxic cyanobacteria increasing with climate change. Water Res 229:119435. https://doi.org/10.1016/j.watres.2022.119435

Zhao S, Chang X, Li J, Zhu Y, Pan X, Hua Z, Li J (2024) The two-way immunotoxicity in native fish induced by exudates of *Microcystis aeruginosa*: Immunostimulation and immunosuppression. J Hazard Mater 461:132554. https://doi.org/10.1016/j.jhazmat.2023.132554

Zhao J, Yuan Q, Lei X, Lieke T, Liu Y, Steinberg CEW, Pan B, Xing B (2025) The double-edged environmental effect of dissolved organic matter in global climate change. Environ Biogeochem Process 1(1):e009. https://doi.org/10.48130/ebp-0025-0009

Zingone A, Escalera L, Aligizaki K, Fernández-Tejedor M, Ismael A, Montresor M, Mozetič P, Taş S, Totti C (2021) Toxic marine microalgae and noxious blooms in the Mediterranean Sea: a contribution to the global HAB status report. Harmful Algae 102:101843. https://doi.org/10.1016/j.hal.2020.101843

Cyanotoxins in Action

2

In 1878, the first written report of animal poisoning by a cyanobacterium from Lake Alexandrina (Murray estuary, Australia) was published (Francis 1878).[1] From British herbarium specimens at the Natural History Museum, London, Metcalf et al. (2012) identified microcystins (MCs) from even earlier times, e.g., in *Phormidium* (*Oscillatoria*) *terebriforme* from 1839, *Dolichospermum* (*Anabaena*) from 1843, *Oscillatoria contexta* from 1844, and *Nostoc commune* from 1850.

A global analysis found documentation indicating geographic expansion of *Microcystis*, with recorded blooms in at least 108 countries, 79 of which have also reported the hepatotoxin MC (Harke et al. 2016). Compared to what is known about the effects of MCs, information on other cyanotoxins and marine toxins is rather incomplete. Therefore, this chapter will focus on the effects of MCs, regardless of whether animals are affected through diet or exposure. In fact, Diez-Quijada et al. (2021) pointed out in a comprehensive review that toxic effects are generally independent of the type of exposure, namely, intraperitoneal injection, immersion, or diet. Therefore, non-dietary studies will not be excluded in the following sections. However, it is worth noting that the toxicities differ significantly in magnitude across the various routes of exposure. For instance, MC-LR is 100 times less toxic by oral ingestion than it is by intraperitoneal injection (Bouaïcha et al. 2019). The cyclic structure of MCs is essential for the toxicity, and opening of the macrocycle at any position makes the molecule much less toxic (Bouaïcha et al. 2019).

Recently, Jacinavicius et al. (2023) compared the toxicity of several cyanobacteria to the development of early life stages of zebrafish. Extracts of the strains *Desertifilum tharense*, *Anagnostidinema amphibium*, and *Nostoc* sp. promoted

[1] Toxicity is the effect, bloom-forming cyanobacteria are well known to the public and many scientists. However, the biological function of cyanotoxins has only been understood for a few decades. Wei et al. (2024) briefly summarize the biological functions of MCs as enhancing the plasticity of toxic cyanobacteria to changing environments. Intracellular MCs function in car-bon/nitrogen metabolism and antioxidation, while extracellular MCs function in cell-to-cell communication and colony formation.

T. Lieke, C. E. W. Steinberg, *The Ecology of Phycotoxins*,
https://doi.org/10.1007/978-3-032-18380-4_2

Fig. 2.1 *Microcystis aeruginosa* blooms (**a**) in Dianchi Lake, Kunming, China (2009) © C.E.W. Steinberg, and (**b**) in Lednice Castle Pond, Czechia (2024) © T. Lieke

morphological abnormalities in zebrafish larvae. In contrast, extracts from *Microcystis aeruginosa* and *Chlorogloeopsis* sp. did not promote such abnormalities. The *Microcystis* result contrasts with an earlier study that found gross malformations and high mortality during embryonic development with various crude aqueous *Microcystis*-containing extracts from field samples (Oberemm et al. 1997). This contradiction remains unexplained.

Examples of extensive *Microcystis aeruginosa* blooms in the environment are presented in Fig. 2.1.

From farmed invertebrates, only a few records of cyanobacteria-caused mass mortalities are available. For instance, Zimba et al. (2006) report a co-occurrence of Pacific white shrimp (*Penaeus vannamei*) mortalities and MC in a southeastern USA shrimp facility. High concentrations of free MC-LR were found in the hepatopancreas of dead shrimps.

In fish, the uptake of cyanotoxins generally occurs through gills and contaminated foods, with zooplankton and phytoplankton as primary sources Banerjee et al. (2021). Hepatotoxicity, neurotoxicity, reproductive toxicity, immunotoxicity, and cytogenotoxicity are some of the major toxicological impairments induced by cyanotoxins. Modes of action, with the exception of epigenetics, are summarized in Fig. 2.2.

Banerjee et al. (2021) present a comprehensive list of freshwater fishes affected by cyanotoxins. Inhibition of key enzymatic activities (such as protein phosphatases, acetylcholinesterase, and antioxidants) related to hyperphosphorylation, oxidative stress, xenobiotic detoxification, and alterations in gene expression patterns might be the cornerstone of different cyanotoxin-induced toxicities in teleosts. Although the mechanisms and pathogenesis of hepatotoxicity and neurotoxicity of different cyanotoxins in fish are well understood, the cytogenotoxicity, reproductive toxicity, and immunotoxicity of cyanotoxins remain relatively poorly understood to date. The response to cyanotoxins generates ROS, reduces the contents of spontaneous antioxidants (glutathione, ascorbic acid, tocopherols, etc.), and downregulates

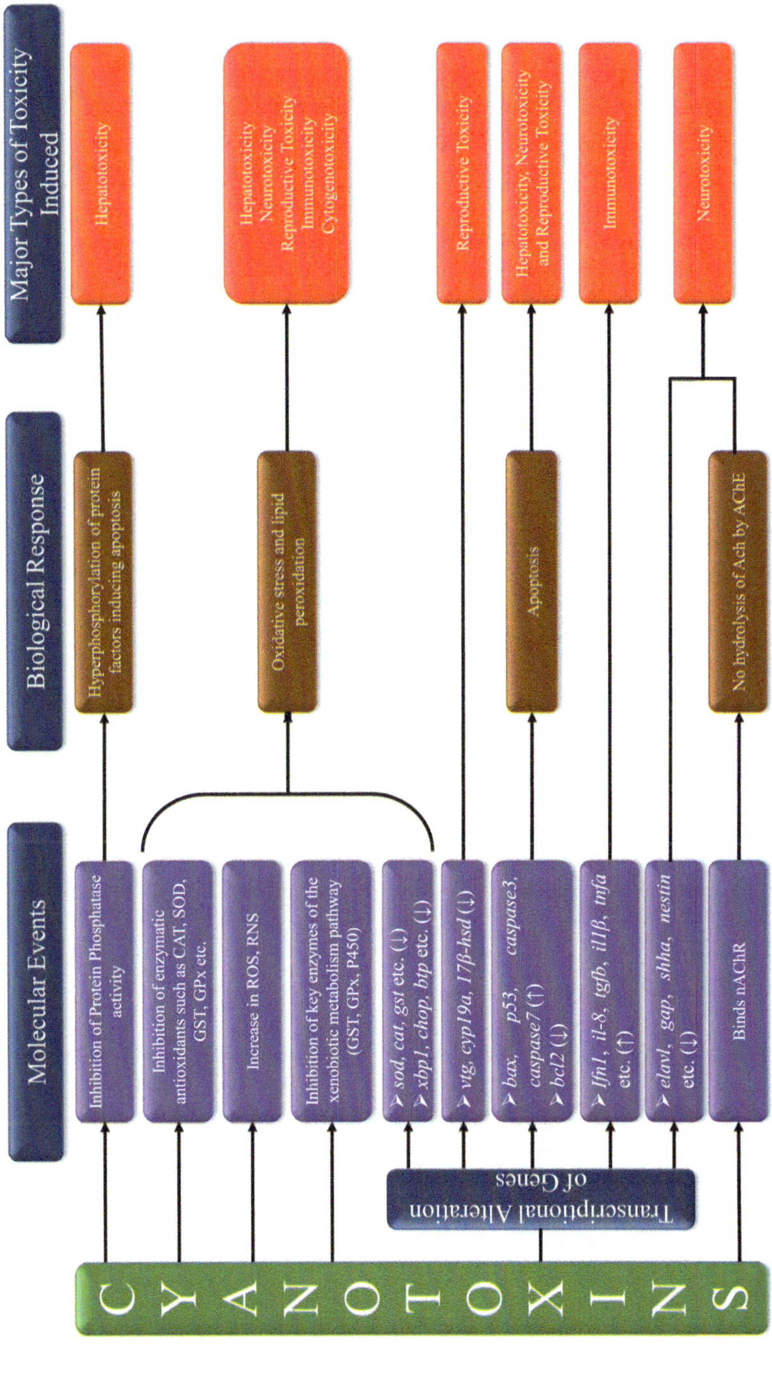

Fig. 2.2 Highlights of different molecular mechanisms responsible for various toxicities induced by cyanotoxins. (From Banerjee et al. (2021), with permission from Elsevier)

antioxidant defense genes (*sod*, *cat*, *gpx*, etc.) associated with oxidative stress. In general, oxygen activation is the initial response when homeostasis is disturbed, regardless of the underlying cause of the disturbance. It serves as a fundamental response to any stress and acts as a trigger for various defensive mechanisms (Steinberg 2012; Lieke et al. 2021). If the activated oxygen is not balanced, the subsequent oxidative stress leads to cyanotoxin-mediated genotoxicity, cytotoxicity, and finally, apoptosis—even in seemingly tolerant fish species, such as Nile tilapia (da Silva et al. 2011). The occurrence of oxidative stress in microcystin-exposed organisms has been demonstrated on several occasions, as noted by Pflugmacher (2004) and Wiegand and Pflugmacher (2005).

In a comprehensive review, Ger et al. (2016) examined the interplay between food quality, toxicity, and tolerance to cyanotoxins in shaping aquatic food webs. In addition to being a public health hazard, cyanobacteria have long been considered poor-quality food for important zooplankton grazers that link phytoplankton to higher trophic levels (reviewed in Steinberg (2022a)). While previous laboratory studies have found negative effects of nutritional constraints and defensive traits (i.e., toxicity and colonial or filamentous morphology) on the fitness of large generalist grazers (i.e., *Daphnia*), cyanobacterial blooms often coexist with high biomass of small-bodied zooplankton in nature. Indeed, recent studies highlight the remarkable diversity and flexibility of zooplankton responses to cyanobacterial prey. However, prolonged exposure to cyanobacteria can shift zooplankton communities toward better-adapted species, select for more tolerant genotypes within a species, and induce traits within the lifetime of individual zooplankton. Recent studies show that epigenetics, in the form of microRNAs, also contribute significantly to the adaptive responses of *D. magna* to toxic cyanobacterial blooms (Lyu et al. 2024) (→Chap. 5). Building on the information from Ger et al. (2016), we will also discuss the role of gut microbiota in cyanotoxin toxicity and tolerance to cyanotoxins.

Consequently, toxicity studies of cyanotoxins, particularly microcystins, comprise this chapter, while tolerance mechanisms are discussed in Chap. 5 "The Tolerance Modes to Phycotoxins."

2.1 Microcystins

In a recent study examining the presence of microcystins in 190 Canadian lakes, MC-LR (Fig. 2.3a) and MC-LA[2] were the most abundant of 14 microcystin congeners (MacKeigan et al. 2023). Consequently, most studies of MCs have been conducted with MC-LR. Microcystins are known to be potent hepatotoxins with genotoxic properties; they are potent inhibitors of the serine/threonine phosphatases PP1 and PP2A (Campos and Vasconcelos 2010). This can occur not only through reversible but also through covalent interactions. The first evidence for covalently bound MCs from non-mammalian studies dates back more than 25 years. Williams et al. (1997a, b) found both covalently bound and free MCs in salmon liver,

[2] MC-LA is an analog of MC-LR with Ala substituted in place of Arg.

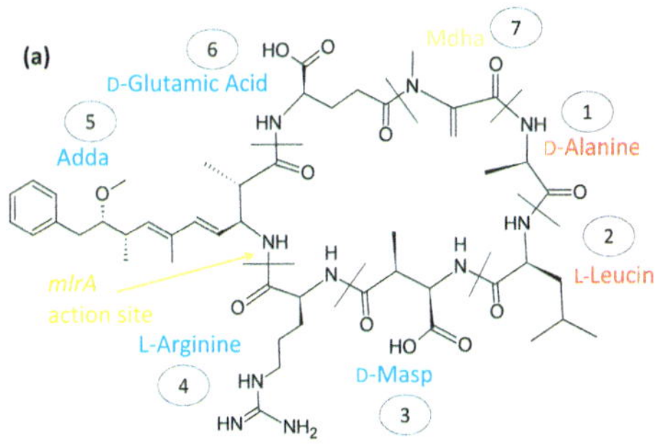

Fig. 2.3 (**a**) Structures of MC-LR, one of the most common microcystin congeners, and (**b**) nodularin. Common amino acids are marked in blue, while different ones are marked in red. D-*Masp* D-methylaspartic acid, *Adda* 3-amino-9-methoxy-2,6,8-trimethyl-10-phenyldeca-4,6-dienoic acid, *Mdha N*-methyldehydroalanine, *Mdhb N*-methyldehydrobutrine. ADDA is responsible for hydrophobic interactions and the inhibition of protein phosphatases 1 and 2A (PP1 and PP2A). The active site of microcystinase (MlrA) is indicated (Krishnan and Mou 2021) (credit to MDPI, amended)

Dungeness crab (*Metacarcinus magister*) larvae, and blue mussel (*Mytilus edulis*). Smith et al. (2010) hypothesized that covalently bound MCs become bioavailable in the consumer's digestive system through the digestion of the attached protein phosphatase, and thus can accumulate at higher trophic levels. However, the findings of Ibelings et al. (2005) do not support the idea of MC biomagnification in a lake food web.

About 20 years ago, Malbrouck and Kestemont (2006) presented a comprehensive review of phenotypic and biochemical effects of MC exposure. In early life stages, exposure to MCs causes dose-dependent effects, including decreased embryonic hatching, reduced survival and growth rates, and histopathological effects (enlarged and opaque yolk sac, small head, curved body and tail, hepatobiliary abnormalities, ultrastructural changes in hepatocytes, and heart rate abnormalities). In juveniles and adults, exposure to MC affects growth rate and osmoregulation, increases serum liver enzyme activities and heart rate, alters behavior, and produces histopathological effects in the liver, intestines, kidneys, heart, spleen, or gills.

Using meta-analyses, Kozlowsky-Suzuki et al. (2012) showed that MC biodilution is the prevailing process in aquatic food webs and is consistent across groups of aquatic consumers, with the exception of zooplankton and zooplanktivorous fish. Nevertheless, the biodilution magnitude varies across consumers and is likely dependent on large differences in consumer feeding mode and physiology. Thus, studies on the toxicokinetics of different variants of MC are clearly needed, as the dynamics of MC uptake and depuration/detoxification are not only species-specific but also dependent on several factors, such as molecular hydrophobicity, environmental variation, behavior, and the availability of alternative food sources. Generally low MC hydrophobicity and poor trophic transfer are additional factors that may support biodilution and hinder biomagnification. These key statements by Kozlowsky-Suzuki et al. (2012) are confirmed in later, more detailed studies, such as those by Larson et al. (2014) or Sotton et al. (2014). This also applies to lower trophic levels in benthic marine ecosystems (Umehara et al. 2017).

Recently, Glidewell et al. (2025) provided further support for the biodilution hypothesis by studying the trophic transfer of MCs from *Daphnia pulex* to the predatory larvae of a damselfly, *Enallagma* sp. MCs are effectively sequestered in the tissues of *D. pulex*, and these sequestered toxins are then transferred to the tissues of *Enallagma* sp. The contribution of gut contents to toxin transfer is negligible, and the concentration of MCs decreases with increasing trophic levels.

Liu et al. (2002) demonstrated that MC-LR can have adverse transgenerational effects in fish exposed to it. Their study on the Chinese loach (*Misgurnus mizolepis*) revealed that embryos are more sensitive to MC-LR exposure at later developmental stages compared to earlier ones. Juveniles, on the other hand, show much greater tolerance than embryos and larvae. The effects of MC-LR are both dose-dependent and stage-specific, with higher mortality rates and developmental abnormalities observed at increased concentrations. Notable phenotypical abnormalities include pericardial edema, tubular heart formation, bradycardia, impaired yolk resorption, reduced head size, body and tail curvature, and abnormal hatching. Later, Chen and Xie (2016) summarized the toxic biochemical and molecular modes of action of MCs (Fig. 2.3a) as follows: Organic anion-transporting polypeptides (OATPs) are responsible for the active transmembrane transport of MCs into cells. The first identified and still one of the most important toxic modes of action is: MCs are highly potent and specific inhibitors of eukaryotic protein serine/threonine phosphatases 1 and 2A (PP1 and PP2A), which causes phosphorylation/dephosphorylation imbalance of key control proteins that regulate cytoskeleton organization, DNA damage repair, cellular proliferation, and cell death. MCs' exposure also induces excessive production of reactive oxygen species (ROS) and, thereby, oxidative stress, leading to cytoskeletal disruption and mitochondrial dysfunction, DNA damage, and necroptosis (Wu et al. 2019b). MCs induce cell apoptosis mediated by ROS in the mitochondria and endoplasmic reticulum (ER) pathways. Cytoskeleton disruption, cell cycle arrest, and mitosis defects are also involved in MC-induced cell death (Malécot et al. 2013; Chen and Xie 2016; Zhang et al. 2022b).

Yoon et al. (2017) highlighted a gender-specific difference in detoxification capacity. Heat maps and changes in metabolite levels show different patterns in male and female zebrafish exposed to low concentrations of MC-LR (20 ppb). The metabolites with increased levels in both genders are arginine, aspartate, glycine, NAD^+, $NADP^+$, pyruvate, threonine, and *myo*-inositol. In contrast, those with decreased levels are betaine, citrate, creatine phosphate, NADPH, phenylalanine, and succinate. Contrary to that, glutathione conjugation for excretion of MC-LR during detoxification is decreased only in male zebrafish, and due to the energy requirement for excretion of MC-LR, the level of energy-related metabolites such as fumarate and creatine is also increased in male zebrafish (Fig. 2.4). It can be assumed that females utilize energy-related metabolites for body maintenance, particularly for reproduction.

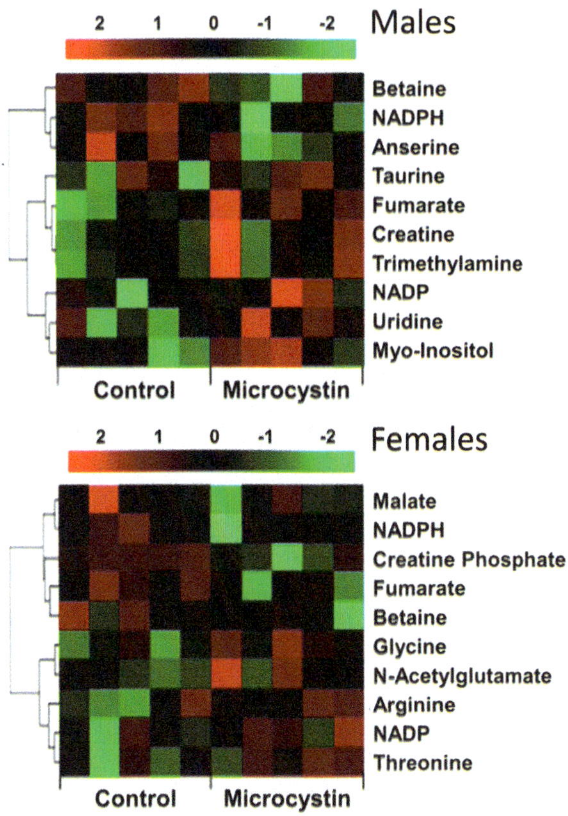

Fig. 2.4 Heatmaps of metabolites from MC-LR-treated male and female zebrafish compared to control. Red represents an increase in metabolite concentrations, as indicated by their log transformation, while green represents a decrease. Each variable was scaled by the mean and standard deviation values. The metabolites were clustered by Pearson coefficient distance (from Yoon et al. (2017), with permission from Wiley & Sons)

In addition to the mitochondrial pathway, another intrinsic apoptotic pathway triggered by ER stress has been identified in MCs toxicity first in cell lines (Xing et al. 2008; Menezes et al. 2013) and later also in different fish species (see below).

Recently, Li et al. (2025) employed the newly developed matrix-assisted laser desorption/ionization mass spectrometry imaging (MALDI-MSI) technique for spatial multi-omics analysis of metabolic heterogeneity in zebrafish exposed to MC-LR, providing a more comprehensive picture of the toxic effects of MC. The intake of MC-LR not only caused oxidative damage to zebrafish but also disrupted the innate immune network of ABC transporters and IgA production, the TCA cycle, and bile acid metabolism pathways in zebrafish. In addition, MSI revealed that MC-LR intake may induce liver injury by disrupting bile acid metabolism by mapping the distribution of metabolites in the whole zebrafish. Spatial multi-omics analysis reveals that MC-LR alters energy metabolism, lipid metabolism, methylation processes, and immunomodulation pathways by modulating the expression of key genes, including *gatm*, *gnmt*, *cyp2p9*, and *tdo2b*.[3] However, for better understanding, each effect will be discussed separately in the following sections.

2.1.1 Hepatotoxins

MC-LR is the most toxic and extensively studied class of MCs, and among all MCs, it poses the greatest threat to human health. In 2010, the International Agency for Research on Cancer classified MC-LR as a possible human carcinogen (Ma et al. 2021b). This paper also emphasizes the action of non-coding RNAs (ncRNAs), which are just beginning to be understood in aquatic animals. NcRNAs are many kinds of RNA molecules that can regulate gene expression at the transcriptional and post-transcriptional levels without being translated into proteins. Briefly, microRNAs (miRNAs) are small single-stranded ncRNAs that can regulate the expression of target genes by binding to the recognition sequences in mRNAs and control many biological processes in eukaryotic organisms, including carcinogenesis.

Furthermore, the role of circular RNAs (circRNAs) and long non-coding RNAs (lncRNAs) in carcinogenesis induced by MC-LR has drawn attention. CircRNA is a special class of ncRNA molecules with a closed ring structure that acts as a miRNA sponge to regulate gene transcription and protein translation (Ebbesen et al. 2017). LncRNAs, which are transcripts longer than 200 nucleotides, play crucial regulatory roles in various biological processes, including epigenetic modification, cell cycle regulation, and cell differentiation (Ma et al. 2021b). The role of epigenetics is crucial when discussing the effects of MCs, as it may influence not only immediate toxicity but also long-term physiological and, therefore, evolutionary responses. Since this topic is crucial in understanding MC-induced changes across generations, we will revisit it in detail below (Sect. 2.1.9).

Combining transcriptomics and proteomics, Qiao et al. (2016a, b) and Le Manach et al. (2016) evaluated the mechanisms of reproductive toxicity in medaka (*Oryzias latipes*) following chronic exposure to MC-LR. Marie 2020 summarized the major results of these studies, which investigated the global proteomic effect induced by environmental/low-dose exposure using 28-day balneation with diluted cyanobacterial extracts on the liver proteome of medaka fish, in addition to anatomical and pathological observations. This study reveals a remarkable sex-dependent response to the different hepatotoxic treatments of Microcystis, in agreement with the clustering based on the amount of upregulated and downregulated proteins (red and green, respectively). Molecular pathways involved in homeostasis and metabolism

[3] *gatm*: glycine amidinotransferase is a mitochondrial enzyme involved in creatine biosynthesis (Joncquel-Chevalier Curt et al. 2015); *gnmt* glycine *N*-methyltransferase is an enzyme catalyzing the synthesis of *N*-methylglycine (sarcosine) from glycine using *S*-adenosylmethionine (SAM) as the methyl donor (Luka et al. 2009); *cyp2p9*: cytochrome P450 family 2 subfamily C member 9 is an enzyme involved in the metabolism, by oxidation, of both xenobiotics, including drugs, and endogenous compounds, include-ing fatty acids; *tdo2b*: tryptophan 2,3-dioxygenase B catalyzing the oxidative cleavage of the L-tryptophan pyrrole ring and converts L-tryptophan to *N*-formyl-L-kynurenine (Steinberg 2022c).

exhibit more modified proteins in the livers of both sexes. These findings suggest that chronic exposure to Microcystis extracts containing MC induces distinct molecular alterations compared to the effects observed under acute or short-term exposure conditions (Marie (2020), and references therein).

A noteworthy aspect of the two following studies was the replication of proteomic results from identical exposure protocols and analytical procedures, which were conducted on fish from the respective parent and descendant generations (Qiao et al. 2016a; Le Manach et al. 2016). The comparison of the dysregulated proteins reveals a similar hepatic alteration pattern under the same cyanotoxin treatments. Remarkably, these data indicate that the MC-containing treatments activate the cellular defense against oxidative stress and the immune response. In addition, proteins implicated in protein modification, lipid and nucleotide metabolism, as well as oviparous-specific proteins, were comparatively downregulated in both analyses (Marie 2020).

Applying a novel chemical proteomic approach to identify proteins that specifically interact with MC-arginine-arginine (MC-RR) in zebrafish liver, Tuo et al. (2016) identified 17 key proteins that play a crucial role in MC-induced liver toxicity. These proteins play critical roles in cytoskeleton assembly, oxidative stress, and the regulation of metabolic disorders. Moreover, in addition to the well-known inhibition of protein phosphate activity, the overall toxicity of MCs was simultaneously modulated by the distribution of MCs in cells and their interactions with other target proteins.

In support, Du et al. (2019) showed that short-term (96 h) exposure to *M. aeruginosa* causes severe histological damage to the liver and reduces its protein phosphatase activity in male zebrafish. The proteomic analysis showed that *M. aeruginosa* interferes with ribosome translation, immune responses, energy metabolism, oxidative phosphorylation, and other biological regulatory processes in zebrafish livers. However, protein expression of Keratin 5 and HSP 90 differs from previous results in zebrafish exposed to pure algal toxin, indicating that *M. aeruginosa* exerts more complex toxic effects than the individual purified toxins, a long-established knowledge (Oberemm et al. 1999).

Of note, even at sublethal but environmentally realistic MC-LR exposure concentrations, Rodrigues et al. (2022) observed severe changes in the liver of the South American fish *Astyanax altiparanae*, including cytoplasmic degradation, displacement, and an increase in nuclear volume and area of hepatocytes. Similarly, but at elevated environmental levels, Mielewczyk et al. (2023) demonstrated that MC-LR exposure results in significant and persistent hypocalcemia, while plasma sodium and branchial activities of sodium/potassium ATPase, proton ATPase, and calcium ATPase enzymes remain unaffected.

Similar toxicity mechanisms of MC-LR were reported by Cao et al. (2024) in the mitten crab (*Eriocheir sinensis*): Involvement of reactive oxygen species (ROS) in hepatopancreatic cytotoxicity, oxidative stress, and the activation of apoptosis through *caspases 3* and *6*.

2.1.1.1 Liver Carcinogens

As one of the first studies, Fonseca et al. (2018) demonstrated the carcinogenic potential of *M. aeruginosa* lysate in zebrafish. Exemplified by the tumor suppressor gene p53, gene expression exhibits a biphasic time pattern, with downregulation of this gene during the first 96 hours and subsequent overexpression after exposure to a moderate lysate concentration. It appears that, following long-term exposure, the zebrafish liver exhibits a compensatory defense mechanism at the molecular level. The *p53* gene is considered one of the most important genes in cell defense. It serves as the primary sensor of genotoxic stress, functioning as a transcriptional transactivator in DNA repair, apoptosis, and tumor suppression pathways, as demonstrated in human carcinoma cells (Fu et al. 2005; Žegura et al. 2008).

This study in zebrafish differs from a previous one by Brzuzan et al. (2009), which showed that intraperitoneal injection of MC-LR induces early mRNA expression of *p53* in whitefish (*Coregonus lavaretus*). This difference may be species-specific. Nevertheless, the response of *p53* to MC-LR indicates that it is involved in promoting cell cycle arrest and apoptosis.

Ma et al. (2021b) summarize the current understanding of hepatotoxicity in an educational graphic (Fig. 2.5).

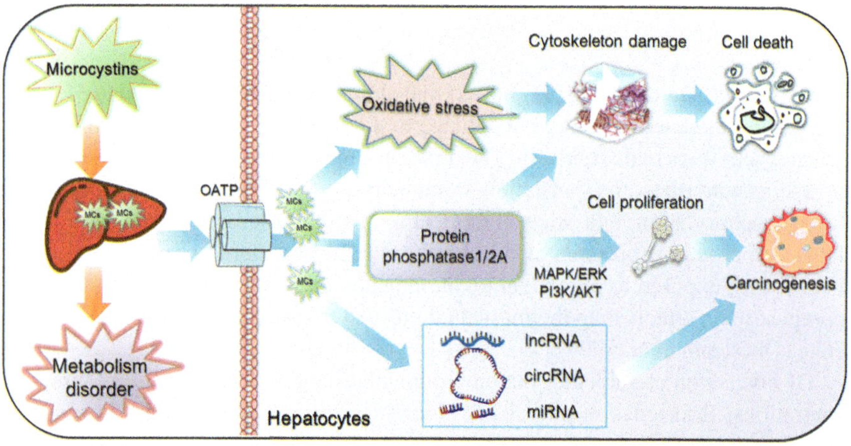

Fig. 2.5 Hepatotoxicity of MC-LR revealed by Omics studies. MC-LR has a hepatotoxic effect mainly by inducing oxidative stress and inhibiting protein phosphatase 1/2A activity. The most prominently affected pathway is the cytoskeleton assembly, which could lead to hepatocyte apoptosis and necrosis. Meanwhile, the inhibition of PP1/2A can regulate cell proliferation by activating the MAPK/ERK pathway and the PI3K/AKT pathway, thereby contributing to the occurrence and development of hepatocellular carcinoma induced by MC-LR. Furthermore, several transcriptomic studies have revealed that various non-coding RNAs (ncRNAs), including microRNA (miRNA), circular RNA (circRNA), and long non-coding RNA (lncRNA), play a crucial role in cell proliferation and hepatocarcinogenesis. Additionally, metabolic results showed that MC-LR could significantly disrupt hepatic metabolic function. Abbreviation: *OATP* organic anion transporting polypeptide, *MAPK* mitogen-activated protein kinase; see Glossary for other abbreviations (from Ma et al. (2021b), credit to Elsevier)

2.1.2 Lipid Homeostasis

Recent studies have shown that MCs also negatively impact lipid metabolism and, consequently, lipid homeostasis. Zhang et al. (2020a) reported that chronic MC-LR exposure induces abnormal lipid metabolism via ER stress in male zebrafish. This is indicated by abnormal hepatic lipid accumulation,[4] which occurs through alterations in the expression of lipid metabolism-related genes (e.g sterol regulatory element-binding protein-1, SREBP), a transcription factor that is a major regulator of fatty acid and lipid biosynthesis.

Disruption of lipid metabolism also occurs in invertebrates. Recently, Duan et al. (2022) showed the responses of lipid metabolism in the hepatopancreas of Pacific white shrimp (*P. vannamei*) to MC-LR exposure. This exposure induces glycerophospholipid remodeling, elevation of ceramides and sphingomyelins, reduction of triglycerides, and changes in lipid metabolism-related indicators. These processes can lead to oxidative stress, membrane dysfunction, and autophagy, which comprehensively promote tissue damage in the hepatopancreas.

Disruption of hepatic fatty acid metabolism is also reported in non-bred animals (Cai et al. 2022a). MC-LR causes mitochondrial DNA damage, leading to mitochondrial dysfunction in the liver (Li et al. 2015a) and eventually to dysregulation of lipid metabolism (Lee et al. 2018). MC-LR-induced lipid accumulation in *Chironomus pallidivittatus* larvae is driven by upregulated fatty acid biosynthesis and elongation via activation of the PPAR signaling pathway. Increased levels of total lipids, total cholesterol, and non-esterified fatty acids in the larvae further support this observation. These results align well with the findings in zebrafish mentioned above, indicating that fatty acid biosynthesis and elongation can promote lipid accumulation.

Confirming the above results of altered lipid metabolism, Yang et al. (2022) showed in the gastropod *Bellamya aeruginosa* that MC exposure can cause hepatic energy consumption associated with various metabolic disturbances involving lipid metabolism, protein catabolism, osmoregulation, glycolysis, glycogenolysis, and the tricarboxylic acid (TCA) cycle. Moreover, the metabolic disturbance worsened as more MCs accumulated over time during treatment with MC-producing cyanobacteria. In particular, the authors observed that MC exposure causes a significant increase in creatine, a variety of amino acids (leucine, isoleucine, valine, threonine, alanine, methionine, glutamate, aspartate, and lysine), carboxylic acids (lactate, acetate, and D-3-hydroxybutyrate), and choline and its derivatives (phosphocholine and glycerophosphocholine), but decreases the energy substance (lipids, glucose, and glycogen) in the hepatopancreas. Overall, the accumulation of MCs leads to oxidative stress-mediated damage to mitochondrial functions. This work confirms a previous proteomic study in *Daphnia magna* showing that differential proteins impaired by *M. aeruginosa* are strongly associated with lipid, carbohydrate, amino acid, and energy metabolism, resulting in reduced fitness (Lyu et al. 2016).

---

[4] The development of non-alcoholic fatty liver disease (steatosis hepatis) caused by microcystins is well documented (Lad et al. 2019).

2.1.3 Neurotoxins

Do MCs simply stop being toxic without a liver to process them? Without any doubt, MCs are harmful even to these animals, as Li et al. (2009b) showed in *Caenorhabditis elegans* and Wu et al. (2016) even in zebrafish embryos. Li et al. (2009b) discovered that MC-LR reduced lifespan, delayed development, lengthened generation time, decreased brood size, suppressed locomotion behavior, and decreased *hsp-16-2-gfp* expression. Subsequently, Ju et al. (2013, 2014) detected the underlying mechanism on a phenotypic level: By applying neurotoxicological tests equivalent to mammalian standard tests, the authors showed that MC-LR is neurotoxic. Saul et al. (2014) refined this mechanism by whole-genome microarray analysis. This analysis revealed that a neuromodulating action is the dominant response. Most of the 201 differentially expressed genes are associated with neurobehavior, neurogenesis, and signaling-associated pathways. In addition, a whole-genome miRNA-microarray highlighted that, in particular, members of the *let-7* family are differentially regulated. These miRNAs are involved in the developmental timing of cell fates, including neurons, and are likely also part of the stress response machinery. Overall, neurological modulation is the primary transcriptional stress response in *C. elegans* exposed to MC-LR.

In cladocerans, DeMott (1999) and Zhu et al. (2013) reported that toxic *Microcystis* inhibits feeding, but they exhibited an unexpected diversity and complexity in the pattern of feeding inhibition. *Daphnia magna* exhibits the strongest inhibition of feeding after 1 h of exposure to toxic food, but substantially recovers after 24 h in the same mixtures. This pattern of inhibition and recovery may balance the benefits of reduced toxin ingestion with the disadvantage of reduced energy intake. In the puddle water flea *Moina macrocopa*, the feeding inhibition likely reflects an avoidance behavior towards the toxic food (Zhu et al. 2013). This can explain why small-sized cladocerans often replace dominant *Daphnia* during cyanobacteria blooms.

Furthermore, the example of *Daphnia longispina* (Table 2.1) shows that the response to MCs is clone-specific rather than species-specific. This implies that contrasting results can be expected in different species when more than one clone is studied.

Even in animals with distinct livers, MCs exhibit a type of neurotoxic action, as they alter the swimming behavior of animals exposed to them. MC-LR altered the chronobiology of zebrafish, favoring nighttime activities. Furthermore, the highest exposure concentration decreased the general motility (Fig. 2.6) and even spawning activity and success (Baganz et al. 1998). The underlying mechanisms of reproductive toxicity have recently been discovered in the medaka (Qiao et al. 2016a) (see below). In sunbleak (*Leucaspius delineatus*), the behavioral shift resulted in a reversal, going from a predominantly diurnal activity to a nocturnal one (Baganz et al. 2004).

Later, a reduction in swimming activity at high exposure concentrations of MC-LR was also observed by Kist et al. (2011) again in zebrafish and Cazenave et al. (2008) in the Rio de la Plata onesided livebearer (*Jenynsia multidentata*). The

Table 2.1 Studies on *Microcystis*/microcystin effects on life history traits of selected cladocerans

Cladoceran species	*Microcystis*/microcystin	Effect	References
Bosmina longirostris	MC-LR	Survival↓ at temperature↑	Lamb et al. (2019)
Ceriodaphnia cornuta	*M. aeruginosa* PCC7820	Survival↓	Liu et al. (2006)
	M. aeruginosa PCC7806	Offspring numbers↓, survival time↓, primipara age↑	Gu et al. (2020)
	M. aeruginosa NPLJ-4	Survival↓, fecundity↓, growth↓	de Aquino Santos et al. (2021)
C. dubia	Collected *M. aeruginosa*	Population growth↔	Alva-Martínez et al. (2001)
		Reproduction↓	Lira et al. (2017)
Daphnia carinata	Collected *M. aeruginosa*	Males↑, ephippia↓	Deng et al. (2010)
	Low conc.	2 resting eggs per ephippium	
	High conc.	Empty ephippia	
	M. aeruginosa FACHB-905	Maternal exposure→ unexposed offspring fitness↓	Jiang et al. (2013)
	M. aeruginosa PCC7806	Offspring resistance↑	Akbar et al. (2017)
D. carinata, Moina (irrasa) affinis	Collected *M. aeruginosa* with MC-LR & MC-RR	*Moina*↑, *Daphnia*↓	Shao et al. (2014)
D. galeata	*M. aeruginosa* PCC7806	Feed ingestion↓	Rohrlack et al. (1999a, b, c)
	M. aeruginosa MRC & MRD	Feed ingestion↓, molting↓: Effective MCs other than MC-LR	Kaebernick et al. (2001)
	MC-producing and non-MC--producing *M. aeruginosa*	Intraspecific variations in sensitivity to MC	Drugă et al. (2016)
	Collected *M. aeruginosa*, field observation	Clone turnover due to sexual reproduction	Ma et al. (2021a)
D. laevis	Collected *M. aeruginosa*	Longevity↑, gross reproduction↓, generation time↓	Nandini et al. (2000)
D. longispina	*M. aeruginosa* HUB5–24, HUB063	Food ingestion↓	Henning et al. (1991)
	M. aeruginosa	Longevity↑	Stangenberg (1968)
D. longispina 2 clones	*M. aeruginosa* PCC7820	Longevity↓, growth↓, reproduction↓	Hietala et al. (1995)
D. lumholtzi	*M. aeruginosa* PCC7806	<50%: Primipara age↑	Semyalo et al. (2009)
	M. aeruginosa UTEX LB 2063	Inferior competitor to *D. Pulex*	Fey and Cottingham (2011)
	MC-LR, cyanobacterial extract; multigenerational (chronic) or maternal exposure	**Chronic**: Offspring survival↓ & reproduction↓ **Maternal**: Offspring recovery only partly in F_2	Dao et al. (2018)
	MCs	Survival↓, fecundity↓, reproduction↓	Vo et al. (2020)

(continued)

Table 2.1 (continued)

Cladoceran species	*Microcystis*/microcystin	Effect	References
D. magna	*M. aeruginosa* various cultures	Survival↓, food intake↓ (no interrelation)	Nizan et al. (1986)
	M. aeruginosa HUB5–24, HUB063	Food ingestion↓, survival↓	Henning et al. (1991)
	M. aeruginosa UTEX 2061, PCC7820	Body growth↓, population growth↓, filtering rate↓	Smith and Gilbert (1995)
	Collected *M. aeruginosa*	Food uptake↓↓	Łotocka (2001)
	MC-LR	Not toxic at 3 mg L^{-1}	Reinikainen et al. (2001)
	M. aeruginosa PCC7806	Toxicity depends on metabolites other than MC	Lürling (2003a)
	M. aeruginosa NIVA-CYA140, NIVA-CYA 228/1	Feed ingestion↓, population growth↓, survival↓	Lürling (2003b); Lürling and van der Grinten (2003)
	M. aeruginosa NIVA-CYA140	♂♂,♀♀: Growth↓; ♀♀: Reproduction↓	Lürling and Beekman (2006)
	MC-LR	Oxidative stress↑	Chen et al. (2005)
	M. Aeruginosa PCC7806	Lipid, carbohydrate, amino acid, energy metabolism↓	Lyu et al. (2016)
	M. aeruginosa NIVA-CYA 228/1	Tolerance of filial generation↑	Gustafsson et al. (2005)
	M. aeruginosa PCC7806	Protease activity↓	Agrawal et al. (2005b)
	M. Aeruginosa PCC7806 WT	Two genes of basic metabolism↑	Schwarzenberger et al. (2009)
		Population growth↓[a], neonate size↓	Sarnelle et al. (2010)
	MC-LR	Longevity↓, growth↓, reproduction↓	Liu et al. (2011)
	Planktothrix agardhii PMC 75.02 extracts	Growth↑, reproduction↓	Hulot et al. (2012)
	MC-RR	Growth↓, reproduction↓	
	MC-LR	Offspring size ↓, offspring survival↑	Ortiz-Rodríguez et al. (2012)
	M. aeruginosa PCC7806	Reproduction↓	De Coninck et al. (2013)
	M. aeruginosa, MC-LR	Growth↓, reproduction↓, digestive enzymes↓, offspring DNA anomaly↑	Wu et al. (2013)
	MC, *M. Aeruginosa* PCC7806	Toxicity depends on metabolites other than MC	Jungmann (1992, 1995); Smutná et al. (2014)
	Low MC-LR concentrations	Median swimming speed↑	Nikitin and Latypova (2014)

	MC-LR + ammonia		Interactive effect on: Time to first clutch↑, size at first clutch↓, number of clutches↓, total offspring per female↓	Yang et al. (2012)
	MC-LR + ammonia	High	Offspring survival↓	Zhu et al. (2015)
		Low	Offspring survival↔	
	Collected *M. aeruginosa* MC-LR, -RR, -YR		Individual growth↓, population growth↓	Chi et al. (2016)
	M. Aeruginosa PCC7806		Offspring resistance↑	Akbar et al. (2017)
			Maturation↓, reproduction↓	Wang et al. (2019b)
			Chloride tolerance↓	Li et al. (2020b)
	MC-LR, MC-LF		Swimming performance↓	Pawlik-Skowrońska and Bownik (2021)
	Anabaenopeptin-B		Swimming performance↓	
	MC-LR, MC-LF + anabaenopeptin-B		Swimming performance↓↓	
D. mitsukuri 2 clones	*M. aeruginosa* PCC7806		Maturation size↓, offspring Number↓, tail spine defense↔	Lu et al. (2021)
D. pulex various clones	*M. aeruginosa* PCC7820		Longevity↓, growth↓, reproduction↓	Hietala et al. (1995, 1997)
D. pulex	MC-LR		Surviral↓	(Pawlik-Skowrońska et al. 2019)
	Microcystis spp. extract		Survival↔	
	M. aeruginosa PCC7820		Survival↓, dead offspring↑	Reinikainen et al. (1994, 1995)
	M. aeruginosa UTEX 2061		Ability to interfere with *Keratella cochlearis*↓	Sartonov (1995)
	M. Aeruginosa PCC7820		Ephippia number↑	Laurén-Määttä et al. (1997)
	M. aeruginosa UTEX 2061, PCC7820		Body growth↓, population growth↓, filtering rate↓	Smith and Gilbert (1995)
	Collected *M. aeruginosa*		Population growth↔	Alva-Martínez et al. (2001)
	M. Aeruginosa PCC7820		Longevity↓	Liu et al. (2005)
	Collected *M. aeruginosa*		Males↑, ephippia↑	Deng et al. (2010)
	Low conc.		2 resting eggs	
	High conc.		Empty ephippia	
	M. aeruginosa UTEX LB2385		Fitness↓ based on energy budget	Asselman et al. (2012)
	M. aeruginosa FACHB-905		Offspring tolerance↑	Li and Jiang et al. (2014)
	M. aeruginosa PCC7820		Empty ephippia↑	Li et al. (2014)
	M. aeruginosa PCC7806		Body size↓, reproduction↓, energy allocation to reproduction↑	Zhou et al. (2020)

(continued)

Table 2.1 (continued)

Cladoceran species	*Microcystis*/microcystin	Effect	References
D. pulicaria	*M. aeruginosa* MRC & MRD	Feed ingestion↓, molting↓: Effective MCs other than MC-LR	Kaebernick et al. (2001)
	M. aeruginosa UTCC436	Reversible impairment of feeding behavior	Ghadouani et al. (2004)
	MC-LR	Nonreversible impairment of feeding behavior	
	M. aeruginosa UTEX2667	Long-term tolerance↑ to dietary toxic cyanobacteria	Sarnelle and Wilson (2005)
	Planktothrix rubescens TCC 29–1, *P. Agardhii* TCC 83–2	*Daphnia*-mediated toxin transfer in the food chain	Oberhaus et al. (2007)
	M. aeruginosa PCC7806	Maturation↓, primipara size↓, Eggs↓, newborns↓	Cerbin et al. (2010)
	Collected *Microcystis* & *Dolichospermum*	Phytoplankton control↔	Chislock et al. (2013)
D. pulicaria 2 clones	*M. aeruginosa* UTEX 2667, MC	Strain-specific growth↓ & population growth↓	Wilson and Hay (2007)
D. similis	MC-LR	Heartbeat rate↓	Herrera et al. (2014)
D. Similoides & *Moina micrura*	*M. Aeruginosa* PCC7806	*Daphnia* ↑, *Moina*↓	Tang et al. (2017)
D. similoides, 2 clones		Body growth↓, molting frequency↓	Lyu et al. (2019)
D. similoides	MC-LR	Reproduction↓	Cao et al. (2014)
	M. aeruginosa F469 & *Raphidiopsis raciborskii* N8	Growth↓, reproduction↓ F469 > N8 reduction	Huang et al. (2015)
D. sinensis	*M. aeruginosa* 7820	Empty ephippia↑	Li et al. (2014)
	Collected *M. aeruginosa*	Maternal exposure → offspring number↓, offspring fitness↑	Peng et al. (2018)
	Collected *M. aeruginosa*	Strain-specific nuclear receptor transcription↓	Xu et al. (2018b)
	M. aeruginosa F469 & *C. raciborskii* N8	Growth↓, survival↓, reproduction↓	Lei et al. (2020)
	M. aeruginosa	*Hox* differently expressed in various clones	Xu et al. (2022)
Daphnia spp.	Microviridin J	Protease activity↓	Rohrlack et al. (2003)
Diaphanosoma brachyurum	Collected *M. aeruginosa*	Assimilation↓↓	Hanazato and Yasuno (1988)
D. celebensis	MC-LR	Oxidative stress↑	Won et al. (2021)
D. mongolianum	*M. aeruginosa*, Madrid University	Growth↔	Nandini et al. (2021)
Eubosmina coregoni	*M. Aeruginosa* HUB5–24, HUB063	Food ingestion↑, survival↔	Henning et al. (1991)

Macrothrix spinosa	*M. aeruginosa* NPLJ-4	Survival↓, fecundity↓, growth↓	de Aquino Santos et al. (2021)
Moina (irrasa) affinis, D. carinata	Collected *M. aeruginosa* with MC-LR & MC-RR	*Moina*↑, *daphnia*↓	Shao et al. (2014)
M. macrocopa	Collected *M. aeruginosa*	Growth↓, reproduction↓, assimilation↓	Hanazato and Yasuno (1984, 1988)
	Collected *M. aeruginosa*	Population growth↓	Alva-Martínez et al. (2001)
	M. aeruginosa PCC7806	Protease↓	Agrawal et al. (2005a)
M. micrura	*M. aeruginosa* PCC7820	Survival↓	Liu et al. (2006)
	Collected *Microcystis* spp.	Reproduction↓	Shakri et al. (2020)
Simocephalus vetulus	*M. aeruginosa* FACHB-905	Clone-specific population increase rate↓	Chen et al. (2015)

[a]Noteworthy, Sarnelle et al. (2010) point out that although *Daphnia* survival, population growth, and neonate size were negatively affected by MC, *Daphnia* populations that have prior experience with toxic cyanobacteria may show positive population growth even at high concentrations of cyanobacterial toxins

↑support/upregulation/increase; ↔ no obvious adverse effect; ↓reduction/downregulation/delay/inhibition; WT wildtype

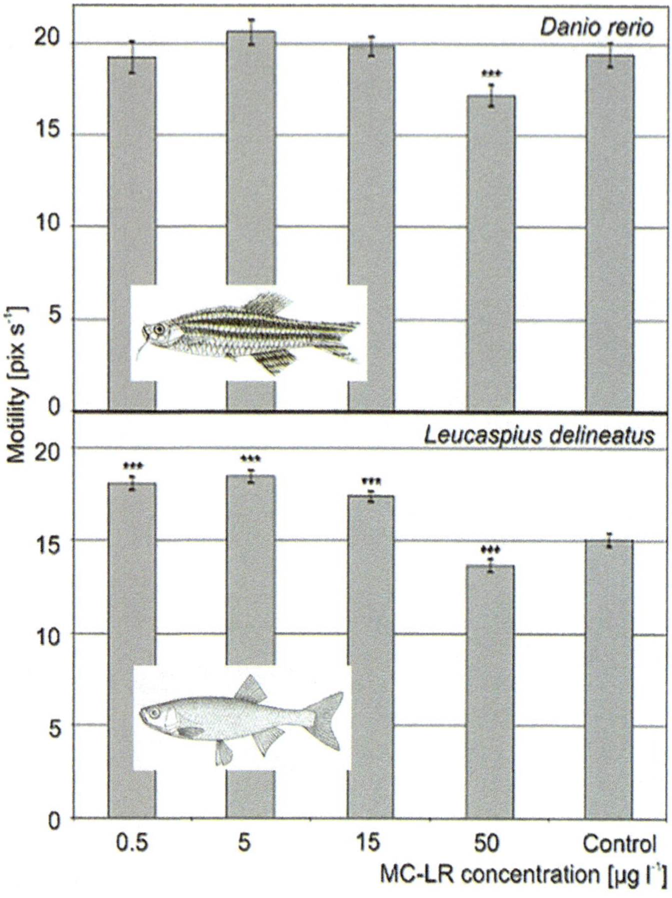

Fig. 2.6 Average motility of *Danio rerio* and *Leucaspius delineatus* during exposure to different concentrations of MC-LR over the whole measuring time per day (23 h per day) and over the whole exposure period. Means and 95% confidence intervals are shown (from Baganz et al. (2004) with permission from Wiley & Sons) (images in public domain, credit to Day (1878) and Seeley (1886))

latter authors also showed that this coincides with the onset of the detoxification activity via glutathione transferase. Therefore, it can be hypothesized that the exposed individuals reallocate their energy, favoring detoxification over swimming activity.

Under MC-LR stress, red claw crayfish (*Cherax quadricarinatus*) exhibit notable behavioral changes. Initially, they display increased activity and restlessness following injection, which subsides within an hour. Subsequently, their swimming activity decreases, and they exhibit less aggressive behavior compared to pre-injection levels. Importantly, crayfish mortality did not occur in all experimental groups (Lu et al. 2023).

The neurotoxic mechanism of MCs is gaining increasing attention. For instance, Hu et al. (2016) provide evidence that MCs can cross the blood-brain barrier via organic anion-transporting polypeptides (Faltermann et al. 2014, 2016; Steiner et al. 2016). Once inside the brain, MCs induce neurostructural, functional, and behavioral changes, impacting neurotransmission, ion channels, signal transduction, oxidative stress, and cytoskeletal integrity. Evidence of the neurotoxic action of MCs in fish also includes accumulation. Qian et al. (2018) reported that *Microcystis aeruginosa* exposure affected the dopaminergic and cholinergic systems in the early life stages of zebrafish by changing the transcription of key nervous system genes.

Consequently, the neurotoxicity of MCs is a multi-pathway process. In addition to the direct neurotoxic mechanism, Hu et al. (2016) also propose an indirect one induced by hepatotoxicity through a medium, i.e., ammonia. Recent whole-genome transcriptomic studies are progressively uncovering the intricate mechanisms underlying cyanotoxin intoxication. Furthermore, they reveal the involvement of miRNAs in the toxic action of MCs.

Building on the behavioral studies referenced earlier and insights from research on hepatopancreas-bearing animals, MCs have also been shown to affect the nervous system in vertebrates, as evidenced by biomolecular studies in zebrafish (Yan et al. 2017). Upon exposure to MC-LR, the expression of genes encoding γ-aminobutyric acid (GABA) type A receptor α1 subunit (*gabra1*), glutamate decarboxylase 1b (*gad1b*), and glutaminase (*glsa*) is upregulated in a concentration-dependent manner. In contrast, the GABA transporter (*gat1*) gene is down-regulated. These changes indicate an impact on the GABA and glutamate metabolic pathways, key components of the GABA neurotransmitter system, ultimately leading to its disruption.

In a companion paper, Wu et al. (2017) found parental transfer of MC-LR induced transgenerational effects in unexposed zebrafish offspring. At environmentally relevant concentrations, MC-LR interferes with the neurotransmitter systems (including the dopamine signaling system, serotonin system, and cholinergic system) and disrupts neuronal development.

Finally, Hinojosa et al. (2019) highlight potential underlying mechanisms: since MCs can cross the blood-brain barrier through organic anion-transporting polypeptides (OATPs), they can induce damage to the nervous system, as demonstrated in a variety of non-mammalian species, including zebrafish (Meier-Abt et al. 2006). Subsequently, Steiner et al. (2016) and Marić et al. (2021) show a congener-specific uptake of MCs, most likely dependent on their basic physicochemical properties, with lipophilicity being the most obvious determinant. The presence of nominally highly toxic MC congeners, such as MC-LR, in cyanobacterial blooms does, therefore, not necessarily translate into a high ecological risk if their ability to be transported by Oatp1d1 is limited. Conversely, nominally less toxic MC variants, like MC-LF or MC-LA, which are taken up by Oatp1d1, can pose significant ecological threats. This is particularly relevant for tissues and organs with high expression of Oatp1d1 or other related transporters with overlapping substrate specificities, where the increased uptake can lead to pronounced neurotoxic and renal effects (Marić et al. 2021).

Further mechanisms of neurotoxicity are involved: Xu et al. (2020) exposed male Nile tilapia to *Microcystis* subchronically to understand the neurotoxicity of *Microcystis* at the proteomic level. The authors found that *Microcystis* impacts proteins related to energy metabolism, the ubiquitin-proteasome system, signal transduction, and transport. Yu et al. (2021) enlarge the catalog of zebrafish brain damages induced by *M. aeruginosa* bloom: The neurotoxicity of this cyanobacterium seriously damages the neurotransmitter conduction in the nervous system and brain information storage and transmission, and makes it more susceptible to neurological diseases, such as Alzheimer's disease, Parkinson's disease, or epileptic encephalopathy.

Cai et al. (2022b) observed abnormal neurobehavior in early life stages of fish after exposure to cyanobacterial exudates. The authors exposed embryos of the golden-line barbel (*Sinocyclocheilus grahami*), an endangered endemic fish in cyanobacteria-dominated Lake Danchi, Kunming, China, to *M. aeruginosa* exudates. These exudates inhibited fish neurobehavior, including touch response, social frequency, swimming distance, and swimming speed (although no exposure figures are given, these findings seem to agree well with those mentioned above in *Leucaspius delineatus* (Baganz et al. 2004)). Additionally, they increased the response to light stimulation. Neurobehavioral suppression results from a decrease in the concentration of excitatory neurotransmitters, such as acetylcholine and dopamine, despite an increase in the number of receptors. *M. aeruginosa* exudates can affect gene and protein expression of neurotransmitters, synthetic and/or degrading enzymes, and receptors. Specifically, neurotransmitter receptors are overexpressed in response to ligand deprivation (Doğanlar et al. 2019). Similarly, exudates increase the expression of five target receptors while decreasing the levels of their respective neurotransmitters. This study demonstrates that the toxicity of *M. aeruginosa* exudates induces changes in the neurobehavior of *S. grahami*, ranging from alterations in gene expression to changes in the production of metabolites related to neurochemical compounds.

Recently, Luan et al. (2024) presented an intriguing approach to mitigate neurotoxic activity in MC-exposed zebrafish. As expected, MC-LR exposure inhibits the swimming speed; however, the presence of *Lactobacillus rhamnosus* mitigates this abnormality. To elucidate the mechanism of how *L. rhamnosus* alleviates MC-LR-induced neurotoxicity, the authors examined the bioaccumulation of MC-LR, changes in neurotransmitters, immune biochemical indicators, and hormone content of the hypothalamic-pituitary-interrenal (HPI) axis in zebrafish along the gut-brain axis. This paper demonstrates that *L. rhamnosus* reverses the abnormal swimming behavior and ultimately alleviates neurotoxicity in zebrafish by modulating intestinal and brain neural signaling, neuroinflammation, and the HPI axis responses—a potential template for broader applications of probiotics in the aquaculture industry.

2.1.4 Immunotoxins

Recently, Lin et al. (2021) combined the classical MC-mediated protein phosphatase inhibition pathway with neurotoxicity in fish. MCs can induce a pro-inflammatory response and lead to immune dysfunction by impairing immunocytes and influencing the secretion of immune molecules. The possible mechanisms of MC-induced immunotoxicity are graphically summarized in Fig. 2.7, implying that MC exposure poses a critical threat to the immune function of fish. Moreover, recent studies have reported that parental exposure to MC-LR can interfere with the immune function of F_1 larvae, indicating that MCs can result in cross-generational immunotoxicity (Lin et al. 2020) (see next Section).

Cytokine gene expression is the parameter most frequently studied in aquatic animals (fishes, shrimps, and bivalves) (Li et al. 2009a; Wei et al. 2009; Li et al. 2013a; Rymuszka and Adaszek 2012a, 2013; Juhel et al. 2015; Chen et al. 2016a; Gao et al. 2017; Li et al. 2019). These studies present evidence that exposure to MCs has an adverse effect on the immune system, leading to immunomodulation. Fish immunity tends to proceed toward an immunostimulatory response at low MC concentrations but toward an immunosuppressive response at high concentrations (Qiao et al. 2013; Juhel et al. 2015; Lin et al. 2017; Qian et al. 2019) (from Diez-Quijada et al. (2021), references added).

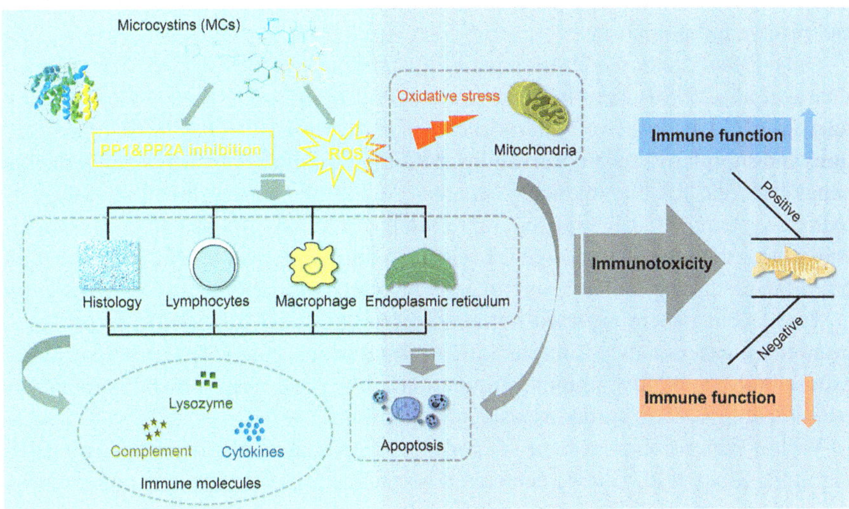

Fig. 2.7 Schematic overview of the immunotoxicity of MCs in fish. MCs are specific inhibitors of serine/threonine PP1 and PP2A, resulting in an imbalance of phosphorylation and dephosphorylation of key proteins, which subsequently causes abnormal cell proliferation and damage to the cytoskeletal organization. MC exposure is also associated with the overproduction of reactive oxygen species (ROS) and oxidative stress in immune organs, leading to apoptosis and immune dysfunction. MC-induced immuno-toxicity greatly depends on exposure routes, time, and concentrations/doses (from Lin et al. (2021), credit to MDPI, Basel)

Wei et al. (2018) confirmed the immunotoxicity of MC-LR in grass carp, specifically in the complement and coagulation cascades. Furthermore, MC-LR can induce apoptosis (Wei et al. 2019), whereby *caspase-3*, *caspase-9*, and the recently in grass carp discovered *Cicaspase-8*-like might play an indispensable role. In addition to modulating caspase genes, the same laboratory has shown that increasing MC-LR doses reduces the hepatic contents of tumor protein P53 (Liu et al. 2021), also known as cellular tumor antigen or guardian of the genome (Toufektchan and Toledo 2018).

In cell cultures, Rymuszka and Adaszek (2012a) reported the expression of both pro- and anti-inflammatory cytokines in carp blood and head kidney leukocytes exposed to MC-stress. The authors demonstrate a crucial role of the disturbance in the expression of pro- and anti-inflammatory cytokines during MC-induced stress. These results confirm that the modulatory action of cyanotoxins is accompanied by different patterns of cytokine expression in immune cells. *il-1β* and *tnf-α* genes are highly expressed in cells incubated with 0.01 mg mL^{-1} MC, but decreased with 0.1 mg mL^{-1} MC. Exposure of blood or head kidney leukocytes to MC-LR results in significant increases in *il-10* and *tgf-β* expression at the higher concentration (0.1 mg mL^{-1}). Overall, the response in head kidney cells appears to be more sensitive to cyanotoxins than that in blood cells. This is because the head kidney is the major hematopoietic organ in fish (a specific bone marrow equivalent). At this point, determining the time course of pro- and anti-inflammatory gene expression would have been important, as the innate immune system's defense always begins with a transient pro-inflammatory stimulation to initiate the anti-inflammatory machinery and resolve inflammation.

Only a few studies address aquatic invertebrates in this regard. For instance, Zhang et al. (2019a) examined the immunological effects of two environmentally relevant MC-LR doses on *Macrobrachium rosenbergii*. Through histological sections and submicroscopic observations, MC-LR exhibits a dose-dependent effect on hepatopancreas toxicity, inducing hepatocyte apoptosis and dysfunction of the antioxidant system. MC-LR exposure represses the immune response of *M. rosenbergii* and affects the expression of genes related to immunity. Under MC-LR exposure, bacterial infections result in a significantly increased mortality rate of *M. rosenbergii*.

Two recent papers reported hormetic, or as they call it, dualistic responses in cyanotoxin-exposed fish. Lin et al. (2017) observed a hormetic response in MC-LR-exposed male zebrafish: immunostimulation at low concentrations and immunosuppression at high concentrations, as depicted in Fig. 2.8.

With a major component of *M. aeruginosa* exudates, phytosphingosine (PHS, Fig. 2.9), Zhao et al. (2024) detected a hormetic response: immunostimulation at low concentration and immunosuppression at high concentration in the endemic *Sinocyclocheilus graham*, with the highest number of genes expressed at low exposure concentrations. PHS induces immunostimulation by increasing leukocyte numbers, IgM, and lysozyme levels in serum and skin mucus, upregulating protease content and activity in skin mucus, increasing pro-immune factors, and directly activating the TLR/NF-κB, FoxO1, and PPARγ pathways. Immunosuppression by PHS is characterized by tissue damage, a lower ACH50 level, and apoptosis and

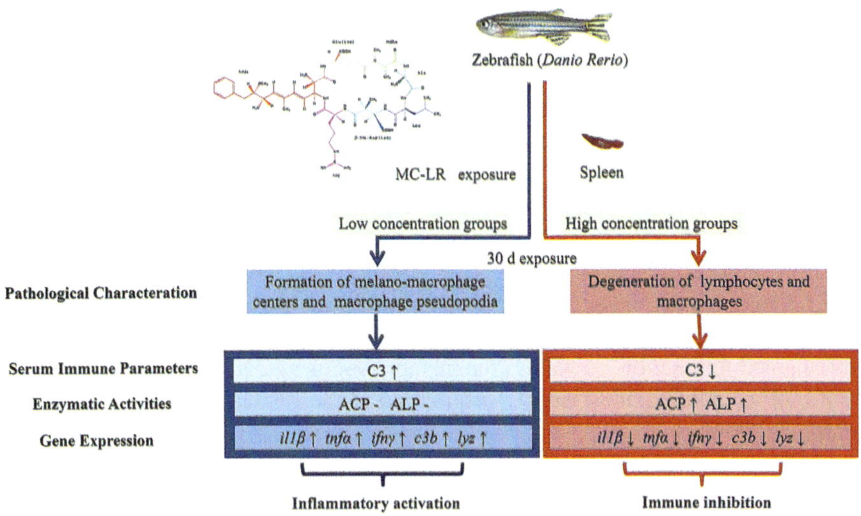

Fig. 2.8 Hormetic immunomodulation of sub-chronic microcystin-LR exposure on the innate--immune defense system in male zebrafish (from Lin et al. (2017), with permission from Elsevier)

Fig. 2.9 Structure of phytosphingosine

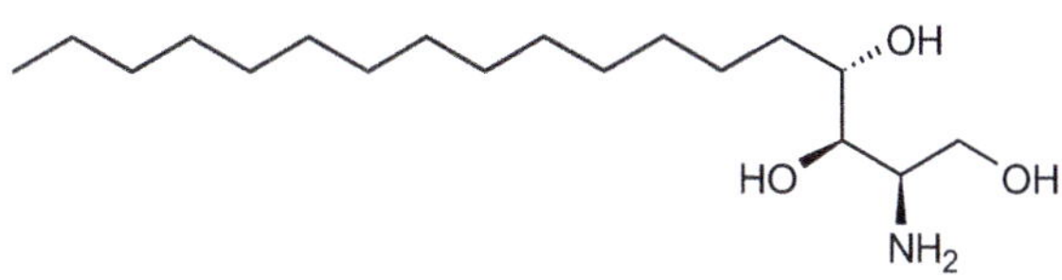

necroptosis of immune cells, as indicated by the activation of the p53 pathway (Liu and Gu 2022). Other cytotoxicity studies show that PHS induces ROS-dependent apoptosis in cells (Li et al. 2022) and inhibits cell proliferation by damaging DNA (Sun et al. 2023).

Hormesis has also been observed in vitro with MC-LR and anatoxin-a: exposure at relatively low concentrations stimulates some immune cell activities in carp blood and head kidney leukocytes (Rymuszka and Adaszek 2012b) and macrophage activity in crucian carp fed cyanobacteria (Qiao et al. 2013). However, higher concentrations of cyanotoxins suppress immune function in both systems studied.

2.1.5 Endoplasmic Stress

Chen et al. (2017b) histopathologically confirmed that MC-LR damages the liver, and the effects are more pronounced in fish chronically exposed to 10 μg L^{-1} than to 1 μg L^{-1} MC-LR. A metabolomic analysis also revealed alterations in hepatic function, which encompassed changes in several metabolic pathways, including those involving small molecules related to energy, glucose, lipids, and amino acid metabolism. This is accompanied by changes in the transcription of genes involved in glycolysis (*pfk1b, peak*), tricarboxylic acid cycle (*pdha1a, sdha*), electron transport chain (*nd1, cox1*), and oxidative phosphorylation (*atp6*). All of these genes were significantly downregulated in the livers of zebrafish exposed to 10 μg L^{-1} MC-LR. For fish exposed to 1 μg L^{-1} MC-LR, levels of mRNA for *pdha1* and *atp6* were elevated. Alterations of transcripts of genes encode proteins essential for protein folding (*pdi, ppcti*) and ER stress (*ern1, xbp-1s, eif2ak3, eif2s1, atf6, atf4, hspa5, hsp90b1, chop*). Abundances of transcripts of *pdi, ppcti, atf6, atf4*, and *hspa5* were significantly greater in fish exposed to 1 μg L^{-1} MC-LR than in the unexposed controls. In individuals exposed to 10 μg L^{-1} MC-LR, elevated levels of mRNA for *pdi, ppcti, ern1, xbp-1s, eif2ak3, atf6, atf4, hspa5*, and *hsp90b1* were observed compared to control animals. The ER is the subcellular organelle responsible for protein folding and maturation, as well as the metabolism of lipids and the overall maintenance of calcium homeostasis. A perturbation of the ER leads to an overload with unfolded proteins.

Furthermore, lactate was significantly elevated in individuals exposed to MC-LR compared to unexposed controls. This indicated a shift toward anaerobic metabolism, which was confirmed by impaired mitochondrial respiration. Proteomics revealed that MC-LR significantly influenced the expression of multiple proteins, including those involved in protein folding and metabolism. Clearly, ER stress contributes to the disruption of lipid metabolism in the liver of zebrafish exposed to MC-LR.

In contrast to MC-sensitive zebrafish, silver carp (*Hypophthalmichthys molitrix*) is known for its high tolerance to MCs, which might be due to the high basic GSH level in the liver or an increased GSH synthesis. To fully understand the mechanism behind the high tolerance of silver carp to MCs, Fu and He (2012) evaluated whether the genes involved in glutathione synthesis were under positive selection in silver carp. In fact, the authors recovered 41 of 44 genes involved in glutathione biosynthesis, and of these, five genes were identified as having undergone positive selection between silver carp and zebrafish. Positive selection pressure on the *gst* gene might be the result of the adaptation of silver carp to eutrophic bodies.

Even transgenerational effects can be anticipated in organisms exposed to cyanotoxins. Lin et al. (2020) found MC-LR both in P_0 gonads and in F_1 embryos and indicated that MC-LR can be transferred directly from the P_0 parental fish to F_1 offspring. The adverse effects on sex hormone levels, sexual development, and fecundity in P_0 generation, along with abnormal development in F1 offspring, are observed. Furthermore, downregulation of antioxidant genes (*cat, mn-sod, gpx1a*) and upregulation of innate immune-related genes (*tlr4a, tnfα, myd88, il1β*) as well as increased contents of pro-inflammatory cytokines (*tnf-α, il-1β, il-6*) are noticed in F_1 offspring without and with being exposed to MC-LR themselves. However, significant differences between the two F_1 embryo treatments demonstrate that continuous MC-LR exposure can result in a higher degree of inflammatory response compared to those without MC-LR exposure. In summary, MC-LR exhibits cross-generational immunotoxic effects in fish by disrupting the antioxidant system and triggering an inflammatory response.

Invertebrates studied so far exhibit nearly identical transgenerational behavior. By applying shotgun proteomics, Shahmohamadloo et al. (2020) identified proteins in *Daphnia magna* parents and neonates that are significantly expressed in the presence of cell-bound MCs at realistic concentrations to freshwater ecosystems. Moreover, the somatic growth rate of *D. magna* demonstrates a significant decrease in the fitness of parents chronically exposed to toxic *Microcystis*. In *D. magna* parents, the authors identified a significant increase in the abundance of proteins with key functions related to reproductive success and development, as well as detoxification of superoxide radicals and motoric activity. These findings indicate the investment of parents in growth, development, and survival in the presence of cell-bound MCs. In *D. magna* neonates, they identified the opposite—a significant decrease in the abundance of proteins with key functions related to apoptosis, metabolism, DNA damage repair, and immunity. These findings indicate that neonates have a decreased ability to support key functions related to their own growth and development in the presence of cell-bound MCs.

Given the stress responses observed in *D. magna* in the present study, there is potential for cell-bound MCs to influence food web dynamics in freshwater ecosystems experiencing harmful algal blooms dominated by Mi*crocystis*. Future work in the area of environmental toxicogenomics is needed to characterize the effects of dietary exposure and accumulation of MCs in *Daphnia* over multiple generations, to assess whether the protein responses observed in this study can be repaired or reestablished to normal in *Daphnia* after exposure to a toxic *Microcystis* bloom.

2.1.6 Reproductive Toxicity

In the hermaphrodite model animal *C. elegans*, MC-LR has been shown to affect reproduction through multiple pathways. For example, Li et al. (2015b) demonstrate that MC-LR exposure induces abnormal sperm morphology and causes severe defects in sperm activation, transactivation, sperm behavior, and competition. Additionally, the expression of *spe-10*, *spe-15*, and *fer-1*[5] is significantly decreased.

In males, spermatogenesis, in which spermatogonia develop and differentiate to form sperm, is vital for healthy reproduction. MC-LR has been found to have deleterious effects on spermatogonia, with various types of non-coding RNAs playing a central role in post-transcriptional regulation (Ma et al. 2021b).

The blood-testis barrier (BTB) is essential for the maintenance of normal male germ cell development, and is formed by basal tight junctions between adjacent Sertoli cells. These cells are "nurse" cells of the testicles that are part of a seminiferous tubule and help in the process of spermatogenesis (Griswold 1998). The mechanism of male reproductive toxicity is to disturb the BTB integrity and spermatogenesis mainly through affecting cell junctions and inducing apoptosis. In addition, cell

[5] *spe-10* is required for spermatids development in spermatogenesis and fertility (Gleason et al. 2006), while *spe-15* is related to spermiogenesis or sperm activation (Kelleher et al. 2000). *fer-1* plays a role in sperm pseudopods extension and motility (Krajacic et al. 2009).

growth and proliferation, histone modification, and cell–cell signaling pathways are also affected by MC-LR (Ma et al. 2021b).

Evidence is accumulating, primarily from exposure studies rather than dietary studies, that cyanotoxins have the potential to cause transgenerational adverse effects in invertebrates and fish. For the sake of completeness, a brief overview of the key findings is provided.

2.1.6.1 Invertebrates

One possibility to ameliorate MC effects is to biotransform MC via glutathione transferase (GST) to a less toxic glutathione conjugate. Carneiro et al. (2017) demonstrated that modulation of different hepatic GST classes in three bivalve species exposed to purified MC-LR is the biochemical basis for their varying sensitivity to this toxin. This pathway is thought to play a key role in *Daphnias*' ability to resist MC and may explain the increased tolerance observed in the offspring of mothers previously exposed to toxic cyanobacteria.

Thus, Ortiz-Rodríguez et al. (2012) conducted multigenerational experiments with *Daphnia magna* by exposing the parental generation to MC for either 1 or 7 days. They then evaluated the enzyme-mediated tolerance of the offspring to MC by measuring the acute effects on biotransformation, antioxidant, and metabolic enzyme activity. Additionally, they performed 21-day chronic tests with the offspring to assess toxicity and growth outcomes. Seven days of exposure of the parental generation to MC induces higher activity of GST and malate dehydrogenase in the offspring and enables them to increase the catalase activity when challenged with MC, whereas 1 day of exposure of the parental generation did not. Offspring from non-exposed and 1-day-exposed mothers suffer decreased survival when exposed to MC compared with offspring from 7-day-exposed mothers; survival is correlated with the elevated activity of GST, malate dehydrogenase, and catalase, indicating maternal transfer of activation factors. However, increased survival occurs at the expense of individual growth. In this study, the transgenerational mechanism remains obscure.

Radersma et al. (2018) confirm that maternal exposure of *Daphnia magna* to MC-producing cyanobacteria can result in increased offspring tolerance to MC. However, the timing of maternal effects, their magnitude relative to the negative fitness consequences for offspring, and weak overall support across studies shed doubt on their adaptive function. The authors, therefore, assume that this maternal effect represents an example of how phenotypic accommodation to stressful conditions can span generations rather than a mechanism that has been selected for its ability to transmit information from parents to offspring.

In contrast, Dao et al. (2018) demonstrated in the tropical *Daphnia lumholtzi* that dissolved MCs at an exposure level of 1 mg L-1 do not affect the life history traits of P_0 *Daphnia*. Instead, continuous toxin exposure impairs the survivorship and delays the maturation and reproduction of the daphnids in F1 and F2 generations.

Simulating aquaculture breeding conditions, Zhang et al. (2019b) exposed male *Macrobrachium rosenbergii* to environmentally relevant concentrations of MC-LR for 1, 2, and 3 weeks. MC-LR enters the testes, lowers testosterone levels in the hemolymph, damages testicular germ cells, mitochondria, and cell junctions, and inhibits testicular development. When studying the sister species *M. nipponense*, Sun et al. (2021) take it a step further by demonstrating that larval health is also affected by exposed males, as evidenced by the upregulation of apoptosis and deterioration of antioxidant capabilities and immunity.

2.1.6.2 Fishes

The daily activity at the time of mating and spawning disappears in female and male zebrafish exposed to 50 μg L^{-1} MC-LR for 6 days. Along with this eliminated activity, no spawned eggs are found. Suppose this response also occurs in native fish species, changes in species composition in eutrophic waters may not only be attributed to losses of spawning sites but also to the direct effects of cyanotoxins on fish reproduction (Baganz et al. 1998).

Zi et al. (2018) tested an endangered fish species (*Sinocyclocheilus grahami*) in Dianchi Lake, Kunming, China, to assess the risks of cell-free exudates produced by the dominant cyanobacterium (*M. aeruginosa*) on embryo development and the molecular mechanisms responsible. Exudates are associated with reduced fertilization and hatching rates, and increased mortality and malformations, typically associated with heart failure. Proteomic analysis revealed that the two most altered proteins are closely associated with heart function.

In a study using female and male medaka, individuals were exposed to 5 μg L^{-1} MC-LR for 30 days (Trinchet et al. 2011). In the testis of treated fish, in addition to abnormalities of cellular division that sometimes occur during normal spermatogenesis, numerous lytic areas occur in the seminiferous tubules, indicating an increase in abnormal cell divisions progressing towards death. MC-LR also has a significant impact on the ovary, particularly the destruction of gonadosomatic tissue, which reduces the area of the gonads and results in a reduction in vitellus storage, potentially due to changes in the relationships between the follicular cells and the oocytes. Such effects following MC-LR treatment, particularly the decrease in yolk, are certainly associated with the significantly lower number of eggs spawned per treated female (Chen et al. 2016b). Some other high-impact studies are listed in Table 2.2.

Similar to the above study on *M. nipponense*, Lin et al. (2020) examined the transgenerational effects of MC-LR in zebrafish. Adult zebrafish pairs are exposed for 60 days, and the embryos (F_1) are hatched without or with continued MC-LR exposures at the same concentrations until 5 days post-fertilization. The heat map of gene transcriptions related to antioxidant capacity and immune function in F_1 larvae is presented in Fig. 2.10. For antioxidant defense genes, the mRNA levels of *cat*, *mn-sod*, *cu/zn-sod*, and *gpx1a* show downregulation. Furthermore, transcript levels of *tlr4a*, *myd88*, and pro-inflammatory *il-1β* and *tnfα* are upregulated. This, together with the downregulation of *lyz* and *c3b* mRNAs in F_1 larvae, indicates a weakening of the innate immune system.

Ma et al. (2021b) summarized the current understanding of reproductive toxicity in an educational graphic (Fig. 2.11).

Notably, Liu et al. (2024a) fed female zebrafish probiotics to mitigate the adverse effects of cyanotoxins. They used the probiotic *Lactobacillus rhamnosus* GG to

Table 2.2 Reproductive toxicity of microcystins in zebrafish model

Gender/life stage	Microcystin; dosis (μg L^{-1})	Observed effect	References
♀, ♂	MC-LR; 1, 5, 20	F_1 growth↓, F_1 immunity↓, F_1 LPO↑	Liu et al. (2014)
♀, ♂	MC-LR; 0, 1, 5, 25	♀♀: T4↓; T3↔; ♂♂: T4↔; T3↔; F_1: Growth↓, THs↓	Cheng et al. (2017)
♀, ♂	MC-RR; 0, 1, 5, 25	F_1 larvae: MC-LR↑, developmental neurotoxicity↑	Wu et al. (2017)
♀, ♂	MC-LR; 0, 0.4, 2, 10	♂♂: E2↑, T↓; F_1 larvae: MC-LR↑, hatchability↓, *cat*↓, *mn-sod*↓, *gpx1a*↓, *tlr4a*↑, *myd88*↑, *tnf-α*↑, *IL-1β*↑, TNF-α↑, IL-1β↑, IL-6↑ (Fig. 2.10)	Lin et al. (2020)
♀♀	MC-LR; 2, 10, 50	Egg production↓, fertilization↓, hatchability↓: At 10: E2↑, VTG↑; at 50: E2↓, VTG↓	Zhao et al. (2015)
♀♀	MC-LR, i.p. 50, 200 μg kg^{-1} BW	Oxidative stress↑, ovary structure↓, ovary function↓	Hou et al. (2014)
♀♀	MC-LR; 1, 10, 100	Meiotic maturation↓ due to phosphorylation disorder in oocytes	Liu et al. (2019)
♀♀	*M. Aeruginosa* FACHB-905, 3.7 × 10^5 cells mL^{-1}	Fertilization↓, hatchability↓, E2↓, T↓, hepatic *vtg1*↓, plasma VTG↑	Liu et al. (2018)
♀♀	MC-LR; 0, 1, 50	E2↑, T↑, VTG↑, reversibility of reproductive impairments	Kawan et al. (2019)
♂♂	MC-LR; 0, 0.3, 3, 30	Growth↓, T → E2↑, hepatic *vtg1*↑ → feminization of ♂♂	Su et al. (2016)
♂♂	MC-RR; 2000 μg kg^{-1} BW	Cytoskeleton ↓, oxidative stress↑	Zhao et al. (2012a)
♂♂	MC-LR; 0, 5, 20	F_1 larvae: Survival↓, hatchability↓, heart rate↓, malformation↑	Zhao et al. (2021)
Developing embryos	MC-RR; 0, 300, 1000, 3000	Growth↓, THs↓: → developmental toxicity	Xie et al. (2015)
Larvae	MC-LR, 500 MC-RR, 500	T3↑, T4↑ T3↔, T4↔	Guo et al. (2024)

↓ reduction/deterioration, ↑ increase, *BW* body weight, E2 17β-estradiol, *VTG* vitellogenin, *T* testosterone, *THs* thyroid hormones, *T3* triiodothyronine, *T4* thyroxine, ***cat*** *catalase*, ***mn-sod*** *manganese superoxide dismutase*, ***gpx1a*** *glutathione peroxidase 1a*, ***tlr4a*** *toll-like receptor 4a;* ***myd88*** *myeloid differentiation factor 88*, pro-inflammatory: ***Tnf-α*** *tumor necrosis factor α*, ***IL-1β*** *interleukin 1β*, *IL-6* interleukin-6

counteract MC-LR-induced oogenesis defects in zebrafish and monitored the effect using transcriptomics. The probiotic diet reversed most of the MC-induced abnormal changes in the transcript levels of *follicle-stimulating hormone receptor* (*fshr*), *luteinizing hormone/choriogonadotropin receptor* (*lhcgr*), aromatase (estrogen synthase, *cyp19a1a*), and cholesterol monooxygenase (*cyp11a1*). Overall, probiotic supplementation modulated the expression of the endocrine system, energy metabolism, and RNA degradation and translation.

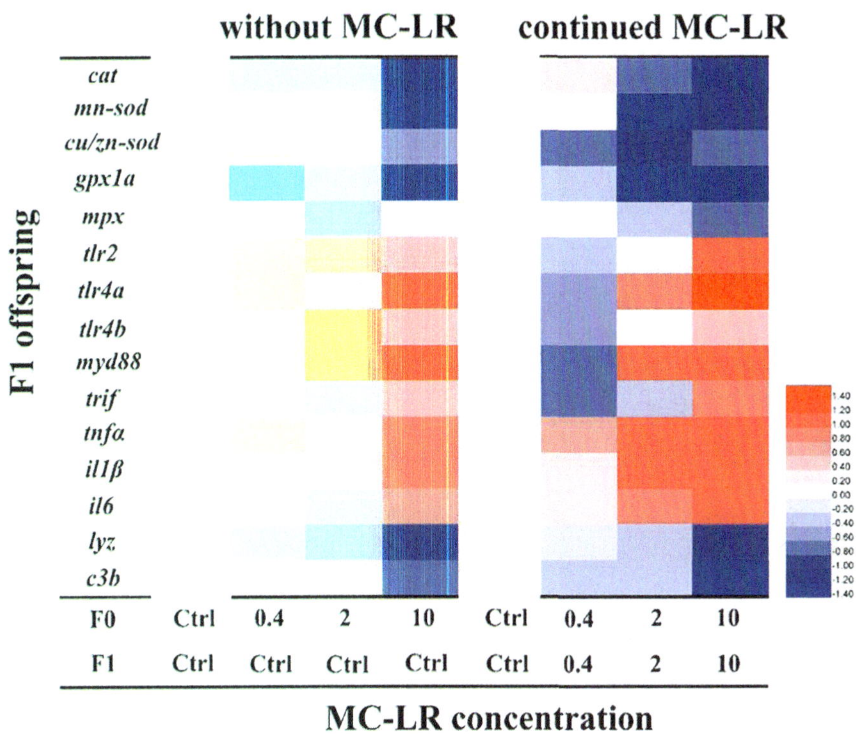

Fig. 2.10 Heat map of gene transcriptions related to antioxidant capacity and innate immune function in F_1 zebrafish larvae without/with continued MC-LR exposure at concentrations of 0, 0.4, 2, and 10 μg L^{-1}. Gene transcripts were assayed in 120 hpf F_1 larvae without/with continued MC-LR exposure relative to F_1 control (mean ± SEM, $N = 3$). The red color indicates upregulation, and the blue color downregulation, and the specific values in the heat map were derived for the fold change of gene expressions converted by log base 2 (from Lin et al. (2020), with permission from the American Chemical Society)

2.1.7 Endocrine Disruption

Recent studies have demonstrated that *M. aeruginosa* also possesses endocrine-disrupting (ED) potential in aquatic animals. Male Nile tilapias exposed to lyophilized *M. aeruginosa* or purified MC-LR have upregulated levels of 17β-estradiol and testosterone. Furthermore, exposure to *Microcystis* alters the expression of genes involved in the hypothalamic-pituitary-gonadal (HPG) axis, disrupting the balance of sex hormones. These findings highlight the environmental significance of *Microcystis* as a potential source of endocrine-disrupting compounds (Chen et al. 2017a). Beyond MC-LR, cylindrospermopsin has also been suggested to act as an ED. Both toxins have ED activity mediated by very different mechanisms, such as estrogenic responses via binding to an estrogen receptor (ER), pathological changes in several organs and cells (testis, ovarian cells), and a decreased gonad-somatic index. Moreover, toxic effects mediated by ROS, changes in transcriptional

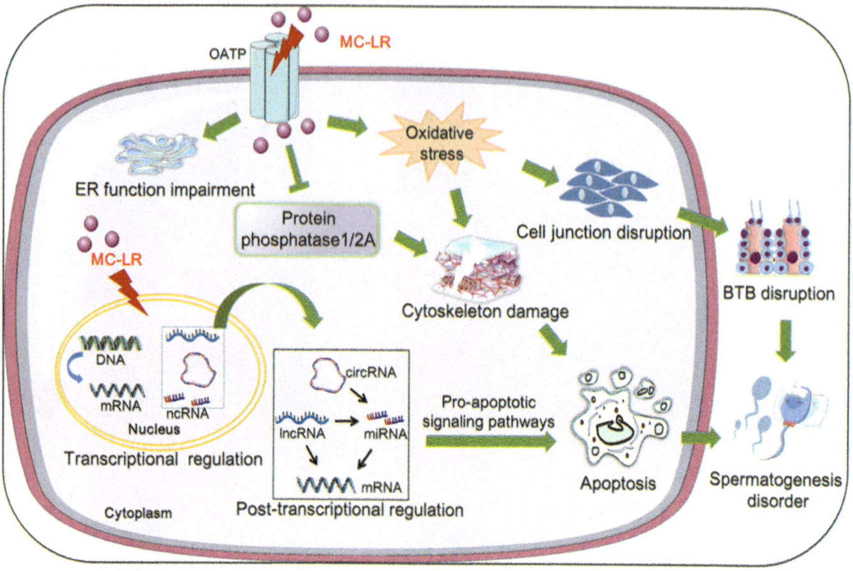

Fig. 2.11 Reproductive toxicity mechanisms of MC-LR revealed by Omics studies. MC-LR can disrupt protein phosphatase activities, induce oxidative stress, and impair the function of the endoplasmic reticulum, destroying the cytoskeleton and cell junctions. In turn, apoptosis of the germ cells occurs, and the integrity of the blood-testis barrier can be compromised, contributing to spermatogenesis disorders. In addition, non-coding RNAs also participate in the male reproductive toxicity of MC-LR through a complicated regulatory network. *OATP* organic anion transporting polypeptide, *ER* endoplasmic reticulum, *BTB* blood-testis barrier, *MAPK* mitogen-activated protein kinase (from Ma et al. (2021b), credit to Elsevier)

responses on several endocrine axes and steroidogenesis-related genes, and changes in hormone levels are reviewed by Casas-Rodriguez et al. (2022) and Zhang et al. (2022a). The latter authors even differentiate the ED effects by organ.

To illustrate the significance of this environmental risk, Table 2.3. summarizes key effects of MC-LR, primarily identified in zebrafish. For additional insights on the impact of MC-LR on daphnids, the reader is referred to Table 2.1. It is worth noting that exposure of *Daphnia* females to Microcystis exudates results in increased offspring production (Xu et al. 2019). Unfortunately, nothing is reported about the viability of the offspring. However, it is unlikely that these effects are "free of charge" in the subsequent generations, as previously observed (Saul et al. 2013; Dao et al. 2018).

Liu et al. (2024b) assessed the estrogenic potency of MC-LR in black-spotted frogs by combining field and laboratory approaches. In 13 bloom areas of Zhejiang province, China, the MC-LR concentrations in water ranged from 0.87 to 8.77 $\mu g\ L^{-1}$ and were correlated with sex hormone profiles in frogs, suggesting possible estrogenic activity of MC-LR. Tadpoles exposed to 1 $\mu g\ L^{-1}$, an environmentally relevant concentration, display a female-biased sex ratio relative to controls. Transcriptomic results reveal that MC-LR induces numerous and complex effects on gene expression across multiple endocrine axes. Additionally, exposure of male adults

Table 2.3 Selected endocrine-disrupting effects in *Daphnia* and fish caused by MC-LR exposure

Species	MC-LR preparation	Effect	References
Daphnia magna	*M. Aeruginosa* exudate from 1×10^4 and 2×10^5 cells mL^{-1}	17β-HSD↑, ecdysone↑, juvenile hormone↑, vitellogenin↑, offspring number↑	Xu et al. (2019)
Danio rerio embryos	*Microcystis aeruginosa, Planktothrix agardhii, Aphanizomenon gracile*	E_2↑, retinoid-like activity↑	Jonas et al. (2015)
Danio rerio larvae	0–1000 μg L^{-1}	*Vgt*↑	Rogers et al. (2011)
Danio rerio ♂, ♀	1, 5, 20 μg L^{-1}	♀: E_2↑, T↑, FSH↑ ♂: Spermatozoa↓, E_2↑, T↑	Liu et al. (2016b)
Danio rerio ♂♂	1–5000 μg L^{-1}	At<30: E_2↑, T↑ non-dose-dependent	Hou et al. (2018)
Danio rerio ♂♂	0, 3, 30 μg L^{-1}	T↓, *gnrh2*↓, *lhb*↓, *ar*↓, *lhr*↓, E_2↑, *cyp19a1b*↑, *cyp19a1a*↑, *vtg1*↑	Lin et al. (2018)
Danio rerio ♂♂	0, 0.3, 3, 30 μg L^{-1}	T/E_2↓, *vtg1*↑	Su et al. (2016)
Danio rerio ♀♀	0–100 μg L^{-1}	Disruption of normal meiotic resumption of meiosis	Liu et al. (2019)
Danio rerio ♀♀	0, 5, 20 μg L^{-1}	Eggs↓, VTG↓, oocyte maturation↑	Zhan et al. (2020)

17β-HSD 17β-hydroxysteroid-dehydrogenase involved in ecdysteroid biosynthesis, *E_2* 17β-estradiol, *ecdysone* Molting hormone, *FSH* Follicle-stimulating hormone, *T* Testosterone, *vitellogenin* Egg yolk precursor, ***gnrh2*** *Gonadotropin releasing hormone 2*, ***lhb*** *Luteinizing hormone subunit beta*, ***ar*** *Androgen receptor*, ***lhr*** *Luteinizing hormone receptor*, ***cyp19a1b, cyp19a1a*** *cytochrome P450 aromatases* (T → E2), ***vtg*** *Vitellogenin*

significantly increases the estradiol (E2)/testosterone (T) ratio by 3.5 times relative to controls. Downregulation of genes related to male reproductive endocrine function has also been reported. MC-LR enhances the expression of specific estrogen receptor (ER) proteins, which induce estrogenic effects by activating the ER pathway and hypothalamic–pituitary–gonadal (HPG) axis. For amphibians, MC-LR is an estrogenic endocrine disruptor at environmentally relevant concentrations, as summarized in Fig. 2.12.

2.1.8 Effects on Gut Microflora

Microcystins also affect the intestinal microflora, which could both enhance and attenuate the toxic effects. The modulation of the microflora is presented here, starting with two invertebrates. In *Chironomus pallidivittatus* larvae, exposure to MC-LR alters the diversity and abundance of the gut microbiota, favoring both pathogenic and MC-degrading bacteria, such as *Microbacterium*, *Sphingomonas*, and *Pseudomonas*. This exposure-induced direct and indirect MC detoxification, combined with the antioxidant response, indicates that these chironomid larvae have the potential to resist MC-LR.

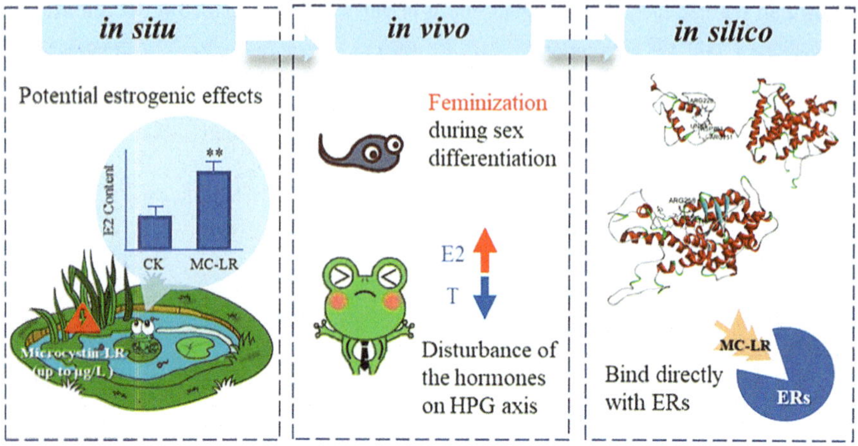

Fig. 2.12 Summary cartoon of feminizing effects at various levels on amphibians at environmentally relevant concentrations (from Liu et al. (2024b), with permission from the American Chemical Society)

Recently, Zhang et al. (2023d) analyzed the effects of MC-LR on the intestinal flora of Louisiana crawfish (*Procambarus clarkii*) and predicted its metabolic functions. In the untreated controls, the phylum *Planctomycetes* and the genus *Pseudorhodobacter* dominate. The relative abundance of the phylum *Firmicutes* is higher in the group treated with 40 μg L^{-1} MC-LR, whereas the relative abundances of the phylum *Planctomycetes* and the genera *Dysgonomonas*, *Brevundimonas*, and *Anaerorhabdus* are higher in the control group. The authors used PICRUSt (Phylogenetic Investigation of Communities by Reconstruction of Unobserved States) to predict the primary functions of the intestinal microbiota. Microbial sequences annotated to ribosome biogenesis, DNA replication proteins, homologous recombination, DNA repair and recombination proteins, DNA replication, and chaperones and folding catalysts are significantly enriched under MC-LR exposure, with the highest levels observed at the highest exposure concentration. Conversely, sequences associated with pathways such as glycine, serine, and threonine metabolism, porphyrin and chlorophyll metabolism, galactose metabolism, amino acid metabolism, valine, leucine, and isoleucine biosynthesis, phenylalanine metabolism, and mineral absorption are significantly reduced following MC-LR exposure.

The remarkably increased proportion of functional genes related to DNA replication, repair, and recombination in the intestinal flora of *P. clarkii* exposed to MC-LR demonstrates that MC-LR exposure has the potential to accelerate intestinal microbial genetic information transfer and may result in disruption of the intestinal flora, which may further lead to abnormal intestinal microbial metabolism and genetics in *P. clarkii*.

Similarly, MC-LR injection also alters the gut microflora in the Chinese mitten crab (*Eriocheir sinensis*). Du et al. (2024) reported that the gut microbial diversity in the MC-LR treatment group is significantly higher than in the control group. The MC-LR group has a significantly higher relative abundance of *Proteobacteria*,

Fusobacteria, and *Actinobacteria* and a significantly lower relative abundance of *Firmicutes*. At the genus level, the relative abundance of *Acinetobacter*, *Flavobacterium*, *Acidovorax*, *Paracoccus*, *Cloacibacterium*, *Gemmobacter*, *Aeromonas*, and *Dechloromonas* is increased in the intestine of the MC-LR-treated group, but the relative abundance of *Lachnospiraceae*_NK4A136_group is decreased. *Lachnospiraceae*_NK4A136_group belongs to the *Firmicutes*, which are generally beneficial bacteria that play a crucial role in maintaining the health of the body.

Similarly, Duan et al. (2024) found that three MC variants (-LR, -YR, -RR) had adverse effects on the histology, physiology, and metabolism of the hepatopancreas and intestinal microbiota of *Penaeus vannamei*. A reduced abundance is observed in the diversity and composition of the gut microbiota, particularly in potentially beneficial bacteria (*Alloprevotella, Bacteroides, Collinsella, Faecalibacterium,* and *Prevotellaceae UCG-001*) (Fig. 2.13), which have a positive correlation with the metabolite berberine. A review of the beneficial effects of berberine on fish nutrition can be found in Wang et al. (2022a), which includes its effect on gut microbiota (Tian et al. 2022). However, the presence of berberine in this study is critical and may be more coincidental than a causal relationship, as its production by non-engineered bacteria and yeasts is not yet understood (Han and Li 2023). In all available aquaculture studies, berberine has been supplemented in the diet, rather than produced by the gut microbiota.[6]

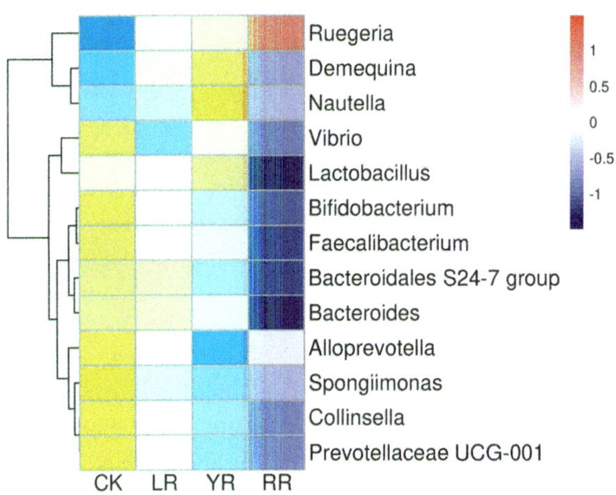

Fig. 2.13 Changes in the composition of the intestinal microbial community of *Penaeus vannamei* following exposure to three MCs variants (MC-LR, MC-YR, and MC-RR) compared to Control (CK): Proportional representation of bacterial genus abundance. Red colors indicate increased bacterial abundance and blue colors indicate decreased bacterial abundance (from Duan et al. (2024), with permission from Elsevier)

[6]This topic would be suitable for public scientific discussion, e.g. in the usual form of 'Letter' and 'Reply' to the Editor, as good journals use to do, as lively debate is an essential part of science and a guarantee of scientific progress. However, the journal in question does not accept 'Letters to the Editor'. See also the polemic in Steinberg (2024).

The change in *Lactobacillus* is equivocal (Fig. 2.13) and cannot be plausibly explained.

Other examples of the modulation of the gut microbiota are based on the "super-organism concept" of Wilson and Sober (1989) and the holo-omics approach proposed by Limborg et al. (2018) to solve scientific and practical problems arising from the traditional fragmented approaches of different disciplines. The authors argue that the adoption of holo-omic approaches would overcome the limited functional insights of current analytical strategies by simultaneously considering the holobiont at multiple molecular levels. This includes unraveling interactions between not only the host genome but also its epigenome, transcriptome, microbial metagenome, and metatranscriptome. Ideally, studies would also include analyses of the associated proteomes and metabolomes, as well as metaproteomes and metametabolomes, to elucidate the functional pathways that control the host phenotype. Therefore, a holobiont metabolomic study in *Oryzias latipes* will conclude this section.

In this study, Foucault et al. (2022) and Gallet et al. (2023) demonstrate how the gut microbiota and holobiont metabolome composition of the medaka (*Oryzias latipes*) are affected by exposure to *Microcystis aeruginosa*. Both gut microbiota and gut metabolome show significant compositional changes after only two days of exposure. The differences in composition in the fish gut are significant. Visually, samples from different time points are scattered along the first axis of the Principal Coordinates Analysis (PCoA), and the two treatments are well separated on the second axis of the PCoA. A dominant symbiont, a member of the *Firmicutes*, decreases in relative abundance, while several genera of *Proteobacteria* and *Actinobacteriota* increase. Changes in microbiota composition occur earlier and more rapidly than changes in metabolome composition. The liver and muscle metabolomes are much less affected than the gut, confirming that the gut and its associated microbiota are at the forefront of exposure. This study highlights that even brief cyanobacterial blooms, which are becoming increasingly frequent, trigger changes in microbiota composition and the metabolome of the holobiont.

Gallet et al. (2023) further note that the decrease in *Firmicutes* abundance induced by *M. aeruginosa* occurs between 10 and 100 $\mu g\ L^{-1}$ Chl*a*, suggesting the existence of a notable tipping point for the responsiveness of the gut microbiome to cyanobacterial bloom intensity. But what happens when the exposure is less than 10 $\mu g\ L^{-1}$ Chl*a*? Could there even be a hormetic response with benefits for the gut microbiota and the host at low exposures? In other words, this tipping point may reflect the two extremes of the gut microbiota: eubiosis and dysbiosis (Iebba et al. 2016), but further experimental evidence is needed.

2.1.9 Microevolution and Epigenetics

In addition to clonal selection, multiple epigenetic pathways associated with MC-LR exposure are increasingly being discovered to explain differences in tolerance to microcystins.

2.1.9.1 Microevolution

Microevolution refers to small-scale evolutionary changes within populations over relatively short timescales, often driven by natural selection, genetic drift, mutation, and gene flow. It is based on clonal selection and contributes to the often-observed high level of genetic differentiation among natural populations of cyclical parthenogenetic species, such as cladocerans (e.g., *Bosmina* spp. or *Moina* spp.), rotifers (e.g., *Branchionus* spp. and *Keratella* spp.), and certain aquatic copepods (such as *Cyclops* spp.) (Papakostas et al. 2013; Adamczuk and Mieczan 2019). The alternating reproductive modes contribute to population structure and genetic diversity (Capaul and Ebert 2003; Zhang and Jiang 2022b). During asexual reproduction, clonal lineages can rapidly expand, allowing beneficial traits to spread quickly in response to environmental pressures, such as toxic algal blooms. Sexual reproduction, which typically occurs seasonally, introduces genetic recombination, increasing genetic diversity and enabling populations to adapt to changing conditions. This balance between clonal selection and genetic recombination makes cyclical parthenogens excellent models for studying rapid evolutionary responses in ecological contexts.

Zhang and Jiang (2022a) explained the tolerance and loss of tolerance of *Ceriodaphnia cornuta* to cyanotoxins from *Microcystis aeruginosa* by forward and reverse evolution of multivariate responses. In particular, clonal selection has a significant influence on evolutionary responses. The *C. cornuta* population under high selection pressure shows higher tolerance to the toxins than those under low selection pressure. When the *C. cornuta* population is fed pure green algae, almost all of the evolved traits are restored to their original state in a six-month experiment. That is, the toxin-tolerant phenotypes are culled from the population in the absence of selection pressure, suggesting a high cost of maintaining tolerance. Genotype dynamics suggest that asymmetric intraspecific competition may be the primary force driving the rapid reversal of evolution. These results indicate that the change in cyanobacterial biomass, from proliferation to decline, leads to rapid forward and reverse evolution of multivariate phenotypic responses in *C. cornuta*. By resurrecting dormant eggs of *C. cornuta* from a sediment core covering different periods of eutrophication, Zhang and Jiang (2022b) confirmed multiple evolutionary responses to toxic cyanobacteria, e.g., microevolution.

The same laboratory compared two other cladocerans (*Daphnia pulex* and *Simocephalus vetulus*) for their tolerance towards *Microcystis*. After exposure, both grazers evolved higher tolerance. The intrinsic population growth rate of *S. vetulus* feeding on cyanobacteria is negatively correlated with that of green algae. However, cyanobacterial selection results in an overall increase in *D. pulex* when both cyanobacteria and green algae are fed (Jiang et al. 2016).

Lyu et al. (2019) performed fitness and transcriptome analyses on two *Daphnia similoides* clones. In the presence of toxic *Microcystis*, the sensitive clone experiences a significant decrease in body growth rate compared to the tolerant clone. There are a total of more than 1800 differentially expressed genes, highlighting the pathways of glutathione metabolism, protein processing in the ER, amino/nucleotide sugar metabolism, and arachidonic acid metabolism (Fig. 2.14). These

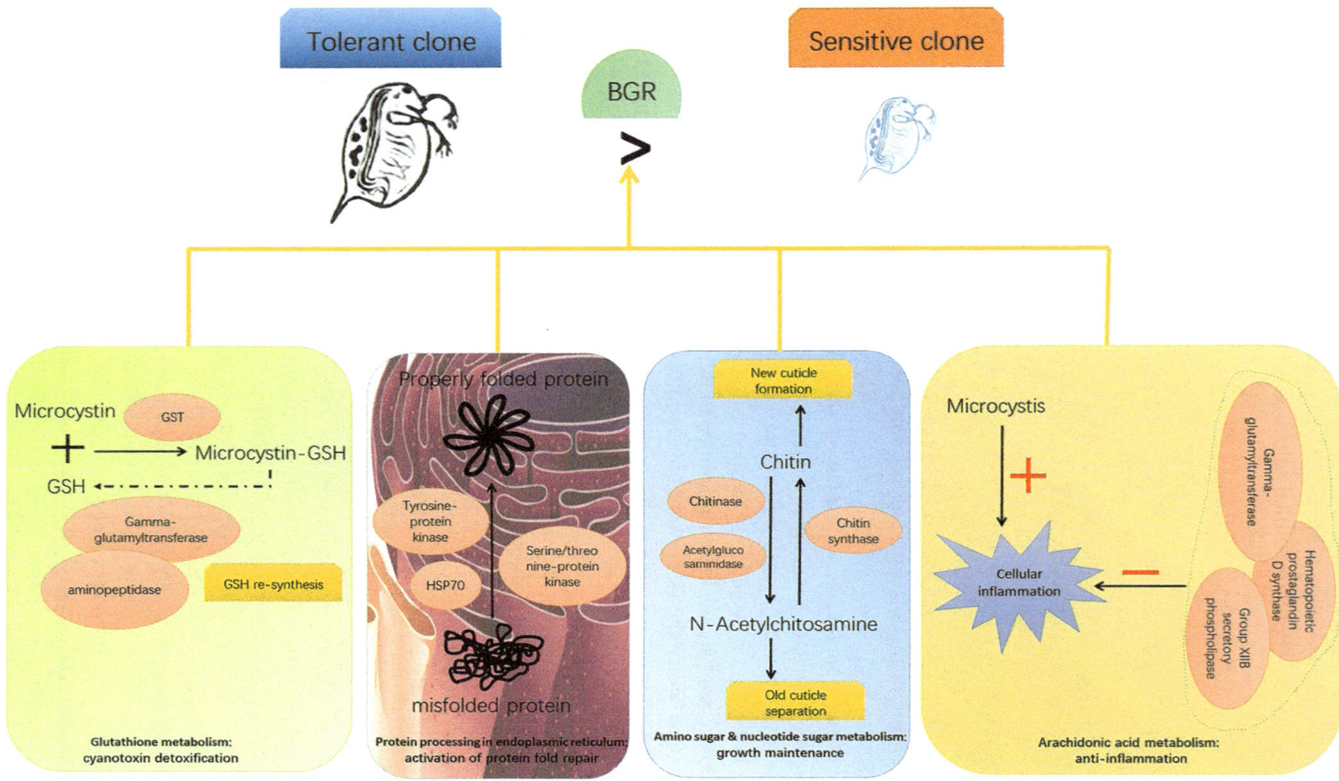

Fig. 2.14 Diagram of function pathways for body growth rates of the tolerant clone compared with the sensitive clone of *Daphnia similoides* (from Lyu et al. (2019), credit to Wiley & Sons)

significantly altered pathways suggest that the strong tolerance of clonal variation is involved in cyanotoxin detoxification, activation of protein fold repair, inflammation mediation, and molting/growth maintenance. Lyu et al. (2019) not only discovered evidence consistent with previous findings (Asselman et al. 2012; Schwarzenberger et al. 2014) but also identified novel candidate pathways that are largely responsible for the tolerance of *D. similoides* to cyanobacteria.

It would have been interesting to examine epigenetic variations of daphnids in these experiments, as other studies have explained rapid adaptation to changing environmental conditions, including *Microcystis* tolerance, through epigenetic mechanisms, such as altered DNA methylation patterns (Asselman et al. 2017; Hawes et al. 2019; Piferrer and Wang 2023; Vogt 2023). Different DNA methylation patterns may, in turn, be responsible for different gene expression patterns and could explain the described differentially expressed genes found in sensitive and tolerant clones.

2.1.9.2 DNA Methylation

Due to the worldwide eutrophication of freshwater bodies, numerous studies have been conducted on the effects of cyanobacteria on key planktonic crustaceans. A selection is presented in Table 2.1 featuring Cladocera in general and *Daphnia* in particular. In stagnant freshwater bodies, *Daphnia* is a keystone species in the transfer of energy and nutrients (Sarnelle 2005).

The most striking result is that intergenerational effects cause the offspring, although smaller, to be more tolerant to cyanobacterial metabolites. These metabolites are not necessarily only MCs (Smutná et al. 2014). Transgenerational stress resistance can be inherited through epigenetic mechanisms, such as modifications of DNA methylation, as demonstrated in *Daphnia magna* in response to stress by natural xenobiotics (Menzel et al. 2011) or salinity changes (Jeremias et al. 2018).

While direct evidence for specific epigenetic mechanisms for cyanotoxin tolerance is still lacking, they are highly probable. Asselman et al. 2017 demonstrated the role of epigenetics in regulating stress responses in *D. magna* when exposed to dietary *M. aeruginosa*. Their study highlighted preferential differential methylation of the amino acids serine and threonine (Fig. 2.15). The authors specifically observed the enrichment of several gene families involved in protein regulation among differentially methylated genes: ribosomal proteins and Rab family proteins. Ribosomal proteins are differentially expressed in response to *M. aeruginosa* exposure, likely due to its effects on protein synthesis in *Daphnia*. In addition, E3 ubiquitin ligases, involved in protein degradation, and Ras-related proteins of the Rab family, involved in protein transport, are also enriched in differentially methylated genes. These results indicate that *Daphnia* invests more energy and resources in dealing with protein regulation and potentially misfolded proteins under *Microcystis* stress, and that, consequently, less energy is available for reproduction and growth. This shift in resource allocation contributes to the reduced reproduction and growth observed under *Microcystis* stress (Asselman et al. (2017) and references therein).

The vertebrate counterpart demonstrates that paternal exposure to MC-LR triggers developmental neurotoxicity in zebrafish offspring through DNA

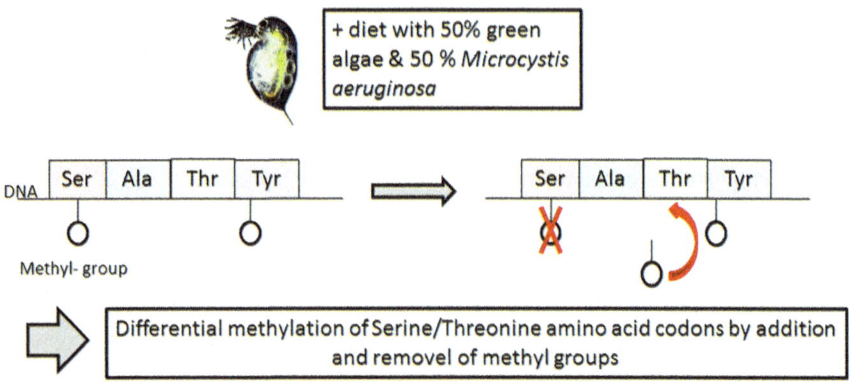

Fig. 2.15 Modulation of specific DNA methylation in *Daphnia magna* on *Microcystis* diet (from Asselman et al. (2017), credit to the American Chemical Society)

methylation-mediated activation of the MAPK pathway. Zhao et al. (2021) reported that male exposure to environmentally relevant concentrations of MC-LR during spermatogenesis induces apoptosis, leads to significant cellular damage, and promotes epigenetic modifications in testicular cells. Furthermore, paternal MC-LR exposure leads to a decrease in locomotor activity and downregulation of specific genes involved in neurodevelopment, triggering the activation of the MAPK pathway through phosphorylation in F_1 zebrafish larvae. An epigenetic programming of the sperm provides a mechanistic link between the paternal age effect and offspring neurodevelopmental disorders.

2.1.9.3 Non-protein Coding RNAs

The transcriptome of a cell comprises a vast array of RNA molecules, both protein-coding and non-coding. Recent advances in sequencing technologies have provided deeper insights into transcriptomic complexity, revealing that a significant portion of the genome is transcribed into a diverse population of non-protein-coding RNAs (ncRNAs). The biological significance of ncRNAs has been largely underestimated (Derrien et al. 2012). In the following, we will focus on microRNAs and long non-coding RNAs.

MicroRNAs

Together with transcription factors, microRNAs are the most abundant gene regulatory factors in multicellular genomes, controlling the expression of genomic information (Hobert 2008). MicroRNAs play their regulatory roles by targeting mRNAs for cleavage or translational repression (Bartel 2004) and may be involved in a variety of toxic effects, both directly and indirectly, through the gut microbiota.

Direct Response

Applying quantitative real-time PCR and miRNA microarray, several studies have provided convincing evidence that miRNAs are involved in hepato- and neurotoxicity, hepatitis and tumorigenicity, as well as reproductive toxicity induced by MCs (Zhao et al. 2012c; Saul et al. 2014; Yang et al. 2018). Brzuzan et al. (2010, 2012) found that, after a lag period of more than 8 hours, *let-7c*, *miR-9b*, *miR-16b*, *miR-34b*, and *miR-122* are significantly upregulated in whitefish (*Coregonus lavaretus*) liver after MCs-exposure. In addition, *mir-10*, *mir-122*, and *mir-92* are likely plasma biomarkers of liver damage caused by MC-LR (Brzuzan et al. 2016).

Zebrafish embryos, exposed to crude MCs extracted from a batch culture of a toxic *Microcystis aeruginosa* strain, confirmed the significance of *miR-122* (Li et al. 2013b). *miR-122* is conserved among vertebrate species. *miR-122* is not present in invertebrates, and no close paralogs of *miR-122* have been detected (Chang et al. 2004). Li et al. (2013b) found that MCs exposure elevated the transcription levels of miR-21 and miR-27b, while the expression of miR-122 and miR-148 was downregulated. It is unclear why *miR-122* is downregulated in microcystin-challenged zebrafish embryos; this finding contrasts with the above finding in whitefish.

In zebrafish embryos exposed to MC-RR, the picture is different. Zhao et al. (2011) reported that the expressions of 31 miRNAs are affected at 72 hpf. Expressions of *miR-430* and *miR-125* families are also significantly changed. The altered expressions of *miR-31* (Steinberg 2022b) and *miR-126* are likely responsible for the loss of vascular integrity. Bioinformatics analysis shows that several expressed miRNAs are predicted to be potential MC-responsive miRNA-target pairs, and that their aberrant expressions are the possible molecular mechanisms for the various developmental defects caused by MC-RR. Zebrafish and whitefish belong to the MC-sensitive fish species. Therefore, the question arises whether or not MC-resistant species, such as silver carp (*Hypophthalmichthys molitrix*), do respond differently. Feng et al. (2017) studied the expression of 7 miRNAs (*let-7b*, *miR-21*, *miR-122*, *miR-27b*, *miR-148*, *miR-125a*, and *miR-143*) that play regulatory roles in various biological processes, including signal transduction, apoptosis, cell cycle, and fatty acid metabolism. In particular, *let-7b*, *miR-125a*, and *miR-143* are tumor suppressor miRNAs.

Actually, *miR-122–5p* in whitefish (*Coregonus lavaretus*) has been identified as a plasma biomarker of liver injury after exposure to MC-LR (Florczyk et al. 2016). Plasma levels of this miRNA are elevated within 8 h of exposure and show high specificity. *MiR-122* is involved in the immune response by regulating interleukin-15,[7] as shown in the orange-spotted grouper (*Epinephelus coioides*) (Liu et al. 2020). Furthermore, *miR-148a-3p* and *miR-92a-3p*, although responding more slowly to MC-LR exposure and having lower specificity, also show potential (Florczyk et al. 2016).

Expression patterns are assessed in the liver, spleen, and kidney of silver carp following 8, 24, and 48 h of exposure to MC-LR via intraperitoneal injection. The results of qPCR analysis show that the expression of the 7 miRNAs is either upregulated or downregulated in response to MC-LR exposure. Interestingly, all three tumor suppressor miRNAs are upregulated in the liver, as demonstrated even 8 h

[7] Interleukin-15 is a cytokine that induces the proliferation of natural killer cells killing virally infected cells (Perera et al. 2012).

after intraperitoneal injection, and their levels increase over time in several instances. This finding contrasts with the whitefish study, which detected a delayed response, and with the zebrafish study, which reported even a downregulation of *miR-125a* (Zhao et al. 2011). This difference in miRNA expression between MC-sensitive and -resistant fish species is obviously part of the underlying biomolecular mechanism. This assumption is supported by findings showing that members of the *miR-125* group are involved in responses to a viral infection (Salazar et al. 2020).

The systemic toxicity in silver carp is predicted due to the upregulation of four miRNAs (*miR-16*, *miR-181a-3p*, *miR-223*, *miR-451*), exemplified with *miR-451* (Fig. 2.16). These miRNAs are the key components of multiple signaling cascades (Feng et al. 2020c). Numerous studies have confirmed that *miR-451* is involved in various biological mechanisms, including differentiation and development, cell cycle and proliferation, cell survival, and apoptosis (Khordadmehr et al. 2019). There is clear evidence for the role of *miR-451* in regulating multiple signaling pathways, including the Wnt signaling, AMPK signaling, and IL-6R-STAT3 pathways (Liu et al. 2016c; Mamoori et al. 2016; Bitarte et al. 2011).

In zebrafish, Ding et al. (2021) demonstrated that even environmentally relevant levels of MC-LR can have a profound impact on the intestine, increasing its permeability, triggering inflammation, disrupting oxidative balance, inducing apoptosis, altering the gut microbiome, and affecting miRNA expression. The authors specifically analyzed *miR-146a*, *miR-155*, *miR-16*, *miR-21*, and *miR-223*, two of which are identical to those analyzed in the silver carp study above. Exposure to MC-LR alters the transcription of *miR-16* (decrease), and *miR-21* and *miR-223* (both increase).

miR-146a plays a crucial role in innate immunity and autoimmune diseases and has been widely investigated as a regulator of the immune response in vertebrates (Boldin et al. 2011). In a preceding paper, *miR-146a* is upregulated by a synthetic xenobiotic (paraquat) in common carp (Ma et al. 2018); however, this miRNA does not respond to a natural xenobiotic (MC-LR).

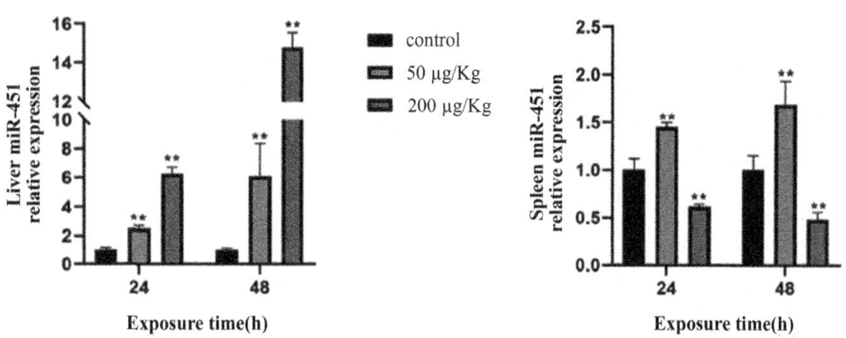

Fig. 2.16 Expression of *miR-451* in liver and spleen of silver carp after MC-LR exposure. * indicates significantly different response from the control group (*$P < 0.05$, **$P < 0.01$) (from Feng et al. (2020c), credit to MDPI)

Recent and increasing numbers of studies prove that *miR-155* plays an important role in the regulation of inflammatory processes and immune pathways (O'Connell et al. 2007); however, it does obviously not respond to MC-LR exposure.

miR-16 participates in various pathways and acts by regulating different genes in biological processes (Ke et al. 2013). *miR-16* levels in the zebrafish gut are markedly decreased by MC-LR. Decreased *miR-16* levels also occur in HepG2 cells (Feng et al. 2020a), indicating that *miR-16* may participate in the toxicity of MC-LR. The same is found in silver carp (*Hypophthalmichthys molitrix*), in concert with three other microRNAs, *miR-181a-3p*, *miR-223*, and *miR-451* (Feng et al. 2020b).

miR-146a plays a critical role in innate immunity and autoimmune diseases and has been extensively studied as a regulator of vertebrate immune response (Boldin et al. 2011; Zhang et al. 2023b) and LC-PUFA biosynthesis (Chen et al. 2018). *miR-155* plays an important role in the regulation of inflammatory processes and immune pathways (O'Connell et al. 2007), and the induction of *miR-155* is associated with high levels of pro-inflammatory cytokines such as IL-1β, IL-6, and IL-8 (Tili et al. 2013; Tang et al. 2020). *miR-16* is involved in various pathways and acts by regulating various genes in biological processes (Ke et al. 2013), such as LC-PUFA biosynthesis (Sun et al. 2020) or antioxidant response (Tang et al. 2013). However, there are no reports on the effect of pollutants on the expression of *miR-16* in fish. In a current study, *miR-16* levels in the zebrafish intestine are significantly decreased by MC-LR. Decreased *miR-16* levels are also reported in HepG2 cells caused by MC-LR (Feng et al. 2020a), suggesting that *miR-16* is involved in the toxicity of MC-LR. As a vital onco-miRNA, *miR-21* also plays a key role in pollutant toxicity (Roy and Sen 2011). *miR-21* levels are significantly elevated in the zebrafish gut after the MC-LR exposure, which is consistent with results in WRL-68 cells and mice treated with MC-LR (Xu et al. 2012; Zhao et al. 2012b). In addition to carcinogenic processes, *miR-21* can be indicative of general inflammation (Olivieri et al. 2012).

miR-223 is a cross-species conserved miRNA involved in hematopoiesis, lipid metabolism, inflammation, and other biological processes (Tang et al. 2013). In the current study, the level of intestinal *miR-223* is significantly downregulated by MC-LR. The correlations shown indicate that the causal relationships still need to be investigated: Are the expressed miRNAs the cause or effect of the transcribed genes?

To the best of our knowledge, two invertebrates have been studied with respect to MC-LR exposure. Zhang et al. (2023a) demonstrated that MC-LR exposure affects the degradation pathway of misfolded proteins in the ER of *Penaeus vannamei* hepatopancreas, and *miR-181-5p* is likely to play an important role in this effect. In addition, Zhang et al. (2023c) highlighted novel aspects of *miR-282–5p*-mediated autophagy in juvenile *Eriocheir sinensis* treated by MC-LR. MC-LR induces autophagy via *miR-282–5p*-targeted PIK3R1. Overexpression of *miR-282–5p* mimics MC-LR-induced autophagy by inhibiting PIK3R1 expression, whereas silencing of *miR-282–5p* inhibits autophagy by promoting PIK3R1 expression.

Toxicity via Gut Microbiota

MicroRNAs also have the ability to influence the gut microbiota (DuPont and DuPont 2011), and altering miRNAs can affect both the growth and composition of the microbiota (Liu et al. 2016a; Teng et al. 2018; Tavasolian and Inman 2021). In the zebrafish study, statistical analysis shows that genes for tight junction proteins, cytokines, oxidative stress, and apoptosis are associated with miRNAs and gut microbiota. The gut microbiota is also positively or negatively related to miRNAs, indicating that damage to the fish gut may contribute to the altered miRNAs and gut microbiota.

Ding et al. (2021) correlated the adverse effects of MC-LR on the zebrafish intestine to gut microbiota and microRNAs. The authors performed a subacute toxicity test on zebrafish exposed to MC-LR (35 μg L^{-1}) for 21 d. MC-LR treatment reduces the mRNA levels of tight junction genes (*claudin-5*, *occludin*, and *zonula occludens-1*) and alters the levels of diamine oxidase and D-lactic, indicating increased intestinal permeability in zebrafish. Furthermore, MC-LR treatment significantly increases the levels of intestinal IL-1β and IL-8 but decreases the levels of IL-10 and TGF-β, indicating that this exposure induces an inflammatory response in the fish intestine. MC-LR exposure also activates superoxide dismutase and catalase, generally upregulates the levels of *p53*, *bax*, *bcl-2*, *caspase-3*, and *caspase-9*, downregulates the levels of *caspase-8,* and causes significant histological injury in the fish intestine. The novelty of the paper is that MC-LR exposure also significantly alters the microbial community in the zebrafish gut and the expression of miRNAs (*miR-146a*, *miR-155*, *miR-16*, *miR-21*, and *miR-223*). A network analysis reveals the correlation of biochemical indicators, miRNAs, and gut microbiota in zebrafish: claudin-5, DAO,[8] IL-10, and TGF-β are positively correlated with *Plesiomonas* and *Tsukamurella*, while CAT is negatively correlated with *Aeromonas* but positively correlated with *miR-21*; apoptosis-related genes are correlated with *Vibrio*, *Methylobacterium*, *Pseudonocardia*, *Clostridium_sensu_stricto_13*, *Paenisporosarcina*, *Pontibacter*, and *Haliangium*; *miR-146a* and *miR-223* are positively correlated with *Geobacter* and *Marmoricola*, respectively, while *miR-21* is negatively correlated with *Aeromonas*; *miR-16* is negatively correlated with *Ambiguous_taxa* and *Fonticella* but positively correlated with *Pseudonocardia*; in addition, *miR-155* is negatively correlated with *Ambiguous_taxa* but positively correlated with *Pseudonocardia.*

Long Non-coding RNAs

Besides microRNAs, another group of non-coding RNAs has been identified as playing a role in the development of MC-toxicity. Using RNA sequencing data, Florczyk et al. (2021) de novo assembled a high-quality whitefish (*Coregonus lavaretus*) liver transcriptome. Whitefish are i.p. treated with 100 μg MC-LR kg^{-1} of body weight for 1, 6, 9, and 14 days. A total of 94 differentially expressed (DE), putatively evolutionarily conserved lncRNAs and also more than 4000 DE

[8] DAO is a specific intracellular enzyme detected in the intestinal mucosal epithelium (Xiao et al. 2019) and may indirectly indicate gut mucosa injury.

putatively novel whitefish lncRNAs were identified. Gene ontology terms analysis shows that, after 1 day of exposure, the co-upregulated pairs are enriched in transcription regulators. For example, this group comprises two subunits of a pro-apoptotic transcription factor, suggesting that these transcripts play a role in regulating the response to severe liver damage (Florczyk et al. 2021). Although the picture of lncRNA action in MC-toxicity is far from complete, this first inventory promises many more interesting discoveries.

Histone Modifications

To date, evidence of MC-mediated histone alterations is available only from studies in mammals. Wang et al. (2020d) demonstrate that histone acetylation may play a crucial role in MC-LR-induced apoptosis and cell cycle disorder in Sprague-Dawley rats. MC-LR increases histone deacetylase activity, thereby decreasing the level of histone acetylation and associated gene expression. This activates the mitochondrial apoptotic pathway and increases the expression of apoptosis-related factors to induce apoptosis. Additionally, MC-LR also causes cell cycle arrest, inducing testicular cell apoptosis in rats. In a companion paper, Yuan et al. (2020) suggest that the increase in apoptosis gene levels induced by MC-LR can be attributed to the methylation of H3K4 in the testicular cells.

2.2 Anatoxin-a

Anatoxin-a (ATX, Fig. 2.17) is a secondary, bicyclic amine alkaloid produced by various cyanobacterial genera, including *Dolichospermum* (formerly *Anabaena*), *Aphanizomenon*, *Planktothrix*, and *Microcoleus*. It is a potent neurotoxin, often referred to as the "Very Fast Death Factor." Unlike microcystins, which primarily target the liver, anatoxin-a acts on the nervous system and can be fatal within minutes to hours. As an acetylcholine mimic, it binds irreversibly to nicotinic acetylcholine receptors, causing continuous nerve stimulation which leads to rapid paralysis, respiratory failure, and death in exposed animals. It was first discovered in the early 1960s in Canada and was isolated in 1972 (Aráoz et al. 2010). Its presence in freshwater ecosystems poses significant risks to wildlife, livestock, and humans, making it a critical concern for water quality management. Nevertheless, studies on the effects of ATX on aquatic animals are still in their infancy.

Fig. 2.17 Structures of anatoxin-a and β-N-methylamino-L-alanine

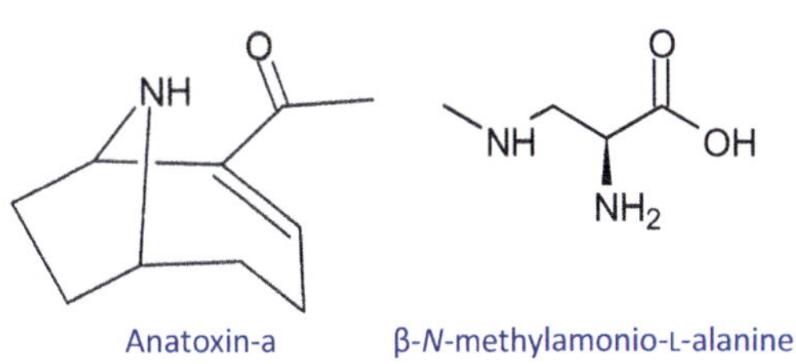

2.2.1 Invertebrates

A few studies have been devoted to ATX intoxication in invertebrates. In one of the first studies, *Dolichospermum* (*Anabaena*) *affinis* affects both survival and fecundity of *Daphnia pulex*, while *D. flos-aquae* and its toxin ATX primarily affect fecundity. The presence of cyanobacteria affects brood size, brood number, time to first reproduction, and interclutch interval. Increasing concentrations of *D. flos-aquae* mainly affect demographic parameters, especially at lower temperatures, while increasing concentrations of *D. affinis* had less effect (Claska and Gilbert 1998).

Later, Pawlik-Skowrońska et al. (2019) showed that the rotifer *Brachionus calyciflorus* is more resistant to pure ATX than *Daphnia pulex*. ATX significantly reduces the survival of this cladoceran, and its effect is stronger than that of MC-LR at the same concentrations. In addition, ATX exposure alters behavioral parameters in *Daphnia magna*, but not at the most commonly detected environmental levels of this cyanotoxin (Bownik and Pawlik-Skowrońska 2019).

Studying three clones of *D. magna*, Schwarzenberger and Martin-Creuzburg (2021) found that it can cope with dietary ATX if the percentage of the neurotoxin-producing cyanobacterium is not too high. Furthermore, the upregulation of nicotine-acetylcholine receptor genes is maternally transferred and leads to higher fitness in the offspring of experienced mothers. Nicotinic acetylcholine receptors respond to the neurotransmitter acetylcholine and to drugs such as the agonist nicotine.

2.2.2 Fishes

Research on the effects of anatoxin-a in fish is even more limited. A study by Colas et al. (2020) suggests that female medaka fish do not appear to accumulate ATX and show substantial recovery within 24 hours after a single subacute oral liquid exposure at the no-observed-adverse-effect level. These observations suggest that if the dose remains below the acute toxicological limit that produces neurotoxicosis with fatal consequences, the organism can recover completely and rapidly, without inducing obvious effects even if the exposure is repeated several times, as shown in mice (Fawell et al. 1999). However, the question of chronic ATX exposure, which can induce more insidious long-term pathologies in the central nervous system (Lombardo and Maskos 2015), remains unanswered.

Of note, in contrast to previous studies that have specifically identified (±)-ATX as a neurotoxin, in zebrafish (*Danio rerio*) and fathead minnow (*Pimephales promelas*) larvae exposed to (±)-ATX, proteomics identifies other potential mechanisms by which this toxin might elicit a toxicological response, e.g., via endocrine disruption and post-translational modifications (Langan et al. 2024). This study highlights the need for rapid expansion of omics-based characterization in future cross-species studies.

2.3 Nodularins

In the bays of the Baltic Sea, the recruitment of Eurasian perch (*Perca fluviatilis*) larvae coincides with the formation of cyanobacterial (*Nodularia*) blooms. Persson et al. (2011) demonstrated that bloom conditions of toxic and non-toxic cyanobacterium *Nodularia* reduce the growth and energy reserves of perch larvae. In this respect, the adverse effects of cyanobacteria are likely to be caused by both toxicity and turbidity, leading to increased energy expenditure due to detoxification and prolonged foraging times, alteration of chemosensory signals, or difficulties in prey handling.

Back to the pure toxins. There is significant homology in structure and function between MCs and nodularins (NODs), and MCs have been studied in greater detail. Therefore, facts from MCs are often extended to NODs (Chen et al. 2013). NOD-R (Fig. 2.3b) is the predominant toxin variant among the ten known nodularin congeners. NODs are cyclic non-ribosomal pentapeptides that contain several uncommon non-proteinogenic amino acids, including *N*-methyl-didehydroaminobutyric acid and the β-amino acid ADDA. These compounds are relatively stable: light, temperature, and microwaves have little effect on degrading them (Chen et al. 2013).

Like MCs, NODs can pose hepatotoxic and carcinogenic threats to animal health (Chen et al. 2013). In addition, NODs have been linked to neurotoxicity (Lehtonen et al. 2003) and the induction of apoptosis (Rymuszka and Sieroslawska 2018). Although NODs have not been classifiable as carcinogens due to the lack of exposure data, a classical study considered them more toxic than MCs (Ohta et al. 1994). At the molecular level, NOD is processed in a complex manner to induce toxic effects. During digestion, NODs diffuse from the small intestine into the liver due to active uptake by an unspecific organic anion transporter in the bile acid carrier transport system. This transporter is expressed in the gastrointestinal tract, kidney, brain, and liver (Žegura et al. 2011). Once in the liver, NOD inhibits three key enzymes, specifically the catalytic units of serine/threonine protein phosphatases: protein phosphatase 1, protein phosphatase 2A, and protein phosphatase 3 (Ohta et al. 1994). These enzymes normally dephosphorylate proteins, a process essential for regulating cellular activities. Dephosphorylation inhibits the normal function of the protein, leading to cellular dysfunction and toxicity. Recently, Chen et al. (2020) mentioned oxidative stress as one toxic mode of NODs; Štern et al. (2019) specify that this oxidative stress is central in the genotoxic mode of action.

The effects of NOD on the reproductive system are not well understood. Therefore, Biswas et al. (2024) studied the toxicological effects of NOD on the reproductive endocrine system of female zebrafish at environmentally realistic concentrations. After 14 days of exposure to nodularin, decreased gonadosomatic index, germinal vesicle breakdown, and decreased follicle-stimulating hormone, luteinizing hormone, 17β-estradiol levels, and increased testosterone levels in female zebrafish (Fig. 2.18) suggested that nodularin may disrupt both oocyte growth and maturation. In support of these data, changes in the expression of several marker genes on the hypothalamic-pituitary-gonadal-liver (HPGL) axis were observed. Transcription levels of genes related to steroidogenesis, including

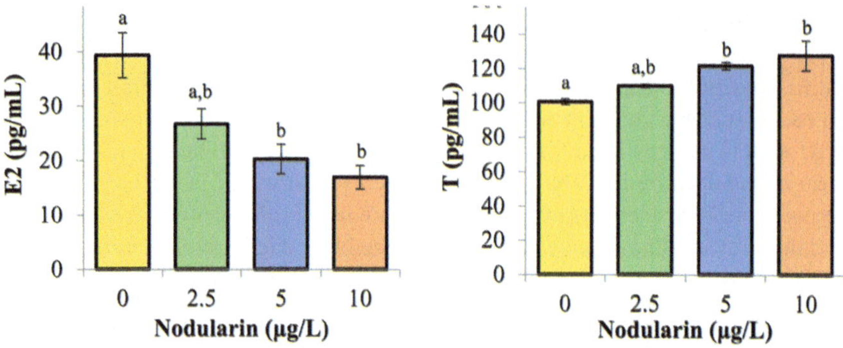

Fig. 2.18 Plasma levels of 17β-estradiol (E2) and testosterone (T) in female zebrafish following exposure to Nodularin for 14 days. The values represent mean ± SE (triplicates). Different small letter above the error bars represents significant difference ($P < 0.05$) (from Biswas et al. (2024), with permission from Elsevier)

cytochrome P450 aromatase (*cyp19a1a*) in the ovary and primary vitellogenin genes (*vtg1*, *vtg2*, and *vtg3*) in the liver, were downregulated, and marker genes for oxidative stress (*sod*, *cat*, and *gpx*) were upregulated on the HPGL axis. Taken together, NOD is a potent endocrine disruptor that induces oxidative stress and reproductive endocrine toxicity in female zebrafish, highlighting the importance of a comprehensive assessment of its environmental risks. Future studies need to evaluate whether the oral route of contamination is similarly effective.

As mentioned above, a novel mode of action of MC and NOD is coming into focus: ER stress. The ER is the site of synthesis and folding of membrane and secretory proteins, which collectively represent a large fraction of the total protein output of a cell (Lin et al. 2008). Disturbance of protein structure and function results in the accumulation of unfolded or misfolded proteins, leading to ER stress (Malhotra and Kaufman 2007). MC-LR exposure to a human hepatoma cell line induces ER-stress and associated unfolded protein response (UPR) (Christen et al. 2013), and it can be assumed that this mechanism also applies to fish. Therefore, the same laboratory compares the effects of NOD and MC-LR in zebrafish with particular reference to eleuthero-embryos and adult liver (Faltermann et al. 2016). Notably, MC-LR and NOD exhibit similar transcriptional effects. They induce changes in mRNA levels of genes associated with ER-stress, and lead to increased *tnf-α* mRNA levels in the adult liver, highlighting a novel (transcriptional) mode of action in fish. However, despite being taken up by eleuthero-embryos, these cyanobacterial toxins do not induce any detectable transcriptional changes. This is likely due to the toxin's organ-specific effects, particularly on the liver and kidneys, which cannot be detected through whole-embryo sampling.

Furthermore, studies demonstrated that early embryos appear to be protected from MC-LR (Oberemm et al. 1997, 1999; Azevedo-Linhares et al. 2018) as long as they have not hatched, as the chorion likely acts as a barrier to cyanotoxins (Wang et al. 2005).

2.4 β-*N*-Methylamino-L-Alanine

In recent years, other cyanotoxins have come into focus that might bioaccumulate in the food chain. For instance, β-*N*-methylamino-L-alanine (BMAA, Fig. 2.17), a neurotoxic non-protein amino acid produced by most cyanobacteria, has been implicated in causing devastating neurodegenerative diseases on the island of Guam in the Pacific Ocean (Garruto and Yase 1986) and has also been detected in stranded dolphins (Davis et al. 2019). Unlike other biomagnifying compounds, BMAA is not lipophilic and, thus, likely does not accumulate in fatty tissues. An alternative mechanism is required to explain BMAA bioaccumulation: Murch et al. (2004) demonstrated that BMAA can bind to proteins and acts as an endogenous toxin reservoir when cleaved from the protein. It can cross the blood–brain barrier and bioaccumulates in neuroproteins (Weiss et al. 1989). In cultured non-nitrogen-fixing cyanobacteria, BMAA production increases in a nitrogen-depleted medium (Downing et al. 2011).

Contrasting BMAA concentrations and bioaccumulation detected in studies conducted by different research groups can result from species specificity, the impact of ecological niches, or the physiological state of the organism. The different selectivity of the analytical procedures used in the determination of BMAA also has to be taken into account. Faassen (2014) judges even more harshly about older studies: For the majority of the reviewed articles, it was unclear whether BMAA was correctly identified, either because inadequate analytical methods were used, or because poor reporting of analyses made it impossible to verify the results. Therefore, this section focuses on studies from recent years.

BMAA has also been detected in freshwater systems. Jiao et al. (2014) were one of the first laboratories to report the occurrence and transfer of BMAA within the aquatic food webs of Gonghu Bay (Lake Taihu, China). During the bloom outbreak, these authors even observed a ten-fold biomagnification of BMAA in the predatory carp *Chanodichthys erythropterus*. In a subsequent survey of six aquaculture ponds around this lake, Wu et al. (2019a) reported an increase of BMAA bioconcentration along the food chain, with the lowest values in herbivorous grass carp (*C. idella*) and the highest in carnivorous (molluskivorous) black carp (*Mylopharyngodon piceus*). Notably, the omnivorous/carnivorous mitten crab (*E. sinensis*) exhibits medium bioconcentration factors. The low BMAA bioconcentration values in *M. nipponense* can be attributed to the less pronounced cyanobacterial blooms in the corresponding ponds. In addition, the BMAA biomagnification in the aquaculture ponds is lower than in the natural environment, which may be due to the addition of artificial feed that further increased the scatter of the data. Overall, BMAA can be found in high concentrations in aquatic animals in many areas of the world where cyanobacteria blooms occur.

In the Baltic Sea, zooplankton is considered a crucial link in the transport of BMAA to organisms at higher trophic levels (Jonasson et al. 2010). BMAA was identified in the soft tissues of mollusks: *Mytilus edulis* (0.115–0.200 $\mu g\ g^{-1}$ DW) and *Ostrea edulis* (0.006–0.140 $\mu g\ g^{-1}$ DW); the highest concentration of this amino acid was determined in turbot brain tissue (1.29 $\mu g\ g^{-1}$ DW). However, recent studies indicate that BMAA is only present in phytoplankton and herbivorous consumers of the pelagic food chain (Zguna et al. 2019). In the study by Błaszczyk et al. (2021), BMAA is not detected in any of the extracts from fish and bivalve mollusks.

As with every new discovery, confirming studies always meet negating ones, provided reliable analytics are applied. Here is another confirming study: Wang et al. (2021a) presented evidence of food web biomagnification of BMAA in a diatom-dominated marine ecosystem in China. The trophic magnification factors for zooplankton, bivalve mollusks, carnivorous crustaceans, and carnivorous gastropod mollusks are ca. 4.58, 30.1, 42.5, and 74.4, respectively. In addition to cyanobacteria, diatoms and dinoflagellates are proposed as producers of BMAA (Jiang et al. 2014; Lage et al. 2014). Actually, Wang et al. (2021a) cultured a total of 56 diatom strains of the genera *Pseudo-nitzschia*, *Thalassiosira*, *Chaetoceros*, *Planktoniella*, and *Minidiscus* isolated from the Chinese coast, among which 21 strains contain BMAA mainly in precipitated bound form at toxin concentrations ranging from 0.11 to 3.95 μg g^{-1} DW.

2.5 Cylindrospermopsin

Cylindrospermopsin (CYN) is a polycyclic uracil derivative containing guanidino and sulfate groups (Fig. 2.19) and is produced by a variety of cyanobacteria. It was first found in a bloom of *Raphidiopsis* (*Cylindrospermopsis*) *raciborskii* in Australia (Bourke et al. 1983). CYN progressively accumulates in the liver, where it binds to DNA, induces DNA fragmentation, and inhibits protein synthesis. It disrupts metabolic processes, leading to oxidative damage, which is linked to cytotoxic effects in multiple organs, including the liver, kidneys, spleen, and heart. Among these, the liver is the most severely affected organ, with CYN significantly reducing hepatocyte CYP450 activity (Sieroslawska and Rymuszka 2015b). In a human cell line,

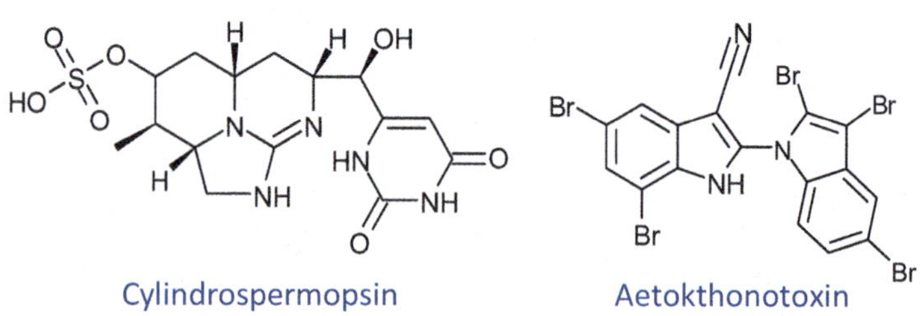

Fig. 2.19 Chemical structures of cylindrospermopsin (CYN) and aetokthonotoxin (AETX, see Sect. 2.6), two cyanotoxins with severe toxicological implications. CYN is a tricyclic guanidine alkaloid featuring a characteristic uracil moiety as the heterocyclic aromatic ring on the right-hand side of its structure. This uracil group contributes to CYN's ability to interfere with nucleic acid metabolism, resulting in the inhibition of protein synthesis and genotoxic effects. AETX is a neurotoxic compound recently identified as the causative agent of vacuolar myelinopathy, a fatal neurological disease in wildlife. Unlike CYN, AETX is a brominated hydrophobic metabolite that structurally differs from the guanidine alkaloids.

Bain et al. (2007) demonstrated that CYN activates the p53 transcription factor, a crucial regulator of cellular stress response. p53 plays a key role in maintaining genomic stability by initiating DNA repair mechanisms when damage is detected. If the damage is irreparable, p53 promotes apoptosis to prevent the propagation of mutated or damaged cells. This tumor suppressor protein is often referred to as the "guardian of the genome" due to its central function in preventing cancerous transformations. In contrast, Lankoff et al. (2007) reported that CYN increases the frequency of necrotic cells and necrotic cell death in a dose- and time-dependent manner, with a minimal impact on apoptosis. Therefore, it has been debated whether CYN has a tumorigenic effect, and although it is still suspected, it has not been clearly proven to be carcinogenic (Bláhová et al. 2009; Yang et al. 2021).

Histopathological changes and DNA damage induced by cylindrospermopsin in different tissues have frequently been observed in fish, such as Nile tilapia (Guzmán-Guillén et al. 2015) or guppy (Rabelo et al. 2021), often following oxidative stress (Sieroslawska and Rymuszka 2015a; Puerto et al. 2014; Silva et al. 2017). Applying a battery of mammalian in vitro tests, Puerto, Prieto et al. (2018) recently concluded that CYN is likely a pro-genotoxic compound, as one of its metabolites, rather than the educt itself, upregulates genes downstream of the tumor-suppressor gene *p53*. Moreover, oxidative stress does not seem to be directly related to genotoxicity. Furthermore, CYN displays immunomodulatory effects as shown in phagocytic cells of common carp: CYN activates the inflammatory machinery by expression of pro-inflammatory cytokines (*il-1β*, *tnf-α*) and, to a lesser extent, the anti-inflammatory *tgf-β* (Sieroslawska et al. 2015). However, ER stress has not yet been identified—the gene encoding the ER chaperone BiP is not modulated (Raška et al. 2019; Vanova et al. 2019). Máthé et al. (Máthé et al. 2017) present a comprehensive set of examples illustrating the cellular effects of CYN in human and animal cells. In addition to this compilation, Moraes et al. (2021) provide novel information and unmistakable evidence demonstrating that CYN directly impairs spermatogenesis in zebrafish testis at environmentally relevant levels of CYN. Furthermore, CYN adversely affects the development of fish embryos, as evidenced by cardiac toxicity in zebrafish (Wang et al. 2020b). Reproductive toxicity on *R. raciborskii* diet is also observed in *Daphnia*; the effective principle, however, does not appear to be CYN (see below).

The same laboratory adds some more adverse effects: CYN disrupts the vascular development of zebrafish embryos and decreases the abundance of blood vessels, resulting in vascular malformations and deletions, and inhibition of common cardinal vein remodeling (Wang et al. 2020a).

In a classic study of mice given purified CYN orally, Humpage and Falconer (2003) observed a hormetic response. Body weights significantly increase at low doses (30 and 60 $\mu g\ kg^{-1}\ day^{-1}$) and decrease at high doses (432 and 657 $\mu g\ kg^{-1}\ day^{-1}$). The authors interpret this non-linear response as compensatory hypertrophy in response to exposure to a blocking agent in cell metabolism, namely CYN. One might assume hormesis behind it, as defined initially by Calabrese and Baldwin (2003): adaptive response. Remarkably, the concept of a hormetic response to CYN exposure is only considered in a few papers, such as Guzmán-Guillén et al. (2017), who found an increase in photosynthetic yield in exposed carrots.

Although they are also exposed to CYN-producing cyanobacteria in the field, aquatic invertebrates affected by this alkaloid are almost entirely neglected in scientific studies. In a standard bioassay, Metcalf et al. (2002) compared the acute toxicities in brine shrimp *Artemia salina* of CYN, MC-LR, and three protein synthesis inhibitors (cycloheximide, chloramphenicol, and tetracycline). They found that CYN is slightly less potent than MC-LR, but more toxic than the protein synthesis inhibitors.

In *Daphnia magna*, Nogueira et al. (2006) identified disruption of epithelial cell adhesion in the gut after feeding CYN-producing cyanobacteria. This effect occurred only on *Raphidiopsis raciborskii* diet, but not on *Chrysosporum* (*Aphanizomenon*) *ovalisporum*. Therefore, the authors assume that *R. raciborskii* may produce other, as yet unidentified active metabolite(s) responsible for disrupting epithelial cell adhesion, in addition to CYN. A follow-up study identifying the substance in question has not been made; however, we know that *R. raciborskii* can also produce saxitoxins and gonyautoxins (Humpage et al. 1994; Stucken et al. 2014; Werner et al. 2020), which may account for the observed adverse effects in *Daphnia*.

In support, Kinnear et al. (2007) showed that *R. raciborskii* cells with CYN are more toxic than the extracted toxin itself in the snail *Melanoides tuberculata*. It is noteworthy that exposure to whole-cell extracts or live algae does not result in significant changes in the behavior or growth rates of adult snails. However, clear changes in the number of hatchlings released from parent snails occur. Exposure to whole-cell extracts increases the number of hatchlings. In contrast, decreases in hatchling numbers occur in exposures to live *R. raciborskii* cells.

There is also increasing evidence that the effects of cyanotoxins on other taxa are not simply allelopathic. CYN can stimulate protein phosphatase expression in some eukaryotic algae (Bar-Yosef et al. 2010), thereby increasing readily available phosphorus (P) resources; this is the enslavement of eukaryotic algae by the cyanobacterium of *Umezakia* (*Aphanizomenon*) *ovalisporum*. In other words, CYN is a poison that releases phosphate. More specifically, CYN induces the production of alkaline phosphatase even when the responding algae are not P-deficient, which can only be regarded as toxicity if the diversion of resources to produce alkaline phosphatase decreases the fitness of the alga (Raven 2010). Overall, this implies that our understanding of anthropocentrically important facets of cyanobacterial HAB biology (bloom proliferation, toxin production) is incomplete if we ignore ecological opportunities arising from assemblage-scale phytoplankton dynamics (Bramburger et al. 2023) (for more details, see Chap. 4).

2.6 Aetokthonotoxin

Recently, another very potent cyanotoxin has been identified: Aetokthonotoxin (AETX). This is a structurally unusual pentabrominated biindole alkaloid (Fig. 2.19) and causes vacuolar myelinopathy (VM). VM is a neurological disease characterized by widespread vacuolization in the white matter of the brain. First diagnosed in 1994 in bald eagles, it has since spread throughout the southeastern United States (Breinlinger et al. 2021).

Breinlinger et al. (2021) showed that it is produced by a cyanobacterium (*Aetokthonos hydrillicola*) growing on the invasive plant *Hydrilla verticillata* in man-made water bodies. AETX biosynthesis relies on the availability of bromide. Seasonal environmental conditions that promote toxin production in *A. hydrillicola* are watershed-specific. AETX is transferred through the food from herbivorous fish and wildlife to birds of prey, such as the bald eagle (Fig. 2.20).

Subsequently, Schwark et al. (2023) found that *Ae. hydrillicola* is more than just an eagle killer: it produces highly toxic dolastatin derivatives. In a cytotoxicity assay, the authors found the crude extract of the cyanobacterium to be much more potent than pure AETX, prompting further investigation. The authors describe the isolation and structure elucidation of aetokthonostatins (AESTs), linear peptides belonging to the dolastatin (→Glossary) compound family and biosynthesized by a gene cluster encoding a nonribosomal peptide synthetase/polyketide synthase. AEST potently affects microtubule dynamics and inhibits reproduction in the nematode *C. elegans*.

2.7 Concluding Considerations

Several ecophysiological considerations conclude this chapter, namely

- Toxin production by picocyanobacteria.
- Toxin production can be accelerated by grazing.
- Potential biodegradation of microcystins.
- A more practical consideration: are cyanobacterial proteins a potential aquafeed?

2.7.1 Picocyanobacteria: A Matter of Concern?

Picocyanobacteria (with cell sizes in the range of 0.2–2.0 µm) are essential organisms in the world's oceans and freshwater ecosystems. *Prochlorococcus* and *Synechococcus* are the two most abundant marine cyanobacteria. They represent a significant fraction of the total primary production of the world oceans and comprise a major fraction of the prey biomass available to phagotrophic protists (Frias-Lopez et al. 2009). Their dominance in phytoplankton biomass is common in both oligotrophic and eutrophic waters (Śliwińska-Wilczewska et al. 2018). Their role is expected to become even more critical with the effects of climate change (Worden and Wilken 2016).

The production of cyanotoxins by picoplanktonic cyanobacteria, along with their ecological significance, remains an underexplored topic—and the few existing studies have primarily focused on marine ecosystems. For instance, Martins et al. (2005) reported that strains of *Synechocystis* and *Synechococcus* from the Portuguese coast

Fig. 2.20 From the cyanobacterium to the bald eagle—toxin transmission through the food chain. *Aetokthonos hydrillicola*, growing in colonies on aquatic vegetation, produces the neurotoxin

(continued)

caused toxic effects in mice. In fact, toxin concentrations can be significant. Wiśniewska et al. (2022) found MC-LR in a laboratory strain of *Synechococcus* up to 420 fg cell^{-1}. Gómez-Leyva et al. (2024) identified another MC variant (MC-YR) in *Synechococcus* strain 2004 at 0.56 μg g^{-1}.

Picocyanobacteria are often described as a non-blooming group (e.g., Stockner et al. (2002)). However, it is becoming increasingly clear that picocyanobacteria are significant contributors to massive blooms in tropical and subtropical coastal areas, and even occur in temperate waters. Significant blooms of picocyanobacteria have been recorded in the Mediterranean Sea (Sorokin et al. 2004; Sorokin and Dallocchio 2008; Sorokin and Zakuskina 2010), Baltic Sea (Kuosa 1991), Black Sea (Uysal 2000), Hungarian lakes (Somogyi et al. 2009), Moroccan ponds (Mezrioui and Oudra 1998; Oudra et al. 2000), San Francisco Bay (Ning et al. 2000), Gulf of Mexico (Wawrik and Paul 2004), Florida Bay and Pensacola Bay (Phlips et al. 1999; Lynch and Phlips 2000; Murrell and Lores 2004; Gardner and McCarthy 2009; Goleski et al. 2010), Caloosahatchee River (southwest Florida (Urakawa et al. 2023; Hancock et al. 2024)), Indian River Lagoon (Florida) (Papacek et al. 2024), Lake Okeechobee (Florida) (Lefler et al. 2023; Krausfeldt et al. 2024), Chesapeake Bay (Weissberger and Glibert 2021), Great South Bay (New York) (Gobler et al. 2004), Seto Inland Sea (Nakamura et al. 1997), Guantánamo Bay, (Cuba) (Hall et al. 2018), and Gippsland Lakes (Australia) (Beardall 2008).[9] Picoplanktonic cyanobacteria blooms occur worldwide; the focus of studies along the east coast of the USA appears to be a matter of financial and technical resources rather than the abundance of harmful algal blooms.

Sorokin et al. (2004) reported that picocyanobacteria can provide up to 98% of phytoplankton production. By making such a large proportion of phytoplankton, picoplanktonic organisms affect not only the composition and quantity of matter but also the flow of energy to higher trophic levels (Parvathi et al. 2014). Furthermore, Motwani and Gorokhova (2013) suggested that picoplankton are crucial components of the mesozooplankton diet, which needs to be considered in food web models and productivity assessments. In surface layers, the number of picocyanobacteria ranges from a few hundred to several thousand and sometimes even a few million cells per mL of water (Sorokin and Dallocchio 2008; Parvathi et al. 2014). The large number of autotrophic picoplankton makes these organisms crucial for the ecological stability of aquatic ecosystems (Sánchez-Baracaldo et al. 2008).

Fig. 2.20 (continued) AETX. Waterbirds, tadpoles, aquatic turtles, snails, and fish consume this contaminated vegetation and develop VM. Predators develop VM when they consume animals that have been grazing on *A. hydrillicola*–covered plants (from Breinlinger et al. (2021); total image credit to the American Association for the Advancement of Science; individual images: Greenfrog tadpole, B. Gratwicke; American coot, G. S. Segler; Grass carp, R. Hagerty; Snail kite, Sirkfish; Painted turtle, U.S. Fish and Wildlife Service; Bald eagle, W. H. Majoros. Images are all under the Creative Commons Attribution Generic License)

[9] This and the next section is adapted from Śliwińska-Wilczewska et al. (2018), credit to MDPI.

Recently, Pagliara et al. (2021) summarized that *Synechococcus* and *Synechocystis* strains may be responsible for the production of microcystin in marine and freshwater systems (Carmichael and Li 2006; Vareli et al. 2012, 2013). However, this assertion is too broad. In particular, Carmichael and Li (2006) found that one of several *Synechococcus* strains was positive for microcystins by ELISA and LC/MS. And often, *Synechococcus* has been identified without cyanotoxins being detected (Silva-Stenico et al. 2011). On the other hand, Vareli et al. (2012) reported that microcystin concentrations in *Mytilus galloprovincialis*, collected from Amvrakikos Gulf, Greece, ranged from 45 ± 2 to 141.5 ± 13.5 ng g^{-1} (ww), while the concentrations in the water ranged from 0.003 to 19.8 ng L^{-1} with *Synechococcus* as the likely major producer. Although microcystins have a low lipophilicity at environmentally relevant pH values (log K_{ow} ~1 (Liang et al. 2011), some bioaccumulation has occurred. The developmental fate of the contaminated mussels was not reported by the authors.

In addition to MC, Gin et al. (2021) discovered that one strain of *Synechococcus* is capable of producing both CYN and ATX. This finding is particularly noteworthy, as the production of these cyanotoxins had never been previously reported. The *Synechococcus* strain produced about 9.0×10^{-5}–6.8×10^{-4} fg CYN $cell^{-1}$ and 4.7×10^{-4}–1.5×10^{-2} fg ATX $cell^{-1}$. Furthermore, Jakubowska and Szeląg-Wasielewska (2015) also listed BMAA and NOD present in picoplanktonic cyanobacteria.

After conducting a survey of New Zealand lakes and using LC-MS, Wood et al. (2017) concluded that cyanotoxins were only detected in eutrophic lowland lakes, with *Synechococcus* being the most abundant taxon. However, the PCR-based methods detected genes involved in microcystin production in many other lakes. Sequencing identified that this was due to the presence of the *mcyE* gene (involved in the production of microcystin) in *Dolichospermum lemmermannii*. (The molecular structure and biosynthetic pathways of microcystins will be explored in greater detail in Chap. 4) Similar results have been obtained by other researchers, demonstrating that the presence of one gene does not necessarily mean that the entire operon is present, or that the toxin is being produced. This highlights the need for molecular techniques to be aligned with other confirmatory analytical tools. Molecular methods offer numerous promising prospects for assessing the diversity of aquatic microbial communities across varying spatial and temporal scales. In addition to providing broad-scale data on diversity, they have tremendous potential to complement current monitoring programs aimed at measuring water quality or protecting human health.

Śliwińska-Wilczewska et al. (2018) concluded their review with a concern that is worth quoting here: *Picocyanobacterial blooms can cause the death of a large proportion of fauna and benthic flora, as well as economic losses. In addition, some picocyanobacteria are capable of producing secondary metabolites, which may also be a source of harmful bioactive compounds. A dense bloom of picocyanobacteria is a new phenomenon that requires further investigation. While the interactive effects of climate change on harmful picocyanobacterial blooms are complex, much of the current knowledge suggests that these processes are likely to increase the magnitude and frequency of these events. It is therefore essential to carefully examine the role of these small organisms in aquatic ecosystems.*

2.7.2 Grazing and Xenobiotics Increase Toxin Production

Cyanobacteria have been shown to increase cyanotoxin production under adverse conditions, such as predator grazing or exposure to xenobiotics, as a management strategy for controlling cyanobacteria.

2.7.2.1 Grazing

As we conclude this ecotoxicological chapter on cyanobacteria, we would like to highlight an issue that has long been ignored: Grazing herbivores increase toxin production in cyanobacteria. The production of Microcystin indicates this.

The defense hypothesis of toxin production in photoautotrophic organisms is supported by a body of work, primarily focused on *Microcystis aeruginosa*, which demonstrated that these organisms have inducible defenses, increasing toxin production in response to the presence of zooplankton and fish. For instance, Jang et al. (2004) showed that potentially toxic *Microcystis* responds to a chemical signal or feeding-related activity of both the phytoplanktivorous silver carp (*Hypophthalmichthys molitrix*) and the omnivorous ginbuna or silver crucian carp (*Carassius langsdorfi*) by producing microcystin, even though neither fish species appeared to feed heavily on *Microcystis*. Fish actively respond to exposure to microcystin-producing cyanobacteria through avoidance behavior (Baganz et al. 1998, 2004) and by minimizing adverse effects (Godlewska et al. 2018). These responses occur regardless of the type of interaction, whether through direct intoxication (Ernst et al. 2007; Sotton et al. 2012), visual deprivation (Gray et al. 2014), or food deprivation (Reichwaldt et al. 2013).

In addition to being toxic, cyanotoxins can serve as infochemicals and contribute to specific responses of the species that produce them. Cerbin et al. (2013) demonstrated that *Aphanizomenon gracile* switches its growth mode in the presence of *Daphnia* and invests more in trichome width to prevent filaments from being broken by grazers, thus minimizing predation pressure. In fact, thicker filaments are more harmful to *Daphnia* than thinner ones (Wejnerowski et al. 2015).

In support, Vilar et al. (2021) reported that infochemicals of *Daphnia gessneri* induce an upregulation of *sxt* genes and consequently an increase in saxitoxin production by the cyanobacterium *Raphidiopsis raciborskii*. Conversely, a loss of growth may occur as a trade-off for increased toxin production, as discussed above for *Alexandrium catenella*. Van Gremberghe et al. (2009) highlight substantial strain-specific variability in the toxin production response of photoautotrophs.

Several studies indicate that there is a mutualistic interaction between the toxic *Microcystis* and *Daphnia*, mediated by infochemicals rather than direct predation. Bojadzija Savic et al. (2019) and Rzymski et al. (2020) suggest that the presence of grazers via exudates can trigger a series of events in the *Microcystis* population, with cells undergoing oxidative stress-induced programmed cell death associated with the release of MC and further secondary metabolites, indicating either a protective or anti-grazer response.

2.7.2.2 Xenobiotics

Lieke et al. (2020) recently reviewed the currently used xenobiotic treatment options for reducing cyanobacterial blooms, including triphenylmethane dyes, copper, acriflavine, and formalin. However, the risks associated with many of these "traditional" chemical treatments—ranging from toxicity to fish to environmental hazards and potential risks to consumers—have led to bans in the aquaculture industry in many countries. Furthermore, this approach is problematic not only due to the direct or indirect toxicity of these treatments but also because suboptimal dosing or insufficient treatment duration can paradoxically lead to increased cyanotoxin production.

Cyanobacterial blooms are often controlled with $CuSO_4$ (Lieke et al. 2020; Watson et al. 2024), but the remaining Cu^{2+} affects cyanobacterial growth and metabolism in a hormetic manner. Low $CuSO_4$ concentrations (<0.1 mg L^{-1}) facilitate algal growth, photosynthesis, and enzyme activity. Conversely, high concentrations of $CuSO_4$ (>0.1 mg L^{-1}) inhibit growth and markedly reduce photosynthesis (Zhang et al. 2024).

Using anthraquinone, present in root exudates of e.g. the reed canary grass (*Phalaris arundinacea*) at a concentration of 100 μg L^{-1}, rather than 1000 μg L^{-1}, Castro-Castellon et al. (2025) found that this natural xenobiotic stimulated *Microcystis* growth by up to 70% compared to controls (dark bars, Fig. 2.21a). The biochemical mechanisms of why anthraquinone, out of the ten tested xenobiotics, induces a hormetic response remain to be elucidated. However, anthraquinone is not unique, as the hormetic effect on *Microcystis* growth has been demonstrated with exposure to other substances, such as the allelochemical ethyl-2-methyl acetoacetate isolated from *Phragmites communis*, pentachlorophenol (pesticide), glyphosate (herbicide), luteolin (flavonoid with algicidal activity), sulfamethoxazole, tetracycline, and ciprofloxacin, spiramycin, ampicillin, and amoxicillin (antibiotics), flame retardants, and lubricants (Hong et al. 2008; Liu et al. 2015a, b, c, 2016d; Li et al. 2020a, 2024; Wang et al. 2019c; Wang et al. 2021b; Xu et al. 2021a; Ye et al. 2020; Zhang et al. 2022c). Exposure to low-dose glyphosate stimulates the growth of *M. aeruginosa* primarily through the upregulation of photosynthesis-related proteins and an increase in photosynthetic activity, rather than by regulating the shikimate pathway (Xu et al. 2021b). And xenobiotic-challenged *Microcystis* produce more microcystin, as shown with amoxicillin (Fig. 2.21b) by stimulated photosynthesis[10] (Liu et al. 2016d), as well as with a mixture of spiramycin and ampicillin (Wang et al. 2019c) and with several concentrations of glyphosate exposure (Fig. 2.22) (Ye et al. 2020).

[10]Microcystin synthetase (mcyB) significantly interconnected with three PSII proteins (psbA, psb28 and psbO), two PSI proteins (psaB and psaC), one carbon fixation protein (cbbL), three photosynthetic pigments (apcB, pcA and pcB) and one photosynthesis-related electron transfer protein (fdx) (Liu et al. 2016d).

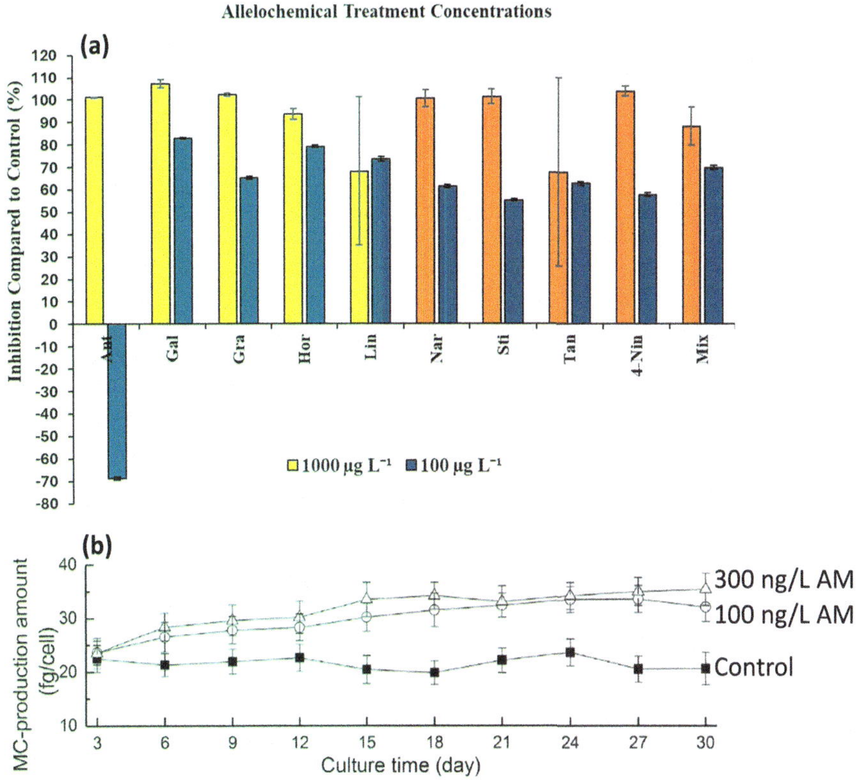

Fig. 2.21 (**a**) Percentage inhibition of *Microcystis* by commercial allelochemicals (mean ± SD) with a concentration of 1000 µg L^{-1} (orange bars) and of 100 µg L^{-1} (blue bars) compared to a *Microcystis* control in the absence of allelochemicals ($N = 6$). Tested allelochemicals: *Ant* anthraquinone, *Gal* gallic acid, *Gra* gramine, *Hor* hordenine, *Lin* linoleic acid, *Nar* naringuin, *Sti* stigmasterol, *Tan* tannic acid, *4-Nin* 4-nitroindol-5-carboxaldehyde, *Mix* allelochemical-mix (from Castro-Castellon et al. (2025), credit to Taylor & Francis). (**b**) Effect of amoxicillin on MC-production of *Microcystis aeruginosa* during the 30 days of exposure. Mean and standard deviation of three replicates are shown (from Liu et al. (2016d), with permission from Elsevier)

2.7.3 The Microcystin Biodegradation

Krishnan and Mou (2021) reviewed the current knowledge on heterotrophic MC biodegradation, primarily based on studies on bacteria. The relatively slow microbial degradation that mimics persistence (Welker et al. 2001) can be significantly facilitated if preceded by catalyzed photolytic degradation, in which dissolved humic substances play a central role as catalyst (Welker and Steinberg 1999, 2000). Depending on the presence of adapted bacteria and abiotic environmental conditions (such as temperature, the presence of photocatalysts, and water hardness), relatively high degradation rates can be observed (Feng et al. 2024). The authors found a half-life of MC-LR of 1 day in summer, under the combined effect of direct

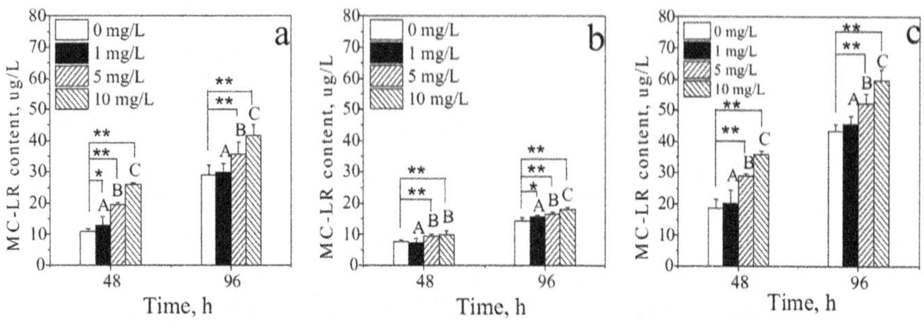

Fig. 2.22 The effects of different concentrations of glyphosate on intracellular MC-LR (**a**), extracellular MR-LR (**b**), and total cell MC-LR (**c**) at 15 °C (* represents $P < 0.05$ and ** represents $P < 0.01$ relative to the control group by ANOVA). Different capital letters above adjacent bars indicate a significant difference ($P < 0.05$) between different concentrations. From Ye et al. (2020), with permission from Springer Nature)

and indirect photodegradation. The identified transformation products revealed that the conversion of MC-LR involved oxidation, hydroxylation, and/or bond cleavage with the addition of Adda and Mdha moieties (Abbreviations, see Fig. 2.3).

Currently, and without initial photolytic breakdown, one complete MC degradation pathway is known, which is encoded by multiple *mlr* genes. Four genes, namely *mlrA*, *mlrB*, *mlrC*, and *mlrD*, are sequentially located in a cluster. *mlrD* encodes a transporter-like protein, whereas *mlr*A, B, and C encode enzymes required for the degradation of MCs (Fig. 2.3a). The enzyme MlrA (microcystinase) initiates degradation by breaking the ring structure of MCs at the peptide bond between Adda and Arg, thereby linearizing the cyclic MC-LR. The linearized MC-LR is only one-hundredth as toxic, or even less, than the cyclic one (Jones et al. 1994; Somdee et al. 2013). Next, a serine peptidase (MlrB) catalyzes the hydrolysis of the linearized MC-LR at the Ala-Leu peptide bond, producing a tetrapeptide (Jones and Orr 1994). Finally, the third enzyme (MlrC) randomly cuts the peptide bonds between amino acid residues, resulting in undetermined peptide fragments and amino acids. Since the initial cleavage is the most critical step in MC-LR degradation, genetic analysis of MC-LR degradation has primarily focused on the analysis of *mlrA* genes (Ho et al. 2006; Yang et al. 2014; Wang et al. 2019a). This *mlr*-dependent degradation is regulated by acylhomoserine lactone (AHL)-mediated quorum sensing (QS), highlighting the role of cell-to-cell communication in controlling toxin breakdown (Zeng et al. 2020). This QS encodes two enzyme systems, namely *novR1/novI1* and *novR2/novI2*, which upregulate the degradation of MC. Identification of these regulators will aid in controlling the biological degradation of MCs.

Studies on bacterially mediated MC degradation are dominated by culture-dependent work (Jones et al. 1994; Park et al. 2001). Nearly a third of current MC-degrading bacterial isolates are affiliated with genera of a single family, i.e., *Sphingomonas*, *Sphingopyxis*, and *Sphingoscinicella* of the *Sphingomonaceae* family (*Alphaproteobacteria*) (Massey and Yang 2020), with the rest belonging to *Gammaproteobacteria* and *Firmicutes*. However, many recent culture-dependent and independent studies have consistently indicated that MC-degrading bacteria are obviously more taxonomically diverse (Mou et al. 2013; Krishnan et al. 2020). MC-degrading bacteria are now known to be distributed among species (approx. 100) of *Microbacterium* of Actinobacteria families (Manage et al. 2009; Lawton et al. 2011); *Bacillus*, *Brevibacillus*, and *Lactobacillus* of *Firmicutes* families (Hu et al. 2012; Kansole and Lin 2016; Nybom et al. 2007); and *Pseudomonas*, *Stenotrophomonas*, *Acidovorax*, and *Sphingomonas* of *Alphaproteobacteria*, *Betaproteobacteria*, and *Gammaproteobacteria* families (Jones et al. 1994; Krishnan et al. 2018).

Many studies have revealed that MC-degrading bacteria may carry *mlr* genes, as explained above (Manage et al. 2009; Mou et al. 2013; Krishnan et al. 2020). Non-*mlr* bacteria might degrade MCs via the xenobiotic degradation pathway (Mou et al. 2013). A recent study has shown the presence of microcystinase, linearized-microcystinase, tetrapeptidease, and PAAase (phenylacetic acidase) as enzymes for the degradation of MC by bacteria (Yang et al. 2020). This study is the first to report the presence of PAAase for the complete degradation of MC using the *mlr* degradation pathway. Non-*mlr* MC-degrading bacterial cultures (Jones et al. 1994) are taxonomically more diverse than *mlr*-based MC degraders. Most non-*mlr*-based degraders are affiliated with *Proteobacteria* (80%), including Alpha-, Beta-, and Gammaproteobacteria classes, followed by *Actinobacteria* (10%) and *Firmicutes* (10%). In contrast, bacteria containing *mlr* genes (Fischer et al. 2005) primarily belong to *Alphaproteobacteria* (60%), with some found in the phylum *Firmicutes* and the class *Gammaproteobacteria* (Park et al. 2001) (extract taken from Krishnan and Mou (2021), credit MDPI, Basel, references added).

Using metabolic profiling, Zhang et al. (2020b) identified functional genes within specialized genomic regions, with the *mlr* gene cluster located nearby. Additionally, the authors identified transposable elements, suggesting that these genomic regions might undergo horizontal gene transfer to acquire new functionalities. This indicates an adaptive force driving the evolution of the genome of these organisms. The recruitment of MC-degrading capacity might enrich their metabolic traits and enhance their environmental adaptation.

Previously, the Mlr pathway was the only enzymatic pathway clearly described for MC biodegradation from a biochemical and molecular perspective (Bourne et al. 1996, 2001; Dziga et al. 2013, 2016); however, other pathways have been elucidated from various intermediate products or mlr-lacking biodegradation activity (Edwards et al. 2008; Mou et al. 2013; Lezcano et al. 2016; Morón-López et al. 2017; Dexter et al. 2021). Nearly two decades ago, a new potent cyanotoxin-degrading bacterium was identified: *Paucibacter toxinivorans* (Rapala et al. 2005). *P. toxinivorans* (*mlr-*) degrade MC-LR in different culture conditions, along with other MC variants, such as -RR and -YR (R: arginine and Y: tyrosine) (Rapala et al. 2005; Lezcano et al. 2016; Morón-López et al. 2017). More recently, Santos et al. (2021) show the degradation of several peptides by strain 2C20 of this microcystin degrader. Not all MCs are completely degraded within 7 days by *P. toxinivorans*, with MC-YR showing the highest degradation among MCs (Fig. 2.23).

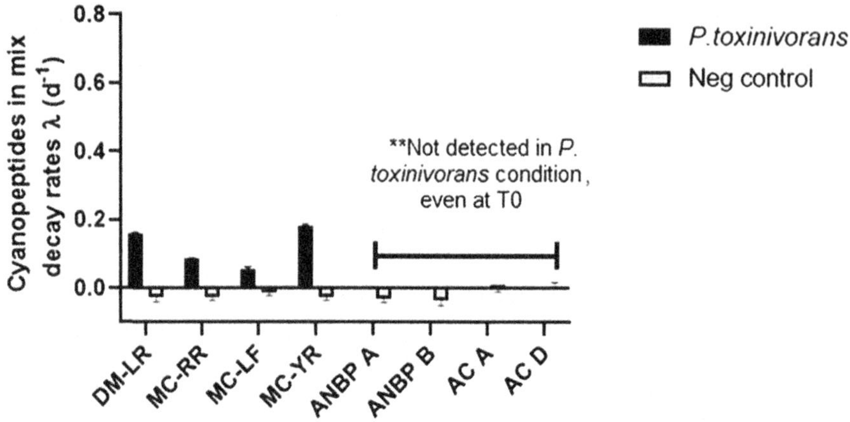

Fig. 2.23 Biodegradation of several cyanobacterial peptides by *Paucibacter toxinivorous* from the mixed condition with an initial concentration of 1 μg mL^{-1}, showing the remaining concentration as exponential decay rates over 7 days of incubation. Non-MCs peptides were not detected even at time 0. Data are expressed as mean and standard deviation ($N = 3$). The graph shows four MC variants designated as DM-LR, -RR, LF, and YR, in addition to other cyanopeptides such as ANPB (anabaenopeptin) A and B, and AC (aerucyclamide) A and D (from Santos et al. (2021), credit to MDPI, Basel)

2.7.4 Cyanobacteria as Potential Aquafeed

Due to their high protein content, with a high proportion of essential amino acids (Han et al. 2015; Chen et al. 2021), cyanobacteria, especially nitrogen-fixing cyanobacteria, represent a promising option for producing biomass with high nutritional value while promoting a more efficient and environmentally friendly use of nitrogen resources (Lucato et al. 2024). In most cases, specific sterols and sometimes long-chain PUFAs need to be added to the aquafeed (von Elert et al. 2003; Martin-Creuzburg et al. 2008), if the toxin content can be managed. Here, in addition to the empirically oriented review by Chen et al. (2021), we will present a few hypotheses on how to address the toxin content.

2.7.4.1 Selective Breeding

Selective breeding may be a way to use even toxin-producing cyanobacteria. Shahmohamadloo et al. (2022) investigated the tissue accumulation and proteomic response of rainbow trout and lake trout to MCs. For both species, MCs accumulate at a higher concentration in the liver than in the muscle, and activate cellular responses related to oxidative stress, apoptosis, DNA repair, and carcinogenicity. However, intraspecific proteomic effects differ between rainbow trout and lake trout, as does interspecific accumulation and retention of MCs in tissues within each species. Interspecific variability in fish tissue MC concentrations may be related to feeding habits and dietary exposure (Wood et al. 2014), while intraspecific variability may result from individual differences in mechanistic responses to toxicity, biotransformation, excretion of MCs (Adamovský et al. 2007), and subsequent epimutations (Anastasiadi and Piferrer 2019). On the latter point, intraspecific variation, which can evolve rapidly in multiple species within an ecosystem as well as in aquaculture, is an essential driver of community structure and ecosystem function (Rudman et al. 2015; Anastasiadi and Piferrer 2019). Overall, well-acclimated or epimutated individuals with low MC accumulation and few adverse effects can serve as a foundation for less sensitive strains of farmed fish and invertebrates that can tolerate cyanobacteria and utilize their proteins.

Studies have shown that estuarine and marine bivalve species such as the Mediterranean mussel (*Mytilus galloprovincialis*) and the Manila clam (*Ruditapes philippinarum*) modulate their detoxification and oxidative stress defense systems after exposure to MCs (Carneiro et al. 2017; Kim et al. 2017). This suggests that selective breeding can also be used in invertebrates for genetic improvement, for example, to breed more disease- or stress-resistant strains (Yáñez et al. 2023; Jiang et al. 2024). The genetic improvement to achieve more MC-resistant invertebrates seems achievable.

2.7.4.2 Prophylactic Nutrition

Prophylactic nutrition includes measures that block the effects of cyanotoxins once an organism has internalized them and may comprise antagomirs, prebiotics, and probiotics.

Antagomirs

The effect of antagomirs on apoptosis-blocking miRNAs in human medicine, particularly in cancer therapy, is gaining focus in research and application (Krützfeldt et al. 2005; Tong and Nemunaitis 2008; Gambari et al. 2016). The current chapter has demonstrated that epigenetic regulation is being identified as a major mechanism in displaying the toxic action of cyanotoxins. Also, knowledge of this regulatory pathway is more advanced for cyanotoxins than for phycotoxins or mycotoxins. Since miRNAs play a central role in cyanobacterial intoxication, the use of antagomirs seems feasible in the long term, provided that the biomolecular and epigenetic basis of the toxic effect in aquatic animals is extended and robust. Krützfeldt, Rajewsky, et al. (2005) coined the term antagomirs, also known as anti-miRs, which are a class of chemically engineered oligonucleotides designed to silence endogenous miRNAs. Furthermore, this laboratory developed an *antagomir-122*. As seen above, *miR-122* also plays a central, not yet fully understood role in the development of MC-LR toxicity in mammals (Xu et al. 2018a; Wang et al. 2020c, 2022b) as well as in MC-sensitive and MC-tolerant fish species (Li et al. 2013b; Florczyk et al. 2016; Feng et al. 2017; Liu et al. 2020); nevertheless, the outcome of dietary administration of *antagomir-122* in MC-containing aquafeed would be not only scientifically interesting, but also forward-looking in terms of aquaculture practice with new protein sources.

Prebiotics

Prebiotics as a prophylactic dietary approach to counteract cyanotoxin exposure have not yet been thoroughly tested. However, a laboratory study on a mixotrophic alga (*Ochromonas* sp.) demonstrated the significant role of chitin in both tolerance and degradation. Zhang et al. (2018) reported that *N*-acetylglucosamine, which is the monomeric unit of the chitin polymer, plays a central role in the resistance of this alga to MCs. Therefore, the question arises whether the tolerance of aquatic animals, especially fish, to cyanobacteria can be improved by an appropriate prebiotic diet containing chitin or other *N*-acetylglucosamine-containing prebiotics. This hypothesis should be tested experimentally.

Dietary astaxanthin, a type of prebiotic, has the potential to enhance the growth of *Penaeus vannamei* by preserving the integrity of the intestinal tract. It also alters the composition of the intestinal microbiota by increasing the relative abundance of the *Firmicutes* phylum, while reducing the abundance of both *Shewanella* and *Vibrio* in MC-stressed individuals (Song et al. 2024). Both of these microbial genera are potentially harmful. Overall, astaxanthin may mitigate the detrimental effects of MC on Pacific white shrimp *(P. vannamei)* aquaculture.

Probiotics

To optimize the gut microbiome for enhanced host tolerance to cyanotoxins, lessons can be learned from human medicine, specifically through microbiome profiling and targeted interventions. Advances in microbiome sequencing and analysis are enabling targeted approaches to probiotic intervention. High-resolution microbiome profiling can identify specific bacterial strains that are depleted or overabundant in the gut. Metagenomic and metabolomic analyses can reveal functional deficits in the microbiome that specific probiotic strains could address. Additionally, integrating host genetic and immune data with microbiome profiles can aid in predicting responses to probiotic interventions. The development of AI-driven algorithms that integrate life history trait data with microbiome profiles holds promise for optimizing probiotic interventions. Machine learning algorithms can be used to predict the most effective probiotic combinations based on specific microbiome profiles and host trait characteristics (D'Urso and Broccolo (2024) with references therein).

MC ingestion leads to gut microbial dysbiosis, characterized by the suppression of metabolically beneficial microbiota and deficiencies in, e.g., bile acids and short-chain fatty acids. It can be hypothesized that dietary probiotics reprogram the gut microbiota and metabolome. Indeed, besides the fact that several probiotic strains can degrade cyanotoxins in the water phase (Nybom et al. 2007, 2008), dietary probiotics are an emerging and effective tool against cyanobacteria and their toxins in the gut of aquatic animals.

Based on a previous study showing that *Lactobacillus plantarum* Ep-M17 has a probiotic activity on *P. vannamei* health and growth (Du et al. 2022), Xu et al. (2023) fed *P. vannamei* MC-LR alone or combined with Ep-M17 for 6 weeks. MC-LR stress leads to slow growth and reduced survival rates in shrimps. However, feeding Ep-M17 significantly increases both the growth and survival rates. In addition, MC-LR stress causes severe tissue damage in the shrimp hepatopancreas and intestine, but Ep-M17 significantly reduces the toxic effects and protects the integrity of these tissues. Additionally, Ep-M17 significantly enhances the activities of antioxidant enzymes and digestive enzymes, and induces higher expression of immune-related genes, thereby promoting digestive and immune responses in shrimps. Furthermore, MC-LR stress disrupts the gut flora, whereas the use of Ep-M17 significantly increases the abundance of immune- and metabolism-related bacteria, inhibiting the growth of pathogenic bacteria to maintain gut flora balance and intestinal health.

In zebrafish, Luan et al. (2024) showed that probiotic *Lactobacillus rhamnosus* ameliorates MC-LR neurotoxicity via the gut-brain axis. MC-LR exposure inhibits the swimming speed of zebrafish, while the presence of *L. rhamnosus* alleviates this abnormality. In addition, *L. rhamnosus* ultimately alleviates neurotoxicity in zebrafish by modulating intestinal and brain neural signaling, as well as neuroinflammation and hypothalamic-pituitary-interrenal axis responses.

Neither of these studies used metabolomics; this aspect of the research still remains to be investigated.

References

Adamczuk M, Mieczan T (2019) Within-species phenotypic diversity enhances resistance to stress – a case study using the polymorphic species *Bosmina longirostris*. Intern Rev Hydrobiol 104(5–6):137–146. https://doi.org/10.1002/iroh.201901985

Adamovský O, Kopp R, Hilscherová K, Babica P, Palíková M, Pašková V, Navrátil S, Maršálek B, Bláha L (2007) Microcystin kinetics (bioaccumulation and elimination) and biochemical responses in common carp (*Cyprinus carpio*) and silver carp (*Hypophthalmichthys molitrix*) exposed to toxic cyanobacterial blooms. Environ Toxicol Chem 26(12):2687–2693. https://doi.org/10.1897/07-213.1

Agrawal MK, Bagchi D, Bagchi SN (2005a) Cysteine and serine protease-mediated proteolysis in body homogenate of a zooplankter, *Moina macrocopa*, is inhibited by the toxic cyanobacterium, *Microcystis aeruginosa* PCC7806. Comp Biochem Physiol B 141(1):33–41. https://doi.org/10.1016/j.cbpc.2005.01.002

Agrawal MK, Zitt A, Bagchi D, Weckesser J, Bagchi SN, von Elert E (2005b) Characterization of proteases in guts of *Daphnia magna* and their inhibition by *Microcystis aeruginosa* PCC 7806. Environ Toxicol 20(3):314–322. https://doi.org/10.1002/tox.20123

Akbar S, Du J, Lin H, Kong X, Sun S, Tian X (2017) Understanding interactive inducible defenses of *daphnia* and its phytoplankton prey. Harmful Algae 66:47–56. https://doi.org/10.1016/j.hal.2017.05.003

Alva-Martínez AF, Sarma SSS, Nandini S (2001) Comparative population dynamics of three species of cladocera in relation to different levels of *Chlorella vulgaris* and *Microcystis aeruginosa*. Crustaceana 74(8):749–764. https://doi.org/10.1163/156854001317015571

Anastasiadi D, Piferrer F (2019) Epimutations in developmental genes underlie the onset of domestication in farmed European sea bass. Mol Biol Evol 36(10):2252–2264. https://doi.org/10.1093/molbev/msz153

Aráoz R, Molgó J, Tandeau de Marsac N (2010) Neurotoxic cyanobacterial toxins. Toxicon 56(5):813–828. https://doi.org/10.1016/j.toxicon.2009.07.036

Asselman J, De Coninck DIM, Glaholt S, Colbourne JK, Janssen CR, Shaw JR, De Schamphelaere KAC (2012) Identification of pathways, gene networks, and paralogous gene families in

Daphnia pulex responding to exposure to the toxic cyanobacterium *Microcystis aeruginosa*. Environ Sci Technol 46(15):8448–8457. https://doi.org/10.1021/es301100j

Asselman J, De Coninck DIM, Beert E, Janssen CR, Orsini L, Pfrender ME, Decaestecker E, De Schamphelaere KAC (2017) Bisulfite sequencing with *daphnia* highlights a role for epigenetics in regulating stress response to *Microcystis* through preferential differential methylation of serine and threonine amino acids. Environ Sci Technol 51(2):924–931. https://doi.org/10.1021/acs.est.6b03870

Azevedo-Linhares M, Souza ATC, Lenz CA, Leite NF, Brito IA, Folle NMT, Garcia JE, Filipak Neto F, Oliveira Ribeiro CA (2018) Microcystin and pyriproxyfen are toxic to early stages of development in *Rhamdia quelen*: an experimental and modelling study. Ecotox Environ Saf 166:311–319. https://doi.org/10.1016/j.ecoenv.2018.09.064

Baganz D, Staaks G, Steinberg C (1998) Impact of the cyanobacteria toxin, microcystin-LR on behaviour of zebrafish, *Danio rerio*. Water Res 32(3):948–952. https://doi.org/10.1016/S0043-1354(97)00207-8

Baganz D, Staaks G, Pflugmacher S, Steinberg CEW (2004) Comparative study of microcystin-LR-induced behavioral changes of two fish species, *Danio rerio* and *Leucaspius delineatus*. Environ Toxicol 19(6):564–570. https://doi.org/10.1002/tox.20063

Bain P, Shaw G, Patel B (2007) Induction of p53-regulated gene expression in human cell lines exposed to the cyanobacterial toxin cylindrospermopsin. J Toxicol Environ Health A 70(19):1687–1693. https://doi.org/10.1080/15287390701434877

Banerjee S, Maity S, Guchhait R, Chatterjee A, Biswas C, Adhikari M, Pramanick K (2021) Toxic effects of cyanotoxins in teleost fish: a comprehensive review. Aquat Toxicol 240:105971. https://doi.org/10.1016/j.aquatox.2021.105971

Bartel DP (2004) MicroRNAs: genomics, biogenesis, mechanism, and function. Cell 116(2):281–297. https://doi.org/10.1016/S0092-8674(04)00045-5

Bar-Yosef Y, Sukenik A, Hadas O, Viner-Mozzini Y, Kaplan A (2010) Enslavement in the water body by toxic *Aphanizomenon ovalisporum*, inducing alkaline phosphatase in phytoplanktons. Curr Biol 20(17):1557–1561. https://doi.org/10.1016/j.cub.2010.07.032

Beardall J (2008) Blooms of *Synechococcus*: An analysis of the problem worldwide and possible causative factors in relation to nuisance blooms in the Gippsland lakes. Abbreviation 2. Monash University,, Clayton, VIC, Australia

Biswas C, Adhikari M, Pramanick K (2024) Toxicological effects of nodularin on the reproductive endocrine system of female zebrafish (*Danio rerio*). Aquat Toxicol 273:107000. https://doi.org/10.1016/j.aquatox.2024.107000

Bitarte N, Bandres E, Boni V, Zarate R, Rodriguez J, Gonzalez-Huarriz M, Lopez I, Javier Sola J, Alonso MM, Fortes P, Garcia-Foncillas J (2011) MicroRNA-451 is involved in the self-renewal, tumorigenicity, and chemoresistance of colorectal cancer stem cells. Stem Cells 29(11):1661–1671. https://doi.org/10.1002/stem.741

Bláhová L, Oravec M, Marsálek B, Sejnohová L, Simek Z, Bláha L (2009) The first occurrence of the cyanobacterial alkaloid toxin cylindrospermopsin in The Czech Republic as determined by immunochemical and LC/MS methods. Toxicon 53(5):519–524. https://doi.org/10.1016/j.toxicon.2009.01.014

Błaszczyk A, Siedlecka-Kroplewska K, Woźniak M, Mazur-Marzec H (2021) Presence of ß-*N*-methylamino-L-alanine in cyanobacteria and aquatic organisms from waters of northern Poland; BMAA toxicity studies. Toxicon 194:90–97. https://doi.org/10.1016/j.toxicon.2021.02.007

Bojadzija Savic G, Edwards C, Briand E, Lawton L, Wiegand C, Bormans M (2019) *Daphnia magna* exudates impact physiological and metabolic changes in *Microcystis aeruginosa*. Toxins 11(7):421. https://doi.org/10.3390/toxins11070421

Boldin MP, Taganov KD, Rao DS, Yang L, Zhao JL, Kalwani M, Garcia-Flores Y, Luong M, Devrekanli A, Xu J, Sun G, Tay J, Linsley PS, Baltimore D (2011) *miR-146a* is a significant brake on autoimmunity, myeloproliferation, and cancer in mice. J Exp Med 208(6):1189–1201. https://doi.org/10.1084/jem.20101823

Bouaïcha N, Miles CO, Beach DG, Labidi Z, Djabri A, Benayache NY, Nguyen-Quang T (2019) Structural diversity, characterization and toxicology of microcystins. Toxins 11(12):714. https://doi.org/10.3390/toxins11120714

Bourke ATC, Hawes RB, Neilson A, Stallman ND (1983) An outbreak of hepato-enteritis (the Palm Island mystery disease) possibly caused by algal intoxication. Toxicon 21:45–48. https://doi.org/10.1016/0041-0101(83)90151-4

Bourne DG, Jones GJ, Blakeley RL, Jones A, Negri AP, Riddles P (1996) Enzymatic pathway for the bacterial degradation of the cyanobacterial cyclic peptide toxin microcystin LR. Appl Environ Microbiol 62(11):4086–4094. https://doi.org/10.1128/aem.62.11.4086-4094.1996

Bourne DG, Riddles P, Jones GJ, Smith W, Blakeley RL (2001) Characterisation of a gene cluster involved in bacterial degradation of the cyanobacterial toxin microcystin LR. Environ Toxicol 16(6):523–534. https://doi.org/10.1002/tox.10013

Bownik A, Pawlik-Skowrońska B (2019) Early indicators of behavioral and physiological disturbances in *Daphnia magna* (Cladocera) induced by cyanobacterial neurotoxin anatoxin-a. Sci Total Environ 695:133913. https://doi.org/10.1016/j.scitotenv.2019.133913

Bramburger AJ, Filstrup CT, Reavie ED, Sheik CS, Haffner GD, Depew DC, Downing JA (2023) Paradox versus paradigm: a disconnect between understanding and management of freshwater cyanobacterial harmful algae blooms. Freshw Biol 68(2):191–201. https://doi.org/10.1111/fwb.14019

Breinlinger S, Phillips TJ, Haram BN, Mareš J, Martínez Yerena JA, Hrouzek P, Sobotka R, Henderson WM, Schmieder P, Williams SM, Lauderdale JD, Wilde HD, Gerrin W, Kust A, Washington JW, Wagner C, Geier B, Liebeke M, Enke H, Niedermeyer THJ, Wilde SB (2021) Hunting the eagle killer: a cyanobacterial neurotoxin causes vacuolar myelinopathy. Science 371 (6536).eaax9050. https://doi.org/10.1126/science.aax9050

Brzuzan P, Woźny M, Ciesielski S, Łuczyński MK, Góra M, Kuźmiński H, Dobosz S (2009) Microcystin-LR induced apoptosis and mRNA expression of p53 and cdkn1a in liver of whitefish (*Coregonus lavaretus* L.). Toxicon 54 (2):170-183. https://doi.org/10.1016/j.toxicon.2009.03.032

Brzuzan P, Woêny M, Wolinska L, Piasecka A, Łuczynski MK (2010) MicroRNA expression in liver of whitefish (*Coregonus lavaretus*) exposed to microcystin-LR. Envirion Biotechnol 6(2):53–60

Brzuzan P, Woźny M, Wolińska L, Piasecka A (2012) Expression profiling *in vivo* demonstrates rapid changes in liver microRNA levels of whitefish (*Coregonus lavaretus*) following microcystin-LR exposure. Aquat Toxicol 122-123:188–196. https://doi.org/10.1016/j.aquatox.2012.07.001

Brzuzan P, Florczyk M, Łakomiak A, Woźny M (2016) Illumina sequencing reveals aberrant expression of MicroRNAs and their variants in whitefish (*Coregonus lavaretus*) liver after exposure to microcystin-LR. PLoS One 11(7):e0158899. https://doi.org/10.1371/journal.pone.0158899

Cai S, Shu Y, Tian C, Wang C, Fang T, Xiao B, Wu X (2022a) Effects of chronic exposure to microcystin-LR on life-history traits, intestinal microbiota and transcriptomic responses in *Chironomus pallidivittatus*. Sci Total Environ 823:153624. https://doi.org/10.1016/j.scitotenv.2022.153624

Cai W, MacIsaac HJ, Xu R, Zhang J, Pan X, Zhang Y, Yang J, Dixon B, Li J, Zi Y, Chang X (2022b) Abnormal neurobehavior in fish early life stages after exposure to cyanobacterial exudates. Ecotox Environ Saf 245:114119. https://doi.org/10.1016/j.ecoenv.2022.114119

Calabrese EJ, Baldwin LA (2003) Toxicology rethinks its central belief. Nature 421(6924):691–692. https://doi.org/10.1038/421691a

Campos A, Vasconcelos V (2010) Molecular mechanisms of microcystin toxicity in animal cells. Int J Mol Sci 11(1):268–287. https://doi.org/10.3390/ijms11010268

Cao H, Lyu K, Xiang F, Yang Z (2014) Life history responses of *Daphnia similoides* simultaneously exposed to microcystin-LR and ammonia and their postexposure recovery. Environ Toxicol Chem 33(11):2497–2505. https://doi.org/10.1002/etc.2701

Cao L, Shao N, Du J, Zhu H, Gao J, Li Q, Sun Y, Hu J, Yin G, Xu G (2024) Involvement of reactive oxygen species (ROS) in the hepatopancreatic cytotoxicity, oxidative stress, and apoptosis induced by microcystin-LR in *Eriocheir sinensis*. Comp Biochem Phys C 276:109801. https://doi.org/10.1016/j.cbpc.2023.109801

Capaul M, Ebert D (2003) Parasite-mediated selection in experimental *Daphnia magna* populations. Evolution 57(2):249–260. https://doi.org/10.1554/0014-3820(2003)057[0249:pmsied]2.0.co;2

Carmichael WW, Li R (2006) Cyanobacteria toxins in the Salton Sea. Saline Systems 2(1):5. https://doi.org/10.1186/1746-1448-2-5

Carneiro M, Antas P, Reis B, Azevedo J, Osório H, Campos A, Vasconcelos V, Martins JC (2017) Modulation of hepatic glutathione transferases isoenzymes in three bivalve species exposed to purified microcystin-LR and *Microcystis* extracts. Toxicon 137:150–157. https://doi.org/10.1016/j.toxicon.2017.07.003

Casas-Rodriguez A, Cameán AM, Jos A (2022) Potential endocrine disruption of cyanobacterial toxins, microcystins and cylindrospermopsin: a review. Toxins 14(12):882. https://doi.org/10.3390/toxins14120882

Castro-Castellon AT, Hughes JMR, Chipps MJ, Hankins NP (2025) Hormesis and antagonism in low-dose *Phalaris* allelochemicals during *Microcystis* control. Inland Waters:1–22. https://doi.org/10.1080/20442041.2024.2426388

Cazenave J, Nores ML, Miceli M, Díaz MP, Wunderlin DA, Bistoni MA (2008) Changes in the swimming activity and the glutathione *S*-transferase activity of *Jenynsia multidentata* fed with microcystin-RR. Water Res 42(4–5):1299–1307. https://doi.org/10.1016/j.watres.2007.09.025

Cerbin S, Kraak MHS, de Voogt P, Visser PM, van Donk E (2010) Combined and single effects of pesticide carbaryl and toxic *Microcystis aeruginosa* on the life history of *Daphnia pulicaria*. Hydrobiologia 643(1):129–138. https://doi.org/10.1007/s10750-010-0130-1

Cerbin S, Wejnerowski Ł, Dziuba M (2013) *Aphanizomenon gracile* increases in width in the presence of *daphnia*. A defence mechanism against grazing? J Limnol 72 (3):505-511.e41. https://doi.org/10.4081/jlimnol.2013.e41

Chang J, Nicolas E, Marks D, Sander C, Lerro A, Buendia MA, Xu C, Mason WS, Moloshok T, Bort R, Zaret KS, Taylor JM (2004) miR-122, a mammalian liver-specific microRNA, is processed from *hcr* mRNA and may downregulate the high affinity cationic amino acid transporter CAT-1. RNA Biol 1(2):106–113. https://doi.org/10.4161/rna.1.2.1066

Chen L, Xie P (2016) Mechanisms of microcystin-induced cytotoxicity and apoptosis. Mini-Rev Med Chem 16(13):1018–1031. https://doi.org/10.2174/1389557516666160219130407

Chen W, Song L, Ou D, Gan N (2005) Chronic toxicity and responses of several important enzymes in *Daphnia magna* on exposure to sublethal microcystin-LR. Environ Toxicol 20(3):323–330. https://doi.org/10.1002/tox.20108

Chen Y, Shen D, Fang D (2013) Nodularins in poisoning. Clin Chim Acta 425:18–29. https://doi.org/10.1016/j.cca.2013.07.005

Chen Y, Zhang L, Jiang X (2015) Fine-scale local adaptation of cyanobacterial tolerance in *Simocephalus vetulus* (Crustacea: Cladocera). J Plankton Res 37(4):764–772. https://doi.org/10.1093/plankt/fbv047

Chen C, Liu W, Wang L, Li J, Chen Y, Jin J, Kawan A, Zhang X (2016a) Pathological damage and immunomodulatory effects of zebrafish exposed to microcystin-LR. Toxicon 118:13–20. https://doi.org/10.1016/j.toxicon.2016.04.030

Chen L, Chen J, Zhang X, Xie P (2016b) A review of reproductive toxicity of microcystins. J Hazard Mater 301:381–399. https://doi.org/10.1016/j.jhazmat.2015.08.041

Chen J, Meng S, Xu H, Zhang Z, Wu X (2017a) Effects of *Microcystis* on hypothalamic-pituitary-gonadal-liver axis in Nile tilapia (*Oreochromis niloticus*). Bull Environ Contam Toxicol 98(4):562–566. https://doi.org/10.1007/s00128-017-2051-4

Chen L, Hu Y, He J, Chen J, Giesy JP, Xie P (2017b) Responses of the proteome and metabolome in livers of zebrafish exposed chronically to environmentally relevant concentrations of microcystin-LR. Environ Sci Technol 51(1):596–607. https://doi.org/10.1021/acs.est.6b03990

Chen C, Zhang J, Zhang M, You C, Liu Y, Wang S, Li Y (2018) miR-146a is involved in the regulation of vertebrate LC-PUFA biosynthesis by targeting elovl5 as demonstrated in rabbitfish *Siganus canaliculatus*. Gene 676:306–314. https://doi.org/10.1016/j.gene.2018.08.063

Chen G, Wang L, Li W, Zhang Q, Hu T (2020) Nodularin induced oxidative stress contributes to developmental toxicity in zebrafish embryos. Ecotox Environ Saf 194:110444. https://doi.org/10.1016/j.ecoenv.2020.110444

Chen L, Giesy JP, Adamovsky O, Svirčev Z, Meriluoto J, Codd GA, Mijovic B, Shi T, Tuo X, Li SC, Pan BZ, Chen J, Xie P (2021) Challenges of using blooms of *Microcystis* spp. in animal feeds: a comprehensive review of nutritional, toxicological and microbial health evaluation. Sci Total Environ 764:142319. https://doi.org/10.1016/j.scitotenv.2020.142319

Cheng H, Yan W, Wu Q, Liu C, Gong X, Hung TC, Li G (2017) Parental exposure to microcystin-LR induced thyroid endocrine disruption in zebrafish offspring, a transgenerational toxicity. Environ Pollut 230:981–988. https://doi.org/10.1016/j.envpol.2017.07.061

Chi VTM, Luu PT, Son DT (2016) Detrimental impacts of toxic *Microcystis aeruginosa* from Vietnam on life history traits of *Daphnia magna*. J Viet Env 9(1):56–61. https://doi.org/10.13141/jve.vol8.no1.pp56-61

Chislock MF, Sarnelle O, Jernigan LM, Wilson AE (2013) Do high concentrations of microcystin prevent *daphnia* control of phytoplankton? Water Res 47(6):1961–1970. https://doi.org/10.1016/j.watres.2012.12.038

Christen V, Meili N, Fent K (2013) Microcystin-LR induces endoplasmatic reticulum stress and leads to induction of NFκB, interferon-alpha, and tumor necrosis factor-alpha. Environ Sci Technol 47(7):3378–3385. https://doi.org/10.1021/es304886y

Claska ME, Gilbert JJ (1998) The effect of temperature on the response of *daphnia* to toxic cyanobacteria. Freshw Biol 39(2):221–232. https://doi.org/10.1046/j.1365-2427.1998.00276.x

Colas S, Duval C, Marie B (2020) Toxicity, transfer and depuration of anatoxin-a (cyanobacterial neurotoxin) in medaka fish exposed by single-dose gavage. Aquat Toxicol 222:105422. https://doi.org/10.1016/j.aquatox.2020.105422

D'Urso F, Broccolo F (2024) Applications of artificial intelligence in microbiome analysis and probiotic interventions—an overview and perspective based on the current state of the art. Appl Sci 14(19). https://doi.org/10.3390/app14198627

da Silva RRP, Pires OR, Grisolia CK (2011) Genotoxicity in *Oreochromis niloticus* (Cichlidae) induced by *Microcystis* spp bloom extract containing microcystins. Toxicon 58(3):259–264. https://doi.org/10.1016/j.toxicon.2011.06.005

Dao T-S, Vo T-M-C, Wiegand C, Bui B-T, Dinh KV (2018) Transgenerational effects of cyanobacterial toxins on a tropical micro-crustacean *Daphnia lumholtzi* across three generations. Environ Pollut 243:791–799. https://doi.org/10.1016/j.envpol.2018.09.055

Davis DA, Mondo K, Stern E, Annor AK, Murch SJ, Coyne TM, Brand LE, Niemeyer ME, Sharp S, Bradley WG, Cox PA, Mash DC (2019) Cyanobacterial neurotoxin BMAA and brain pathology in stranded dolphins. PLoS One 14(3):e0213346. https://doi.org/10.1371/journal.pone.0213346

Day F (1878) The fishes of India; being a natural history of the fishes known to inhabit the seas and fresh waters of India, Burma, and Ceylon, vol 2 volumes. Bernard Quaritch, London

de Aquino Santos AS, Vilar MCP, Amorim CA, Molica RJR, do Nascimento Moura A (2021) Exposure to toxic *Microcystis* via intact cell ingestion and cell crude extract differently affects small-bodied cladocerans. Environ Sci Pollut Res https://doi.org/10.1007/s11356-021-17402-9

De Coninck DIM, Janssen CR, De Schamphelaere KAC (2013) An investigation of the inter-clonal variation of the interactive effects of cadmium and *Microcystis aeruginosa* on the reproductive performance of *Daphnia magna*. Aquat Toxicol 140-141:425–431. https://doi.org/10.1016/j.aquatox.2013.07.005

DeMott WR (1999) Foraging strategies and growth inhibition in five daphnids feeding on mixtures of a toxic cyanobacterium and a green alga. Freshw Biol 42(2):263–274. https://doi.org/10.1046/j.1365-2427.1999.444494.x

Deng D, Zhang S, Li Y, Meng X, Yang W, Li Y, Li X (2010) Effects of *Microcystis aeruginosa* on population dynamics and sexual reproduction in two *daphnia* species. J Plankton Res 32(10):1385–1392. https://doi.org/10.1093/plankt/fbq069
Derrien T, Guigó R, Johnson R (2012) The long non-coding RNAs: a new (p)layer in the "dark matter". Front Gen 2(107). https://doi.org/10.3389/fgene.2011.00107
Dexter J, McCormick AJ, Fu P, Dziga D (2021) Microcystinase—a review of the natural occurrence, heterologous expression, and biotechnological application of MlrA. Water Res 189:116646. https://doi.org/10.1016/j.watres.2020.116646
Diez-Quijada L, Benítez-González MM, Puerto M, Jos A, Cameán AM (2021) Immunotoxic effects induced by microcystins and cylindrospermopsin: a review. Toxins 13(10):711. https://doi.org/10.3390/toxins13100711
Ding W, Shangguan Y, Zhu Y, Sultan Y, Feng Y, Zhang B, Liu Y, Ma J, Li X (2021) Negative impacts of microcystin-LR and glyphosate on zebrafish intestine: linked with gut microbiota and microRNAs? Environ Pollut 286:117685. https://doi.org/10.1016/j.envpol.2021.117685
Doğanlar ZB, Güçlü H, Öztopuz Ö, Türkön H, Dogan A, Uzun M, Doğanlar O (2019) The role of melatonin in oxidative stress, DNA damage, apoptosis and angiogenesis in fetal eye under preeclampsia and melatonin deficiency stress. Curr Eye Res 44(10):1157–1169. https://doi.org/10.1080/02713683.2019.1619778
Downing S, Banack SA, Metcalf JS, Cox PA, Downing TG (2011) Nitrogen starvation of cyanobacteria results in the production of *β-N*-methylamino-L-alanine. Toxicon 58(2):187–194. https://doi.org/10.1016/j.toxicon.2011.05.017
Drugă B, Turko P, Spaak P, Pomati F (2016) Cyanobacteria affect fitness and genetic structure of experimental *daphnia* populations. Environ Sci Technol 50(7):3416–3424. https://doi.org/10.1021/acs.est.5b05973
Du B, Liu G, Ke M, Zhang Z, Zheng M, Lu T, Sun L, Qian H (2019) Proteomic analysis of the hepatotoxicity of *Microcystis aeruginosa* in adult zebrafish (*Danio rerio*) and its potential mechanisms. Environ Pollut 254:113019. https://doi.org/10.1016/j.envpol.2019.113019
Du Y, Xu W, Wu T, Li H, Hu X, Chen J (2022) Enhancement of growth, survival, immunity and disease resistance in *Litopenaeus vannamei*, by the probiotic, *lactobacillus plantarum* ep-M17. Fish Shellfish Immunol 129:36–51. https://doi.org/10.1016/j.fsi.2022.08.066
Du J, Cao L, Gao J, Nie Z, Li Q, Sun Y, Shao N, Hu J, Zhou L, Yin G, Xu G (2024) Effects of microcystin-LR on immune function, lipid metabolism and intestinal microbial structure in *Eriocheir sinensis*. Aquac Rep 35:101994. https://doi.org/10.1016/j.aqrep.2024.101994
Duan Y, Zeng S, Lu Z, Dan X, Mo Z, Xing Y, Zhang J, Li Y (2022) Responses of lipid metabolism and lipidomics in the hepatopancreas of Pacific white shrimp *Litopenaeus vannamei* to microcystin-LR exposure. Sci Total Environ 820:153245. https://doi.org/10.1016/j.scitotenv.2022.153245
Duan Y, Nan Y, Xiao M, Yang Y (2024) Toxicity of three microcystin variants on the histology, physiological and metabolism of hepatopancreas and intestinal microbiota of *Litopenaeus vannamei*. Comp Biochem Phys C 280:109904. https://doi.org/10.1016/j.cbpc.2024.109904
DuPont AW, DuPont HL (2011) The intestinal microbiota and chronic disorders of the gut. Nat Rev Gastroenterol Hepatol 8(9):523–531. https://doi.org/10.1038/nrgastro.2011.133
Dziga D, Wasylewski M, Wladyka B, Nybom S, Meriluoto J (2013) Microbial degradation of microcystins. Chem Res Toxicol 26(6):841–852. https://doi.org/10.1021/tx4000045
Dziga D, Zielinska G, Wladyka B, Bochenska O, Maksylewicz A, Strzalka W, Meriluoto J (2016) Characterization of enzymatic activity of MlrB and MlrC proteins involved in bacterial degradation of cyanotoxins microcystins. Toxins 8(3). https://doi.org/10.3390/toxins8030076
Ebbesen KK, Hansen TB, Kjems J (2017) Insights into circular RNA biology. RNA Biol 14(8):1035–1045. https://doi.org/10.1080/15476286.2016.1271524
Edwards C, Graham D, Fowler N, Lawton LA (2008) Biodegradation of microcystins and nodularin in freshwaters. Chemosphere 73(8):1315–1321. https://doi.org/10.1016/j.chemosphere.2008.07.015
Ernst B, Hoeger SJ, O'Brien E, Dietrich DR (2007) Physiological stress and pathology in European whitefish (*Coregonus lavaretus*) induced by subchronic exposure to environmentally relevant

densities of *Planktothrix rubescens*. Aquat Toxicol 82(1):15–26. https://doi.org/10.1016/j.aquatox.2007.01.007

Faassen EJ (2014) Presence of the neurotoxin BMAA in aquatic ecosystems: what do we really know? Toxins 6(3):1109–1138. https://doi.org/10.3390/toxins6031109

Faltermann S, Zucchi S, Kohler E, Blom JF, Pernthaler J, Fent K (2014) Molecular effects of the cyanobacterial toxin cyanopeptolin (CP1020) occurring in algal blooms: global transcriptome analysis in zebrafish embryos. Aquat Toxicol 149:33–39. https://doi.org/10.1016/j.aquatox.2014.01.018

Faltermann S, Grundler V, Gademann K, Pernthaler J, Fent K (2016) Comparative effects of nodularin and microcystin-LR in zebrafish: 2. Uptake and molecular effects in eleuthero-embryos and adult liver with focus on endoplasmic reticulum stress. Aquat Toxicol 171:77–87. https://doi.org/10.1016/j.aquatox.2015.12.001

Fawell JK, Mitchell RE, Hill RE, Everett DJ (1999) The toxicity of cyanobacterial toxins in the mouse. II anatoxin-a Hum Exper Toxicol 18(3):168–173. https://doi.org/10.1177/096032719901800306

Feng Y, Ma J, Xiang R, Li X (2017) Alterations in microRNA expression in the tissues of silver carp (*Hypophthalmichthys molitrix*) following microcystin-LR exposure. Toxicon 128:15–22. https://doi.org/10.1016/j.toxicon.2017.01.016

Feng Y, Chen X, Ding W, Ma J, Zhang B, Li X (2020a) MicroRNA-16 participates in the cell cycle alteration of HepG2 cells induced by MC-LR. Ecotox Environ Saf 192:110295. https://doi.org/10.1016/j.ecoenv.2020.110295

Feng Y, Chen X, Ma J, Zhang B, Li X (2020b) Aberrant expressional profiling of known microRNAs in the liver of silver carp (*Hypophthalmichthys molitrix*) following microcystin-LR exposure based on samllRNA sequencing. Toxins 12(1). https://doi.org/10.3390/toxins12010041

Feng Y, Chen X, Ma J, Zhang B, Li X (2020c) Aberrant expressional profiling of known microRNAs in the liver of silver carp (*Hypophthalmichthys molitrix*) following microcystin-LR exposure based on smallRNA sequencing. Toxins 12(1). https://doi.org/10.3390/toxins12010041

Feng J, Kiki C, Li X, Sun Q, Zhao F (2024) Deciphering the natural attenuation of cyanotoxins: dissipation, transformation pathways, and genotoxicity. Environ Health. https://doi.org/10.1021/envhealth.4c00149

Fey SB, Cottingham KL (2011) Linking biotic interactions and climate change to the success of exotic *Daphnia lumholtzi*. Freshw Biol 56(11):2196–2209. https://doi.org/10.1111/j.1365-2427.2011.02646.x

Fischer WJ, Altheimer S, Cattori V, Meier PJ, Dietrich DR, Hagenbuch B (2005) Organic anion transporting polypeptides expressed in liver and brain mediate uptake of microcystin. Toxicol Appl Pharmacol 203(3):257–263. https://doi.org/10.1016/j.taap.2004.08.012

Florczyk M, Brzuzan P, Krom J, Woźny M, Łakomiak A (2016) miR-122-5p as a plasma biomarker of liver injury in fish exposed to microcystin-LR. J Fish Dis 39(6):741–751. https://doi.org/10.1111/jfd.12406

Florczyk M, Brzuzan P, Woźny M (2021) De novo profiling of long non-coding RNAs involved in MC-LR-induced liver injury in whitefish: discovery and perspectives. Int J Mol Sci 22(2):1–22.941. https://doi.org/10.3390/ijms22020941

Fonseca VB, Sopezki MDS, Yunes JS, Zanette J (2018) Effect of a toxic *Microcystis aeruginosa* lysate on the mRNA expression of proto-oncogenes and tumor suppressor genes in zebrafish. Ecotox Environ Saf 161:729–734. https://doi.org/10.1016/j.ecoenv.2018.06.007

Foucault P, Gallet A, Duval C, Marie B, Duperron S (2022) Gut microbiota and holobiont metabolome composition of the medaka fish (*Oryzias latipes*) are affected by a short exposure to the cyanobacterium *Microcystis aeruginosa*. Aquat Toxicol 253:106329. https://doi.org/10.1016/j.aquatox.2022.106329

Francis G (1878) Poisonous Australian lake. Nature 18(444):11–12. https://doi.org/10.1038/018011d0

Frias-Lopez J, Thompson A, Waldbauer J, Chisholm SW (2009) Use of stable isotope-labelled cells to identify active grazers of picocyanobacteria in ocean surface waters. Environ Microbiol 11(2):512–525. https://doi.org/10.1111/j.1462-2920.2008.01793.x

Fu B, He S (2012) Transcriptome analysis of silver carp (*Hypophthalmichthys molitrix*) by paired-end RNA sequencing. DNA Res 19(2):131–142. https://doi.org/10.1093/dnares/dsr046

Fu W-y, Chen J-p, Wang X-m, Xu L-h (2005) Altered expression of p53, Bcl-2 and Bax induced by microcystin-LR in vivo and in vitro. Toxicon 46(2):171–177. https://doi.org/10.1016/j.toxicon.2005.03.021

Gallet A, Halary S, Duval C, Huet H, Duperron S, Marie B (2023) Disruption of fish gut microbiota composition and holobiont's metabolome during a simulated *Microcystis aeruginosa* (cyanobacteria) bloom. Microbiome 11(1):108. https://doi.org/10.1186/s40168-023-01558-2

Gambari R, Brognara E, Spandidos DA, Fabbri E (2016) Targeting oncomiRNAs and mimicking tumor suppressor miRNAs: new trends in the development of miRNA therapeutic strategies in oncology (review). Int J Oncol 49(1):5–32. https://doi.org/10.3892/ijo.2016.3503

Gao J, Zuo H, Yang L, He JH, Niu S, Weng S, He J, Xu X (2017) Long-term influence of cyanobacterial bloom on the immune system of *Litopenaeus vannamei*. Fish Shellfish Immunol 61:79–85. https://doi.org/10.1016/j.fsi.2016.12.015

Gardner WS, McCarthy MJ (2009) Nitrogen dynamics at the sediment–water interface in shallow, sub-tropical Florida bay: why denitrification efficiency may decrease with increased eutrophication. Biogeochemistry 95(2):185–198. https://doi.org/10.1007/s10533-009-9329-5

Garruto RM, Yase Y (1986) Neurodegenerative disorders of the western Pacific: the search for mechanisms of pathogenesis. Trends Neurosci 9:368–374. https://doi.org/10.1016/0166-2236(86)90116-5

Ger KA, Urrutia-Cordero P, Frost PC, Hansson L-A, Sarnelle O, Wilson AE, Lürling M (2016) The interaction between cyanobacteria and zooplankton in a more eutrophic world. Harmful Algae 54:128–144. https://doi.org/10.1016/j.hal.2015.12.005

Ghadouani A, Pinel-Alloul B, Plath K, Codd GA, Lampert W (2004) Effects of *Microcystis aeruginosa* and purified microcystin-LR on the feeding behavior of *Daphnia pulicaria*. Limnol Oceanogr 49(3):666–679. https://doi.org/10.4319/lo.2004.49.3.0666

Gin KY-H, Sim ZY, Goh KC, Kok JWK, Te SH, Tran NH, Li W, He Y (2021) Novel cyanotoxin-producing *Synechococcus* in tropical lakes. Water Res 192:116828. https://doi.org/10.1016/j.watres.2021.116828

Gleason EJ, Lindsey WC, Kroft TL, Singson AW, L'Hernault SW (2006) *Spe-10* encodes a DHHC-CRD zinc-finger membrane protein required for endoplasmic reticulum/Golgi membrane morphogenesis during *Caenorhabditis elegans* spermatogenesis. Genetics 172(1):145–158. https://doi.org/10.1534/genetics.105.047340

Glidewell D, Beyer JE, Hambright KD (2025) Microcystins bioaccumulate but do not biomagnify in an experimental aquatic food chain. Harmful Algae 141:102768. https://doi.org/10.1016/j.hal.2024.102768

Gobler CJ, Boneillo GE, Debenham CJ, Caron DA (2004) Nutrient limitation, organic matter cycling, and plankton dynamics during an *Aureococcus anophagefferens* bloom. Aquat Microb Ecol 35(1):31–43. https://doi.org/10.3354/ame035031

Godlewska M, Balk H, Kaczkowski Z, Jurczak T, Izydorczyk K, Długoszewski B, Jaskulska A, Gągała-Borowska I, Mankiewicz-Boczek J (2018) Night fish avoidance of *Microcystis* bloom revealed by simultaneous hydroacoustic measurements of both organisms. Fish Res 207:74–84. https://doi.org/10.1016/j.fishres.2018.05.025

Goleski JA, Koch F, Marcoval MA, Wall CC, Jochem FJ, Peterson BJ, Gobler CJ (2010) The role of zooplankton grazing and nutrient loading in the occurrence of harmful cyanobacterial blooms in Florida bay, USA. Estuar Coasts 33(5):1202–1215. https://doi.org/10.1007/s12237-010-9294-1

Gómez-Leyva Y, Torrecillas A, Aboal M (2024) Cyanotoxins in epipelic and epiphytic cyanobacteria from a hypersaline coastal lagoon, an environmental hazard in climate warming times and a potential source of new compounds. Mar Drugs 22(8). https://doi.org/10.3390/md22080334

Gray SM, Bieber FME, Mcdonnell LH, Chapman LJ, Mandrak NE (2014) Experimental evidence for species-specific response to turbidity in imperilled fishes. Aquat Conserv 24(4):546–560. https://doi.org/10.1002/aqc.2436

Griswold MD (1998) The central role of Sertoli cells in spermatogenesis. Semin Cell Dev Biol 9(4):411–416. https://doi.org/10.1006/scdb.1998.0203

Gu L, Qin S, Zhu S, Lu N, Sun Y, Zhang L, Huang Y, Lyu K, Chen Y, Yang Z (2020) *Microcystis aeruginosa* affects the inducible anti-predator responses of *Ceriodaphnia cornuta*. Environ Pollut 259:113952. https://doi.org/10.1016/j.envpol.2020.113952

Guo X-c, Lu S-y, Zhang S-n, Xie P, Li G-y, Shi Z-q, Zhou Y-t, Wang Y-m (2024) Combined inhibitory effects of microcystin-LR and microcystin-RR on growth and development in zebrafish larvae. Comp Biochem Phys C 277:109824. https://doi.org/10.1016/j.cbpc.2023.109824

Gustafsson S, Rengefors K, Hansson L-A (2005) Increased consumer fitness following transfer of toxin tolerance to offspring via maternal effects. Ecology 86(10):2561–2567. https://doi.org/10.1890/04-1710

Guzmán-Guillén R, Prieto AI, Moreno I, Vasconcelos VM, Moyano R, Blanco A, Cameán Fernandez AM (2015) Cyanobacterium producing cylindrospermopsin cause histopathological changes at environmentally relevant concentrations in subchronically exposed tilapia (*Oreochromis niloticus*). Environ Toxicol 30(3):261–277. https://doi.org/10.1002/tox.21904

Guzmán-Guillén R, Campos A, Machado J, Freitas M, Azevedo J, Pinto E, Almeida A, Cameán AM, Vasconcelos V (2017) Effects of *Chrysosporum* (*Aphanizomenon*) *ovalisporum* extracts containing cylindrospermopsin on growth, photosynthetic capacity, and mineral content of carrots (*Daucus carota*). Ecotoxicology 26(1):22–31. https://doi.org/10.1007/s10646-016-1737-4

Hall NS, Litaker RW, Kenworthy WJ, Vandersea MW, Sunda WG, Reid JP, Slone DH, Butler S (2018) Consortial brown tide – picocyanobacteria blooms in Guantánamo Bay. Cuba Harmful Algae 73:30–43. https://doi.org/10.1016/j.hal.2018.01.003

Han J, Li S (2023) De novo biosynthesis of berberine and halogenated benzylisoquinoline alkaloids in *Saccharomyces cerevisiae*. Communications Chemistry 6(1):27. https://doi.org/10.1038/s42004-023-00821-9

Han S, Zhou Q, Xu Y, Vanogtrop F, Guo Q, Liu G, Yan S (2015) Valuable ingredients and feed toxicity evaluation of *Microcystis aeruginosa* acidolysis product in mice. Exp Biol Med (Maywood) 240(10):1333–1339. https://doi.org/10.1177/1535370214563894

Hanazato T, Yasuno M (1984) Growth, reproduction and assimilation of *Moina macrocopa* fed on *Microcystis* and/or *chlorella*. Jap J Ecol 34(2):195–202. https://doi.org/10.18960/seitai.34.2_195

Hanazato T, Yasuno M (1988) Assimilation of *Diaphanosoma brachyurum* and *Moina macrocopa* on *Microcystis*. Jpn J Limnol 49(1):37–41. https://doi.org/10.3739/rikusui.49.37

Hancock TL, Dahedl EK, Kratz MA, Urakawa H (2024) *Synechococcus dominance* induced after hydrogen peroxide treatment of *Microcystis* bloom in the Caloosahatchee River. Florida Environ Pollut 345:123508. https://doi.org/10.1016/j.envpol.2024.123508

Harke MJ, Steffen MM, Gobler CJ, Otten TG, Wilhelm SW, Wood SA, Paerl HW (2016) A review of the global ecology, genomics, and biogeography of the toxic cyanobacterium, *Microcystis* spp. Harmful Algae 54:4–20. https://doi.org/10.1016/j.hal.2015.12.007

Hawes NA, Amadoru A, Tremblay LA, Pochon X, Dunphy B, Fidler AE, Smith KF (2019) Epigenetic patterns associated with an ascidian invasion: a comparison of closely related clades in their native and introduced ranges. Sci Rep 9(1):14275. https://doi.org/10.1038/s41598-019-49813-7

Henning M, Hertel H, Wall H, Kohl JG (1991) Strain-specific influence of *Microcystis aeruginosa* on food ingestion and assimilation of some cladocerans and copepods. Int Rev Hydrobiol 76(1):37–45. https://doi.org/10.1002/iroh.19910760105

Herrera N, Palacio J, Echeverri F, Ferrão-Filho A (2014) Effects of a cyanobacterial bloom sample containing microcystin-LR on the ecophysiology of *Daphnia similis*. Toxicol Rep 1:909–914. https://doi.org/10.1016/j.toxrep.2014.10.017

Hietala J, Reinikainen M, Walls M (1995) Variation in life history responses of *daphnia* to toxic *Microcystis aeruginosa*. J Plankton Res 17(12):2307–2318. https://doi.org/10.1093/plankt/17.12.2307

Hietala J, Laurén-Määttä C, Walls M (1997) Life history responses of *daphnia* clones to toxic *Microcystis* at different food levels. J Plankton Res 19(7):917–926. https://doi.org/10.1093/plankt/19.7.917

Hinojosa MG, Gutiérrez-Praena D, Prieto AI, Guzmán-Guillén R, Jos A, Cameán AM (2019) Neurotoxicity induced by microcystins and cylindrospermopsin: a review. Sci Total Environ 668:547–565. https://doi.org/10.1016/j.scitotenv.2019.02.426

Ho L, Meyn T, Keegan A, Hoefel D, Brookes J, Saint CP, Newcombe G (2006) Bacterial degradation of microcystin toxins within a biologically active sand filter. Water Res 40(4):768–774. https://doi.org/10.1016/j.watres.2005.12.009

Hobert O (2008) Gene regulation by transcription factors and microRNAs. Science 319(5871):1785–1786. https://doi.org/10.1126/science.1151651

Hong Y, Hu HY, Li FM (2008) Physiological and biochemical effects of allelochemical ethyl 2-methyl acetoacetate (EMA) on cyanobacterium *Microcystis aeruginosa*. Ecotox Environ Saf 71(2):527–534. https://doi.org/10.1016/j.ecoenv.2007.10.010

Hou J, Li L, Xue T, Long M, Su Y, Wu N (2014) Damage and recovery of the ovary in female zebrafish i.p.-injected with MC-LR. Aquat Toxicol 155:110–118. https://doi.org/10.1016/j.aquatox.2014.06.010

Hou J, Su Y, Lin W, Guo H, Li L, Anderson DM, Li D, Tang R, Chi W, Zhang X (2018) Estrogenic potency of MC-LR is induced via stimulating steroidogenesis: *In vitro* and *in vivo* evidence. Environ Pollut 240:615–622. https://doi.org/10.1016/j.envpol.2018.04.140

Hu L, Zhang F, Liu C, Wang M (2012) Biodegradation of microcystins by *bacillus* sp. strain EMB. Energy Procedia 16:2054–2059. https://doi.org/10.1016/j.egypro.2012.01.312

Hu Y, Chen J, Fan H, Xie P, He J (2016) A review of neurotoxicity of microcystins. Environ Sci Pollut Res 23(8):7211–7219. https://doi.org/10.1007/s11356-016-6073-y

Huang H, Lei L, Peng L, Han B (2015) Effect of *Microcystis aeruginosa* and *Cylindrospermopsis raciborskii* on the growth and reproduction of *Daphnia similoides* (large sub-tropical cladoceran). Chin J App Eviron Biol 21(2):366–371. https://doi.org/10.3724/SP.J.1145.2014.09020

Hulot FD, Carmignac D, Legendre S, Yépémian C, Bernard C (2012) Effects of microcystin–producing and microcystin–free strains of *Planktothrix agardhii* on long–term population dynamics of *Daphnia magna*. Ann Limnol 48(3):337–347. https://doi.org/10.1051/limn/2012023

Humpage AR, Falconer IR (2003) Oral toxicity of the cyanobacterial toxin cylindrospermopsin in male Swiss albino mice: determination of no observed adverse effect level for deriving a drinking water guideline value. Environ Toxicol 18(2):94–103. https://doi.org/10.1002/tox.10104

Humpage AR, Rositano J, Bretag AH, Brown R, Baker PD, Nicholson BC, Steffensen DA (1994) Paralytic shellfish poisons from Australian cyanobacterial blooms. Aust J Mar Freshwat Res 45(5):761–771. https://doi.org/10.1071/MF9940761

Ibelings BW, Bruning K, de Jonge J, Wolfstein K, Pires LMD, Postma J, Burger T (2005) Distribution of microcystins in a lake foodweb: no evidence for biomagnification. Microb Ecol 49(4):487–500. https://doi.org/10.1007/s00248-004-0014-x

Iebba V, Totino V, Gagliardi A, Santangelo F, Cacciotti F, Trancassini M, Mancini C, Cicerone C, Corazziari E, Pantanella F, Schippa S (2016) Eubiosis and dysbiosis: the two sides of the microbiota. New Microbiol 39(1):1–12

Jacinavicius FR, Geraldes V, Fernandes K, Crnkovic CM, Gama WA, Pinto E (2023) Toxicological effects of cyanobacterial metabolites on zebrafish larval development. Harmful Algae 125:102430. https://doi.org/10.1016/j.hal.2023.102430

Jakubowska N, Szeląg-Wasielewska E (2015) Toxic picoplanktonic cyanobacteria--review. Mar Drugs 13 (3):1497–1518. https://doi.org/10.3390/md13031497

Jang MH, Ha K, Lucas MC, Joo GJ, Takamura N (2004) Changes in microcystin production by *Microcystis aeruginosa* exposed to phytoplanktivorous and omnivorous fish. Aquat Toxicol 68(1):51–59. https://doi.org/10.1016/j.aquatox.2004.02.002

Jeremias G, Barbosa J, Marques SM, De Schamphelaere KAC, Van Nieuwerburgh F, Deforce D, Gonçalves FJM, Pereira JL, Asselman J (2018) Transgenerational inheritance of DNA hypomethylation in *Daphnia magna* in response to salinity stress. Environ Sci Technol 52(17):10114–10123. https://doi.org/10.1021/acs.est.8b03225

Jiang X, Yang W, Zhao S, Liang H, Zhao Y, Chen L, Li R (2013) Maternal effects of inducible tolerance against the toxic cyanobacterium *Microcystis aeruginosa* in the grazer *Daphnia carinata*. Environ Pollut 178:142–146. https://doi.org/10.1016/j.envpol.2013.03.017

Jiang L, Eriksson J, Lage S, Jonasson S, Shams S, Mehine M, Ilag LL, Rasmussen U (2014) Diatoms: a novel source for the neurotoxin BMAA in aquatic environments. PLoS One 9(1):e84578. https://doi.org/10.1371/journal.pone.0084578

Jiang X, Gao H, Zhang L, Liang H, Zhu X (2016) Rapid evolution of tolerance to toxic *Microcystis* in two cladoceran grazers. Sci Rep 6:25319. https://doi.org/10.1038/srep25319

Jiang K, Chen C, Jiang G, Chi Y, Xu C, Kong L, Yu H, Liu S, Li Q (2024) Genetic improvement of oysters: current status, challenges, and prospects. Rev Aquac 16(2):796–817. https://doi.org/10.1111/raq.12868

Jiao Y, Chen Q, Chen X, Wang X, Liao X, Jiang L, Wu J, Yang L (2014) Occurrence and transfer of a cyanobacterial neurotoxin β-methylamino-L-alanine within the aquatic food webs of Gonghu Bay (Lake Taihu, China) to evaluate the potential human health risk. Sci Total Environ 468-469:457–463. https://doi.org/10.1016/j.scitotenv.2013.08.064

Jonas A, Scholz S, Fetter E, Sychrova E, Novakova K, Ortmann J, Benisek M, Adamovsky O, Giesy JP, Hilscherova K (2015) Endocrine, teratogenic and neurotoxic effects of cyanobacteria detected by cellular *in vitro* and zebrafish embryos assays. Chemosphere 120:321–327. https://doi.org/10.1016/j.chemosphere.2014.07.074

Jonasson S, Eriksson J, Berntzon L, Spáčil Z, Ilag LL, Ronnevi L-O, Rasmussen U, Bergman B (2010) Transfer of a cyanobacterial neurotoxin within a temperate aquatic ecosystem suggests pathways for human exposure. Proc Natl Acad Sci USA 107(20):9252–9257. https://doi.org/10.1073/pnas.0914417107

Joncquel-Chevalier Curt M, Voicu P-M, Fontaine M, Dessein A-F, Porchet N, Mention-Mulliez K, Dobbelaere D, Soto-Ares G, Cheillan D, Vamecq J (2015) Creatine biosynthesis and transport in health and disease. Biochimie 119:146–165. https://doi.org/10.1016/j.biochi.2015.10.022

Jones GJ, Orr PT (1994) Release and degradation of microcystin following algicide treatment of a *Microcystis aeruginosa* bloom in a recreational lake, as determined by HPLC and protein phosphatase inhibition assay. Water Res 28(4):871–876. https://doi.org/10.1016/0043-1354(94)90093-0

Jones GJ, Bourne DG, Blakeley RL, Doelle H (1994) Degradation of the cyanobacterial hepatotoxin microcystin by aquatic bacteria. Nat Toxins 2(4):228–235. https://doi.org/10.1002/nt.2620020412

Ju J, Ruan Q, Li X, Liu R, Li Y, Pu Y, Yin L, Wang D (2013) Neurotoxicological evaluation of microcystin-LR exposure at environmental relevant concentrations on nematode *Caenorhabditis elegans*. Environ Sci Pollut Res 20(3):1823–1830. https://doi.org/10.1007/s11356-012-1151-2

Ju J, Saul N, Kochan C, Putschew A, Pu Y, Yin L, Steinberg CEW (2014) Cyanobacterial xenobiotics as evaluated by a *Caenorhabditis elegans* neurotoxicity screening test. Int J Environ Res Publ Hlth 11(5):4589–4606. https://doi.org/10.3390/ijerph110504589

Juhel G, Ramsay RM, Davenport J, O'Halloran J, Culloty SC (2015) Effect of the microcystin-producing cyanobacterium, *Microcystis aeruginosa*, on immune functions of the zebra mussel *Dreissena polymorpha*. J Shellfish Res 34(2):433–442. https://doi.org/10.2983/035.034.0227

Jungmann D (1992) Toxic compounds isolated from *Microcystis* PCC786 that are more active against *daphnia* than two microcystins. Limnol Oceanogr 37(8):1777–1783

Jungmann D (1995) Isolation, purification, and characterization of new *daphnia*-toxic compound from axenic *Microcystis flos-aquae* strain PCC7806. J Chem Ecol 21(11):1665–1676. https://doi.org/10.1007/BF02033668

Kaebernick M, Rohrlack T, Christoffersen K, Neilan BA (2001) A spontaneous mutant of microcystin biosynthesis: genetic characterization and effect on *daphnia*. Environ Microbiol 3(11):669–679. https://doi.org/10.1046/j.1462-2920.2001.00241.x

Kansole MMR, Lin T-F (2016) Microcystin-LR biodegradation by *bacillus* sp.: reaction rates and possible genes involved in the degradation. Water 8(11):508

Kawan A, Zhang T, Liu W, Mukhtar H, Zhan C, Zhang X (2019) Recovery of reproductive function of female zebrafish from the toxic effects of microcystin-LR exposure. Aquat Toxicol 214:105240. https://doi.org/10.1016/j.aquatox.2019.105240

Ke Y, Zhao W, Xiong J, Cao R (2013) Downregulation of miR-16 promotes growth and motility by targeting HDGF in non-small cell lung cancer cells. FEBS Lett 587(18):3153–3157. https://doi.org/10.1016/j.febslet.2013.08.010

Kelleher JF, Mandell MA, Moulder G, Hill KL, L'Hernault SW, Barstead R, Titus MA (2000) Myosin VI is required for asymmetric segregation of cellular components during *C. Elegans* spermatogenesis. Curr Biol 10(23):1489–1496. https://doi.org/10.1016/s0960-9822(00)00828-9

Khordadmehr M, Jigari-Asl F, Ezzati H, Shahbazi R, Sadreddini S, Safaei S, Baradaran B (2019) A comprehensive review on *miR-451*: a promising cancer biomarker with therapeutic potential. J Cell Physiol 234(12):21716–21731. https://doi.org/10.1002/jcp.28888

Kim YD, Kim WJ, Shin YK, Lee DH, Kim YJ, Kim JK, Rhee JS (2017) Microcystin-LR bioconcentration induces antioxidant responses in the digestive gland of two marine bivalves *Crassostrea gigas* and *Mytilus edulis*. Aquat Toxicol 188:119–129. https://doi.org/10.1016/j.aquatox.2017.05.003

Kinnear SHW, Duivenvoorden LJ, Fabbro LD (2007) Sublethal responses in *Melanoides tuberculata* following exposure to *Cylindrospermopsis raciborskii* containing cylindrospermopsin. Harmful Algae 6(5):642–650. https://doi.org/10.1016/j.hal.2007.01.004

Kist LW, Piato AL, da Rosa JGS, Koakoski G, Barcellos LJG, Yunes JS, Bonan CD, Bogo MR (2011) Acute exposure to microcystin-producing cyanobacterium *Microcystis aeruginosa* alters adult zebrafish (*Danio rerio*) swimming performance parameters. J Toxicol 2011. https://doi.org/10.1155/2011/280304

Kozlowsky-Suzuki B, Wilson AE, Ferrão-Filho AS (2012) Biomagnification or biodilution of microcystins in aquatic foodwebs? Meta-analyses of laboratory and field studies. Harmful Algae 18:47–55. https://doi.org/10.1016/j.hal.2012.04.002

Krajacic P, Hermanowski J, Lozynska O, Khurana TS, Lamitina T (2009) C. Elegans dysferlin homolog fer-1 is expressed in muscle, and fer-1 mutations initiate altered gene expression of muscle enriched genes. Physiol Genomics 40 (1):8-14. https://doi.org/10.1152/physiolgenomics.00106.2009

Krausfeldt LE, Shmakova E, Lee HW, Mazzei V, Loftin KA, Smith RP, Karwacki E, Fortman PE, Rosen BH, Urakawa H, Dadlani M, Colwell RR, Lopez JV (2024) Microbial diversity, genomics, and phage–host interactions of cyanobacterial harmful algal blooms. mSystems 9 (7). https://doi.org/10.1128/msystems.00709-23

Krishnan A, Mou X (2021) A brief review of the structure, cytotoxicity, synthesis, and biodegradation of microcystins. Water 13(16):2147. https://doi.org/10.3390/w13162147

Krishnan A, Zhang Y-Q, Mou X (2018) Isolation and characterization of microcystin-degrading bacteria from Lake Erie. Bull Environ Contam Toxicol 101(5):617–623. https://doi.org/10.1007/s00128-018-2468-4

Krishnan A, Zhang Y, Balaban M, Seo Y, Mou X (2020) Taxonomic and genotypical heterogeneity of microcystin degrading bacterioplankton in western Lake Erie. Harmful Algae 98:101895. https://doi.org/10.1016/j.hal.2020.101895

Krützfeldt J, Rajewsky N, Braich R, Rajeev KG, Tuschl T, Manoharan M, Stoffel M (2005) Silencing of microRNAs in vivo with 'antagomirs'. Nature 438(7068):685–689. https://doi.org/10.1038/nature04303

Kuosa H (1991) Picoplanktonic algae in the northern Baltic Sea: seasonal dynamics and flagellate grazing. Mar Ecol Prog Ser 73:269–276. https://doi.org/10.3354/meps073269

Lad A, Su RC, Breidenbach JD, Stemmer PM, Carruthers NJ, Sanchez NK, Khalaf FK, Zhang S, Kleinhenz AL, Dube P, Mohammed CJ, Westrick JA, Crawford EL, Palagama D, Baliu-Rodriguez D, Isailovic D, Levison B, Modyanov N, Gohara AF, Malhotra D, Haller ST, Kennedy DJ (2019) Chronic low dose oral exposure to microcystin-LR exacerbates hepatic injury in a murine model of non-alcoholic fatty liver disease. Toxins 11(9):486. https://doi.org/10.3390/toxins11090486

Lage S, Costa PR, Moita T, Eriksson J, Rasmussen U, Rydberg SJ (2014) BMAA in shellfish from two Portuguese transitional water bodies suggests the marine dinoflagellate *Gymnodinium catenatum* as a potential BMAA source. Aquat Toxicol 152:131–138. https://doi.org/10.1016/j.aquatox.2014.03.029

Lamb MC, Kimmel DG, Field EK (2019) The effects of temperature on *Bosmina longirostris* susceptibility to microcystin-LR acute toxicity. PLoS One 14(7):e0219342. https://doi.org/10.1371/journal.pone.0219342

Langan LM, Lovin LM, Taylor RB, Scarlett KR, Kevin Chambliss C, Chatterjee S, Thad Scott J, Brooks BW (2024) Proteome changes in larval zebrafish (*Danio rerio*) and fathead minnow (*Pimephales promelas*) exposed to (±) anatoxin-a. environ Int:108514. https://doi.org/10.1016/j.envint.2024.108514

Lankoff A, Wojcik A, Lisowska H, Bialczyk J, Dziga D, Carmichael WW (2007) No induction of structural chromosomal aberrations in cylindrospermopsin-treated CHO-K1 cells without and with metabolic activation. Toxicon 50(8):1105–1115. https://doi.org/10.1016/j.toxicon.2007.07.021

Larson D, Ahlgren G, Willén E (2014) Bioaccumulation of microcystins in the food web: a field study of four Swedish lakes. Inland Waters 4(1):91–104. https://doi.org/10.5268/IW-4.1.627

Laurén-Määttä C, Hietala J, Andwalls M (1997) Responses of *Daphnia pulex* populations to toxic cyanobacteria. Freshw Biol 37(3):635–647. https://doi.org/10.1046/j.1365-2427.1997.00189.x

Lawton LA, Welgamage A, Manage PM, Edwards C (2011) Novel bacterial strains for the removal of microcystins from drinking water. Water Sci Technol 63(6):1137–1142. https://doi.org/10.2166/wst.2011.352

Le Manach S, Khenfech N, Huet H, Qiao Q, Duval C, Marie A, Bolbach G, Clodic G, Djediat C, Bernard C, Edery M, Marie B (2016) Gender-specific toxicological effects of chronic exposure to pure microcystin-LR or complex *Microcystis aeruginosa* extracts on adult medaka fish. Environ Sci Technol 50(15):8324–8334. https://doi.org/10.1021/acs.est.6b01903

Lee M-C, Park JC, Lee J-S (2018) Effects of environmental stressors on lipid metabolism in aquatic invertebrates. Aquat Toxicol 200:83–92. https://doi.org/10.1016/j.aquatox.2018.04.016

Lefler FW, Barbosa M, Zimba PV, Smyth AR, Berthold DE, Laughinghouse HD (2023) Spatiotemporal diversity and community structure of cyanobacteria and associated bacteria in the large shallow subtropical Lake Okeechobee (Florida, United States). Front Microbiol 14:1219261. https://doi.org/10.3389/fmicb.2023.1219261

Lehtonen KK, Kankaanpää H, Leiniö S, Sipiä VO, Pflugmacher S, Sandberg-Kilpi E (2003) Accumulation of nodularin-like compounds from the cyanobacterium *Nodularia spumigena* and changes in acetylcholinesterase activity in the clam *Macoma balthica* during short-term laboratory exposure. Aquat Toxicol 64(4):461–476. https://doi.org/10.1016/S0166-445X(03)00101-2

Lei L, Huang H, Peng L, Yang Y, Xiao L, Han BP (2020) Life-history responses of *Daphnia sinensis* simultaneously exposed to *Microcystis aeruginosa* and *Cylindrospermopsis raciborskii*. Ecotoxicology 29(6):771–779. https://doi.org/10.1007/s10646-020-02220-5

Lezcano MÁ, Morón-López J, Agha R, López-Heras I, Nozal L, Quesada A, El-Shehawy R (2016) Presence or absence of *mlr* genes and nutrient concentrations co-determine the microcystin biodegradation efficiency of a natural bacterial community. Toxins 8(11). https://doi.org/10.3390/toxins8110318

Li Q, Jiang X (2014) Offspring tolerance to toxic *Microcystis aeruginosa* in *Daphnia pulex* shaped by maternal food availability and age. Fundam Appl Limnol 185(3–4):315–319. https://doi.org/10.1127/fal/2014/0704

Li Y, Sun B, Wu H, Nie P (2009a) Effects of pure microcystin-LR on the transcription of immune related genes and heat shock proteins in larval stage of zebrafish (*Danio rerio*). Aquaculture 289(1–2):154–160. https://doi.org/10.1016/j.aquaculture.2008.12.029

Li Y, Wang Y, Yin L, Pu Y, Wang D (2009b) Using the nematode *Caenorhabditis elegans* as a model animal for assessing the toxicity induced by microcystin-LR. J Environ Sci 21(3):395–401. https://doi.org/10.1016/S1001-0742(08)62282-2

Li H, Cai Y, Xie P, Li G, Hao L, Xiong Q (2013a) Identification and expression profiles of IL-8 in bighead carp (*Aristichthys nobilis*) in response to microcystin-LR. Arch Environ Contam Toxicol 65(3):537–545. https://doi.org/10.1007/s00244-013-9910-8

Li X, Ma J, Fang Q, Li Y (2013b) Transcription alterations of microRNAs, cytochrome P4501A1 and 3A65, and AhR and PXR in the liver of zebrafish exposed to crude microcystins. Toxicon 73:17–22. https://doi.org/10.1016/j.toxicon.2013.07.002

Li F, Deng D, Zhang X, Ji G, Huang Q (2014) Combined effects of four *Microcystis aeruginosa* strains and *Scenedesmus obliquus* concentrations on population dynamics and resting

egg formation of two *daphnia* species. Limnology 15(3):271–279. https://doi.org/10.1007/s10201-014-0433-4
Li X, Zhao Q, Zhou W, Xu L, Wang Y (2015a) Effects of chronic exposure to microcystin-LR on hepatocyte mitochondrial DNA replication in mice. Environ Sci Technol 49(7):4665–4672. https://doi.org/10.1021/es5059132
Li Y, Zhang M, Chen P, Liu R, Liang G, Yin L, Pu Y (2015b) Effects of microcystin-LR exposure on spermiogenesis in nematode *Caenorhabditis elegans*. Int J Mol Sci 16(9):22927–22937. https://doi.org/10.3390/ijms160922927
Li X, Li J, Meng F, Yao L (2019) Hepatotoxicity and immunotoxicity of MC-LR on silver carp. Ecotox Environ Saf 169:28–32. https://doi.org/10.1016/j.ecoenv.2018.10.110
Li J, Hu J, Cao L, Yuan Y (2020a) Growth, physiological responses and microcystin-production/-release dynamics of *Microcystis aeruginosa* exposed to various luteolin doses. Ecotox Environ Saf 196:110540. https://doi.org/10.1016/j.ecoenv.2020.110540
Li Y, Zhu Y, Ma L, Huang J, Sun Y, Zhang L, Lyu K, Yang Z (2020b) Toxic *Microcystis* reduces tolerance of *daphnia* to increased chloride, and low chloride alleviates the harm of toxic microcystis to *daphnia*. Chemosphere 260:127594. https://doi.org/10.1016/j.chemosphere.2020.127594
Li J, Wen J, Sun C, Zhou Y, Xu J, MacIsaac HJ, Chang X, Cui Q (2022) Phytosphingosine-induced cell apoptosis via a mitochondrially mediated pathway. Toxicology 482:153370. https://doi.org/10.1016/j.tox.2022.153370
Li J, Zhang L, Li Q, Zhang S, Zhang W, Zhao Y, Zheng X, Fan Z (2024) Hormetic effect of a short-chain PFBS on *Microcystis aeruginosa* and its molecular mechanism. J Hazard Mater 467:133596. https://doi.org/10.1016/j.jhazmat.2024.133596
Li J, Li L, Liang W, Li L, Wang R, Wang Z, Ma C (2025) Spatial multi-omics analysis of metabolic heterogeneity in zebrafish exposed to microcystin-LR and its disinfection byproducts. Water res:123599. https://doi.org/10.1016/j.watres.2025.123599
Liang G, Xie P, Chen J, Yu T (2011) Comparative studies on the pH dependence of DOW of microcystin-RR and -LR using LC-MS. Scientific World J 11:20–26. https://doi.org/10.1100/tsw.2011.17
Lieke T, Meinelt T, Hoseinifar SH, Pan B, Straus DL, Steinberg CEW (2020) Sustainable aquaculture requires environmental-friendly treatment strategies for fish diseases. Rev Aquac 12(2):943–965. https://doi.org/10.1111/raq.12365
Lieke T, Steinberg CEW, Bittmann S, Behrens S, Hoseinifar SH, Meinelt T, Knopf K, Kloas W (2021) Fulvic acid accelerates hatching and stimulates antioxidative protection and the innate immune response in zebrafish larvae. Sci Total Environ 796:148780. https://doi.org/10.1016/j.scitotenv.2021.148780
Limborg MT, Alberdi A, Kodama M, Roggenbuck M, Kristiansen K, Gilbert MTP (2018) Applied hologenomics: feasibility and potential in aquaculture. Trends Biotechnol 36(3):252–264. https://doi.org/10.1016/j.tibtech.2017.12.006
Lin JH, Walter P, Yen TSB (2008) Endoplasmic reticulum stress in disease pathogenesis. Annu Rev Pathol 3:399–425. https://doi.org/10.1146/annurev.pathmechdis.3.121806.151434
Lin W, Hou J, Guo H, Qiu Y, Li L, Li D, Tang R (2017) Dualistic immunomodulation of sub-chronic microcystin-LR exposure on the innate-immune defense system in male zebrafish. Chemosphere 183:315–322. https://doi.org/10.1016/j.chemosphere.2017.05.079
Lin W, Guo H, Li Y, Wang L, Zhang D, Hou J, Wu X, Li L, Li D, Zhang X (2018) Single and combined exposure of microcystin-LR and nitrite results in reproductive endocrine disruption via hypothalamic-pituitary-gonadal-liver axis. Chemosphere 211:1137–1146. https://doi.org/10.1016/j.chemosphere.2018.08.049
Lin W, Guo H, Wang L, Zhang D, Wu X, Li L, Qiu Y, Yang L, Li D, Tang R (2020) Parental transfer of microcystin-LR-induced innate immune dysfunction of zebrafish: a cross-generational study. Environ Sci Technol 54(2):1014–1023. https://doi.org/10.1021/acs.est.9b04953
Lin W, Hung TC, Kurobe T, Wang Y, Yang P (2021) Microcystin-induced immunotoxicity in fishes: a scoping review. Toxins 13(11):765. https://doi.org/10.3390/toxins13110765
Lira VS, Moreira IC, Tonello PS, Henriques Vieira AA, Fracácio R (2017) Evaluation of the ecotoxicological effects of *Microcystis aeruginosa* and *Cylindrospermopsis raciborskii*

on *Ceriodaphnia dubia* before and after treatment with ultrasound. Water Air Soil Pollut 228(1):49. https://doi.org/10.1007/s11270-016-3209-0

Liu Y, Gu W (2022) p53 in ferroptosis regulation: the new weapon for the old guardian. Cell Death Differ 29(5):895–910. https://doi.org/10.1038/s41418-022-00943-y

Liu Y, Song L, Li X, Liu T (2002) The toxic effects of microcystin-LR on embryo-larval and juvenile development of loach, *Misguruns mizolepis* Gunthe. Toxicon 40(4):395–399. https://doi.org/10.1016/S0041-0101(01)00173-8

Liu Y, Xie P, Chen F, Wu X (2005) Effect of combinations of the toxic cyanobacterium *Microcystis aeruginosa* PCC7820 and the green alga *Scenedesmus* on the experimental population of *Daphnia pulex*. Bull Environ Contam Toxicol 74(6):1186–1191. https://doi.org/10.1007/s00128-005-0706-z

Liu Y, Xie P, Wu XP (2006) Effects of toxic and nontoxic *Microcystis aeruginosa* on survival, population-increase, and feeding of two small cladocerans. Bull Environ Contam Toxicol 77(4):566–573. https://doi.org/10.1007/s00128-006-1101-0

Liu L, Li K, Chen T, Dai X, Jiang M, Diana JS (2011) Effects of *Microcystis aeruginosa* on life history of water flea *Daphnia magna*. Chin J Oceanol Limnol 29(4):892–897. https://doi.org/10.1007/s00343-011-0518-4

Liu W, Qiao Q, Chen Y, Wu K, Zhang X (2014) Microcystin-LR exposure to adult zebrafish (*Danio rerio*) leads to growth inhibition and immune dysfunction in F_1 offspring, a parental transmission effect of toxicity. Aquat Toxicol 155:360–367. https://doi.org/10.1016/j.aquatox.2014.07.011

Liu Y, Chen S, Chen X, Zhang J, Gao B (2015a) Interactions between *Microcystis aeruginosa* and coexisting amoxicillin contaminant at different phosphorus levels. J Hazard Mater 297:83–91. https://doi.org/10.1016/j.jhazmat.2015.04.064

Liu Y, Chen X, Zhang J, Gao B (2015b) Hormesis effects of amoxicillin on growth and cellular biosynthesis of *Microcystis aeruginosa* at different nitrogen levels. Microb Ecol 69(3): 608–617. https://doi.org/10.1007/s00248-014-0528-9

Liu Y, Wang F, Chen X, Zhang J, Gao B (2015c) Influence of coexisting spiramycin contaminant on the harm of *Microcystis aeruginosa* at different nitrogen levels. J Hazard Mater 285: 517–524. https://doi.org/10.1016/j.jhazmat.2014.11.027

Liu S, da Cunha AP, Rezende RM, Cialic R, Wei Z, Bry L, Comstock LE, Gandhi R, Weiner HL (2016a) The host shapes the gut microbiota via fecal microRNA. Cell Host Microbe 19(1):32–43. https://doi.org/10.1016/j.chom.2015.12.005

Liu W, Chen C, Chen L, Wang L, Li J, Chen Y, Jin J, Kawan A, Zhang X (2016b) Sex-dependent effects of microcystin-LR on hypothalamic-pituitary-gonad axis and gametogenesis of adult zebrafish. Sci Rep 6(1):22819. https://doi.org/10.1038/srep22819

Liu X, Zhang A, Xiang J, Lv Y, Zhang X (2016c) miR-451 acts as a suppressor of angiogenesis in hepatocellular carcinoma by targeting the IL-6R-STAT3 pathway. Oncol Rep 36(3):1385–1392. https://doi.org/10.3892/or.2016.4971

Liu Y, Chen S, Zhang J, Gao B (2016d) Growth, microcystin-production and proteomic responses of *Microcystis aeruginosa* under long-term exposure to amoxicillin. Water Res 93:141–152. https://doi.org/10.1016/j.watres.2016.01.060

Liu G, Ke M, Fan X, Zhang M, Zhu Y, Lu T, Sun L, Qian H (2018) Reproductive and endocrine-disrupting toxicity of *Microcystis aeruginosa* in female zebrafish. Chemosphere 192:289–296. https://doi.org/10.1016/j.chemosphere.2017.10.167

Liu W, Zhan C, Zhang T, Zhang X (2019) Microcystin-LR influences the *in vitro* oocyte maturation of zebrafish by activating the MAPK pathway. Aquat Toxicol 215:105261. https://doi.org/10.1016/j.aquatox.2019.105261

Liu X, Hao Y, Peng L, Liu Y, Wei N, Liang Q (2020) MiR-122 is involved in immune response by regulating Interleukin-15 in the orange-spotted grouper (*Epinephelus coioides*). Fish Shellfish Immunol 106:404–409. https://doi.org/10.1016/j.fsi.2020.08.021

Liu L, Zhou Y, Ruan JM, Liang XM, Lin CG, He L, Wei LL (2021) Molecular cloning of *tp53inp1* in grass carp and its expression in response to microcystins (in Chinese). Acta Hydrobiol Sin 45(4):722–727. https://doi.org/10.7541/2021.2020.101

Liu H, Yang Q, Li G, Hung TC, Zuo J, Luan N, Liu X, Wu Q (2024a) Probiotic *lactobacillus rhamnosus* modulates MCLR-induced oogenesis disorders in zebrafish: evidence from the transcriptome. Sci Total Environ 951:175532. https://doi.org/10.1016/j.scitotenv.2024.175532

Liu Z, Zhang Y, Jia X, Hoskins TD, Lu L, Han Y, Zhang X, Lin H, Shen L, Feng Y, Zheng Y, Hu C, Zhang H (2024b) Microcystin-LR induces estrogenic effects at environmentally relevant concentration in black-spotted pond frogs (*Pelophylax nigromaculatus*): *In situ, in vivo, in vitro*, and *in silico* investigations. Environ Sci Technol 58(22):9559–9569. https://doi.org/10.1021/acs.est.4c01322

Lombardo S, Maskos U (2015) Role of the nicotinic acetylcholine receptor in Alzheimer's disease pathology and treatment. Neuropharmacology 96:255–262. https://doi.org/10.1016/j.neuropharm.2014.11.018

Łotocka M (2001) Toxic effect of cyanobacterial blooms on the grazing activity of *Daphnia magna* Straus. Oceanologia 43(4):441–453

Lu N, Sun Y, Wei J, Gu L, Zhang L, Yang Z, Huang Y (2021) Toxic *Microcystis aeruginosa* alters the resource allocation in *daphnia mitsukuri* responding to fish predation cues. Environ Pollut 278:116918. https://doi.org/10.1016/j.envpol.2021.116918

Lu YP, Zhang XX, Zheng PH, Li JT, Li JJ, Li T, Wang X, Wang DM, Xian JA, Zhang ZL, Wang AL (2023) Effects of microcystin-LR on behavior, histopathology, oxidative stress, non-specific immunity and gene expression of red claw crayfish (*Cherax quadricarinatus*). Aquac Rep 33:101805. https://doi.org/10.1016/j.aqrep.2023.101805

Luan N, Zuo J, Niu Q, Yan W, Hung T-C, Liu H, Wu Q, Wang G, Deng P, Ma X, Qin J, Li G (2024) Probiotic *lactobacillus rhamnosus* alleviates the neurotoxicity of microcystin-LR in zebrafish (*Danio rerio*) through the gut-brain axis. Sci Total Environ 908:168058. https://doi.org/10.1016/j.scitotenv.2023.168058

Lucato V, Sut S, Abiusi F, Mathys A, Dall'Acqua S, Sforza E (2024) From nitrogen to protein: harnessing the power of nitrogen-fixing cyanobacteria for protein-rich biomass production. ACS Sustain Chem Eng 12 (20):7693–7702. https://doi.org/10.1021/acssuschemeng.3c07548

Luka Z, Mudd SH, Wagner C (2009) Glycine *N*-methyltransferase and regulation of *S*-adenosylmethionine levels. J Biol Chem 284(34):22507–22511. https://doi.org/10.1074/jbc.R109.019273

Lürling M (2003a) *Daphnia* growth on microcystin-producing and microcystin-free *Microcystis aeruginosa* in different mixtures with the green alga *Scenedesmus obliquus*. Limnol Oceanogr 48(6):2214–2220. https://doi.org/10.4319/lo.2003.48.6.2214

Lürling M (2003b) Effects of microcystin-free and microcystin-containing strains of the cyanobacterium *Microcystis aeruginosa* on growth of the grazer *Daphnia magna*. Environ Toxicol 18(3):202–210. https://doi.org/10.1002/tox.10115

Lürling M, Beekman W (2006) Growth of *Daphnia magna* males and females fed with the cyanobacterium *Microcystis aeruginosa* and the green alga *Scenedesmus obliquus* in different proportions. Acta Hydrochim Hydrobiol 34(4):375–382. https://doi.org/10.1002/aheh.200500634

Lürling M, van der Grinten E (2003) Life-history characteristics of *daphnia* exposed to dissolved microcystin-LR and to the cyanobacterium *Microcystis aeruginosa* with and without microcystins. Environ Toxicol Chem 22(6):1281–1287. https://doi.org/10.1897/1551-5028(2003)022<1281:LHCODE>2.0.CO;2

Lynch TC, Phlips E (2000) Filtration of the bloom-forming cyanobacteria *Synechococcus* by three sponge species from Florida bay, U.S.A. Bull Mar Sci 67:923–936

Lyu K, Meng Q, Zhu X, Dai D, Zhang L, Huang Y, Yang Z (2016) Changes in iTRAQ-based proteomic profiling of the cladoceran *Daphnia magna* exposed to microcystin-producing and microcystin-free *Microcystis aeruginosa*. Environ Sci Technol 50(9):4798–4807. https://doi.org/10.1021/acs.est.6b00101

Lyu K, Gu L, Wang H, Zhu X, Zhang L, Sun Y, Huang Y, Yang Z (2019) Transcriptomic analysis dissects the mechanistic insight into the *daphnia* clonal variation in tolerance to toxic *Microcystis*. Limnol Oceanogr 64(1):272–283. https://doi.org/10.1002/lno.11038

Lyu K, Wu Y, Li J, Yang Z (2024) MicroRNA *mir-210* modulates the water flea *Daphnia magna* response to cyanobacterial stress. Environ Sci Technol 58(42):18520–18530. https://doi.org/10.1021/acs.est.4c04190

Ma J, Li Y, Wu M, Zhang C, Che Y, Li W, Li X (2018) Serum immune responses in common carp (*Cyprinus carpio* L.) to paraquat exposure: the traditional parameters and circulating microRNAs. Fish Shellfish Immunol 76:133–142. https://doi.org/10.1016/j.fsi.2018.02.046

Ma X, Deng Z, Blair D, Bi Y, Hu W, Yin M (2021a) Cyanobacterial bloom associated with a complete turnover of a *daphnia* population in a warm-temperate eutrophic lake in eastern China. Freshw Biol. https://doi.org/10.1111/fwb.13858

Ma Y, Liu H, Du X, Shi Z, Liu X, Wang R, Zhang S, Tian Z, Shi L, Guo H, Zhang H (2021b) Advances in the toxicology research of microcystins based on omics approaches. Environ Int 154:106661. https://doi.org/10.1016/j.envint.2021.106661

MacKeigan PW, Zastepa A, Taranu ZE, Westrick JA, Liang A, Pick FR, Beisner BE, Gregory-Eaves I (2023) Microcystin concentrations and congener composition in relation to environmental variables across 440 north-temperate and boreal lakes. Sci Total Environ 884:163811. https://doi.org/10.1016/j.scitotenv.2023.163811

Malbrouck C, Kestemont P (2006) Effects of microcystins on fish. Environ Toxicol Chem 25(1):72–86. https://doi.org/10.1897/05-029R.1

Malécot M, Guével B, Pineau C, Holbech BF, Bormans M, Wiegand C (2013) Specific proteomic response of *Unio pictorum* mussel to a mixture of glyphosate and microcystin-LR. J Proteome Res 12(11):5281–5292. https://doi.org/10.1021/pr4006316

Malhotra JD, Kaufman RJ (2007) Endoplasmic reticulum stress and oxidative stress: a vicious cycle or a double-edged sword? Antioxid Redox Signal 9(12):2277–2294. https://doi.org/10.1089/ars.2007.1782

Mamoori A, Gopalan V, Smith RA, Lam AK-Y (2016) Modulatory roles of microRNAs in the regulation of different signalling pathways in large bowel cancer stem cells. Biol Cell 108(3):51–64. https://doi.org/10.1111/boc.201500062

Manage PM, Edwards C, Singh BK, Lawton LA (2009) Isolation and identification of novel microcystin-degrading bacteria. Appl Environ Microbiol 75(21):6924–6928. https://doi.org/10.1128/AEM.01928-09

Marić P, Ahel M, Maraković N, Lončar J, Mihaljević I, Smital T (2021) Selective interaction of microcystin congeners with zebrafish (*Danio rerio*) Oatp1d1 transporter. Chemosphere 283:131155. https://doi.org/10.1016/j.chemosphere.2021.131155

Marie B (2020) Disentangling of the ecotoxicological signal using 'omics' analyses, a lesson from the survey of the impact of cyanobacterial proliferations on fishes. Sci Total Environ 736:139701. https://doi.org/10.1016/j.scitotenv.2020.139701

Martin-Creuzburg D, von Elert E, Hoffmann KH (2008) Nutritional constraints at the cyanobacteria-*Daphnia magna* interface: the role of sterols. Limnol Oceanogr 53(2):456–468. https://doi.org/10.4319/lo.2008.53.2.0456

Martins R, Pereira P, Welker M, Fastner J, Vasconcelos VM (2005) Toxicity of culturable cyanobacteria strains isolated from the Portuguese coast. Toxicon 46(4):454–464

Massey IY, Yang F (2020) A mini review on microcystins and bacterial degradation. Toxins 12(4):268. https://doi.org/10.3390/toxins12040268

Máthé C, Hamvas M, Garda T, Beyer D, Vasas G (2017) Cellular effects of cylindrospermopsin (cyanobacterial alkaloid toxin) and its potential medical consequences. Curr Med Chem 24(1):91–109. https://doi.org/10.2174/0929867323666161028153814

Meier-Abt F, Mokrab Y, Mizuguchi K (2006) Organic anion transporting polypeptides of the OATP/*SLCO* superfamily: identification of new members in nonmammalian species, comparative modeling and a potential transport mode. J Membr Biol 208(3):213–227. https://doi.org/10.1007/s00232-005-7004-x

Menezes C, Alverca E, Dias E, Sam-Bento F, Pereira P (2013) Involvement of endoplasmic reticulum and autophagy in microcystin-LR toxicity in Vero-E6 and HepG2 cell lines. Toxicol In Vitro 27(1):138–148. https://doi.org/10.1016/j.tiv.2012.09.009

Menzel S, Bouchnak R, Menzel R, Steinberg CEW (2011) Dissolved humic substances initiate DNA-methylation in cladocerans. Aquat Toxicol 105(3–4):640–642. https://doi.org/10.1016/j.aquatox.2011.08.025

Metcalf JS, Lindsay J, Beattie KA, Birmingham S, Saker ML, Törökné AK, Codd GA (2002) Toxicity of cylindrospermopsin to the brine shrimp *Artemia salina*: comparisons with protein synthesis inhibitors and microcystins. Toxicon 40(8):1115–1120. https://doi.org/10.1016/S0041-0101(02)00105-8

Metcalf JS, Beattie KA, Purdie EL, Bryant JA, Irvine LM, Codd GA (2012) Analysis of microcystins and microcystin genes in 60-170-year-old dried herbarium specimens of cyanobacteria. Harmful Algae 15:47–52. https://doi.org/10.1016/j.hal.2011.11.004

Mezrioui N-E, Oudra B (1998) Dynamics of picoplankton and microplankton flora in the experimental wastewater stabilization ponds in the arid region of Marrakech, Morocco and cyanobacteria effect on *Escherichia coli* and *vibrio cholerae* survival. In: Wong Y-S, Tam NFY (eds) Wastewater treatment with algae. Springer, Berlin/Heidelberg, pp 165–188. https://doi.org/10.1007/978-3-662-10863-5_10

Mielewczyk DA, Glover CN, Klaczek CE, Goss GG, Saari GN (2023) Sub-chronic exposure to waterborne extracellular microcystin-LR impairs calcium homeostasis in rainbow trout. Ecotox Environ Saf 251:114542. https://doi.org/10.1016/j.ecoenv.2023.114542

Moraes ACN, Fallah HP, Magalhães VF, Habibi HR (2021) Cylindrospermopsin directly disrupts spermatogenesis in isolated male zebrafish testis. Gen Comp Endocrinol 313:113891. https://doi.org/10.1016/j.ygcen.2021.113891

Morón-López J, Nieto-Reyes L, El-Shehawy R (2017) Assessment of the influence of key abiotic factors on the alternative microcystin degradation pathway(s) (*mlr*$^{-}$): a detailed comparison with the mlr route (*mlr*$^{+}$). Sci Total Environ 599–600:1945–1953. https://doi.org/10.1016/j.scitotenv.2017.04.042

Motwani NH, Gorokhova E (2013) Mesozooplankton grazing on picocyanobacteria in the Baltic Sea as inferred from molecular diet analysis. PLoS One 8(11):e79230. https://doi.org/10.1371/journal.pone.0079230

Mou X, Lu X, Jacob J, Sun S, Heath R (2013) Metagenomic identification of bacterioplankton taxa and pathways involved in microcystin degradation in Lake Erie. PLoS One 8(4):e61890. https://doi.org/10.1371/journal.pone.0061890

Murch SJ, Cox PA, Banack SA (2004) A mechanism for slow release of biomagnified cyanobacterial neurotoxins and neurodegenerative disease in Guam. Proc Natl Acad Sci USA 101(33):12228–12231. https://doi.org/10.1073/pnas.0404926101

Murrell MC, Lores EM (2004) Phytoplankton and zooplankton seasonal dynamics in a subtropical estuary: importance of cyanobacteria. J Plankton Res 26(3):371–382. https://doi.org/10.1093/plankt/fbh038

Nakamura Y, Suzuki K, Suzuki S-y, Hiromi J (1997) Production of *Oikopleura dioica* (Appendicularia) following a picoplankton 'bloom' in a eutrophic coastal area. J Plankton Res 19(1):113–124. https://doi.org/10.1093/plankt/19.1.113

Nandini S, Sarma SSS, Ramírez-García P (2000) Life table demography and population growth of *Daphnia laevis* (Cladocera, Anomopoda) under different densities of *Chlorella vulgaris* and *Microcystis aeruginosa*. Crustaceana 73(10):1273–1286. https://doi.org/10.1163/156854000505254

Nandini S, Miracle MR, Vicente E, Sarma SSS (2021) Strain-related differences in bacterivory and demography of *Diaphanosoma mongolianum* (Cladocera) in relation to diet and previous exposure to cyanobacteria in nature. Aquat Ecol 55(4):1225–1239. https://doi.org/10.1007/s10452-021-09892-z

Nikitin O, Latypova V (2014) Behavioral response of *Daphnia magna* (Crustacea, Cladocera) to low concentration of microcystin. In: International multidisciplinary scientific GeoConference surveying geology and mining ecology management. SGEM, pp 85–92

Ning X, Cloern JE, Cole BE (2000) Spatial and temporal variability of picocyanobacteria *Synechococcus* sp. San Francisco Bay Limnol Oceanogr 45(3):695–702. https://doi.org/10.4319/lo.2000.45.3.0695

Nizan S, Dimentman C, Shilo M (1986) Acute toxic effects of the cyanobacterium *Microcystis aeruginosa* on *Daphnia magna*. Limnol Oceanogr 31(3):497–502. https://doi.org/10.4319/lo.1986.31.3.0497

Nogueira ICG, Lobo-da-Cunha A, Vasconcelos VM (2006) Effects of *Cylindrospermopsis raciborskii* and *Aphanizomenon ovalisporum* (cyanobacteria) ingestion on *Daphnia magna* midgut and associated diverticula epithelium. Aquat Toxicol 80(2):194–203. https://doi.org/10.1016/j.aquatox.2006.08.008

Nybom SMK, Salminen SJ, Meriluoto JAO (2007) Removal of microcystin-LR by strains of metabolically active probiotic bacteria. FEMS Microbiol Lett 270(1):27–33. https://doi.org/10.1111/j.1574-6968.2007.00644.x

Nybom SM, Salminen SJ, Meriluoto JA (2008) Specific strains of probiotic bacteria are efficient in removal of several different cyanobacterial toxins from solution. Toxicon 52(2):214–220. https://doi.org/10.1016/j.toxicon.2008.04.169

O'Connell RM, Taganov KD, Boldin MP, Cheng G, Baltimore D (2007) MicroRNA-155 is induced during the macrophage inflammatory response. Proc Natl Acad Sci USA 104(5):1604–1609. https://doi.org/10.1073/pnas.0610731104

Oberemm A, Fastner J, Steinberg CEW (1997) Effects of microcystin-LR and cyanobacterial crude extracts on embryo-larval development of zebrafish (*Danio rerio*). Water Res 31(11):2918–2921. https://doi.org/10.1016/S0043-1354(97)00120-6

Oberemm A, Becker J, Codd GA, Steinberg C (1999) Effects of cyanobacterial toxins and aqueous crude extracts of cyanobacteria on the development of fish and amphibians. Environ Toxicol 14(1):77–88. https://doi.org/10.1002/(SICI)1522-7278(199902)14:1<77::AID-TOX11>3.0.CO;2-F

Oberhaus L, Gélinas M, Pinel-Alloul B, Ghadouani A, Humbert J-F (2007) Grazing of two toxic *Planktothrix* species by *Daphnia pulicaria*: potential for bloom control and transfer of microcystins. J Plankton Res 29(10):827–838. https://doi.org/10.1093/plankt/fbm062

Ohta T, Sueoka E, Iida N, Komori A, Suganuma M, Nishiwaki R, Tatematsu M, Kim SJ, Carmichael WW, Fujiki H (1994) Nodularin, a potent inhibitor of protein phosphatases 1 and 2A, is a new environmental carcinogen in male F344 rat liver. Cancer Res 54(24):6402–6406

Olivieri F, Spazzafumo L, Santini G, Lazzarini R, Albertini MC, Rippo MR, Galeazzi R, Abbatecola AM, Marcheselli F, Monti D, Ostan R, Cevenini E, Antonicelli R, Franceschi C, Procopio AD (2012) Age-related differences in the expression of circulating microRNAs: miR-21 as a new circulating marker of inflammaging. Mech Ageing Dev 133(11–12):675–685. https://doi.org/10.1016/j.mad.2012.09.004

Ortiz-Rodríguez R, Dao TS, Wiegand C (2012) Transgenerational effects of microcystin-LR on *Daphnia magna*. J Exp Biol 215(16):2795–2805. https://doi.org/10.1242/jeb.069211

Oudra B, El Andaloussi M, Franca S, Barros P, Martins R, Oufdou K, Sbiyyaa B, Loudiki M, Mezrioui N, Vasconcelos V (2000) Harmful cyanobacterial toxic blooms in waste stabilisation ponds. Water Sci Technol 42(10–11):179–186. https://doi.org/10.2166/wst.2000.0637

Pagliara P, De Benedetto GE, Francavilla M, Barca A, Caroppo C (2021) Bioactive potential of two marine picocyanobacteria belonging to *Cyanobium* and *Synechococcus* genera. Microorganisms 9(10):2048. https://doi.org/10.3390/microorganisms9102048

Papacek JR, Inglett PW, Phlips EJ, Lasi MA (2024) Nitrogen and phosphorus uptake kinetics in cultures of two novel picoplankton groups responsible for a recent bloom event in a subtropical estuary (Indian River lagoon, Florida). Front Mar Sci 11:1256901. https://doi.org/10.3389/fmars.2024.1256901

Papakostas S, Michaloudi E, Triantafyllidis A, Kappas I, Abatzopoulos TJ (2013) Allochronic divergence and clonal succession: two microevolutionary processes sculpturing population structure of *Brachionus rotifers*. Hydrobiologia 700(1):33–45. https://doi.org/10.1007/s10750-012-1217-7

Park H-D, Sasaki Y, Maruyama T, Yanagisawa E, Hiraishi A, Kato K (2001) Degradation of the cyanobacterial hepatotoxin microcystin by a new bacterium isolated from a hypertrophic lake. Environ Toxicol 16(4):337–343. https://doi.org/10.1002/tox.1041

Parvathi A, Zhong X, Pradeep Ram AS, Jacquet S (2014) Dynamics of auto- and heterotrophic picoplankton and associated viruses in Lake Geneva. Hydrol Earth Syst Sci 18(3):1073–1087. https://doi.org/10.5194/hess-18-1073-2014

Pawlik-Skowrońska B, Bownik A (2021) Cyanobacterial anabaenopeptin-B, microcystins and their mixture cause toxic effects on the behavior of the freshwater crustacean *Daphnia magna* (Cladocera). Toxicon 198:1–11. https://doi.org/10.1016/j.toxicon.2021.04.023

Pawlik-Skowrońska B, Toporowska M, Mazur-Marzec H (2019) Effects of secondary metabolites produced by different cyanobacterial populations on the freshwater zooplankters *Brachionus calyciflorus* and Daphnia pulex. Environ Sci Pollut Res 26(12):11793–11804. https://doi.org/10.1007/s11356-019-04543-1

Peng S, Deng D, He P, Xu X, Zhang C, Cao J, Liu Q, Zhang T (2018) Effects of *Microcystis aeruginosa* on the life history traits and SOD activity of *Daphnia similoides sinensis*. Environ Sci Pollut Res 25(30):30696–30707. https://doi.org/10.1007/s11356-018-3040-9

Perera P-Y, Lichy JH, Waldmann TA, Perera LP (2012) The role of interleukin-15 in inflammation and immune responses to infection: implications for its therapeutic use. Microbes Infect 14(3):247–261. https://doi.org/10.1016/j.micinf.2011.10.006

Persson KJ, Stenroth P, Legrand C (2011) Effects of the filamentous cyanobacterium *Nodularia* on fitness and feeding behavior of young-of-the-year (YOY) Eurasian perch (*Perca fluviatilis*). Toxicon 57(7–8):1033–1040. https://doi.org/10.1016/j.toxicon.2011.04.007

Pflugmacher S (2004) Promotion of oxidative stress in the aquatic macrophyte *Ceratophyllum demersum* during biotransformation of the cyanobacterial toxin microcystin-LR. Aquat Toxicol 70(3):169–178. https://doi.org/10.1016/j.aquatox.2004.06.010

Phlips EJ, Badylak S, Lynch TC (1999) Blooms of the picoplanktonic cyanobacterium *Synechococcus* in Florida bay, a subtropical inner-shelf lagoon. Limnol Oceanogr 44(4):1166–1175. https://doi.org/10.4319/lo.1999.44.4.1166

Piferrer F, Wang H (eds) (2023) Epigenetics in aquaculture. Wiley, Chichester. https://doi.org/10.1002/9781119821946

Puerto M, Jos A, Pichardo S, Moyano R, Blanco A, Cameán AM (2014) Acute exposure to pure cylindrospermopsin results in oxidative stress and pathological alterations in tilapia (*Oreochromis niloticus*). Environ Toxicol 29(4):371–385. https://doi.org/10.1002/tox.21764

Qian H, Liu G, Lu T, Sun L (2018) Developmental neurotoxicity of *Microcystis aeruginosa* in the early life stages of zebrafish. Ecotox Environ Saf 151:35–41. https://doi.org/10.1016/j.ecoenv.2017.12.059

Qian H, Zhang M, Liu G, Lu T, Sun L, Pan X (2019) Effects of different concentrations of *Microcystis aeruginosa* on the intestinal microbiota and immunity of zebrafish (*Danio rerio*). Chemosphere 214:579–586. https://doi.org/10.1016/j.chemosphere.2018.09.156

Qiao Q, Liang H, Zhang X (2013) Effect of cyanobacteria on immune function of crucian carp (*Carassius auratus*) via chronic exposure in diet. Chemosphere 90(3):1167–1176. https://doi.org/10.1016/j.chemosphere.2012.09.025

Qiao Q, Le Manach S, Huet H, Duvernois-Berthet E, Chaouch S, Duval C, Sotton B, Ponger L, Marie A, Mathéron L, Lennon S, Bolbach G, Djediat C, Bernard C, Edery M, Marie B (2016a) An integrated omic analysis of hepatic alteration in medaka fish chronically exposed to cyanotoxins with possible mechanisms of reproductive toxicity. Environ Pollut 219:119–131. https://doi.org/10.1016/j.envpol.2016.10.029

Qiao Q, Le Manach S, Sotton B, Huet H, Duvernois-Berthet E, Paris A, Duval C, Ponger L, Marie A, Blond A, Mathéron L, Vinh J, Bolbach G, Djediat C, Bernard C, Edery M, Marie B (2016b) Deep sexual dimorphism in adult medaka fish liver highlighted by multi-omic approach. Sci Rep 6(1):32459. https://doi.org/10.1038/srep32459

Rabelo JCS, Hanusch AL, de Jesus LWO, Mesquita LA, Franco FC, Silva RA, Sabóia-Morais SMT (2021) DNA damage induced by cylindrospermopsin on different tissues of the biomonitor fish *Poecilia reticulata*. Environ Toxicol 36(6):1125–1134. https://doi.org/10.1002/tox.23111

Radersma R, Hegg A, Noble DWA, Uller T (2018) Timing of maternal exposure to toxic cyanobacteria and offspring fitness in *Daphnia magna*: implications for the evolution of anticipatory maternal effects. Ecol Evol 8(24):12727–12736. https://doi.org/10.1002/ece3.4700

Rapala J, Berg KA, Lyra C, Niemi RM, Manz W, Suomalainen S, Paulin L, Lahti K (2005) *Paucibacter toxinivorans* gen. Nov., sp. nov., a bacterium that degrades cyclic cyanobacterial hepatotoxins microcystins and nodularin. Int J Syst Evol Microbiol 55(4):1563–1568. https://doi.org/10.1099/ijs.0.63599-0

Raška J, Čtveráčková L, Dydowiczová A, Sovadinová I, Bláha L, Babica P (2019) Cylindrospermopsin induces cellular stress and activation of ERK1/2 and p38 MAPK pathways in adult human liver stem cells. Chemosphere 227:43–52. https://doi.org/10.1016/j.chemosphere.2019.03.131

Raven JA (2010) Cyanotoxins: a poison that frees phosphate. Curr Biol 20(19):R850–R852. https://doi.org/10.1016/j.cub.2010.08.012

Reichwaldt ES, Song H, Ghadouani A (2013) Effects of the distribution of a toxic *Microcystis* bloom on the small scale patchiness of zooplankton. PLoS One 8(6):e66674. https://doi.org/10.1371/journal.pone.0066674

Reinikainen M, Ketola M, Walls M (1994) Effects of the concentrations of toxic *Microcystis aeruginosa* and an alternative food on the survival of *Daphnia pulex*. Limnol Oceanogr 39(2):424–432. https://doi.org/10.4319/lo.1994.39.2.0424

Reinikainen M, Ketola M, Jantunen M, Walts M (1995) Effects of *Microcystis aeruginosa* exposure and nutritional status on the reproduction of *Daphnia pulex*. J Plankton Res 17(2):431–436. https://doi.org/10.1093/plankt/17.2.431

Reinikainen M, Meriluoto JAO, Spoof L, Harada KI (2001) The toxicities of a polyunsaturated fatty acid and a microcystin to *Daphnia magna*. Environ Toxicol 16(5):444–448. https://doi.org/10.1002/tox.10003

Rodrigues NB, Pitol DL, Tocchini de Figueiredo FA, Tenfen das Chagas Lima AC, Burdick Henry T, Mardegan Issa JP, de Aragão Umbuzeiro G, Pereira BF (2022) Microcystin-LR at sublethal concentrations induce rapid morphology of liver and muscle tissues in the fish species *Astyanax altiparanae* (Lambari). Toxicon 211:70–78. https://doi.org/10.1016/j.toxicon.2022.03.006

Rogers ED, Henry TB, Twiner MJ, Gouffon JS, McPherson JT, Boyer GL, Sayler GS, Wilhelm SW (2011) Global gene expression profiling in larval zebrafish exposed to microcystin-LR and microcystis reveals endocrine disrupting effects of cyanobacteria. Environ Sci Technol 45(5):1962–1969. https://doi.org/10.1021/es103538b

Rohrlack T, Dittmann E, Henning M, Börner T, Kohl JG (1999a) Role of microcystins in poisoning and food ingestion inhibition of *Daphnia galeata* caused by the cyanobacterium *Microcystis aeruginosa*. Appl Environ Microbiol 65(2):737–739. https://doi.org/10.1128/aem.65.2.737-739.1999

Rohrlack T, Henning M, Kohl JG (1999b) Does the toxic effect of *Microcystis aeruginosa* on *Daphnia galeata* depend on microcystin ingestion rate? Arch Hydrobiol 146(4):385–395. https://doi.org/10.1127/archiv-hydrobiol/146/1999/385

Rohrlack T, Henning M, Kohl JG (1999c) Mechanisms of the inhibitory effect of the cyanobacterium *Microcystis aeruginosa* on *Daphnia galeata*'s ingestion rate. J Plankton Res 21(8):1489–1500. https://doi.org/10.1093/plankt/21.8.1489

Rohrlack T, Christoffersen K, Hansen PE, Zhang W, Czarnecki O, Henning M, Fastner J, Erhard M, Neilan BA, Kaebernick M (2003) Isolation, characterization, and quantitative analysis of microviridin J, a new *Microcystis* metabolite toxic to *daphnia*. J Chem Ecol 29(8):1757–1770. https://doi.org/10.1023/A:1024889925732

Roy S, Sen CK (2011) MiRNA in innate immune responses: novel players in wound inflammation. Physiol Genomics 43(10):557–565. https://doi.org/10.1152/physiolgenomics.00160.2010

Rudman SM, Rodriguez-Cabal MA, Stier A, Sato T, Heavyside J, El-Sabaawi RW, Crutsinger GM (2015) Adaptive genetic variation mediates bottom-up and top-down control in an aquatic ecosystem. Proc Royal Soc B Biol Sci 282(1812):20151234. https://doi.org/10.1098/rspb.2015.1234

Rymuszka A, Adaszek Ł (2012a) Pro- and anti-inflammatory cytokine expression in carp blood and head kidney leukocytes exposed to cyanotoxin stress – an *in vitro* study. Fish Shellfish Immunol 33(2):382–388. https://doi.org/10.1016/j.fsi.2012.05.021

Rymuszka A, Adaszek Ł (2012b) Pro- and anti-inflammatory cytokine expression in carp blood and head kidney leukocytes exposed to cyanotoxin stress—an *in vitro* study. Fish Shellfish Immunol 33(2):382–388. https://doi.org/10.1016/j.fsi.2012.05.021

Rymuszka A, Adaszek Ł (2013) Cytotoxic effects and changes in cytokine gene expression induced by microcystin-containing extract in fish immune cells—an *in vitro* and *in vivo* study. Fish Shellfish Immunol 34(6):1524–1532. https://doi.org/10.1016/j.fsi.2013.03.364

Rymuszka A, Sieroslawska A (2018) Comparative studies on the cytotoxic effects induced by nodularin in primary carp leukocytes and the cells of the fish CLC line. Toxicon 148:7–15. https://doi.org/10.1016/j.toxicon.2018.04.001

Rzymski P, Klimaszyk P, Jurczak T, Poniedziałek B (2020) Oxidative stress, programmed cell death and microcystin release in *Microcystis aeruginosa* in response to *daphnia* grazers. Front Microbiol 11:1201. https://doi.org/10.3389/fmicb.2020.01201

Salazar C, Galaz M, Ojeda N, Marshall SH (2020) Detection of cellular miRNAs in plasma of *Salmo salar* during an ISAV infection. Aquac Rep 17:100320. https://doi.org/10.1016/j.aqrep.2020.100320

Sánchez-Baracaldo P, Handley BA, Hayes PK (2008) Picocyanobacterial community structure of freshwater lakes and the Baltic Sea revealed by phylogenetic analyses and clade-specific quantitative PCR. Microbiology 154(11):3347–3357. https://doi.org/10.1099/mic.0.2008/019836-0

Santos AA, Soldatou S, de Magalhães VF, Azevedo SMFO, Camacho-Muñoz D, Lawton LA, Edwards C (2021) Degradation of multiple peptides by microcystin-degrader *Paucibacter toxinivorans* (2C20). Toxins 13(4):265. https://doi.org/10.3390/TOXINS13040265

Sarnelle O (2005) *Daphnia* as keystone predators: effects on phytoplankton diversity and grazing resistance. J Plankton Res 27(12):1229–1238. https://doi.org/10.1093/plankt/fbi086

Sarnelle O, Wilson AE (2005) Local adaptation of *Daphnia pulicaria* to toxic cyanobacteria. Limnol Oceanogr 50(5):1565–1570. https://doi.org/10.4319/lo.2005.50.5.1565

Sarnelle O, Gustafsson S, Hansson L-A (2010) Effects of cyanobacteria on fitness components of the herbivore *daphnia*. J Plankton Res 32(4):471–477. https://doi.org/10.1093/plankt/fbp151

Sartonov A (1995) Effects of *Microcystis aeruginosa* on interference competition between *Daphnia pulex* and *Keratella cochlearis*. Hydrobiologia 307(1–3):117–126. https://doi.org/10.1007/BF00032003

Saul N, Pietsch K, Stürzenbaum SR, Menzel R, Steinberg CEW (2013) Hormesis and longevity with tannins: free of charge or cost-intensive? Chemosphere 93(6):1005–1008. https://doi.org/10.1016/j.chemosphere.2013.05.069

Saul N, Chakrabarti S, Stürzenbaum SR, Menzel R, Steinberg CEW (2014) Neurotoxic action of microcystin-LR is reflected in the transcriptional stress response of *Caenorhabditis elegans*. Chem Biol Interact 223:51–57. https://doi.org/10.1016/j.cbi.2014.09.007

Schwark M, Yerena JAM, Röhrborn K, Hrouzek P, Divoká P, Štenclová L, Delawská K, Enke H, Vorreiter C, Wiley F, Sippl W, Sobotka R, Saha S, Wilde SB, Mareš J, Niedermeyer THJ (2023) More than just an eagle killer: the freshwater cyanobacterium *Aetokthonos hydrillicola* produces highly toxic dolastatin derivatives. Proc Natl Acad Sci USA 120(40):e2219230120. https://doi.org/10.1073/pnas.2219230120

Schwarzenberger A, Martin-Creuzburg D (2021) *Daphnia*'s adaptive molecular responses to the cyanobacterial neurotoxin anatoxin-α are maternally transferred. Toxins 13(5):326. https://doi.org/10.3390/toxins13050326

Schwarzenberger A, Courts C, von Elert E (2009) Target gene approaches: gene expression in *Daphnia magna* exposed to predator-borne kairomones or to microcystin-producing and microcystin-free *Microcystis aeruginosa*. BMC Genomics 10:527. https://doi.org/10.1186/1471-2164-10-527

Schwarzenberger A, Sadler T, Motameny S, Ben-Khalifa K, Frommolt P, Altmuller J, Konrad K, von Elert E (2014) Deciphering the genetic basis of microcystin tolerance. BMC Genomics 15:776. https://doi.org/10.1186/1471-2164-15-776

Seeley HG (1886) Fresh-water fishes of Europe: a history of their genera, species, structure, habits, and distribution. Cassell & Company, Limited, London

Semyalo R, Rohrlack T, Larsson P (2009) Growth and survival responses of a tropical *daphnia* (*Daphnia lumholtzi*) to cell-bound microcystins. J Plankton Res 31(8):827–835. https://doi.org/10.1093/plankt/fbp032

Shahmohamadloo RS, Simmons DBD, Sibley PK (2020) Shotgun proteomics analysis reveals sublethal effects in *Daphnia magna* exposed to cell-bound microcystins produced by *Microcystis aeruginosa*. Comp Biochem Phys D 33:100656. https://doi.org/10.1016/j.cbd.2020.100656

Shahmohamadloo RS, Ortiz Almirall X, Simmons DBD, Poirier DG, Bhavsar SP, Sibley PK (2022) Fish tissue accumulation and proteomic response to microcystins is species-dependent. Chemosphere 287:132028. https://doi.org/10.1016/j.chemosphere.2021.132028

Shakri A, Yusoff FM, Ismail IS, Toda T (2020) Reduced reproductive capacity in *Moina micrura* Kurz, 1875 exposed to toxic *Microcystis* spp. Asian Fish Sci 33(1):42–49. https://doi.org/10.33997/j.afs.2020.33.1.005

Shao YQ, Deng DG, Meng MR, Zhang XL, Li F (2014) Effects of colonial *Microcystis aeruginosa* and interspecific competition on the population dynamics and resting egg formation of two cladocerans. J Freshw Ecol 29(2):213–223. https://doi.org/10.1080/02705060.2013.873882

Sieroslawska A, Rymuszka A (2015a) Cylindrospermopsin induces oxidative stress and genotoxic effects in the fish CLC cell line. J Appl Toxicol 35(4):426–433. https://doi.org/10.1002/jat.3040

Sieroslawska A, Rymuszka A (2015b) Effects of cylindrospermopsin on a common carp leucocyte cell line. J Appl Toxicol 35(1):83–89. https://doi.org/10.1002/jat.2990

Sieroslawska A, Rymuszka A, Adaszek Ł (2015) Effects of cylindrospermopsin on the phagocytic cells of the common carp (*Cyprinus carpio* L.). J Appl Toxicol 35(11):1406–1414. https://doi.org/10.1002/jat.3118

Silva RC, Liebel S, de Oliveira HHP, Ramsdorf WA, Garcia JRE, Azevedo SMFO, Magalhães VF, Oliveira Ribeiro CA, Filipak Neto F (2017) Cylindrospermopsin effects on cell viability and redox milieu of Neotropical fish *Hoplias malabaricus* hepatocytes. Fish Physiol Biochem 43(5):1237–1244. https://doi.org/10.1007/s10695-017-0367-y

Silva-Stenico ME, Silva CS, Lorenzi AS, Shishido TK, Etchegaray A, Lira SP, Moraes LA, Fiore MF (2011) Non-ribosomal peptides produced by Brazilian cyanobacterial isolates with antimicrobial activity. Microbiol Res 166(3):161–175. https://doi.org/10.1016/j.micres.2010.04.002

Śliwińska-Wilczewska S, Maculewicz J, Barreiro Felpeto A, Latała A (2018) Allelopathic and bloom-forming picocyanobacteria in a changing world. Toxins 10(1):48. https://doi.org/10.3390/toxins10010048

Smith AD, Gilbert KK (1995) Relative susceptibilities of rotifers and cladocerans to *Microcystis aeruginosa*. Arch Hydrobiol 132(3):309–336. https://doi.org/10.1127/archiv-hydrobiol/132/1995/309

Smith JL, Schulz KL, Zimba PV, Boyer GL (2010) Possible mechanism for the foodweb transfer of covalently bound microcystins. Ecotoxicol Environ Saf 73(5):757–761. https://doi.org/10.1016/j.ecoenv.2009.12.003

Smutná M, Babica P, Jarque S, Hilscherová K, Maršálek B, Haeba M, Bláha L (2014) Acute, chronic and reproductive toxicity of complex cyanobacterial blooms in *Daphnia magna* and the role of microcystins. Toxicon 79 (0):11–18. https://doi.org/10.1016/j.toxicon.2013.12.009

Somdee T, Thunders M, Ruck J, Lys I, Allison M (2013) Page R (2013) degradation of [Dha^7]MC-LR by a microcystin degrading bacterium isolated from Lake Rotoiti. New Zealand Int Sch Res Notices 1:596429. https://doi.org/10.1155/2013/596429

Somogyi B, Felföldi T, Vanyovszki J, Ágyi Á, Márialigeti K, Vörös L (2009) Winter bloom of picoeukaryotes in Hungarian shallow turbid soda pans and the role of light and temperature. Aquat Ecol 43(3):735–744. https://doi.org/10.1007/s10452-009-9269-0

Song G, Zhao Y, Lu J, Liu Z, Quan J, Zhu L (2024) Effects of astaxanthin on growth performance, gut structure, and intestinal microorganisms of *Penaeus vannamei* under microcystin-LR stress. Animals 14(1):58. https://doi.org/10.3390/ani14010058

Sorokin YI, Dallocchio F (2008) Dynamics of phosphorus in the Venice lagoon during a picocyanobacteria bloom. J Plankton Res 30(9):1019–1026. https://doi.org/10.1093/plankt/fbn059

Sorokin YI, Zakuskina OY (2010) Features of the Comacchio ecosystem transformed during persistent bloom of picocyanobacteria. J Oceanogr 66(3):373–387. https://doi.org/10.1007/s10872-010-0033-9

Sorokin PY, Sorokin YI, Boscolo R, Giovanardi O (2004) Bloom of picocyanobacteria in the Venice lagoon during summer–autumn 2001: ecological sequences. Hydrobiologia 523(1):71–85. https://doi.org/10.1023/B:HYDR.0000033096.14267.43

Sotton B, Guillard J, Bony S, Devaux A, Domaizon I, Givaudan N, Crespeau F, Huet H, Anneville O (2012) Impact of toxic cyanobacterial blooms on Eurasian perch (*Perca fluviatilis*): experimental study and In situ observations in a Peri-alpine lake. PLoS One 7(12):e52243. https://doi.org/10.1371/journal.pone.0052243

Sotton B, Guillard J, Anneville O, Maréchal M, Savichtcheva O, Domaizon I (2014) Trophic transfer of microcystins through the lake pelagic food web: evidence for the role of zooplankton as a vector in fish contamination. Sci Total Environ 466-467:152–163. https://doi.org/10.1016/j.scitotenv.2013.07.020

Stangenberg M (1968) Toxic effects of *Microcystis aeruginosa* kg. Extracts on *Daphnia longispina* O. F. Müller and *Eucypris virens* Jurine. Hydrobiologia 32 (1–2):81–87. https://doi.org/10.1007/BF00179540

Steinberg CEW (2012) Stress Ecology–Environmental Stress as Ecological Driving Force and Key Player in Evolution. Springer, Dordrecht. https://doi.org/10.1007/978-94-007-2072-5

Steinberg CEW (2022a) Sterols, phospholipids, and wax esters—'*stay healthy, avoid cholesterol*'. In: Aquatic animal nutrition: organic macro- and micro-nutrients. Springer International Publishing, Cham, pp 791–820. https://doi.org/10.1007/978-3-030-87227-4_31

Steinberg CEW (2022b) Sulfur amino acids—'*much more than easy fuel*'. In: Aquatic animal nutrition: organic macro- and micro-nutrients. Springer International Publishing, Cham, pp 163–192. https://doi.org/10.1007/978-3-030-87227-4_9

Steinberg CEW (2022c) The versatile amino acid: tryptophan—'*more controlling than fueling*'. In: Aquatic animal nutrition: organic macro- and micro-nutrients. Springer International Publishing, Cham, pp 117–137. https://doi.org/10.1007/978-3-030-87227-4_7

Steinberg CEW (2024) Aquatic Animal Nutrition: Plant Preparations—'*Ever Tried. Ever Failed. Try Again.*'. In: Aquatic animal nutrition – plant preparations. Springer Nature Switzerland AG, Cham, pp 1–47. https://doi.org/10.1007/978-3-031-70938-8_1

Steiner K, Zimmermann L, Hagenbuch B, Dietrich D (2016) Zebrafish Oatp-mediated transport of microcystin congeners. Arch Toxicol 90(5):1129–1139. https://doi.org/10.1007/s00204-015-1544-3

Štern A, Rotter A, Novak M, Filipič M, Žegura B (2019) Genotoxic effects of the cyanobacterial pentapeptide nodularin in HepG2 cells. Food Chem Toxicol 124:349–358. https://doi.org/10.1016/j.fct.2018.12.019

Stockner J, Callieri C, Cronberg G (2002) Picoplankton and other non-bloom-forming cyanobacteria in lakes. In: Whitton BA, Potts M (eds) The ecology of cyanobacteria: their diversity in time and space. Springer, Netherlands, Dordrecht, pp 195–231. https://doi.org/10.1007/0-306-46855-7_7

Stucken K, John U, Cembella A, Soto-Liebe K, Vásquez M (2014) Impact of nitrogen sources on gene expression and toxin production in the diazotroph *Cylindrospermopsis raciborskii* CS-505 and non-diazotroph *Raphidiopsis brookii* D9. Toxins 6(6):1896–1915. https://doi.org/10.3390/toxins6061896

Su Y, Li L, Hou J, Wu N, Lin W, Li G (2016) Life-cycle exposure to microcystin-LR interferes with the reproductive endocrine system of male zebrafish. Aquat Toxicol 175:205–212. https://doi.org/10.1016/j.aquatox.2016.03.018

Sun J, Chen C, You C, Liu Y, Ma H, Monroig Ó, Tocher DR, Wang S, Li Y (2020) The miR-15/16 cluster is involved in the regulation of vertebrate LC-PUFA biosynthesis by targeting *pparγ* as demonostrated in rabbitfish *Siganus canaliculatus*. Mar Biotechnol 22(4):475–487. https://doi.org/10.1007/s10126-020-09969-0

Sun S, Zheng C, Shi X (2021) Effect of paternal exposure to microcystin-LR on testicular dysfunction, reproduction, and offspring immune response in the oriental river prawn (*Macrobrachium nipponense*). Aquaculture 534:736332. https://doi.org/10.1016/j.aquaculture.2020.736332

Sun C, Chang X, MacIsaac HJ, Wen J, Zhao L, Dai Z, Li J (2023) Phytosphingosine inhibits cell proliferation by damaging DNA in human cell lines. Ecotox Environ Saf 256:114840. https://doi.org/10.1016/j.ecoenv.2023.114840

Tang XL, Xu MJ, Li ZH, Pan Q, Fu JH (2013) Effects of vitamin E on expressions of eight microRNAs in the liver of Nile tilapia (*Oreochromis niloticus*). Fish Shellfish Immunol 34(6):1470–1475. https://doi.org/10.1016/j.fsi.2013.03.353

Tang H, Hou X, Xue X, Chen R, Zhu X, Huang Y, Chen Y (2017) *Microcystis aeruginosa* strengthens the advantage of *Daphnia similoides* in competition with *Moina micrura*. Sci Rep 7(1):10245. https://doi.org/10.1038/s41598-017-10844-7

Tang X, Fu J, Tan X, Shi Y, Ye J, Guan W, Shi Y, Xu M (2020) The miR-155 regulates cytokines expression by SOSC1 signal pathways of fish in vitro and in vivo. Fish Shellfish Immunol 106:28–35. https://doi.org/10.1016/j.fsi.2020.07.042

Tavasolian F, Inman RD (2021) Gut microbiota-microRNA interactions in ankylosing spondylitis. Autoimmun Rev 20(6):102827. https://doi.org/10.1016/j.autrev.2021.102827

Teng Y, Ren Y, Sayed M, Hu X, Lei C, Kumar A, Hutchins E, Mu J, Deng Z, Luo C, Sundaram K, Sriwastva MK, Zhang L, Hsieh M, Reiman R, Haribabu B, Yan J, Jala VR, Miller DM, Van Keuren-Jensen K, Merchant ML, McClain CJ, Park JW, Egilmez NK, Zhang HG (2018) Plant-derived exosomal microRNAs shape the gut microbiota. Cell Host Microbe 24(5):637–652. e638. https://doi.org/10.1016/j.chom.2018.10.001

Tian JJ, Jin YQ, Yu EM, Sun JH, Xia Y, Zhang K, Li ZF, Gong WB, Wang GJ, Xie J (2022) Intestinal farnesoid X receptor mediates the effect of dietary berberine on lipid accumulation in grass carp (*Ctenopharyngodon idella*). Aquaculture 553:738055. https://doi.org/10.1016/j.aquaculture.2022.738055

Tili E, Michaille J-J, Croce CM (2013) MicroRNAs play a central role in molecular dysfunctions linking inflammation with cancer. Immunol Rev 253(1):167–184. https://doi.org/10.1111/imr.12050

Tong AW, Nemunaitis J (2008) Modulation of miRNA activity in human cancer: a new paradigm for cancer gene therapy? Cancer Gene Ther 15(6):341–355. https://doi.org/10.1038/cgt.2008.8

Toufektchan E, Toledo F (2018) The guardian of the genome revisited: p53 downregulates genes required for telomere maintenance, DNA repair, and centromere structure. Cancer 10(5):135. https://doi.org/10.3390/cancers10050135

Trinchet I, Djediat C, Huet H, Dao SP, Edery M (2011) Pathological modifications following sub-chronic exposure of medaka fish (*Oryzias latipes*) to microcystin-LR. Reprod Toxicol 32(3):329–340. https://doi.org/10.1016/j.reprotox.2011.07.006

Tuo X, Chen J, Zhao S, Xie P (2016) Chemical proteomic analysis of the potential toxicological mechanisms of microcystin-RR in zebrafish (*Danio rerio*) liver. Environ Toxicol 31(10):1206–1216. https://doi.org/10.1002/tox.22128

Umehara A, Takahashi T, Komorita T, Orita R, Choi JW, Takenaka R, Mabuchi R, Park HD, Tsutsumi H (2017) Widespread dispersal and bio-accumulation of toxic microcystins in benthic marine ecosystems. Chemosphere 167:492–500. https://doi.org/10.1016/j.chemosphere.2016.10.029

Urakawa H, Steele JH, Hancock TL, Dahedl EK, Schroeder ER, Sereda JV, Kratz MA, García PE, Armstrong RA (2023) Interaction among spring phytoplankton succession, water discharge patterns, and hydrogen peroxide dynamics in the Caloosahatchee River in Southwest Florida. Harmful Algae 126:102434. https://doi.org/10.1016/j.hal.2023.102434

Uysal Z (2000) Pigments, size and distribution of *Synechococcus* spp. in the Black Sea. J Mar Syst 24(3):313–326. https://doi.org/10.1016/S0924-7963(99)00092-5

Van Gremberghe I, Vanormelingen P, Van der Gucht K, Mancheva A, D'Hondt S, De Meester L, Vyverman W (2009) Influence of *daphnia* infochemicals on functional traits of *Microcystis* strains (*cyanobacteria*). Hydrobiologia 635(1):147–155. https://doi.org/10.1007/s10750-009-9907-5

Vanova T, Raška J, Babica P, Sovadinova I, Kunova Bosakova M, Dvorak P, Bláha L, Rotrekl V (2019) Freshwater cyanotoxin cylindrospermopsin has detrimental stage-specific effects on hepatic differentiation from human embryonic stem cells. Toxicol Sci 168(1):241–251. https://doi.org/10.1093/toxsci/kfy293

Vareli K, Zarali E, Zacharioudakis GSA, Vagenas G, Varelis V, Pilidis G, Briasoulis E, Sainis I (2012) Microcystin producing cyanobacterial communities in Amvrakikos gulf (Mediterranean Sea, NW Greece) and toxin accumulation in mussels (*Mytilus galloprovincialis*). Harmful Algae 15:109–118. https://doi.org/10.1016/j.hal.2011.12.005

Vareli K, Jaeger W, Touka A, Frillingos S, Briasoulis E, Sainis I (2013) Hepatotoxic seafood poisoning (HSP) due to microcystins: a threat from the ocean? Mar Drugs 11(8):2751–2768. https://doi.org/10.3390/md11082751

Vilar MCP, Rodrigues TFCP, Silva LO, Pacheco ABF, Ferrão-Filho AS, Azevedo SMFO (2021) Ecophysiological aspects and *sxt* genes expression underlying induced chemical defense in STX-producing *Raphidiopsis raciborskii* (cyanobacteria) against the zooplankter *Daphnia gessneri*. Toxins 13(6):406. https://doi.org/10.3390/toxins13060406

Vo TMC, Bui BT, Wiegand C, Dinh KV, Dao TS (2020) Responses of a tropical micro-crustacean, *Daphnia lumholtzi*, upon exposures to dissolved toxins and living cells of cyanobacteria. Environ Technol Innov 19:100973. https://doi.org/10.1016/j.eti.2020.100973

Vogt G (2023) Environmental adaptation of genetically uniform organisms with the help of epigenetic mechanisms—an insightful perspective on ecoepigenetics. Epigenomes 7(1):1. https://doi.org/10.3390/epigenomes7010001

von Elert E, Martin-Creuzburg D, Le Coz JR (2003) Absence of sterols constrains carbon transfer between cyanobacteria and a freshwater herbivore (*Daphnia galeata*). Proc R Soc B Biol Sci 270(1520):1209–1214. https://doi.org/10.1098/rspb.2003.2357

Wang PJ, Chien MS, Wu FJ, Chou HN, Lee SJ (2005) Inhibition of embryonic development by microcystin-LR in zebrafish, *Danio rerio*. Toxicon 45(3):303–308. https://doi.org/10.1016/j.toxicon.2004.10.016

Wang J, Wang C, Li Q, Shen M, Bai P, Li J, Lin Y, Gan N, Li T, Zhao J (2019a) Microcystin-LR degradation and gene regulation of microcystin-degrading *Novosphingobium* sp. THN1 at different carbon concentrations. Front Microbiol 10(July):1750. https://doi.org/10.3389/fmicb.2019.01750

Wang Y, Qin S, Li Y, Wu G, Sun Y, Zhang L, Huang Y, Lyu K, Chen Y, Yang Z (2019b) Combined effects of ZnO nanoparticles and toxic *Microcystis* on life-history traits of *Daphnia magna*. Chemosphere 233:482–492. https://doi.org/10.1016/j.chemosphere.2019.05.269

Wang Z, Chen Q, Zhang J, Dong J, Ao Y, Wang M, Wang X (2019c) Long-term exposure to antibiotic mixtures favors microcystin synthesis and release in Microcystis aeruginosa with different morphologies. Chemosphere 235:344–353. https://doi.org/10.1016/j.chemosphere.2019.06.192

Wang L, Chen G, Xiao G, Han L, Wang Q, Hu T (2020a) Cylindrospermopsin induces abnormal vascular development through impairing cytoskeleton and promoting vascular endothelial cell apoptosis by the rho/ROCK signaling pathway. Environ Res 183:109236. https://doi.org/10.1016/j.envres.2020.109236

Wang L, Wang Q, Xiao G, Chen G, Han L, Hu T (2020b) Adverse effect of cylindrospermopsin on embryonic development in zebrafish (*Danio rerio*). Chemosphere 241:125060. https://doi.org/10.1016/j.chemosphere.2019.125060

Wang R, Liu X, Wu J, Liu H, Wang W, Chen X, Yuan L, Wang Y, Du X, Ma Y, Losiewicz MD, Zhang X, Zhang H (2020c) Role of microRNA-122 in microcystin-leucine arginine-induced dysregulation of hepatic iron homeostasis in mice. Environ Toxicol 35(8):822–830. https://doi.org/10.1002/tox.22918

Wang Y, Liu H, Liu X, Zhang X, Wu J, Yuan L, Du X, Wang R, Ma Y, Chen X, Cheng X, Zhuang D, Zhang H (2020d) Histone acetylation plays an important role in MC-LR-induced apoptosis and cycle disorder in SD rat testicular cells. Chemosphere 241:125073. https://doi.org/10.1016/j.chemosphere.2019.125073

Wang C, Yan C, Qiu J, Liu C, Yan Y, Ji Y, Wang G, Chen H, Li Y, Li A (2021a) Food web biomagnification of the neurotoxin β-*N*-methylamino-L-alanine in a diatom-dominated marine ecosystem in China. J Hazard Mater 404:124217. https://doi.org/10.1016/j.jhazmat.2020.124217

Wang W, Jiang M, Sheng Y (2021b) Glyphosate accelerates the proliferation of *Microcystis aeruginosa*, a dominant species in cyanobacterial blooms. Environ Toxicol Chem 40(2):342–351. https://doi.org/10.1002/etc.4942

Wang L, Sagada G, Wang C, Gao C, Wang B, Shao Q, Yan Y (2022a) Berberine in fish nutrition: impact on hepatoenteric health, antioxidative and immune status. Front Mar Sci 9:967748. https://doi.org/10.3389/fmars.2022.967748

Wang R, Liu H, Du X, Ma Y, Tian Z, Zhang S, Shi L, Guo H, Zhang H (2022b) MicroRNA-122 overexpression promotes apoptosis and tumor suppressor gene expression induced by microcystin-leucine arginine in mouse liver. Int J Environ Health Res 32(10):2123–2134. https://doi.org/10.1080/09603123.2021.1946489

Watson SE, Taylor CH, Bell V, Bellamy TR, Hooper AS, Taylor H, Jouault M, Kille P, Perkins RG (2024) Impact of copper sulphate treatment on cyanobacterial blooms and subsequent water quality risks. J Environ Manag 366:121828. https://doi.org/10.1016/j.jenvman.2024.121828

Wawrik B, Paul JH (2004) Phytoplankton community structure and productivity along the axis of the Mississippi River plume in oligotrophic Gulf of Mexico waters. Aquat Microb Ecol 35(2):185–196. https://doi.org/10.3354/ame035185

Wei L, Sun B, Chang M, Liu Y, Nie P (2009) Effects of cyanobacterial toxin microcystin-LR on the transcription levels of immune-related genes in grass carp *Ctenopharyngodon idella*. Environ Biol Fish 85(3):231–238. https://doi.org/10.1007/s10641-009-9488-1

Wei L, Liu Y, Zhong S, Wu H, Ruan J, Liu M, Zhou Q, Zhong Q (2018) Transcriptome analysis of grass carp provides insights into the immune-related genes and pathways in response to MC-LR induction. Aquaculture 488:207–216. https://doi.org/10.1016/j.aquaculture.2018.01.016

Wei L, He L, Fu J, Liu Y, Ruan J, Liu L, Zhong Q (2019) Molecular characterization of caspase-8-like and its expression induced by microcystin-LR in grass carp (*Ctenopharygodon idella*). Fish Shellfish Immunol 89:727–735. https://doi.org/10.1016/j.fsi.2019.04.026

Wei N, Hu C, Dittmann E, Song L, Gan N (2024) The biological functions of microcystins. Water res:122119. https://doi.org/10.1016/j.watres.2024.122119

Weiss JH, Koh J-y, Choi DW (1989) Neurotoxicity of *β*-*N*-methylamino-L-alanine (BMAA) and *β*-*N*-oxalylamino-L-alamine (BOAA) on cultured cortical neurons. Brain Res 497(1):64–71. https://doi.org/10.1016/0006-8993(89)90970-0

Weissberger EJ, Glibert PM (2021) Diet of the eastern oyster, *Crassostrea virginica*, growing in a eutrophic tributary of Chesapeake Bay, Maryland, USA. Aquac Rep 20:100655. https://doi.org/10.1016/j.aqrep.2021.100655

Wejnerowski L, Cerbin S, Dziuba MK (2015) Thicker filaments of *Aphanizomenon gracile* are more harmful to *daphnia* than thinner *Cylindrospermopsis raciborskii*. Zool stud 54 (JAN). https://doi.org/10.1186/s40555-014-0084-5

Welker M, Steinberg C (1999) Indirect photolysis of cyanotoxins: one possible mechanism for their low persistence. Water Res 33(5):1159–1164. https://doi.org/10.1016/S0043-1354(98)00318-2

Welker M, Steinberg C (2000) Rates of humic substance photosensitized degradation of microcystin-LR in natural waters. Environ Sci Technol 34(16):3415–3419. https://doi.org/10.1021/es991274t

Welker M, Jones GJ, Steinberg CEW (2001) Release and persistence of microcystins in natural waters. In: Chorus I (ed) Cyanotoxins—occurrence, effects, controlling factors. Springer, Heidelberg, pp 83–101

Werner VR, Tucci A, da Silva LM, Yunes JS, Neuhaus EB, Berthold DE, Laughinghouse HD IV (2020) Morphological, ecological and toxicological aspects of *Raphidiopsis raciborskii* (cyanobacteria) in a eutrophic urban subtropical lake in southern Brazil. Iheringia, Série Botânica 75:e2020018. https://doi.org/10.21826/2446-82312020v75e2020018

Wiegand C, Pflugmacher S (2005) Ecotoxicological effects of selected cyanobacterial secondary metabolites a short review. Toxicol Appl Pharmacol 203 (3 SPEC. ISS.):201-218. https://doi.org/10.1016/j.taap.2004.11.002

Williams DE, Craig M, Dawe SC, Kent ML, Holmes CFB, Andersen RJ (1997a) Evidence for a covalently bound form of microcystin-LR in salmon liver and dungeness crab larvae. Chem Res Toxicol 10(4):463–469. https://doi.org/10.1021/tx9601519

Williams DE, Dawe SC, Kent ML, Andersen RJ, Craig M, Holmes CFB (1997b) Bioaccumulation and clearance of microcystins from salt water mussels, *Mytilus edulis*, and *in vivo* evidence for covalently bound microcystins in mussel tissues. Toxicon 35(11):1617–1625. https://doi.org/10.1016/S0041-0101(97)00039-1

Wilson AE, Hay ME (2007) A direct test of cyanobacterial chemical defense: variable effects of microcystin-treated food on two *Daphnia pulicaria* clones. Limnol Oceanogr 52(4):1467–1479. https://doi.org/10.4319/lo.2007.52.4.1467

Wilson DS, Sober E (1989) Reviving the superorganism. J Theor Biol 136(3):337–356. https://doi.org/10.1016/S0022-5193(89)80169-9

Wiśniewska K, Śliwińska-Wilczewska S, Savoie M, Lewandowska AU (2022) Quantitative and qualitative variability of airborne cyanobacteria and microalgae and their toxins in the coastal zone of the Baltic Sea. Sci Total Environ 826:154152. https://doi.org/10.1016/j.scitotenv.2022.154152

Won E-J, Kim D, Yoo J-W, In S, Shin K-H, Lee Y-M (2021) Oxidative stress responses in brackish water flea exposed to microcystin-LR and algal bloom waters from Nakdong River. Republic of Korea Mar Pollut Bull 162:111868. https://doi.org/10.1016/j.marpolbul.2020.111868

Wood JD, Franklin RB, Garman G, McIninch S, Porter AJ, Bukaveckas PA (2014) Exposure to the cyanotoxin microcystin arising from interspecific differences in feeding habits among fish and shellfish in the james river estuary. Virginia Environ Sci Technol 48(9):5194–5202. https://doi.org/10.1021/es403491k

Wood SA, Maier MY, Puddick J, Pochon X, Zaiko A, Dietrich DR, Hamilton DP (2017) Trophic state and geographic gradients influence planktonic cyanobacterial diversity and distribution in New Zealand lakes. FEMS Microbiol Ecol 93(2). https://doi.org/10.1093/femsec/fiw234

Worden AZ, Wilken S (2016) A plankton bloom shifts as the ocean warms. Science 354(6310):287–288. https://doi.org/10.1126/science.aaj1751

Wu H, Jiang M, Zhu ZG, Zhao Y, Qu R, Liu JJ (2013) Toxic response of *Daphnia magna* to *Microcystis aeruginosa*. Asian J Chem 25(6):2965–2969. https://doi.org/10.14233/ajchem.2013.13452

Wu Q, Yan W, Liu C, Li L, Yu L, Zhao S, Li G (2016) Microcystin-LR exposure induces developmental neurotoxicity in zebrafish embryo. Environ Pollut 213:793–800. https://doi.org/10.1016/j.envpol.2016.03.048

Wu Q, Yan W, Cheng H, Liu C, Hung TC, Guo X, Li G (2017) Parental transfer of microcystin-LR induced transgenerational effects of developmental neurotoxicity in zebrafish offspring. Environ Pollut 231:471–478. https://doi.org/10.1016/j.envpol.2017.08.038

Wu X, Wu H, Gu X, Zhang R, Ye J, Sheng Q (2019a) Biomagnification characteristics and health risk assessment of the neurotoxin BMAA in freshwater aquaculture products of Taihu Lake Basin, China. Chemosphere 229:332–340. https://doi.org/10.1016/j.chemosphere.2019.04.210

Wu YL, He Y, Shi JJ, Zheng TX, Lin XJ, Lin X (2019b) Microcystin-LR promotes necroptosis in primary mouse hepatocytes by overproducing reactive oxygen species. Toxicol Appl Pharmacol 377:114626. https://doi.org/10.1016/j.taap.2019.114626

Xiao L, Cui T, Liu S, Chen B, Wang Y, Yang T, Li T, Chen J (2019) Vitamin a supplementation improves the intestinal mucosal barrier and facilitates the expression of tight junction proteins in rats with diarrhea. Nutrition 57:97–108. https://doi.org/10.1016/j.nut.2018.06.007

Xie L, Yan W, Li J, Yu L, Wang J, Li G, Chen N, Steinman AD (2015) Microcystin-RR exposure results in growth impairment by disrupting thyroid endocrine in zebrafish larvae. Aquat Toxicol 164:16–22. https://doi.org/10.1016/j.aquatox.2015.04.014

Xing ML, Wang XF, Xu LH (2008) Alteration of proteins expression in apoptotic FL cells induced by MCLR. Environ Toxicol 23(4):451–458. https://doi.org/10.1002/tox.20355

Xu L, Qin W, Zhang H, Wang Y, Dou H, Yu D, Ding Y, Yang L, Wang Y (2012) Alterations in microRNA expression linked to microcystin-LR-induced tumorigenicity in human WRL-68 cells. Mutat Res Genet Toxicol Environ Mutagen 743(1):75–82. https://doi.org/10.1016/j.mrgentox.2011.12.024

Xu L, Li T, Ding W, Cao Y, Ge X, Wang Y (2018a) Combined seven miRNAs for early hepatocellular carcinoma detection with chronic low-dose exposure to microcystin-LR in mice. Sci Total Environ 628-629:271–281. https://doi.org/10.1016/j.scitotenv.2018.02.021

Xu X, Zhang Y-N, Peng S, Wu J, Deng D, Zhou Z (2018b) Effects of *Microcystis aeruginosa* on the expression of nuclear receptor genes in *Daphnia similoides sinensis*. Ecotox Environ Saf 150:344–352. https://doi.org/10.1016/j.ecoenv.2017.12.033

Xu R, Jiang Y, MacIsaac HJ, Chen L, Li J, Xu J, Wang T, Zi Y, Chang X (2019) Blooming cyanobacteria alter water flea reproduction via exudates of estrogen analogues. Sci Total Environ 696:133909. https://doi.org/10.1016/j.scitotenv.2019.133909

Xu H, Ding Y, Qiu X, Shimasaki Y, Oshima Y (2020) Quantitative proteomic analysis of Nile tilapia brain following sub-chronically exposed to toxic cyanobacteria *Microcystis aeruginosa*. J Fac Agr, Kyushu Univ 65(2):283–288. https://doi.org/10.5109/4103892

Xu S, Jiang Y, Liu Y, Zhang J (2021a) Antibiotic-accelerated cyanobacterial growth and aquatic community succession towards the formation of cyanobacterial bloom in eutrophic lake water. Environ Pollut 290:118057. https://doi.org/10.1016/j.envpol.2021.118057

Xu S, Liu Y, Zhang J, Gao B (2021b) Proteomic mechanisms for the combined stimulatory effects of glyphosate and antibiotic contaminants on *Microcystis aeruginosa*. Chemosphere 267:129244. https://doi.org/10.1016/j.chemosphere.2020.129244

Xu X, Cao Y, Qi H, Deng D, Zhang YN, Wu J, Peng S, Zhou Z (2022) Effects of toxic *Microcystis aeruginosa* on the expression of Hox genes in *Daphnia similoides sinensis*. Ecol Evol 12(3):e8685. https://doi.org/10.1002/ece3.8685

Xu W, Hu X, Li H, Tian X, Ouyang Z, Du Y, Chen J (2023) Effects of *lactobacillus plantarum* ep-M17 on growth, immunity and intestinal microbiota of *Penaeus vannamei* under microcystin-LR stress. Aquat Toxicol 265:106763. https://doi.org/10.1016/j.aquatox.2023.106763

Yan W, Li L, Li G, Zhao S (2017) Microcystin-LR induces changes in the GABA neurotransmitter system of zebrafish. Aquat Toxicol 188:170–176. https://doi.org/10.1016/j.aquatox.2017.05.006

Yáñez JM, Barría A, López ME, Moen T, Garcia BF, Yoshida GM, Xu P (2023) Genome-wide association and genomic selection in aquaculture. Rev Aquac 15(2):645–675. https://doi.org/10.1111/raq.12750

Yang Z, Lü K, Chen Y, Montagnes DJS (2012) The interactive effects of ammonia and microcystin on life-history traits of the cladoceran *Daphnia magna*: synergistic or antagonistic? PLoS One 7(3):e32285. https://doi.org/10.1371/journal.pone.0032285

Yang F, Zhou Y, Sun R, Wei H, Li Y, Yin L, Pu Y (2014) Biodegradation of microcystin-LR and-RR by a novel microcystin-degrading bacterium isolated from Lake Taihu. Biodegradation 25(3):447–457. https://doi.org/10.1007/s10532-013-9673-y

Yang S, Chen L, Wen C, Zhang X, Feng X, Yang F (2018) MicroRNA expression profiling involved in MC-LR-induced hepatotoxicity using high-throughput sequencing analysis. J Toxicol Environ Health A 81(5):89–97. https://doi.org/10.1080/15287394.2017.1415580

Yang F, Huang F, Feng H, Wei J, Massey IY, Liang G, Zhang F, Yin L, Kacew S, Zhang X, Pu Y (2020) A complete route for biodegradation of potentially carcinogenic cyanotoxin microcystin-LR in a novel indigenous bacterium. Water Res 174:115638. https://doi.org/10.1016/j.watres.2020.115638

Yang Y, Yu G, Chen Y, Jia N, Li R (2021) Four decades of progress in cylindrospermopsin research: the ins and outs of a potent cyanotoxin. J Hazard Mater 406:124653. https://doi.org/10.1016/j.jhazmat.2020.124653

Yang W, Ye Y, Lu K, Zheng Z, Zhu J (2022) NMR-based metabolomic responses of freshwater gastropod *Bellamya aeruginosa* to MC-producing and non MC-producing *Microcystis aeruginosa*. J Oceanol Limnol 40(1):260–272. https://doi.org/10.1007/s00343-021-0336-2

Ye J, Guan Y, Wu L, Wang C, Chen J, Zhou S, Xu C (2020) Effects of glyphosate on microcystin-LR production and release from *Microcystis aeruginosa* at different temperatures. Environ Sci Pollut Res. https://doi.org/10.1007/s11356-020-10185-5

Yoon D, Choi H, Noh H, Ok YJ, Oh S, Ma S, Yoon C, Kim S (2017) Toxicological assessment of microcystin-LR to zebrafish (*Danio rerio*) using metabolomics. Bull Korean Chem Soc 38(4):459–465. https://doi.org/10.1002/bkcs.11112

Yu Y, Zhang Q, Liu G, Deng Y, Kang J, Zhang F, Lu T, Sun L, Qian H (2021) Proteomic analysis of zebrafish brain damage induced by *Microcystis aeruginosa* bloom. Sci Total Environ 795:148865. https://doi.org/10.1016/j.scitotenv.2021.148865

Yuan L, Liu H, Liu X, Zhang X, Wu J, Wang Y, Du X, Wang R, Ma Y, Chen X, Petlulu P, Cheng X, Zhuang D, Guo H, Zhang H (2020) Epigenetic modification of H3K4 and oxidative stress are involved in MC-LR-induced apoptosis in testicular cells of SD rats. Environ Toxicol 35(2):277–291. https://doi.org/10.1002/tox.22865

Žegura B, Zajc I, Lah TT, Filipič M (2008) Patterns of microcystin-LR induced alteration of the expression of genes involved in response to DNA damage and apoptosis. Toxicon 51(4):615–623. https://doi.org/10.1016/j.toxicon.2007.11.009

Žegura B, Štraser A, Filipič M (2011) Genotoxicity and potential carcinogenicity of cyanobacterial toxins—a review. Mutat Res/Rev Mutat Res 727(1):16–41. https://doi.org/10.1016/j.mrrev.2011.01.002

Zeng Y-H, Cai Z-H, Zhu J-M, Du X-P, Zhou J (2020) Two hierarchical LuxR-LuxI type quorum sensing systems in *Novosphingobium* activate microcystin degradation through transcriptional regulation of the *mlr* pathway. Water Res 183:116092. https://doi.org/10.1016/j.watres.2020.116092

Zguna N, Karlson AML, Ilag LL, Garbaras A, Gorokhova E (2019) Insufficient evidence for BMAA transfer in the pelagic and benthic food webs in the Baltic Sea. Sci Rep 9(1):10406. https://doi.org/10.1038/s41598-019-46815-3

Zhan C, Zhang F, Liu W, Zhang X (2020) Microcystin-LR promotes zebrafish (*Danio rerio*) oocyte (*in vivo*) maturation by activating ERK1/2-MPF signaling pathways, and cAMP is involved in this process. Environ Pollut 259:113843. https://doi.org/10.1016/j.envpol.2019.113843

Zhang H, Jiang X (2022a) Forward and reverse evolution of multivariate responses to cyanobacteria in experimental waterflea populations. Limnol Oceanogr 67(11):2418–2430. https://doi.org/10.1002/lno.12210

Zhang H, Jiang X (2022b) Resurrection of dormant zooplankton grazers reveals multiple evolutionary responses to toxic cyanobacteria. Limnol Oceanogr 67(9):2000–2011. https://doi.org/10.1002/lno.12183

Zhang L, Lyu K, Wang N, Gu L, Sun Y, Zhu X, Wang J, Huang Y, Yang Z (2018) Transcriptomic analysis reveals the pathways associated with resisting and degrading microcystin in *Ochromonas*. Environ Sci Technol 52(19):11102–11113. https://doi.org/10.1021/acs.est.8b03106

Zhang Y, Shi Q, Wei W, Xu F, Nie F, Yang H (2019a) Effects of microcystin-LR on the immune dysfunction and ultrastructure of hepatopancreas in giant freshwater prawn *Macrobrachium rosenbergii*. Fish Shellfish Immunol 89:586–594. https://doi.org/10.1016/j.fsi.2019.04.039

Zhang Y, Zhuang H, Yang H, Xue W, Wang L, Wei W (2019b) Microcystin-LR disturbs testicular development of giant freshwater prawn *Macrobrachium rosenbergii*. Chemosphere 222:584–592. https://doi.org/10.1016/j.chemosphere.2019.01.146

Zhang D, Lin W, Liu Y, Guo H, Wang L, Yang L, Li L, Li D, Tang R (2020a) Chronic microcystin-LR exposure induces abnormal lipid metabolism via endoplasmic reticulum stress in male zebrafish. Toxins 12(2):107. https://doi.org/10.3390/toxins12020107

Zhang X, Yang F, Chen L, Feng H, Yin S, Chen M (2020b) Insights into ecological roles and potential evolution of Mlr-dependent microcystin-degrading bacteria. Sci Total Environ 710:136401. https://doi.org/10.1016/j.scitotenv.2019.136401

Zhang S, Liu H, Du X, Chen X, Petlulu P, Tian Z, Shi L, Zhang B, Yuan S, Guo X, Wang Y, Guo H, Zhang H (2022a) A new identity of microcystins: environmental endocrine disruptors? An evidence-based review. Sci Total Environ 851:158262. https://doi.org/10.1016/j.scitotenv.2022.158262

Zhang Y, Gao J, Nie Z, Zhu H, Du J, Cao L, Shao N, Sun Y, Su S, Xu G, Xu P (2022b) Microcystin-LR induces apoptosis in juvenile *Eriocheir sinensis* via the mitochondrial pathway. Ecotox Environ Saf 238:113528. https://doi.org/10.1016/j.ecoenv.2022.113528

Zhang Y, Gao Q, Liu S-s, Tang L, Li X-G, Sun H (2022c) Hormetic dose-response of halogenated organic pollutants on *Microcystis aeruginosa*: joint toxic action and mechanism. Sci Total Environ 829:154581. https://doi.org/10.1016/j.scitotenv.2022.154581

Zhang D, Huang L, Jia Y, Zhang S, Bi X, Dai W (2023a) Integrated analysis of mRNA and microRNA expression profiles in hepatopancreas of *Litopenaeus vannamei* under acute exposure to MC-LR. Front Gen 14:1088191. https://doi.org/10.3389/fgene.2023.1088191

Zhang H, Chen D, Ji Q, Yang M, Ding R (2023b) miR-146a-5p promotes the inflammatory response in PBMCs induced by microcystin-leucine-arginine. J Inflamm Res 16:1979–1993. https://doi.org/10.2147/jir.s403945

Zhang Y, Gao J, Cao L, Du J, Xu G, Xu P (2023c) Microcystin-LR-induced autophagy via miR-282–5p/PIK3R1 pathway in *Eriocheir sinensis* hepatopancreas. Ecotox Environ Saf 267:115661. https://doi.org/10.1016/j.ecoenv.2023.115661

Zhang Y, Li Z, Tian X, Xu P, Sun K, Ren N (2023d) Acute toxic effects of microcystin-LR on crayfish (*Procambarus clarkii*): insights from antioxidant system, histopathology and intestinal flora. Environ Sci Pollut Res. https://doi.org/10.1007/s11356-023-26171-6

Zhang L, Wu Y, Lei P, Xie WM, Liu JE, Meng H, Li S, He H, Wang G, Zhang L (2024) Dual role of Cu^{2+} in the physiological growth and metabolism of *Microcystis aeruginosa*. ACS EST Water 4(10):4636–4645. https://doi.org/10.1021/acsestwater.4c00676

Zhao Y, Xiong Q, Xie P (2011) Analysis of microRNA expression in embryonic developmental toxicity induced by MC-RR. PLoS One 6(7):e22676. https://doi.org/10.1371/journal.pone.0022676

Zhao S, Xie P, Li G, Jun C, Cai Y, Xiong Q, Zhao Y (2012a) The proteomic study on cellular responses of the testes of zebrafish (*Danio rerio*) exposed to microcystin-RR. Proteomics 12(2):300–312. https://doi.org/10.1002/pmic.201100214

Zhao Y, Xie P, Fan H (2012b) Genomic profiling of MicroRNAs and proteomics reveals an early molecular alteration associated with tumorigenesis induced by MC-LR in mice. Environ Sci Technol 46(1):34–41. https://doi.org/10.1021/es201514h

Zhao Y, Xie P, Fan H (2012c) Genomic profiling of microRNAs and proteomics reveals an early molecular alteration associated with tumorigenesis induced by MC-LR in mice. Environ Sci Technol 46(1):34–41. https://doi.org/10.1021/es201514h

Zhao Y, Xie L, Yan Y (2015) Microcystin-LR impairs zebrafish reproduction by affecting oogenesis and endocrine system. Chemosphere 120:115–122. https://doi.org/10.1016/j.chemosphere.2014.06.028

Zhao S, Xu J, Zhang W, Yan W, Li G (2021) Paternal exposure to microcystin-LR triggers developmental neurotoxicity in zebrafish offspring via an epigenetic mechanism involving MAPK pathway. Sci Total Environ 792:148437. https://doi.org/10.1016/j.scitotenv.2021.148437

Zhao S, Chang X, Li J, Zhu Y, Pan X, Hua Z, Li J (2024) The two-way immunotoxicity in native fish induced by exudates of *Microcystis aeruginosa*: Immunostimulation and immunosuppression. J Hazard Mater 461:132554. https://doi.org/10.1016/j.jhazmat.2023.132554

Zhou Q, Lu N, Gu L, Sun Y, Zhang L, Huang Y, Chen Y, Yang Z (2020) *Daphnia* enhances relative reproductive allocation in response to toxic *Microcystis*: changes in the performance of parthenogenetic and sexual reproduction. Environ Pollut 259:113890. https://doi.org/10.1016/j.envpol.2019.113890

Zhu J, Lu K, Sun S, Zhang K (2013) Foraging inhibition in two cladocerans feeding on *Microcystis aeruginosa*. Clean (Weinh) 41(7):645–650. https://doi.org/10.1002/clen.201100643

Zhu X, Wang Q, Zhang L, Liu J, Zhu C, Yang Z (2015) Offspring performance of *Daphnia magna* after short-term maternal exposure to mixtures of microcystin and ammonia. Environ Sci Pollut Res 22(4):2800–2807. https://doi.org/10.1007/s11356-014-3520-5

Zi J, Pan X, MacIsaac HJ, Yang J, Xu R, Chen S, Chang X (2018) Cyanobacteria blooms induce embryonic heart failure in an endangered fish species. Aquat Toxicol 194:78–85. https://doi.org/10.1016/j.aquatox.2017.11.007

Zimba PV, Camus A, Allen EH, Burkholder JM (2006) Co-occurrence of white shrimp, *Litopenaeus vannamei*, mortalities and microcystin toxin in a southeastern USA shrimp facility. Aquaculture 261(3):1048–1055. https://doi.org/10.1016/j.aquaculture.2006.08.037

Marine Phycotoxins in Action

3

3.1 Introduction

The ocean, a vast and seemingly endless expanse of life, harbors organisms both essential and perilous. Among them are marine algae—microscopic architects of aquatic ecosystems, oxygen producers, and, at times, silent harbingers of toxicity. While most marine microalgae support the foundation of marine food webs, a select few wield biochemical weaponry capable of poisoning entire ecosystems and disrupting fisheries. However, the impact of these toxic blooms extends beyond the marine environment. Human exposure occurs through the consumption of contaminated shellfish and finfish, as well as through direct contact with toxin-laden water or airborne aerosols. The effects range from mild gastrointestinal distress to severe neurological impairment, and in extreme cases, even death.

Of the approximately 5000 identified species of marine microalgae, fewer than 2% are currently recognized as harmful or toxic. However, this percentage is rising, and the consequences are becoming increasingly evident. The frequency and global distribution of toxic algal incidents have surged in recent decades, increasing reports of human intoxications from previously undocumented algal sources. This unsettling trend coincides with mounting evidence of large-scale ecological disturbances driven by eutrophication, pollution, and global warming—factors that create ideal conditions for harmful algal blooms (HABs) to flourish. Consequently, the extent to which human activities have contributed to this increase is a significant question (Hallegraeff 1993; Dolah 2000; Landsberg 2002). As the ocean changes, so too does the behavior of these toxic algae, posing a challenge to researchers in keeping pace with emerging threats.

This chapter delves into the complex world of toxic marine algae, exploring their ecological roles, the biochemical nature of their toxins, and the growing risks they pose in a rapidly changing world. Understanding these dynamics is crucial, not only for safeguarding public health and marine biodiversity but also for predicting and mitigating future toxic outbreaks.

T. Lieke, C. E. W. Steinberg, *The Ecology of Phycotoxins*,
https://doi.org/10.1007/978-3-032-18380-4_3

3.2 Mechanistic Overview

Three major groups of marine biotoxins are particularly significant (Leal and Cristiano 2022): paralytic shellfish poisoning toxins (PSP toxins, including saxitoxin and its derivatives), diarrheic shellfish poisoning toxins (DSP toxins, including okadaic acid), and amnesic shellfish poisoning toxins (ASP toxins, including domoic acid, DA). The chemistry of PSP toxins is discussed in Botana and López (2015).

Between 1970 and 2019, PSP and DSP were the most commonly reported syndromes worldwide, accounting for 46% and 42%, respectively, of all events associated with harmful algal blooms (Mardones et al. 2020). Beasley et al. (2023) provide a comprehensive overview of phycotoxins, with an emphasis on their impact on human toxicology.

Marine toxins are highly potent substances that exert rapid effects even at minute concentrations. Bian et al. (2024)[1] provide a comparative overview (Fig. 3.1). Most of these toxins exhibit remarkable selectivity for important physiological targets, including ion channels, enzymes, pumps, and receptors on cell membranes. Voltage-gated sodium channels (VGSCs) play a crucial role in the generation and transmission of action potentials in excitable cells such as neurons and muscle cells (Xu et al. 2019). The toxins that primarily target VGSCs include paralytic shellfish poisoning toxins (PSP toxins), tetrodotoxin (TTX), ciguatoxin (CTX), and brevetoxin (BTX). Among these, CTXs and BTXs specifically activate site 5 of VGSCs, whereas saxitoxin (STX) and TTX bind to site 1 to inhibit ion conduction (Fig. 3.1). The symptoms observed in vivo are consistent with the effects of these toxins that target VGSCs. Neuromuscular complications, resulting from the inhibition of Na^+ influx, are characteristic of PSP, posing a risk to respiration due to muscle unresponsiveness. TTXs exert their toxic effects through a shared mechanism of action; hence, poisoning manifests similar symptoms (Katikou et al. 2022).

In contrast to PSP toxins and TTXs, CTXs and BTXs specifically bind to site 5 of VGSCs (Fig. 3.1). This binding induces an active conformation of the channel, promoting an increased influx of Na^+ ions into cells. The resulting repetitive action potential firing disrupts normal ion homeostasis. Beyond sodium channel activation, CTXs have also been demonstrated to partially inhibit the voltage-gated potassium channels, further amplifying membrane excitability (Hidalgo et al. 2002).

Under normal conditions, the Na^+/K^+-ATPase maintains cellular ion homeostasis by actively transporting Na^+ and K^+ ions against their respective gradients, thus ensuring proper ion distribution between the intracellular and extracellular environments. Palytoxin (PLTXs) target the Na^+/K^+-ATPase pump, transforming it into a non-selective ion channel. This structural modification leads to a continuously open state, allowing the uncontrolled flow of cations along their concentration gradients. The disruption of this tightly regulated mechanism by PLTX can lead to severe cellular dysfunction and tissue damage (Fig. 3.1) (Rossini and Bigiani 2011; Farabegoli et al. 2018).

[1] Credit to Elsevier, excerpt appropriately referenced.

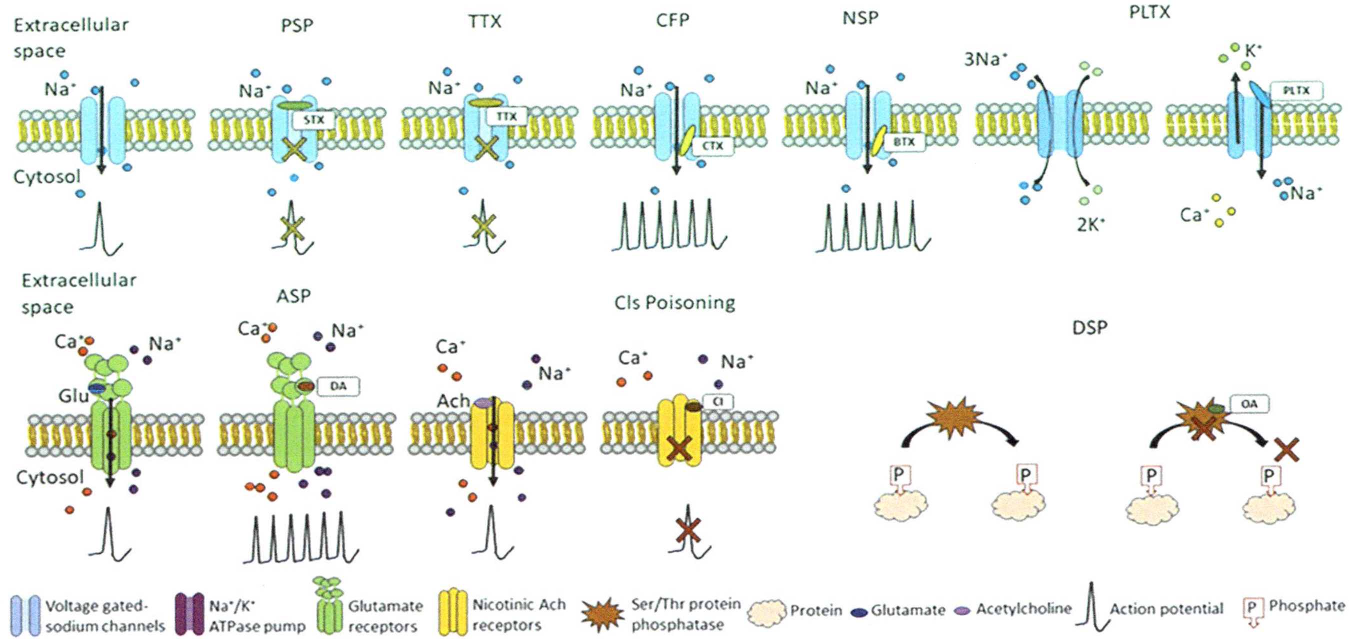

Fig. 3.1 Action mechanism of marine toxins (from Bian et al. (2024), credit to Elsevier). *ASP* amnesic shellfish poisoning, *BTX* brevetoxin, *CFP* ciguatera fish poisoning, *CIs* cyclic imines, *CTX* ciguatoxin, *DA* domoic acid, *DSP* diarrhetic shellfish poisoning, *NSP* neurotoxic shellfish poisoning, *OA* okadaic acid, *PLTX* palytoxins, *PSP* paralytic shellfish poisoning, *STX* saxitoxin, *TTX* tetrodotoxin

Ionotropic glutamate receptors (iGluRs) are key mediators of excitatory neurotransmission in the central nervous system and serve as primary targets for several naturally occurring marine and fungal toxins. Among these, DA, a potent neurotoxin produced by certain diatom species (e.g., *Pseudo-nitzschia*), is a well-characterized agonist of iGluRs. By binding to these receptors—particularly the AMPA and kainate subtypes—DA induces prolonged receptor activation, maintaining the channels in an open state. This facilitates an excessive influx of Ca^{2+} and Na^{+} ions into neurons, inducing sustained depolarization, neuronal excitability, and ultimately excitotoxicity (Fig. 3.1) (Pope et al. 2024).

Cyclic imines (CIs), such as spirolides, gymnodimines, and pinnatoxins, are a class of marine toxins known for their ability to inhibit nicotinic acetylcholine receptors (nAChRs). They act as reversible antagonists, blocking ion flow through these channels and thereby disrupting acetylcholine-mediated neurotransmission at the neuromuscular junction (Araoz et al. 2011). This interference impairs muscle contraction initiated by motor neurons. Following oral exposure, affected animals may exhibit a range of neurological and muscular symptoms, including tremors, reduced mobility, hind limb paralysis, difficulty jumping, and respiratory distress—symptoms that can ultimately lead to death in severe cases (Delcourt et al. 2019; Sosa et al. 2020). In contrast, okadaic acid (OA), another prominent marine toxin, exerts its toxic effects by inhibiting serine/threonine protein phosphatases, particularly PP1 and PP2A. This inhibition leads to the hyperphosphorylation of numerous cellular proteins, thereby disrupting essential cellular processes, including signal transduction, cytoskeletal dynamics, and cell cycle regulation. The resulting cellular dysfunction has been linked to gastrointestinal toxicity, tumor promotion, and various other disease states.

Complementing the human perspective, our chapter focuses on recent studies of the effects of marine algal toxins on aquatic, often farmed, animals, the underlying mechanistic pathways, and potential remedies.

Toxic microalgae and cyanobacteria may bloom in eutrophicated waters, both marine and freshwater, introducing their various phycotoxins into the food webs as they are ingested during feeding. Only certain copepods, cladocerans, rotifers, and some bivalves can feed selectively and avoid ingesting toxic algal cells (Vanderploeg et al. 2001; Kubanek et al. 2007; Tang et al. 2014; Turner 2014; Ray et al. 2016). For example, the brine shrimp *Artemia* does not discriminate between toxic and non-toxic algae, suffering severely from this lack of selectivity, as shown by Neves et al. (2017) (Fig. 3.2). Redundancy analysis applied to survival and behavioral data of *Artemia salina* clearly distinguishes toxic treatments (negative coordinates on the first axis) from nontoxic ones (positive coordinates on the first axis). The positive direction of the vectors for swimming activity, water column position, and survival confirms their positive correlation with the samples from the non-toxic treatment (*Tetraselmis* sp). Brine shrimps exposed to toxic dinoflagellates exhibit reduced swimming activity, ultimately leading to immobility. Swimming behavior has a direct impact on zooplankton dispersal, prey and predator encounters, and predator susceptibility. These traits influence the grazing behavior of zooplankton species (excluding *Artemia*) on harmful algal species and their role in transferring

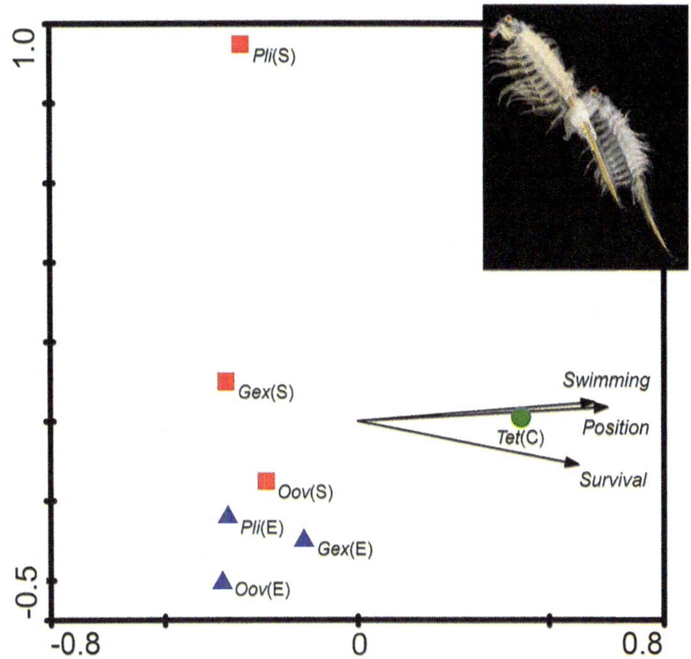

Fig. 3.2 Redundancy analysis of survival and behavioral responses of *Artemia salina* exposed to toxic and non-toxic treatments. Biplot of survival and behavioral data of *A. salina* exposed to the non-toxic chlorophyte *Tetraselmis* sp. (Tet(C), ●) and toxic dinoflagellates: *Prorocentrum lima* in exponential phase (Pli(E), ▲), *P. lima* in stationary phase (Pli(S), ■), *Ostreopsis* cf. *ovata* in exponential phase (Oov(E), ▲), *O.* cf. *ovata* in stationary phase (Oov(S), ■), *Gambierdiscus excentricus* in exponential phase (Gex(E), ▲), and *G. excentricus* in stationary phase (Gex(S), ■) (from Neves et al. (2017), credit to the Public Library of Science; image credit to Hans Hillewaert, Wikimedia)

phycotoxins through the food web. Toxic benthic dinoflagellates have been shown to significantly reduce the survival rates of *Artemia salina* during acute exposure. When exposed to the exponential growth phase of the algae, brine shrimp experienced 100% mortality across all three tested dinoflagellate species.

In contrast, exposure to dinoflagellates in the stationary phase resulted in species-specific mortality rates: 100% for *Prorocentrum lima*, 91% for *Ostreopsis* cf. *ovata*, and 75% for *Gambierdiscus excentricus*. Notably, the cell-free medium did not affect the survival of *A. salina* during acute exposure. Overall, adverse effects related to the toxicity of benthic dinoflagellates are expected to affect their non-selective primary grazers, making them more vulnerable to predation and vectors of toxins through marine food webs.

In addition to the reduced swimming speed commonly observed in zooplankton after ingesting toxic algae—thereby increasing their risk of predation—Lasley-Rasher et al. (2016) demonstrated that the swimming behavior of the copepod *Temora longicornis* is also negatively affected by exposure to the toxic dinoflagellate *Alexandrium catenella* (formerly *A. fundyense*), which is easily ingested. Interestingly, *A. catenella* exposure triggered an increase in swimming speed and trajectory linearity in *T. longicornis*, effects that persisted even after the copepods were removed from the toxic environment. The authors estimated that these behavioral changes could result in a 25–56% increase in encounter rates between copepods and their predators, thereby heightening predation risk despite no immediate mortality.

Typically, sublethal exposure to toxic food algae reduces the fecundity of herbivores. For example, Kozlowsky-Suzuki et al. (2009) compared the oocyte development in the marine copepods *Acartia tonsa* and *T. longicornis*. Toxic diets adversely affected reproduction in both copepods, but through different mechanisms. In particular, the egg production of *A. tonsa* follows the order *Rhodomonas* spp. > *Alexandrium minutum* > *Prorocentrum lima*, *Prymnesium parvum*, and *Nodularia spumigena*, whereas the oocyte production of *T. longicornis* is significantly higher with *Rhodomonas* sp. than with any of the other algae. Oocyte production and development are affected by *A. minutum* and *N. spumigena*. While the adverse effect of *N. spumigena* seems to be related to gonadal atresia, likely caused by severe food restriction (starvation), the adverse impact of *A. minutum* is caused either by a direct effect of saxitoxins or by nutritional deficiencies in oocyte production. The survival of *A. tonsa* ranges from 82 ± 24% in *Rhodomonas* sp. to 58 ± 23% in *P. lima* and does not differ significantly between the food species. In contrast, significant differences in the survival of *T. longicornis* are observed between diets, with a higher survival on *Rhodomonas* sp. and *A. minutum* (>80%) than on *P. parvum* (56 ± 7%) and *P. lima* (35 ± 9%) (Fig. 3.3).

Conversely, the control of toxic phytoplankton by non-selective grazers has been reported as well, as noted by Kozlowsky-Suzuki et al. (2006) for several zooplankton species. The selectivity on *Dinophysis acuta* in natural plankton assemblages progressively decreased as food availability increased for *T. longicornis* and *C. typicus*. A similar response was not observed for *A. bifilosa*, which displayed an even less selective behavior. Grazing by specific copepods can, therefore, potentially contribute as a significant loss factor for *D. acuta* if it is abundant at the onset of blooms.

Currently, around 200 species across diverse groups—including dinoflagellates, diatoms, raphidophytes, prymnesiophytes, silicoflagellates, ciliates, and cyanobacteria—are recognized, suspected, or considered capable of producing toxins harmful to a wide range of aquatic and terrestrial organisms. This number continues to rise as new toxin-producing species and previously uncharacterized toxic compounds are identified (Vale et al. 2009; Ruocco et al. 2020; Long et al. 2021; Ternon et al. 2022). Many of these toxins, produced during harmful algal blooms (HABs), represent naturally occurring but emerging environmental contaminants with the capacity to cause severe poisoning and mortality in humans and wildlife, primarily through the consumption of contaminated seafood (Manivasagan and Kim 2015). The global frequency and severity of HABs have increased markedly over the past four decades. This trend has been driven by multiple anthropogenic and environmental factors, including nutrient enrichment (eutrophication), rising sea surface

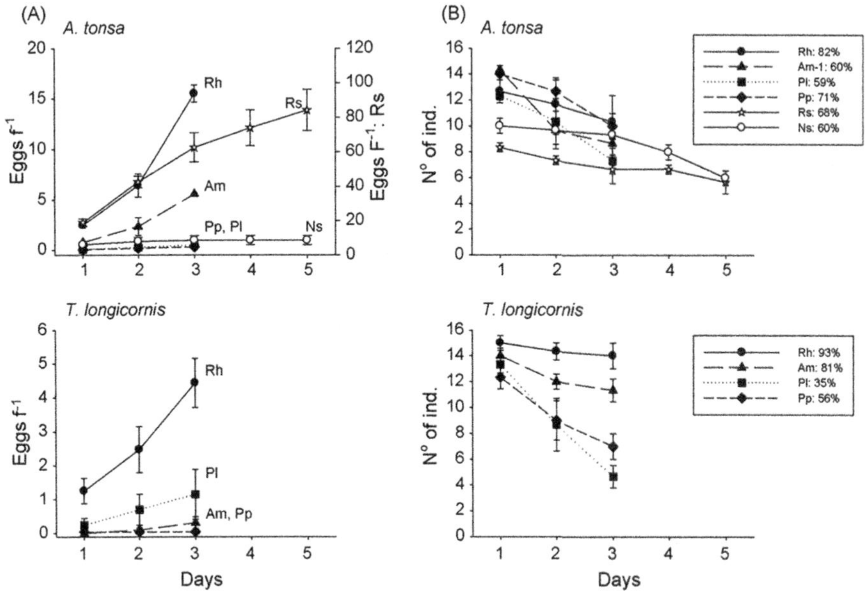

Fig. 3.3 (**a**) Cumulative egg production (eggs f^{-1}) and (**b**) survival (%) of *Acartia tonsa* and *Temora longicornis* during the 3–5 day experiments (mean ± S.E.). Note the different scale of the *A. tonsa* egg production on *R. salina* on the secondary y-axis. Survival rates are indicated in the figure. **Rh** *Rhodomonas* sp., **Rs** *R. salina*, **Am** *Alexandrium minutum*, **At** *A. tamarense*, **Pl** *Prorocentrum lima*, **Pp** *Prymnesium parvum*, **Ns** *Nodularia spumigena* (from Kozlowsky-Suzuki et al. (2009), with permission from Elsevier)

temperatures linked to global warming, the spread of invasive or non-native species through international shipping routes, climate-induced range shifts of native species, and changes in ocean circulation patterns (McGowan 2016). Despite growing attention to their public health and ecological impacts, the evolutionary and ecological roles of algal toxins remain poorly understood and continue to be a subject of scientific debate.

The toxins produced by eukaryotic microalgae can be classified into several major groups based on their structural characteristics. The most prominent of these are as follows:

(a) *Linear and macrocyclic polyethers*, which include compounds such as okadaic acid, the closely related dinophysistoxins, karlotoxins, and prymnesins
(b) *Ladder-frame polyethers*, which include compounds such as brevetoxins and ciguatoxins. Furthermore
(c) *Macrocyclic imines*, e.g., spirolides and gymnodimine
(d) *Tetrahydropurines*, e.g., saxitoxin and its analogs
(e) *Toxic secondary amines*, including domoic acid (Anderson et al. 2012)

However, these authors, as well as Rasmussen et al. (2016a), have raised serious concerns about the validity of many of the so-called ichthyotoxins. Despite numerous reports of algal blooms causing massive fish kills on a global scale, only a limited number of compounds have been identified as true ichthyotoxins. These include the brevetoxin produced by *Karenia brevis*, karlotoxin by *Karlodinium veneficum*, gymnocin from certain strains of *Karenia mikimotoi*, and brevisulcatic acid from *Karenia brevisulcata*.

Specifically, polyether toxins described as prymnesins with hemolytic and cytolytic activities have been isolated from the fish-killing haptophyte *Prymnesium parvum* (Parnas 1963; Ulitzur 1973; Gordon and Colorni 2008). However, these compounds have not been definitively linked to either fish kills or allelochemical interactions. Some researchers claim that reactive oxygen species (ROS) are the primary cause of fish kills by algae such as *Chattonella*[2] *marina*, but this has not yet been substantiated by fish bioassays with chemically generated superoxide or hydrogen peroxide. Furthermore, it is questionable whether this will ever be confirmed, as oxygen activation is the primordial response to all abiotic and biotic "stressors," including food digestion—not just potential toxins (Steinberg 2012). Other research groups have investigated polyunsaturated fatty acids (PUFAs) as potential toxic agents, but the results have been highly variable. Some of this variability is now thought to result from the presence of previously uncharacterized lipid degradation products, which may have confounded earlier findings (Anderson et al. (2012) and references therein). Indeed, more recent studies have questioned this mechanism as the only possible explanation (Li et al. 2018).

The following sections delve into the effects of representative compounds from each of the major chemical groups mentioned above, with a particular focus on their impact following oral ingestion. By exploring their modes of action, toxicokinetics, and physiological consequences, we uncover how these naturally occurring marine toxins can disrupt biological systems—sometimes with severe outcomes for both wildlife and humans.

3.3 Okadaic Acid

Okadaic acid (OkA, Fig. 3.4, for the toxicological profile, see Sosa and Tubaro (2016)) and its analogues—the dinophysistoxins—as related derivatives, are potent marine toxins produced by various species of *Prorocentrum*, *Dinophysis*, and *Alexandrium* (dinoflagellates). These lipophilic toxins are well documented for their tendency to accumulate in marine organisms, particularly in sponges and shellfish (Reguera et al. 2014; Cheng et al. 2016; Lee et al. 2016). OkA is a primary causative agent of diarrhetic shellfish poisoning (DSP), a foodborne illness marked by gastrointestinal distress in humans.

[2] For an insightful review of the life cycle, physiology, ecology, and red tide occurrence of the fish-killing raphidophyte genus *Chattonella*, see Imai and Yamaguchi (2012).

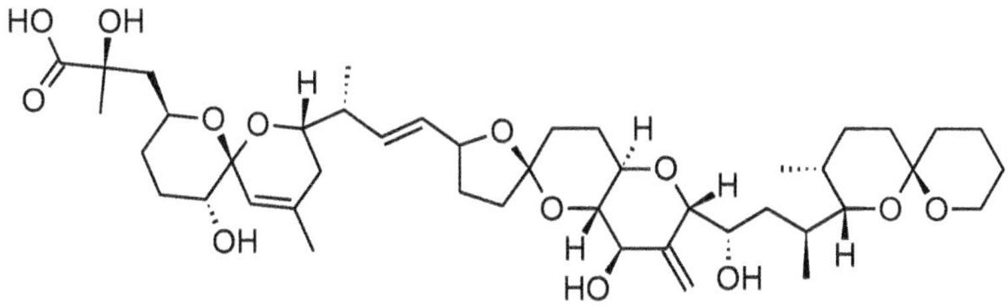

Fig. 3.4 Structure of okadaic acid (OkA)

Its mode of toxicity stems from its role as a powerful inhibitor of serine/threonine protein phosphatase, leading to abnormal phosphorylation of cellular proteins and a cascade of downstream cellular disruptions (Valdiglesias et al. 2013).

Beyond the public health threat, DSP outbreaks result in significant economic damage to the global shellfish and oyster industries due to harvesting bans and product recalls (Conte 1984; Messieh and El-Sabh 1990; Ajani et al. 2013). It is noteworthy that only photosynthetic (including mixotrophic) species have been associated with DSP.

Additional pathways of toxicity have been described for OkA. One such pathway involves the downregulation of neuropeptide Y (*NPY*) expression (Valdiglesias et al. 2013) and the modulation of NPY secretion by neuroblastoma cells (Louzao et al. 2015a, b), suggesting an additional neurotoxic mode of action beyond its established effects on protein phosphatases. NPY is a highly conserved neuropeptide that regulates numerous physiological processes, including appetite control, stress response, circadian rhythm, and neuroprotection. Disruption of NPY signaling by OkA could therefore have wide-ranging effects on central nervous system function and contribute to the neurological impacts of the toxin. The second mechanism is genotoxicity: gill cells of the clam *Ruditapes decussatus* show increased DNA breaks in the presence of lower concentrations of OkA, whereas control cells and those exposed to higher concentrations of OkA exhibit less DNA damage (Florez-Barros et al. 2011). Additionally, significant increases in DNA fragmentation occur in hemolymph and hepatopancreas cells of blue mussel (*Mytilus edulis*) and Pacific oyster (*Magallana* (*Crassostrea*) *gigas*)—even at low exposure concentrations of OkA. Although OkA is relatively hydrophilic, Wei et al. (2019) demonstrated in *Perna viridis* that even different CYP450 isoforms, typically involved in the metabolism of hydrophobic compounds, are involved in the early metabolism of OkA.

In blue mussels exposed to *Prorocentrum minimum*, hemocytes exhibit immune responses, including diapedesis, aggregation, and encapsulation, within the digestive tract. These cellular defenses are triggered without any known microalgal toxins directly affecting circulating hemocytes, as the primary immune response is initiated outside the epithelial barriers of the stomach and intestine. Upon encountering *P. minimum* cells in the gut, otherwise healthy hemocytes exit circulation and subsequently exhibit impaired function in the intestinal environment. Consequently, bacterial proliferation occurs in the stomach and intestine as hemocytes are diverted to respond to ingested *P. minimum*, becoming functionally impaired and unable to control microbial growth effectively (Galimany et al. 2008).

A transcriptomic study in the Pacific oyster (*M. gigas*) exposed to another *Prorocentrum* sp., *P. lima*, showed clear modulation of mRNA expression for genes involved in cell cycle regulation and the immune system. This overexpression can be related to DNA damage and cell cycle perturbations, likely indicating a genotoxic effect, as well as the activation of the innate immune system as a first line of defense (Romero-Geraldo et al. 2014). The potential training of innate immunity will be discussed in Chap. 5 and in the Concluding Remarks.

A subsequent study showed that four of the six tested stress genes were upregulated after a certain time lag. These genes include glutamine synthetase (GS), glutathione transferase (GST), heat shock protein 70 (HSP70), and CuZn superoxide dismutase (SOD). However, heat shock protein 90 (HSP90) and melanogenic peroxidase (POX) genes were unaffected.

This strong impact on the expression of first-line defense genes indicates exposure to various (environmental, chemical, or biological) factors, explaining the mass mortality observed in oyster farms, especially in newly seeded organisms (3–5 mm) (Romero-Geraldo and Hernández-Saavedra 2014). This conclusion is supported by a subsequent histological and immunological study (Romero-Geraldo et al. 2016), which showed that exposure to diarrheic shellfish poisoning from *P. lima* cells decreased the clearance rate, induced structural loss, reduced the tubule area of the digestive gland, and upregulates the expression of genes involved in the cell cycle (*p21*, *cafp55*, *p53*) (Fig. 3.5), cytoskeleton (*tub*, *act*), and inflammatory process (*casp1*). These findings suggest that toxic cells themselves trigger inflammatory tissue processes, disrupt the cell cycle and cytoskeleton, and pose a risk to the integrity of oysters. Overall, these results suggest that additional stressors, such as viral or bacterial infections, may not even be necessary to trigger mortality in oysters. Toxic exposure alone appears sufficient to compromise immune defenses and induce lethal outcomes, particularly in vulnerable juvenile stages.

In addition to transcriptomics, Wang et al. (2021a, b) employed proteomics to reveal further toxicity mechanisms of OkA in the Chinese scallop (*Azumapecten* (*Chlamys*) *farreri*). Although the concentration used is well below that tolerated by the scallops, OkA caused significant changes in their physiology and proteome. OkA not only induced oxidative stress and altered stress gene expression, but also significantly altered biological processes such as amino acid biosynthesis, energy metabolism, cytoskeletal functions, and phagosome formation.

Several recent studies on shellfish species, namely the oyster *Magallana* (*Crassostrea*) *ariakensis* (Zou et al. 2015), the Yesso scallop *Mizuhopecten* (*Patinopecten*) *yessoensis* (Cheng et al. 2016), and four bivalve species (*Ruditapes philippinarum*, *Anadara kagoshimensis* (*Scapharca subcrenata*), *Perna viridis*, *Tegillarca granosa*) (Huang et al. 2014, 2015a, b), suggest different potential defense mechanisms at the biomolecular level that may act individually or in concert to enhance the resistance of the exposed mussel community. Zou et al. (2015) showed that exposure of *P. lima* to *C. ariakensis* caused differential expression levels and changes in the content of glutathione transferase (GST) isoforms. The

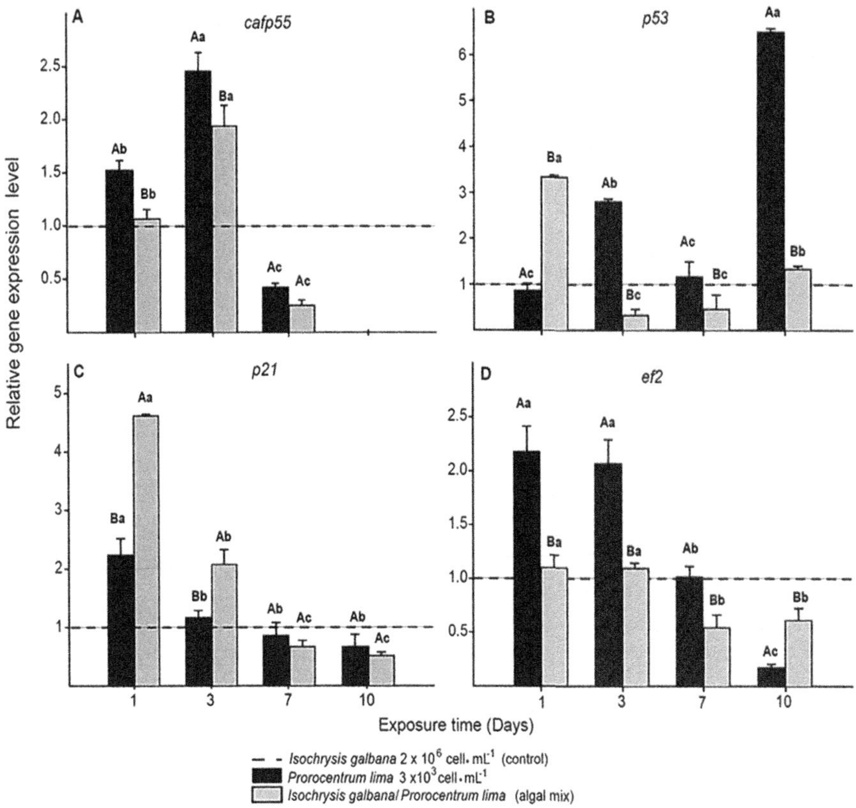

Fig. 3.5 Relative expression of cell cycle regulation genes (**cafp55** chromatin assembly factor 1 p55 subunit; **p53** cycle regulator p53 protein, major tumor suppressor gene; **p21** cycle regulator p21 protein; **ef2** elongation factor 2) in *Magallana (Crassostrea) gigas* exposed to *Isochrysis galbana* (control, – –), *Prorocentrum lima* (black columns), and algal mix (grey columns). ef1-α and gapdh served as endogenous controls. Bars represent mean ± SD. Values in column denoted by different capital letters differ significantly between treatments, where values at each exposure time denoted by different lowercase letters differed significantly (from Romero-Geraldo et al. (2016), with permission from Elsevier)

different GST gene expressions suggest these isozymes play distinct physiological roles in the detoxification of DSP toxins in *M. ariakensis*. Changes in GST expression and activity indicate that GSTs may play a crucial role in protecting cells from oxidative stress induced by *P. lima* and associated DSP toxins. Another potential mechanism of resistance development is described by Huang et al. (2014, 2015a, b): the involvement of ABC transporters, including P-gp, in bivalves resistant to OkA. Although the genes encoding the exporters are modulated, it remains unclear how the exporters contribute to resistance.

In the Yesso scallop, Cheng et al. (2016) identified a set of 61 *hsp70* genes, of which up to 34 are inducible upon OkA stress. This finding aligns with genomic studies showing that bivalves such as the Pacific oyster (*M. gigas*) (Zhang et al.

2012), the Akoya pearl oyster, *Pinctada fucata* (Takeuchi et al. 2016), and an invasive mussel *Xenostrobus securis* (*Limnoperna fortunei*) (Uliano-Silva et al. 2014), contain a significantly increased number of *hsp70* genes in their genomes compared to humans and other model organisms. This suggests the crucial role of HSP70 members in enabling bivalves to cope with and resist stress (Zhang et al. 2012).

Further studies of OkA effects are briefly summarized in Table 3.1.

In one of the first epigenetic studies, Huang et al. (2021) demonstrated that these biomolecular mechanisms might play a central role in the response of the green mussel (*Perna viridis*) to toxic marine algae. The authors identified 90 mature microRNAs (miRNAs) and 33 potential novel miRNAs. After exposure to *P. lima*, the miRNAs *miR-71_5*, *miR-750_1*, and *novel_mir4* were significantly upregulated. This upregulation was associated with the downregulation of target genes involved in apoptosis, the cytoskeleton, and the immune response (Fig. 3.6).

In addition to benthic filter-feeding invertebrates, planktonic invertebrates are also affected by *Prorocentrum donghaiense* blooms. For example, Lin et al. (2014) reported that the copepod *Calanus sinicus* showed higher mortality and lower egg production rates in situ in bloom areas than in non-bloom areas. The model organism *Artemia salina* responds to OkA exposure by increasing proteolysis of damaged proteins and energy expenditure (Table 3.1) (Gong et al. 2021).

The impact of OkA contamination on higher trophic levels within the aquatic food webs remains poorly understood, with only a few studies addressing how bioaccumulation or trophic transfer may affect predators. One example is a gene expression profiling study in zebrafish, which demonstrated that high-dose OkA can cause severe liver damage by inhibiting protein phosphatases PP1 and PP2A, and by regulating the expression of key cell defense systems (e.g., GST, ABC transporters, and HSPs) (Zhang et al. 2014) (Table 3.1).

Despite its potential to accumulate in fish, a study conducted in Hong Kong observed trophic dilution of marine lipophilic phycotoxins, including OkA, with a trophic magnification factor of 0.109 (Fig. 3.7) (Li et al. 2021). This study analyzed 17 representative marine lipophilic phycotoxins (MLPs), including azaspiracids (AZAs), brevetoxins (BTXs), gymnodimine (GYM), spirolides (SPXs), okadaic acid (OA), dinophysistoxins (DTXs), pectenotoxins (PTXs), yessotoxins (YTXs), and ciguatoxins (CTXs). Whether this study accurately represents bioaccumulation across marine food webs remains to be determined.

3.4 Karlotoxins

Karlotoxins are another fascinating toxin group of linear polyether compounds (Fig. 3.8b) produced by the dinoflagellate *Karlodinium veneficum* (Fig. 3.8a). These potent toxins are known for their ichthyotoxic effects and their role as anti-gazing compounds, protecting the producing algae from predation (Adolf et al. 2007). The primary mechanism of action involves the disruption of membrane permeability in vertebrate cells, leading to cellular swelling, lysis, and cell death (Deeds et al. 2015). By 2010, these toxins had been detected both in natural and clonal isolates

Table 3.1 Short survey of toxicity studies of okadaic acid (OkA) in marine animals

Species	OkA source	Observed effect	References
Invertebrates			
Anomalocardia flexuosa	*Prorocentrum lima*	Fast detoxification, no oxidative stress	Leite et al. (2021a, b)
Argopecten irradians	OkA	Oxidative stress↑, innate immunity genes↑, detoxification↑	Chi et al. (2016, 2017, 2018)
Artemia salina	*P. lima*, OkA	Proteolysis of damaged proteins↑, energy expenditure↑	Gong et al. (2021)
Cerastoderma (*Cardium*) *edule*	*Dinophysis* spp.	Modulation of genes involved in: antioxidation, detoxification, transmembrane transport, apoptosis, cytoskeleton, immunity	Domínguez--Pérez et al. (2021)
Crassostrea tulipa (*gasar*)	*P. lima*, *Ostreopsis* cf. *ovata*	Innate and specific immunity↓: *O.o* > *P.l.*	Faustino et al. (2021)
Laeonereis acuta	*P.* cf. *lima*	Oxidative stress↑	Pires et al. (2021)
Magallana (*Crassostrea*) *gigas*	*P. lima*, *P. minimum*, *P. rhathymum*	Modulation (mainly upregulation) of genes of immune system, stress response, contractile systems, cytoskeleton	Matus Hernández and Hernández Saavedra (2021)
	P. lima	*abcb1*↑	Ye et al. (2022)
Magallana (*Crassostrea*) *ariakensis*		8 *gst* isoenzymes differently regulated in gills and digestive gland	Zou et al. (2015)
Mytilus edulis	*Dinophysis acuta*	Clearance rate↓, toxin accumulation↑	Nielsen et al. (2020)
M. galloprovincialis	OkA from *P. concavum*	Transcription of stress proteins↑ and proteins involved in cellular synthesis↑, apoptosis↑	Manfrin et al. (2010)
	P. lima	Modulation of genes involved in carbohydrate metabolism, energy production, genome integrity, defense, immunity	Prego-Faraldo et al. (2018)
		Upregulation of transmembrane transporter genes; modulation of signal transduction, cytoskeleton, post-translational modification, protein turnover, chaperone functions	Martins et al. (2020)
	Dinophysis acuminata	Digestive gland: transcription of MXR-related genes↑	Martínez--Escauriaza et al. (2021)
Mytilus (*coruscus*) *unguiculatus*	*P. lima*	*cat*↑, LPO↑, *sod*↓, *caspase7*↑, *abc1*↑ *gst-ω*↑, *gpx*↑	Tang et al. (2022) Ye et al. (2022)
Octopus vulgaris	Contaminated *Perna perna*	OkA accumulation↑	Mafra et al. (2015)

(continued)

Table 3.1 (continued)

Species	OkA source	Observed effect	References
Palaemon carinicauda	*Prorocentrum minimum*	Crustin↑, proPO↑, lysozyme↑, pathogen resist↓	Mu et al. (2023)
Perna perna	*P. lima*	Phagocytosis↑, ROS↑, immune response↓	Neves et al. (2019)
P. viridis		Several *abc*↑, *cyp3a*-like↑, *gstσ3*↑, cytoskeleton↑, apoptosis inhibiting protein↑	Dou et al. (2020)
		Nrf2-dependent antioxidant defense↑	He et al. (2019)
		MXR activity↑	Huang et al. (2014)
		microRNAs↑ targeting genes of apoptosis, cytoskeleton, immune response (→Fig. 3.6)	Huang et al. (2021)
		Cytoskeletal disorganization, oxidative stress, metabolic dysfunction, and ubiquitination/proteasome activity	Huang et al. (2015a, b)
Ruditapes philippinarum *Scapharca subcrenata*		MXR activity↑	Huang et al. (2015a, b)
Tegillarca granosa		*gpx*↑, *sod*↓	Ye et al. (2022)
Fishes			
Danio rerio	OkA	*mgst1*↓, ABC transporters↓, *hsp90*↓, *hsp40* homolog↑	Zhang et al. (2014)
Dicentrarchus labrax	*P. lima* PL2V	Hyperactivities (jumps, fast let-right turns, surface swims) poor feeding reflexes, abstinence from feeding	Ajuzie (2008)
Diplodus cervinus	OkA	Impaired swimming performance	Corriere et al. (2020)
Oryzias latipes	*P. (arenarium) lima*	Survival↓	Escoffier et al. (2007)
	Crude extracts of P.l.	Survival↓	
	P. emarginatum	Survival↔	
Seriola rivoliana	OkA	Embryo toxicity (mortality of eggs)↑, protein phosphatase↓	Le Du et al. (2017)
Sparus aurata		Oxidative stress↑	Souid et al. (2018)

↑ increase; ↓reduction; ***abc*** *ATB-binding cassettes* (exporters); ***cat*** *catalase*; ***cyp3a*** *cytochrome P450 3a* (biotransformation I); (*m*)***gst*** (*microsomal*) *glutathione transferase* (biotransformation II); ***gpx*** *glutathione peroxidase*; ***Nrf2*** *nuclear factor erythroid 2-related factor 2* (antioxidant); *ROS* reactive oxygen species, *MXR* multixenobiotic resistance; ***sod*** *superoxide dismutase*

from estuarine waters of the U.S. East Coast, including Maryland, North and South Carolina, Georgia, Florida, and Mississippi, as well as from diverse ecosystems across the globe, such as New Zealand coastal waters, Norwegian fjords, and the English Channel. They were also found in water samples during fish kills in a

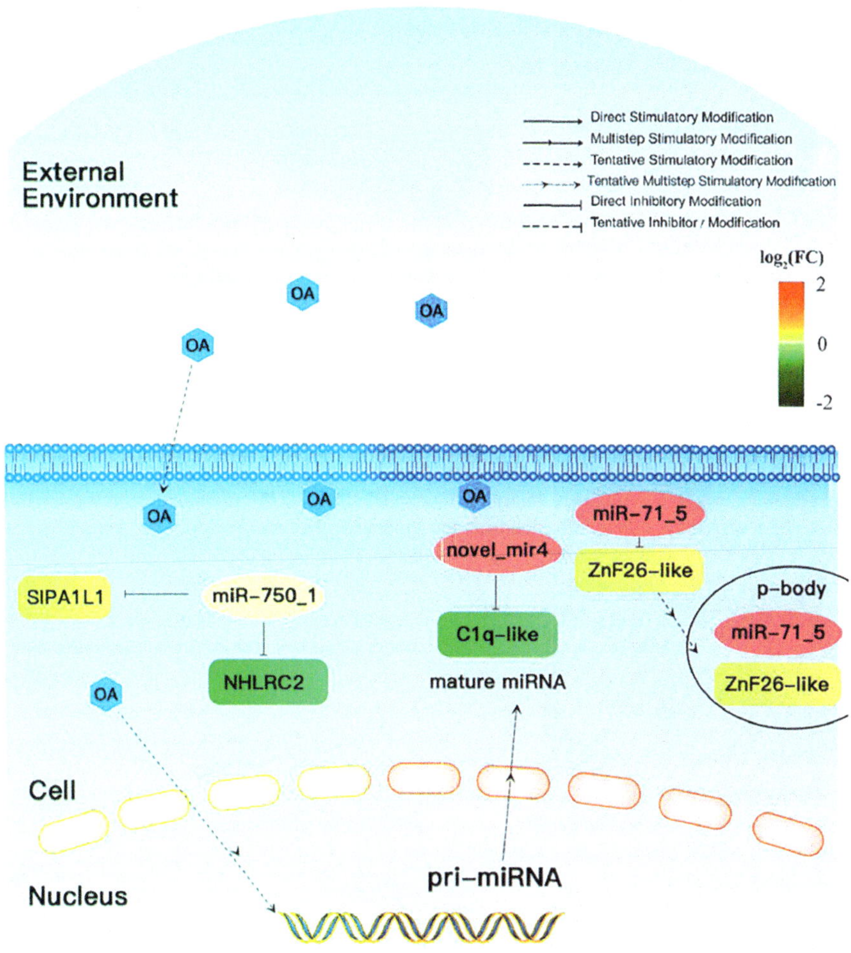

Fig. 3.6 Proposed response mechanism in the digestive gland of *Perna viridis* after exposure to *Prorocentrum lima* (from Huang et al. (2021), with permission from Elsevier). **NHLRC2** is involved in apoptosis; **SIPA1L1** is involved in cytoskeletal reorganization; **C1q-like** is related to the complement system and immune stress; **ZnF26-like** has zinc finger protein function

brackish pond in South Carolina, a tributary of the Chesapeake Bay, and the Swan River in Perth, Australia. In all these instances, high densities of *Karlodinium veneficum* (Fig. 3.8a) were found (Peng et al. (2010) and references therein). Adding to the complexity, a growing number of karlotoxin congeners are being discovered through advanced analytical techniques, revealing a surprising chemical diversity within this toxin group (Cai et al. 2016).

Purified KmTx2 has been proven to kill fish in a dose-dependent manner. Mortality in zebrafish and sheepshead minnow juveniles was observed at concentrations of 0.1–0.5 μg mL^{-1}. This aligns with the concentration range of KmTx2 found

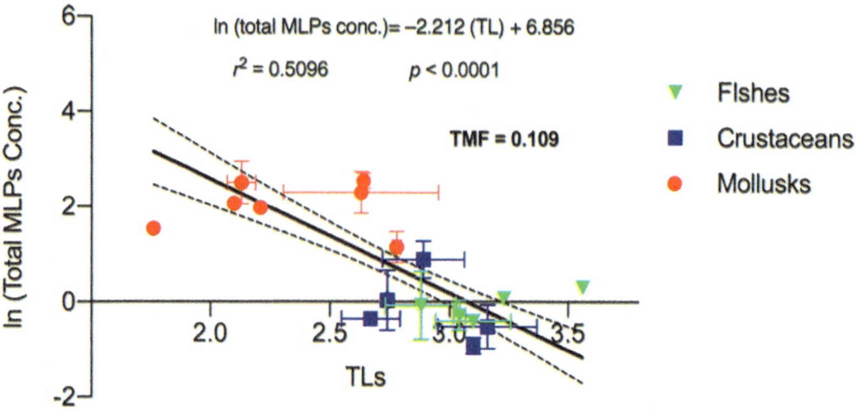

Fig. 3.7 Trophic transfer of concentrations of total marine lipophilic phycotoxins (MLPs) (ng g^{-1}, ww). The solid line represents the linear regression, while the upper and lower dashed lines represent their average 95% confidence intervals. TLs trophic levels. TMF trophic magnification factor (from Li et al. (2021), with permission from the American Chemical Society)

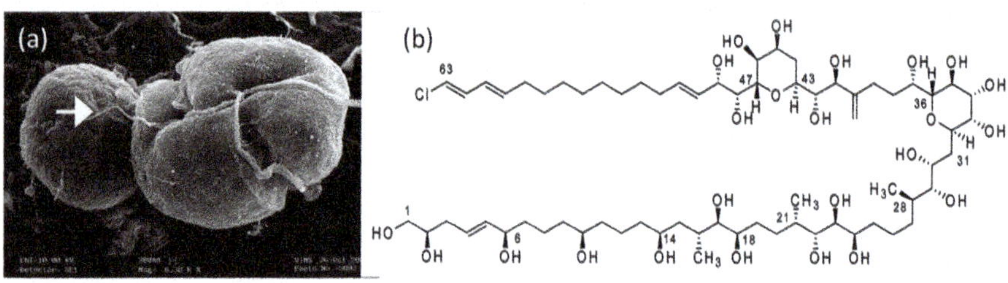

Fig. 3.8 (**a**) SEM micrograph of *Karlodinium veneficum* feeding on *Rhodomonas* sp. (arrowed, from Place et al. (2012), with permission from Elsevier. (**b**) karlotoxin 2 (KmTx2) (from Peng et al. (2010), with permission from the American Chemical Society)

in water samples during fish kills (0.1–0.8 μg mL^{-1}). The pathology observed in the laboratory, specifically epithelial sloughing in the gills of treated fish (Fig. 3.9b), mirrors the gill damage noted in fish kills observed in natural settings (Fig. 3.9c).

In laboratory experiments, Stoecker et al. (2008) compared the effects of *Prorocentrum minimum* and *Karlodinium veneficum* on the survival and development of eastern oyster (*Crassostrea virginica*) embryos and larvae. At 2×10^4 cells mL^{-1} *K. veneficum* caused significant mortality of oyster embryos within one day, with almost no embryos developing into D-hinge larvae. This detrimental effect persists even when an alternative food source (*Isochrysis* sp.) is provided. Additionally, significant mortality was observed when larvae were exposed to *K. veneficum* at concentrations of 10^4 cells mL^{-1} (corresponding to approximately 5 ng mL^{-1} of karlotoxin) (Fig. 3.10). Even organ development in the developing spat appears to be reduced by the ichthyotoxic dinoflagellate (Brownlee et al. 2008). Consequently, embryo and larval survival and maturation may be significantly

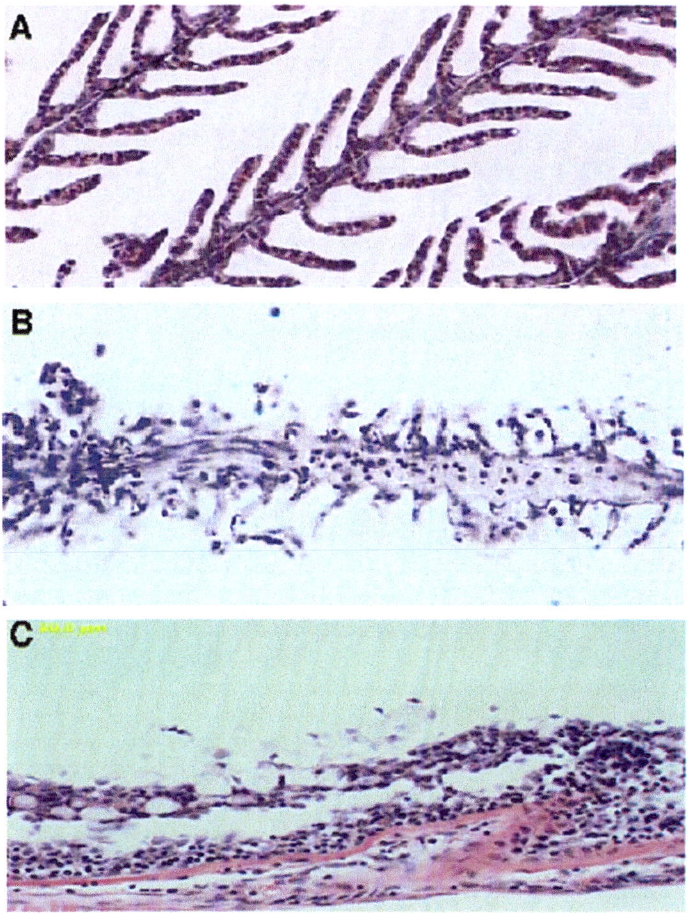

Fig. 3.9 Histological sections (5–6 μm thickness) of zebrafish (*Danio rerio*) juveniles stained with hematoxylin and eosin after exposure. (**a**) Control gill arch. (**b**) KmTx2 exposure (0.5 μg mL^{-1} for 1 h). The loss of intra-lamellar spacing and secondary lamellar fusion should be noted. (**c**) Gut epithelia of a striped killifish (*Fundulus majalis*) exposed in situ to karlotoxin (~0.2 μg mL^{-1}) during a fish-kill event (from Peng et al. (2010), with permission from the American Chemical Society)

reduced when eastern oyster spawning coincides with high *K. veneficum* bloom densities.

Interestingly, Yuan et al. (2023) demonstrated a protective role for the essential long-chain PUFA docosahexaenoic acid (DHA) in the mussel *Perna viridis*. When exposed to the diarrhetic shellfish toxin-producing dinoflagellate *Prorocentrum lima*, DHA acted as a functional detoxification agent. It significantly promoted the esterification of diarrhetic toxins within the hepatopancreas, effectively neutralizing their bioactivity. In addition, DHA supplementation activated the Nrf2 signaling pathway, a key regulator of cellular antioxidant defense. This was reflected in

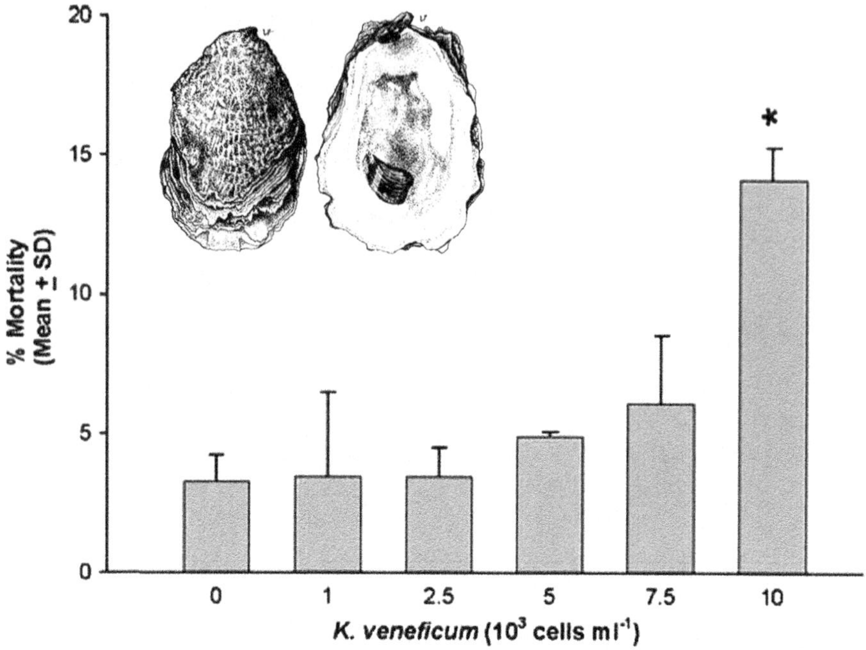

Fig. 3.10 Mortality of 1-day-old *Crassostrea virginica* larvae after 2 days of exposure to *Karlodinium veneficum* (1×10^3 to 10^4 cells mL^{-1}) in the presence of 2×10^4 *Isochrysis* cells mL^{-1}. Control treatment (0) exposed to seawater with added *Isochrysis*. *K. veneficum* concentration significantly affected mortality. * indicates a significant difference ($P < 0.05$) (from Stoecker et al. (2008), with permission from Springer Nature; drawing credit to FAO)

increased expression of Nrf2-associated genes and elevated enzyme activity, ultimately leading to a marked reduction in tissue damage caused by the toxins. These findings open up promising avenues for nutritional or environmental mitigation strategies against shellfish poisoning in aquaculture settings.

3.5 Prymnesins

One of the most prominent cases of prymnesin poisoning occurred in the Oder River (Poland/Germany) in the summer of 2022, resulting in an overwhelming kill of aquatic life, caused by a bloom of the brackish algae *Prymnesium parvum* (Wiederhold et al. 2023; Mora et al. 2024). A similar, though less devastating, die-off in terms of biomass loss and ecological impact occurred in summer 2024 in the Gleiwitz Canal and the Dzierżno Duże Reservoir, Poland.

Prymnesins (Fig. 3.11) are a class of phycotoxins produced by *P. parvum* (Manning and La Claire 2010), algae belonging to the genus *Chrysochromulina*, and the dinoflagellate *Margalefidinium* (*Cochlodinium*) *polykrikoides* (Kudela et al. 2008). These toxins are highly toxic to fish and are known to cause mass fish kills

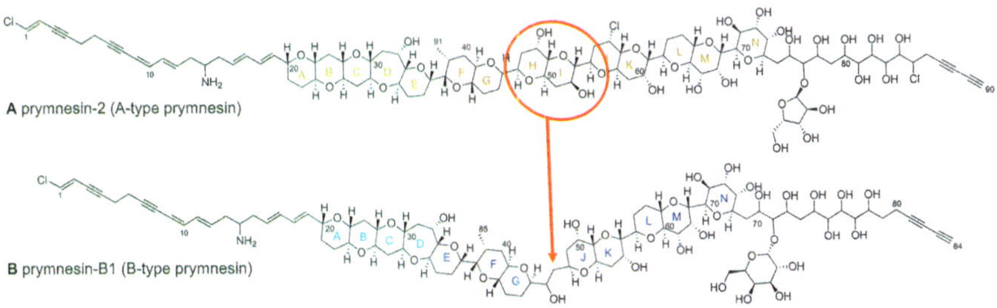

Fig. 3.11 The structures of A- and B-type prymnesins. (**a**) Structure of an A-type prymnesin, here prymnesin-2, note the two aromatic rings labeled H and I, these have been replaced in the B-type by a single carbon-carbon bond; (**b**) Structure of a B-type prymnesin, here prymnesin-B1, notice the lack of ring-system H, I. (modified from Binzer et al. (2019), credit to Elsevier)

worldwide (Table 3.2). The ecological interest in harmful *P. parvum* blooms has increased due to their apparent global increase in incidence (Hallegraeff 1993).

Prymnesins are composed of a large polyether polycyclic core with multiple conjugated double and triple bonds, chlorine and nitrogen heteroatoms, and sugar moieties including L-xylose (Sasaki et al. 2004; Manning and La Claire 2010). *P. parvum* has been attributed to fish kills for more than a century, but clear evidence has only recently been provided, e.g., by Wagstaff et al. (2021). *P. parvum* blooms occur in marine waters, brackish waters (Rasmussen et al. 2016b), and salinized freshwaters, such as the Oder River.

Binzer et al. (2019) screened 26 strains of *P. parvum* with a wide geographical distribution for the production of prymnesins. All strains screened produce prymnesins, indicating that these toxins are significant for the organism. The prymnesins can be classified into three types (A-, B-, and C-type prymnesins) based on the length of their carbon backbone, with each algal strain producing only one type. Phylogenetic analyses using internal transcribed spacer (ITS) sequences revealed a monophyletic origin of all prymnesin types. Clades can, therefore, be defined based on the specific prymnesin compound produced. It is proposed that the C-type prymnesin represents the ancestral state from which both A- and B-type prymnesin-producing strains evolved. This may reflect an evolutionary selective pressure to produce increasingly toxic compounds, as evidenced by prymnesin-2 (an A-type prymnesin) being more harmful than prymnesin-B1 (a B-type prymnesin), as evaluated in a rainbow trout gill cell assay (Rasmussen et al. 2016a, b). This is emphasized by the recent findings of Kuhl et al. (2024), which revealed a 20 kbp deletion in the polyketide synthase (PKS) gene, affecting the number of ketide synthase domains in the corresponding protein. The absence alters the synthesis pathway of the toxin, resulting in those mentioned above structurally different backbones for type B compared to type A prymnesin.

Brooks et al. (2011) also highlighted *Prymnesium parvum* as an emerging threat to inland waters, noting its presence at several freshwater and brackish locations in the U.S.A. To complement these findings, Table 3.2 presents a review of fish kills associated with *P. parvum*. In some cases, multiple ichthyotoxic algae co-occurred

Table 3.2 Fish kills associated with *Prymnesium parvum*

Fish species	Region	References
135 metric tons of caged salmonids	Sandsfjorden, Western Norway	Johnsen et al. (2010)
Cyprinodon hybrids, Lucania parva, Fundulus zebrinus, Gambusia affinis, Cyprinella lutrensis, Dorosoma cepedianum	Pecos River, Texas, USA	Rhodes and Hubbs (1992)
Cyprinus carpio	Various sites in China	Guo et al. (1996)
Cyprinus carpio, Dorosoma cepedianum	Lake Granbury, Texas, USA	Valenti Jr et al. (2010)
Dorosoma sp., *Pimephales* sp., *Ictiobus* sp., *Ameiurus* sp., *Lepisosteus* sp., *Pomoxis* sp.	Lakes Possum Kingdom, Granbury, Whitney, Texas, USA	Roelke et al. (2007, 2011)
Esox lucius, Perca fluviatilis, Rutilus rutilus, Abramis brama, Anguilla anguilla	River Thurne, Hickling Broad, GB	Holdway et al. (1978)
Esox lucius, among others	Hickling Broad, GB	Wagstaff et al. (2021)
Hypophthalmichthys nobilis, Ctenopharyngodon idella, Cyprinus carpio, Silurus glanis, Ameiurus nebulosus, Carassius carassius	Fish ponds, east Hungary	Vasas et al. (2012)
Ictalurus furcatus	Brazos River, Texas, USA	VanLandeghem et al. (2013)
Menidia beryllina	Galveston Bay, Gulf of Mexico	Lundgren et al. (2015)
Micropterus dolomieu, Morone saxatilis, Ictalurus punctatus, Ictalurus furcatus	33 Texas water bodies, USA	Southard et al. (2010)
Micropterus salmoides, Ictalurus punctatus, Morone hybrids	Elizabeth City, North Carolina, USA	Barkoh et al. (2004)
Morone chrysops, Pomoxis annularis, Micropterus salmoides, Lepomis macrochirus, Carpiodes carpio, Aplodinotus grunniens, Ictalurus punctatus, I. furcatus, Pylodictis olivaris	Upper Colorado River, Texas, USA	VanLandeghem et al. (2013)
Morone saxatilis	Lake Texoma, Texas-Oklahoma, USA	Hambright et al. (2010)
Morone saxatilis, Morone hybrids, Ictalurus punctatus, Oncorhynchus mykiss, Cyprinus carpio (koi)	Dundee State Fish Hatchery, Texas, USA	Barkoh et al. (2003)
Rutilus rutilus, Abramis brama, Blicca bjoerkna, Scardinius erythrophthalmus, Alburnus alburnus, Carassius carassius, Esox lucius, Perca fluviatilis, Acerina cernua, Lota lota	Brackish Lake Vargsundet, Åland, SW Finland	Lindholm et al. (1999)
Salmo salar, Oncorhynchus mykiss	Fjord system with freshwater inflow, Norway	Kaartvedt et al. (1991)
Various fish species, 300–350 t in total + bivalves and *Viviparus viviparus*	River Odra, summer 2022	Absalon et al. (2023), Wiederhold et al. (2023)

(continued)

Table 3.2 (continued)

Fish species	Region	References
Not mentioned	Various stations in Sweden	Willén and Mattsson (1997)
	Brackish ponds, The Netherlands	Liebert and Deerns (1920)
	Brackish coastal ponds, Schleswig--Holstein, Germany	Lenz (1933)
	Pond on the North Sea coast, Schleswig--Holstein, Germany	Dietrich and Hesse (1990)
	Brackish fish (breeding) ponds, Israel	Reich and Aschner (1947), Shilo and Shilo (1953)
	Fish pond, Denmark	Otterstrøm and Steeman Nielsen (1940)
	Lake Koronia, Greece	Genitsaris et al. (2009)
	Lake Karla, Greece	Oikonomou et al. (2012)

(e.g., Lindholm et al. (1999)), making it challenging to identify the primary cause of fish mortality. As a result, these fish kills are listed in different tables to account for the various potential algal toxins involved.

A type of putative resistance to toxic *P. parvum* has been reported for rotifers within a Texas reservoir (Davis et al. 2015). Three-day in-lake experiments were conducted in two reservoirs: Lakes Somerville and Whitney. Lake Whitney experiences *P. parvum* blooms, whereas Lake Somerville does not. The experiments focused on the rotifer-dominated zooplankton size fraction and natural phytoplankton assemblages inoculated with *P. parvum*. Fish-lethal levels of toxicity were only occasionally observed during these experiments, suggesting low toxicity levels in the treatments. Rotifers in both lakes selectively fed on *P. parvum*. Interestingly, rotifers in Lake Somerville appeared to benefit from this selective grazing, while those in Lake Whitney did not. The disparity may be attributed to the historical exposure of Lake Somerville rotifers to low levels of toxins produced by *P. parvum*, which potentially allowed them to develop resistance and contribute to the suppression of algal blooms. However, specific tolerance mechanisms were not analyzed in this study.

Overall, understanding toxin production by *P. parvum* remains complex, with the identities and functions of its chemicals unclear (Roelke et al. 2016). The long-held understanding that prymnesins are the primary toxins of *P. parvum* has been challenged by recent studies that have failed to detect prymnesins in toxic *P. parvum* samples. Instead, some researchers proposed several fatty acids as the toxic agents, though this was later contradicted by findings of prymnesins in the same strains. Another study reported fatty acid amides as the primary toxic compounds, relegating fatty acids to a minor role. Subsequent research did not support these findings, suggesting various methodological artifacts might have confounded earlier conclusions. Indeed, the lack of consensus regarding the chemicals produced by *P. parvum* may stem from non-standardized culture techniques, resulting in different levels or types of toxins produced among strains. Furthermore, from inconsistent sample handling protocols, various approaches to characterizing "toxicity," and limitations in analytical instrumentation (Roelke et al. (2016) and references therein).

Regardless of these shortcomings, the ichthyotoxin produced during the massive fish kill in the Oder River in the summer of 2022 was clearly identified (Table 3.2) (Absalon et al. 2023; Wiederhold et al. 2023). A single phylotype, identified through ITS and 18S rRNA sequencing, formed the HAB, and prymnesin B1 (Fig. 3.11) was detected in both cultures and water samples (Mora et al. 2024).

3.6 Brevetoxins and Ciguatera-Causing Agents

Brevetoxins and ciguatoxins are lipid-soluble, highly potent neurotoxins found in marine environments. These dinoflagellate toxin groups are ladder-frame polyethers (Fig. 3.12), a distinctive structural class characterized by a series of interconnected ether rings arranged in a linear or partially branched fashion, resembling the rungs of a ladder.

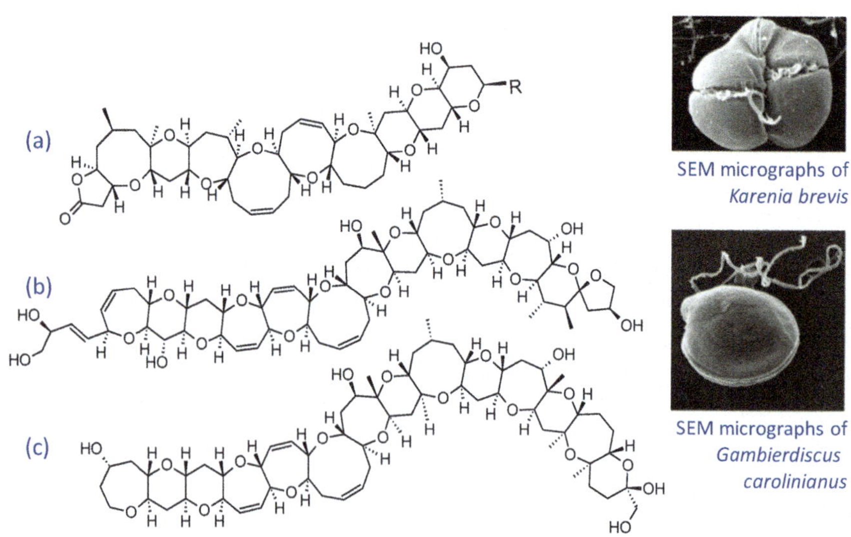

Fig. 3.12 Examples of polyether toxins: (**a**) brevetoxin A, the matrix for PbTx-1, PbTx-7, and PbTx-10; (**b, c**) backbones of Pacific (P) and Caribbean (C) ciguatoxins-1 (P-CTX-1, C-CTX-1). SEM images credit to the Florida Fish and Wildlife Conservation Commission & National Centers for Coastal Ocean Science of the US NOAA

3.6.1 Brevetoxins

Brevetoxins (PbTxs, BTXs) are a group of cyclic polyether compounds (Fig. 3.12a) produced primarily by the dinoflagellate, *Karenia brevis* (image in Fig. 3.12). For a toxicological profile, see Brovedani et al. (2016). *Karenia brevis* was formerly named *Gymnodinium breve* or *Ptychodiscus brevis*, explaining the abbreviation "PbTxs") (Watkins et al. 2008; Fleming et al. 2011). Notably, some raphidophytes, such as *Heterosigma akashiwo*, can also produce PbTxs (Khan et al. 1997).

Brevetoxins accumulate in the gut and hepatopancreas of affected organisms and bind to voltage-gated sodium channels in neurons, leading to the disruption of normal neurological processes (Fig. 3.1). A recent study, however, has shown that not all PbTxs bind to voltage-gated sodium channels. Instead, aldehyde-containing PbTxs are more immunotoxic (McCall et al. 2022).

A novel, structurally more complex marine toxin, brevisulcenal-F (KBT-F, from *Karenia brevisulcata* toxin), was recently isolated from the dinoflagellate *Karenia brevisulcata*. A red tide of *K. brevisulcata* in Wellington Harbour, New Zealand, in 1998 was extremely toxic to fish and marine invertebrates. An extract of *K. brevisulcata* exhibited potent lethality and cytotoxicity in mice, and laboratory cultures of *K. brevisulcata* produced several novel lipid-soluble toxins. One such toxin, KBT-F, was isolated and identified as a ladder-frame polyether scaffold with an LD_{50} (mouse i.p.) of 0.032 mg kg^{-1} (Hamamoto et al. 2012). This discovery highlights that many marine toxins remain to be discovered.

Landsberg 2002 reviewed the effects of PbTxs on aquatic animals that fed on *K. brevis*. When the copepod *Calanus pacificus* consumes *K. brevis*, it exhibits a marked increase in heart rate and loss of motor control. More specifically, Waggett et al. (2012) showed in the copepod *Acartia tonsa* that *K. brevis* affects grazer populations through multiple synergistic mechanisms: (1) reduced ingestion rates, (2) reduced egg production, and (3) increased copepod mortality through a combination of toxicity and nutritional deficiencies.

Other species are also acutely affected by exposure to *K. brevis*. Mollusks, such as the banded tulip (*Cinctura hunteria*), crown conch (*Melongena corona*), and letter olive (*Oliva sayana*), lose muscle control and are unable to right themselves, with mortality rates ranging from 55.5% to 69.0% (Landsberg (2002) and references therein). Additionally, bay scallop (*Argopecten irradians*) recruitment was significantly affected by the North Carolina red tide of *K. brevis* in 1987 (Summerson and Peterson 1990).

Conversely, several species are unaffected by varying concentrations of *K. brevis*. This includes eastern oysters (*Crassostrea virginica*), hooked mussels (*Ischadium recurvum*), the annelid *Polydora websteri*, the barnacle *Amphibalanus eburneus*, the crab *Eurypanopeus depressus*, blue crabs (*Callinectes sapidus*), and stone crabs (*Menippe mercenaria*). A more recent biomolecular analysis has provided insights into the potential mechanism behind the relatively high resistance of oysters to PbTx exposure. Mello et al. (2012) incubated *M. gigas* hemocytes with PbTx-2, a specific PbTx congener, and observed increased expression levels of *hsp70* and *cyp356a1* transcripts, as well as a tendency towards increased *fabp*

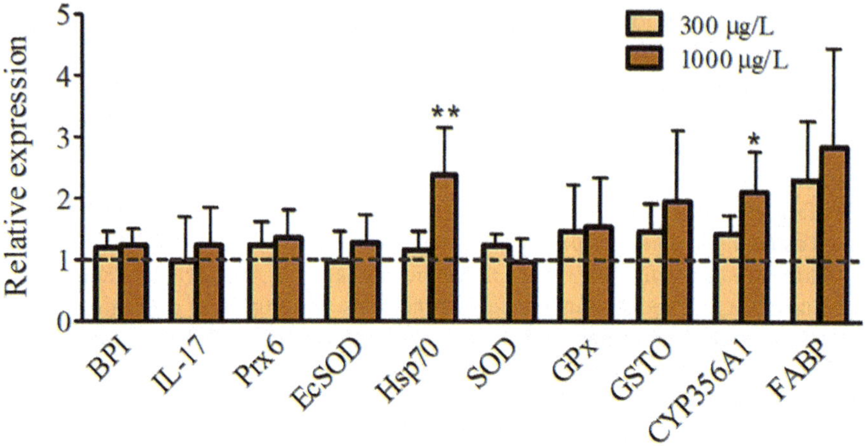

Fig. 3.13 Relative expression of the 10 gene transcripts (normalized to *gapdh*) in *Magallana* (*Crassostrea*) *gigas* hemocytes exposed for 4 h to 300 or 1000 μg L^{-1} of PbTx-2. The dashed line represents the mean value of the control groups (vehicle) for each gene. Bars represent mean + standard deviation. * $P < 0.05$; ** $P < 0.01$ represent significant differences between each gene compared to the corresponding control group (from Mello et al. (2012), credit to MDPI)

expression. The expression of other genes (*bpi*, *il-17*, *gsto*, *ecsod*, *prx6*, *sod*, and *gpx*) remained unaltered (Fig. 3.13). These results indicate that the lack of cytotoxic effects of PbTx-2 in *M. gigas* hemocytes, even at high concentrations, is attributed to the activation of protective mechanisms associated with detoxification (*cyp356a1* and possibly *fabp*) and general stress defense (*hsp70*), rather than immune or antioxidant-related genes (*bpi*, *il-17*, *gsto*, *ecsod*, *prx6*, *sod*, and *gpx*). Nevertheless, additional research, both in vitro and in vivo, is required to substantiate the precise impact of PbTxs on the immune system of cultured or wild bivalves, especially at the gene expression and proteomic levels.

The persistence of toxicity is species-dependent and varies with the concentration of the bloom as well as the feeding and excretion rates of the bivalves. For example, in Pacific oysters, a three-day depuration period resulted in a significant reduction in PbTx concentrations, bringing them close to levels that are acceptable for human consumption in accordance with the relevant regulatory limits (20 MU per 100 g) (Baden 1983). However, in coquinas (*Donax variabilis*) that have accumulated the toxin, the PbTx composition is different from that in *K. brevis*, likely due to metabolic conversion processes (Landsberg 2002).

Suspension-feeding bivalves are the primary vectors for the transfer of several major groups of phycotoxins through the food web. The trophic transfer of PbTxs is complex and often more intricate than initially understood. Unexplained fish kills and other animal mortalities in areas where red tide is endemic are increasingly being linked to post-bloom exposure of biota to PbTxs. Persistent blooms can trigger a cascade of environmental changes, leading to widespread die-offs of benthic communities and causing short-term declines in local fish populations (Fig. 3.14). The long-term impact on these populations remains uncertain, raising concerns

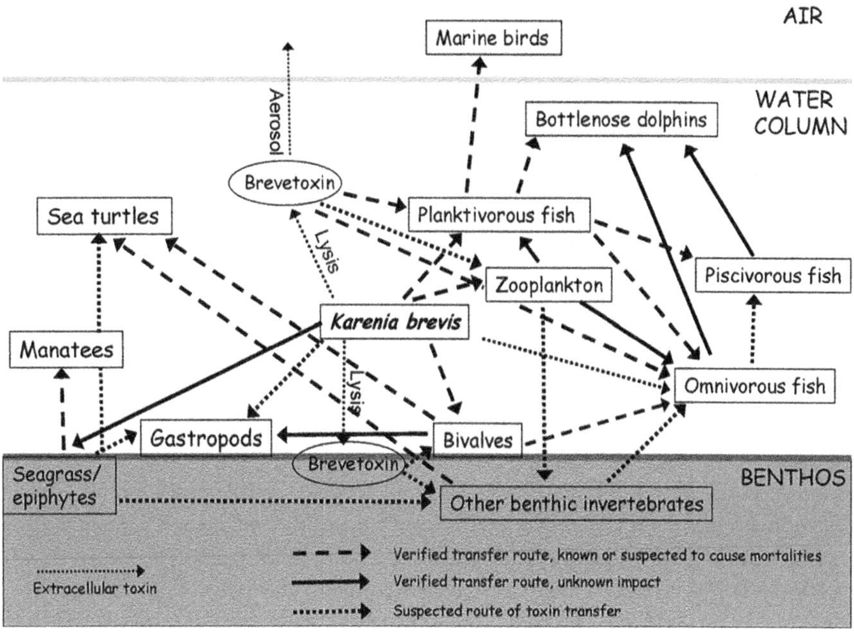

Fig. 3.14 Schematic of the transfer and effects of brevetoxins in the marine environment (from Landsberg et al. (2009), with permission from Elsevier)

about their ability to recover from the persistent effects of red tide (Landsberg et al. 2009).

Not only adults of the Eastern oyster (*Crassostrea virginica*) but also their offspring are susceptible to exposure to *K. brevis*. Rolton et al. (2016) found PbTx in both spermatozoa and oocytes, with estimated amounts of 18% and 26% of the adult PbTx load, respectively. Larvae derived from these gametes showed significantly higher mortality and smaller size during the first 6 days post-fertilization. This may be due to the presence of PbTx in oocyte lipid droplets utilized by the larvae during embryonic and lecithotrophic larval development. However, providing a non-contaminated food source to the larvae appeared to mitigate the early adverse effects of neonatal PbTx exposure.

To identify further transgenerational mechanisms, Gonzalez-Romero et al. (2017) investigated the effects of Florida red tide on two types of mechanisms involved in epigenetic memory in Eastern oysters: histone variants and DNA methylation. Oysters were exposed to increasing concentrations of *K. brevis*, revealing the presence of the histone variants *h2a.x*, *h2a.z*, and *macroh2a* genes in this organism. These histone variants may play a role in maintaining genome integrity during responses to the genotoxic effects of brevetoxins. Furthermore, an increase in *h2a.x* phosphorylation (γ*h2a.x*, a marker of DNA damage) and a decrease in global DNA methylation were observed as the *K. brevis* exposure progressed.

To provide a broad overview of the diverse effects of brevetoxins and their potential ecological role, selected recent studies are summarized in Table 3.3. To illustrate the general importance of brevetoxicosis, studies involving mammals and birds are also included in this table. The ecological effects listed highlight the primordial function of brevetoxins in ecosystems, or at least our state of understanding of these potential defensive chemicals.

In contrast, a subsequent study in the Republic of Kiribati, conducted by Zhu et al. (2024), found that biomagnification of ciguatoxins occurs in the food web with a trophic magnification factor of 2.90. The risks posed by OkA and dinophysistoxins are not negligible in this case. Therefore, future studies are needed to determine the representativeness of these findings.

3.6.2 Ciguatera-Causing Agents

Ciguatera fish poisoning is a foodborne illness caused by the consumption of reef fish contaminated with toxins produced by marine dinoflagellates. The primary toxins responsible for ciguatera are ciguatoxins, which have a similar structure and mode of action as the brevetoxins. Unsurprisingly, the clinical manifestation is therefore quite similar—although often more severe and longer-lasting for ciguatera (Bagnis et al. 1979; Poli et al. 2000; Kirkpatrick et al. 2004). Despite these similarities, these toxins present obvious differences in their impacts on fish. Ciguatoxins are well known to accumulate in fish by trophic transfer in tropical fish food webs (Naar et al. 2007), and have not been documented in association with fish mortalities.

However, ciguatoxins are not the only culprits—other substances, including maitotoxins and palytoxins, have also been identified as contributing to the occurrence of ciguatera and will be discussed within this chapter.

3.6.2.1 Ciguatoxins[3]

Ciguatoxins (CTXs, Fig. 3.12b, c) are a class of approximately 20 lipophilic toxic polycyclic polyethers. They are produced by the epiphytic dinoflagellate genus *Gambierdiscus* (Fig. 3.12), which also produces maitotoxin and gambieric acid. CTX is also produced by the dinoflagellate *Ostreopsis lenticularis*. CTXs are temperature stable over a broad range and thus persist in cooked or frozen fish. They are known to cause ciguatera in humans when consumed with contaminated fish, causing gastrointestinal, neurological, and cardiovascular symptoms that typically last weeks or months (Bagnis et al. 1980). The occurrence of toxic *Gambierdiscus* and *Fukuyoa* spp. is not restricted to tropical and subtropical oceans, as previously thought. A surprisingly high diversity of *Gambierdiscus* has been reported, for example, from Japanese coastal areas (Nishimura et al. 2013). *G. caribaeus* has also been found in the temperate waters of Jeju Island, Korea (Jeong et al. 2012), and

[3]Also see https://www.ecdc.europa.eu/en/infectious-disease-topics/ciguatera-fish-poisoning/facts-ciguatera-fish-poisoning

Table 3.3 Ecological and ecotoxic effects of brevetoxin as reported in recent studies

Toxin/producer	Affected species or community	Effect	References
Ecological effects			
Karenia brevis	*Asterionellopsis glacialis, Thalassiosira pseudonana* Sensitivity: *A. glacialis* < *T. pseudonana*	Membrane integrity↓, photosynthesis↓, energy metabolism↓, cellular protection↓, osmoregulation↓, oxidative stress↑	Poulson-Ellestad et al. (2014a, b), Poulin et al. (2018)
	Phytoplankton in a mesocosm	Species-specific response from strong sensitivity to resistance	Poulson et al. (2010)
	Offshore phytoplankton		Poulson-Ellestad et al. (2014a, b)
	Trieres (*Odontella*) *chinensis*	Hyperexcitability ↔, voltage--activated currents ↓	Kitchen et al. (2018)
	Natural microbial communities	Bacterial abundance↓ & production↓	Meyer et al. (2014)
		Bactericidal effects against sensitive species↑	Sipler et al. (2014)
		Aromatic compounds↑, lipids↑	Patin et al. (2020)
Ostreopsis cf. *ovata*	Benthic diatoms	Stimulation of toxin production by cell-cell interaction	Ternon et al. (2018)
Ecotoxicological effects			
Karenia brevis	*Trichechus manatus latirostris, Tursiops truncatus*	Mortality↑	Flewelling et al. (2005)
		Multi-species mortality↑	Fire et al. (2015)
	Cynoscion nebulosus, Cynoscion arenarius, Sciaenops ocellatus	Juvenile recruitment↓, subadult & adult abundance↔	(Flaherty and Landsberg 2011)
	Caretta caretta	Mortality, unresponsiveness↑, paresis↑, circling↑	Fauquier et al. (2013a)
	Phalacrocorax auritus, Pelecanus occidentalis, Ardea herodias, Gavia immer	Disorientation↑, inability to stand↑, ataxia↑, seizures↑	Fauquier et al. (2013b)
	Brachionus ibericus	Feeds on *K. brevis* in mixed algal cultures	Kubanek et al. (2007)
	Brachionus plicatilis	Rejects *K. brevis*	
	Perna viridis	Most likely: growth↓, survival↓, juvenile recruitment↓	McFarland et al. (2016)
	Crassostrea virginica	Sperm viability↓, fertilization success↓, embryonic & larval survival↓	Rolton et al. (2015)
Brevetoxin	*Fundulus heteroclitus*	St: early phase↑, long-term↓ NDD: early phase↓, long-term↑	Salierno and Kane (2019)
Saxitoxin		St↓, NDD↑	

(continued)

Table 3.3 (continued)

Toxin/producer	Affected species or community	Effect	References
Brevetoxins	*Scophthalmus maximus*	Trophic transfer of Mn and Zn↔	Pouil et al. (2018)
PSP toxin, likely from *Pyrodinium bahamense*	*Chelonia mydas*, *Lepidochelys olivacea*	Mortality↑	Amaya et al. (2018)

↑increase/support; ↓decrease/inhibition; ↔no apparent effect; *NDD* nearest neighbor distance, *PSP toxin* paralytic shellfish poisoning toxin, *St* shoaling time

both *Fukuyoa* and *Gambierdiscus* have been found in the subtropical northern region of New Zealand and the Kermadec Islands, and in the more temperate waters of New South Wales (Australia) (Chinain et al. 2021). The new occurrences of *Gambierdiscus* spp. in temperate areas have coincided with the detection of toxic fish and ciguatera fish poisoning incidents in non-endemic regions. As a result, ciguatera fish poisoning by *Gambierdiscus* now appears to be a global problem (Wang et al. 2022; Li et al. 2023). Currently, ciguatera fish poisoning is the most frequently reported seafood-toxin illness worldwide, acquired from contaminated coral reef fish (Friedman et al. 2008).

The pharmacology of CTXs is summarized in Vale et al. (2015); it is characterized by their ability to cause persistent activation of voltage-gated sodium channels, increase neuronal excitability and neurotransmitter release, impair synaptic vesicle recycling, and induce cell swelling. These effects, combined with the ability to block voltage-gated potassium channels at high doses, are thought to underlie the complex symptoms associated with ciguatera (Nicholson and Lewis 2006).

Effects of Ciguatoxins

Tanyag et al. (2021) demonstrated that CTX accumulation in reef fishes is independent of trophic position in the West Philippine and Sulu Seas. CTXs are present in over 400 fish species, many of which are of commercial importance. This includes herbivores such as rabbitfish (*Siganus guttatus*) and carnivores such as barracuda (*Sphyraena barracuda*), parrotfish (*Scarus quoyi*), grouper (*Plectropomus leopardus*), moray eel (*Gymnothorax isingteena*), and snapper (*Lutjanus campechanus*). Zhu et al. (2022) found a high frequency of ciguatoxic carnivores (snappers and groupers) in the Marakei and Kiritimati Islands, with concentrations that were not safe for human consumption. Similarly, Sanchez-Henao et al. (2021) and Ramos-Sosa et al. (2022) showed that Caribbean C-CTX1 can accumulate in the muscle tissue of goldfish (*Carassius auratus*) and induce toxic symptoms, including hepatic metabolic disorders. Unfortunately, these studies used only phenotypic symptoms (hyperactivity) as an indicator of the detoxification and recovery capacity of this species, rather than using biochemical or biomolecular markers. Consequently, the underlying mechanisms of detoxification remain unclear. In the herbivorous *Naso brevirostris* fed *G. polynesiensis*, Clausing et al. (2024) identified CTX oxidation products indicative of phase I biotransformation activity (Steinberg 2012). However, biotransformation is insufficient to remove the entire CTX load, and the most significant proportion (44% of the total tissue burden) was found in the muscles, which will be consumed by humans. As discussed later, oxidation by CYP enzymes is only one of several ways in which an individual's tolerance to toxins can be increased.

In an earlier study, Naar et al. (2007) showed that CTXs are potent ichthyotoxic neurotoxins with a greater affinity for sodium channels than brevetoxins. Specifically, Pacific ciguatoxin-1 (P-CTX-1) is the most prevalent and potent CTX congener present in fishes from the Pacific Ocean. In a recent study, Mak et al. (2017) micro-injected P-CTX-1 into the yolk sack of 2-day-old marine medaka larvae (*Oryzias melastigma*) to simulate maternal toxin transfer. Exposure at sub-ppb levels (below 1 ng P-CTX-1 per g, with larvae weighing approximately 1 mg) resulted in adverse behavioral changes, altered physiological performance, and reduced survivability as early as 24 h post-injection. P-CTX-1 decreased the heartbeat rate and locomotion of the exposed larvae, likely due to a series of physiological processes and morphological changes, including pericardial edema, failure of swim bladder inflation, and spinal curvature. The exposed marine medaka also exhibited a reduction in the responsiveness to external stimuli, as well as a delay and paralysis in their reactions, which could increase their vulnerability to predation. P-CTX-1 was effectively distributed from the yolk sac to all parts of the fish body, including the head and trunk, within 24 h post-exposure. Notably, repeated low-dose P-CTX-1 exposure resulted in larval mortality comparable to that of a single high-dose exposure. Overall, the results indicated that P-CTX-1 can induce dose-dependent effects on the physiological and behavioral performance of larval marine medaka.

Yan et al. (2017) provide further insight into the toxic effects of P-CTX-1 on the development and molecular responses of marine medaka embryos (6 h post-fertilization) and eleuthero-embryos (within 12 h post-hatching). This study demonstrated that exposure to P-CTX-1 caused significant adverse effects on fish embryonic development, including hatching failure, immune dysfunction, a green-colored gallbladder, hemorrhaging, caudal fin malformation, spinal deformities, bradycardia, and hyperkinetic twitching. Overall, P-CTX-1 is more detrimental to newly hatched larvae than to embryos or ongrowing larvae. Increased expression of key functional genes [*telomerase reverse transcriptase* (stress response), *il-1β*, *tnf-α* (pro-inflammation), *cyclooxygenase 2* (cardiac development and immune toxicity), *tumor protein 53* (apoptosis and stress response), *cytochrome P450 family 26 subfamily B member 1* (bone development)] confirms the observed toxic effects.

Yan et al. (2020) demonstrated that dietary exposure of marine medaka (*Oryzias melastigma*) to P-CTX-1 impaired its reproductive performance. Fish that ingested more than 1.16 pg day^{-1} for 21 days exhibited abnormal behaviors, including diarrhea, abnormal swimming, loss of appetite, and decreased egg production. After 7-day exposure to P-CTX-1 at a dose of 1.93 pg day^{-1}, significant gender-specific differences in reproductive performance and reduced hatching rate of the offspring are observed. Notably, higher numbers of eggs were produced by females in the treatment pairs where only the males were exposed to P-CTX-1—the underlying reasons for this observation warrant further investigation.

Additionally, Li et al. (2022) not only identified several mechanisms of toxicity of the dinoflagellate *G. caribaeus* on *O. melastigma*, but also a tolerance mechanism. The fish, fed *Artemia* nauplii supplemented with *G. caribaeus*, exhibited reduced reproduction, inhibited growth, germ cell differentiation, and embryonic development, as well as altered neural activity and swimming behavior. Consumption of *G. caribaeus* significantly accelerated glycolysis, the tricarboxylic acid cycle, gluconeogenesis, and glycogen synthesis. It induced an immune stress response, upregulated the expression of pro-inflammatory factors, and caused liver tissue damage. Importantly, *O. melastigma* upregulated exporters under *G. caribaeus* exposure, namely *abcc3*, *slc22a6*, *slc16*, and *slc22a6* (phase III biotransformation (Steinberg 2012)), indicating that this fish can at least partially protect itself from toxin damage.

Comparing the results of ciguatera poisoning and biomagnification across different geographical regions of harmful algal blooms revealed significant variations between the Pacific, Atlantic, and Indian Oceans. One reason for the apparent inconsistency may be the different compositions of *Gambierdiscus* species in the biofilms, as recently described by Habibi et al. (2021).

Food Web Biomagnification?

The main ecochemical prerequisites for bioaccumulation and biomagnification are the lipophilicity and persistence of the chemicals in question. Mathematical modeling of biomagnification has shown that chemicals with a log *P* value (the logarithm of the partition coefficient between n-octanol and water) below 1.75 have no potential for biomagnification (Armitage and Gobas 2007). The log *P* value is 2.5 for P-CTX-1 and 3.0 for C-CTX-1 (NCBI PubChem Compound). Therefore, both congeners have the potential for biomagnification. The lack of biomagnification observed for the Caribbean congener may be due to its higher biodegradability or the presence of expanded genes encoding for xenobiotic degradation (Steinberg 2012). This question remains for future studies.

Regardless of the theoretical approach that CTXs have the prerequisite for biomagnification, the critical question remains: Are CTXs transferred up the food web? Indeed, CTXs can be found in various invertebrates, such as clams (giant Pacific clam, *Tridacna maxima*) (Roué et al. 2016) or sea stars (*Ophidiaster ophidianus* and *Marthasterias glacialis*) (Silva et al. 2015). However, studies on their presence in sharks or mullets have been inconclusive. Analysis of CTX extracts from the muscle and liver of sharks, encompassing nine species across four locations, revealed no detectable levels of the three most common Australian CTXs (Meyer et al. 2016). The benthic herbivore *Mugil cephalus* also shows little retention potential for CTXs. Ledreux et al. (2014) found no evidence of accumulation of the most abundant congener, oxocene CTX. Conversely, the less abundant oxopene CTXs may be retained, and their gradual transformation in fish over time increases their potency. Together with sequestration in the liver, this might lead to functional biomagnification in fish of higher trophic levels.

Contrarily, stable isotope studies have shown trophic transfer. For example, Winter and Tosteson (1992) reported that the isotope ratios of fish found in the stomach contents of ciguatoxic barracuda were significantly different from those found in the stomachs of unexposed barracuda. The isotope ratios of the toxic dinoflagellate *Ostreopsis lenticularis*, a suspected vector of ciguatera in the Caribbean, were found to be extremely light, particularly for nitrogen. The elevated nitrogen ratios observed in herbivorous (e.g., doctor fish) and carnivorous (e.g., squirrel fish) species within the stomach contents of CTX-exposed barracudas indicate that the dinoflagellate was a component of the food chain leading to these barracudas.

Mak et al. (2013) conducted a ^{15}N isotope analysis on (CTXs) and found that among the different CTX congeners, Pacific ciguatoxins (P-CTXs) are the most toxic. In coral reef ecosystems where ciguatera is present, P-CTXs were detected in 54% of herbivorous, 72% of omnivorous, and 76% of carnivorous fish, as well as in other marine organisms like the spiny lobster (*Panulirus penicillatus*) and an octopus. Interestingly, no significant correlation was found between P-CTX concentrations and lipid content in the three main predatory fish species examined, suggesting that P-CTX accumulation occurs independently of lipid content.

The ^{15}N and log P-CTX-1 concentrations in fish and invertebrate samples from Marakei, Kiribati, exhibited weak but statistically significant positive correlations (Fig. 3.15). However, analogous relationships were not observed for P-CTX-2 and P-CTX-3, suggesting that biomagnification of CTXs may result from biotransformation of P-CTX-2 and P-CTX-3 to P-CTX-1. In contrast to the above findings, Darius et al. (2007) reported that herbivores in Nuku Hiva, French Polynesia, exhibited elevated CTX levels relative to higher trophic level species. This finding is unexpected, given that herbivorous fish are the initial participants in the CTX transfer chain, and lipophilic components are expected to be biomagnified.

For the omnivorous pinfish (*Lagodon rhomboides*), Bennett and Robertson (2021) showed that C-CTX-1 bioaccumulates at low, environmentally relevant doses and that total ciguatoxicity persists for at least 3.3 months. Furthermore, the authors found that depuration largely depends on fish growth, which acts as a pseudo-elimination factor. These findings highlight that field-determined concentrations of C-CTX may not accurately reflect the mass balance of CTX that is bioavailable in food webs. Juvenile carnivorous lionfish (*Pterois volitans*) show a slow excretion process (Leite et al. 2021a, b). Notably, no acute signs of intoxication were observed throughout the experiment, indicating that *P. volitans* may have some degree of resistance to CTXs.

In a review of CTX processes and pathways in aquatic food webs, Yang et al. (2016) present a region-specific synthesis of the seemingly contradictory results: in most endemic areas, carnivores, such as groupers and other large fish, accumulate higher toxin concentrations than their herbivorous and smaller counterparts. However, in some other areas, such as French Polynesia, the opposite pattern is observed, challenging the biomagnification hypothesis. As noted above, one reason for this apparent inconsistency may be differences in the species composition of CTX producers (Habibi et al. 2021).

Costa et al. (2021) confirmed several unresolved issues regarding the biomagnification of CTXs, including biotransformation. Analyzing fish species from different trophic levels of the Selvagens Islands (NE Atlantic, Madeira, Portugal), the

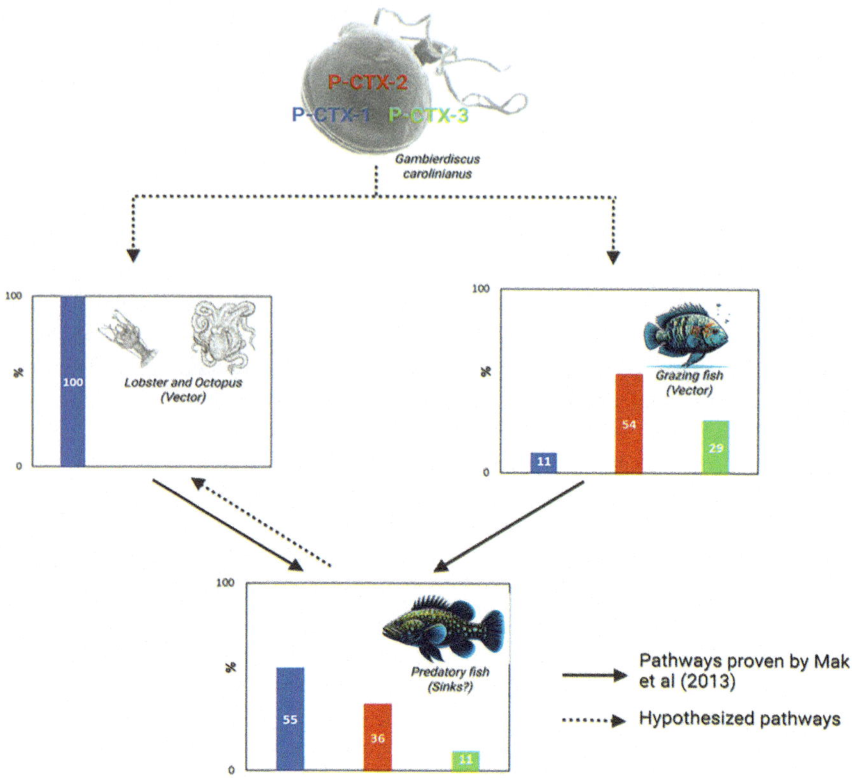

Fig. 3.15 Pathways of Pacific ciguatoxins (P-CTXs) in the food web system in Marakei, Kiribati. (Adapted from Mak et al. (2013), with permission from the American Chemical Society, SEM image credit to the Florida Fish and Wildlife Conservation Commission & National Centers for Coastal Ocean Science of the US NOAA, images of spiny lobster and *Octopus* credit to FAO

authors show that CTX-like toxicity was not detectable in primary consumers, specifically the Bermuda chub (*Kyphosus sectatrix*) and the scribbled leatherjacket filefish (*Aluterus scriptus*). However, toxicity occurred in fish species with an intermediate position in the food web, such as the parrotfish (*Sparisoma cretense*), zebra seabream (*Diplodus cervinus*), and barred hogfish (*Bodianus scrofa*) (Fig. 3.16).

Higher up the food web, grey triggerfish (*Balistes capriscus*) showed reduced CTX-like activity in cell-based assays, but toxicity is consistently present in blacktail comber (*Serranus atricauda*). Among top predators, the yellowmouth barracuda (*Sphyraena viridensis*) and amberjacks (*Seriola* spp.) exhibited highly variable results, ranging from undetectable to very high levels of toxicity (Costa et al. 2021). In a later study, both amberjack (*Seriola* spp.) and dusky grouper (*Epinephelus marginatus*) also showed high CTX-like activity in the cell-based assay (Ramos-Sosa et al. 2023).

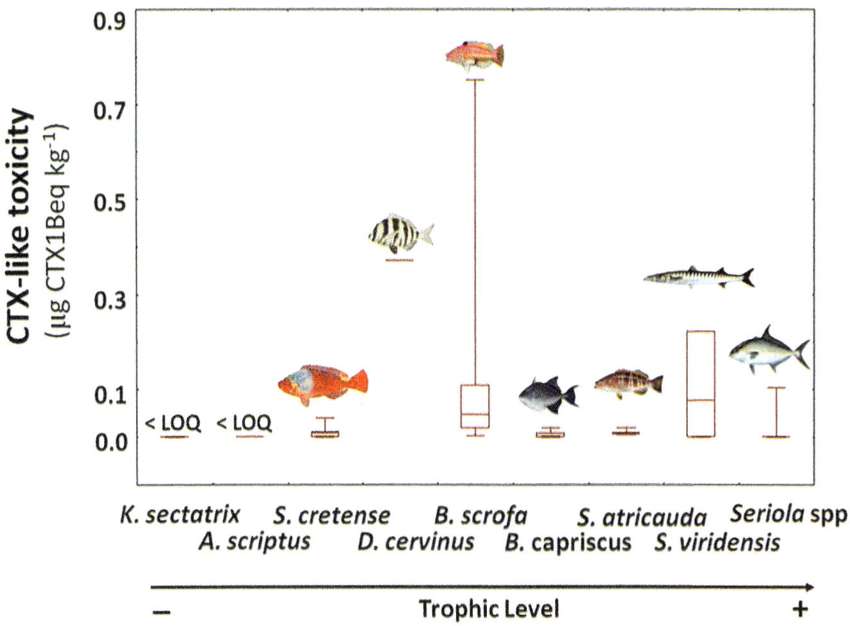

Fig. 3.16 CTX-like toxicity (µg CTX1B equivalent kg^{-1}) determined by neuroblastoma (Neuro-2a) Cell-Based Assay (CBA) in fish species collected from the marine food web of the Selvagens Islands during 5–7 September 2018. The data is presented as median values, along with the 25th and 75th percentiles, minimum and maximum values, with a total sample size of $N = 55$ (from Costa et al. (2021), credit to MDPI, Basel). Full fish names can be found in the text.

3.6.2.2 Further Ciguatera-Causing Agents

Another marine toxin associated with ciguatera-like poisoning is palytoxin (PLTX, Fig. 3.17). This extremely potent compound has been discovered in a near-fatal human poisoning case following the consumption of smoked shortfin squid (*Decapterus macrosoma*), as reported by Kodama et al. (1989). Although structurally and mechanistically distinct from ciguatoxins, PLTX can produce similar clinical symptoms, highlighting the diverse chemical nature of marine toxins capable of causing ciguatera-type syndromes.

PLTX is a polyketide (Moore and Bartolini 1981) and is considered one of the most toxic non-protein substances known in terms of toxicity in mice after intraperitoneal injection, second only to maitotoxin (Wiles et al. 1974). PLTX has been identified as a potent tumor promoter and cytotoxic molecule, causing actin filament distortion and triggering cell death or apoptosis (Louzao et al. 2011). The chemistry and pharmacology of PLTX are described in Ciminiello et al. (2015) and Louzao et al. (2015a, b), and its toxicological profile is presented in Pelin et al. (2016). The basic toxicological mechanism is illustrated in Fig. 3.17.

PLTX is mainly found in *Palythoa* soft corals, dinoflagellates (*Ostreopsis* spp.), and cyanobacteria (*Trichodesmium*). Initially associated with Indo-Pacific waters, PLTX now has a global distribution, with PLTX-like molecules identified on the

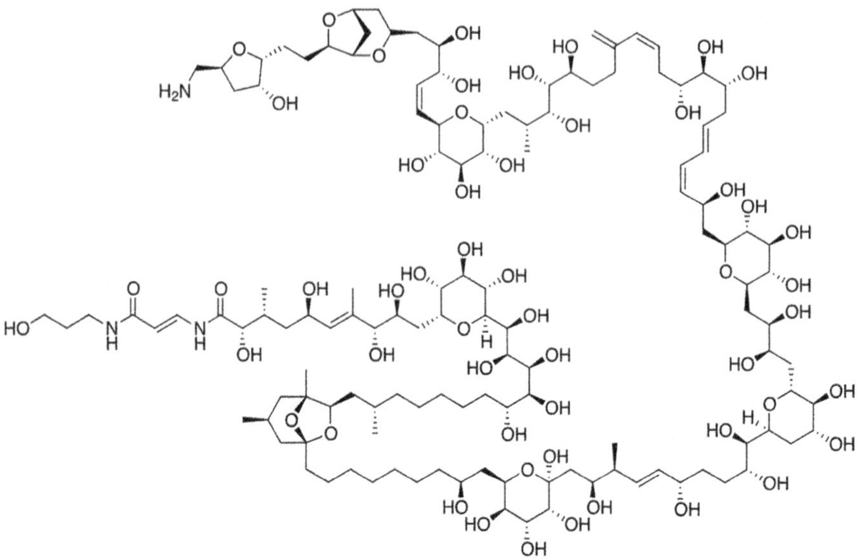

Fig. 3.17 Palytoxin, a toxin isolated from the soft coral *Palythoa toxica* and a metabolite of the marine dinoflagellate *Ostreopsis* spp. (credit to Charlsey, WikiMedia)

Mediterranean and Atlantic coasts (Louzao et al. 2015a, b; Accoroni and Totti 2016). PLTX blocks the Na^+/K^+-ATPase (Habermann 1989) in almost all excitable tissues, inducing passive ion conductance that triggers K^+ efflux, Na^+ influx, and massive membrane depolarization and tissue contraction (Wu 2009). Cases of human death due to PLTX have been reported after the consumption of contaminated crabs in the Philippines, sea urchins in Brazil, and fish in Japan (Wang 2008). Furthermore, PLTX has been identified as a tumor promoter by disrupting cellular signaling cascades (Wattenberg 2011). Overall, PLTX may act on more than one site in excitable cells, functioning as an agonist for low conductance channels that conduct Na^+/K^+, Ca^{2+}, and H^+ ions (Wang 2008).

In an early study, Gleibs et al. (1995) examined the distribution of PLTX in a Caribbean coral reef, finding that PLTX is more widespread in marine ecosystems than expected. In addition to *Palythoa* species, the toxin is also found in *Zoanthus* spp., competitors of *Palythoa* in the reef. Moreover, animals such as crustaceans (*Platypodiella* spp.) that live in close association with *Palythoa* colonies or polychaetes (*Hermodice carunculata*) that feed on *Palythoa* sequester PLTX. In general, herbivorous and omnivorous species show higher toxin levels than carnivorous ones (Biré et al. 2013).

Given the high levels of the toxin present in the bodies of the accumulating animals, it can be deduced that they have developed considerable tolerance or resistance to the toxin. This phenomenon has also been observed in other marine animals in which PLTX has been detected, including fish species (*Aluterus scriptus, Melichthys vidua, D. macrosoma, Epinephelus* sp., *Tetraodon* sp., *Scarus ovifrons*) (Hashimoto et al. 1969; Fukui et al. 1987; Kodama et al. 1989; Taniyama et al.

2001, 2002, 2003), sea urchins (*Paracentrotus lividus*) (Sardo et al. 2021), crabs (*Lophozozymus pictor*, *Demania cultripes*) (Yasumoto et al. 1986), shellfish (mussels: *Mytilus galloprovincialis*, clams: *Venus verrucosa*, hairy mussels: *Modiolus barbatus*) (Aligizaki et al. 2008; Gorbi et al. 2012; Accoroni et al. 2022), sea anemones (*Radianthus crispa*) (Mahnir et al. 1992), and other cnidarians (Zlotnick et al. 1995). However, the mechanism of tolerance remains unclear, making its role as a deterrent to potential predators questionable.

The tolerance of marine animals to the toxin allows its sequestration, enabling PLTX to accumulate in the food chain. Thus, PLTX can accumulate in the food chain, as illustrated in Fig. 3.18. Therefore, symptoms of ciguatera that differ from those regularly observed may be due to PLTX-like toxins rather than ciguatoxin and its congeners (Swift and Swift 1993; Amzil et al. 2012; Sud et al. 2013).

Few studies consider marine invertebrates as more than passive PLTX bioaccumulators. Louzao et al. (2010) found adverse effects of PLTX on the metabolic activity of mantle and hepatopancreas cells of non-adapted *Mytilus galloprovincialis*, demonstrating that PLTX is highly cytotoxic to this bivalve. These results are particularly important for marine aquaculture, as they demonstrate the ability of PLTX to compromise the integrity of bivalves not adapted to the presence of this toxin. This finding is further confirmed by Carella et al. (2015) and Sardo et al. (2021), who observed activation of the inflammatory response in exposed mussels or even death.

Sea urchins also appear to be susceptible to PLTX. In *P. lividus*, Sardo et al. (2021) found that direct exposure to *Ostreopsis* cf. *ovata* cultures caused partial or

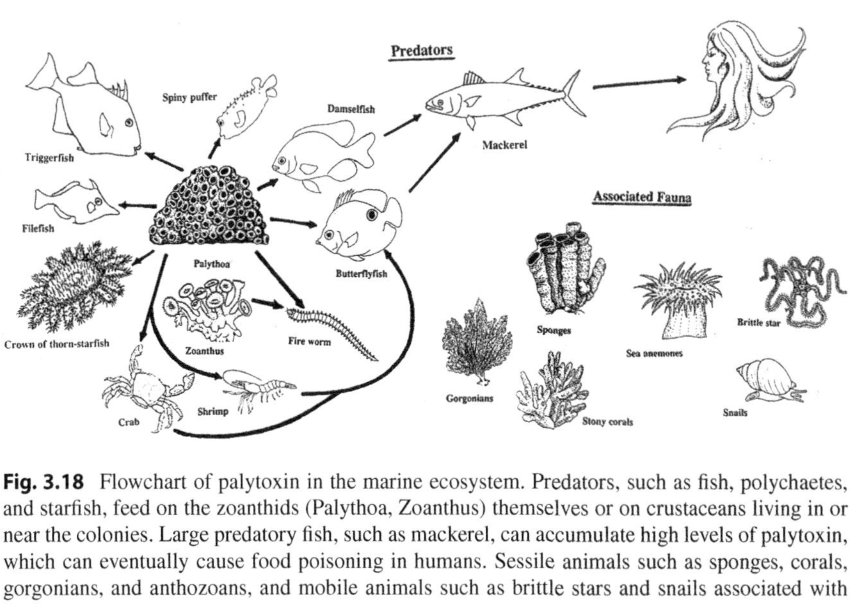

Fig. 3.18 Flowchart of palytoxin in the marine ecosystem. Predators, such as fish, polychaetes, and starfish, feed on the zoanthids (Palythoa, Zoanthus) themselves or on crustaceans living in or near the colonies. Large predatory fish, such as mackerel, can accumulate high levels of palytoxin, which can eventually cause food poisoning in humans. Sessile animals such as sponges, corals, gorgonians, and anthozoans, and mobile animals such as brittle stars and snails associated with zoanthid colonies often sequester palytoxin (from Mebs (1998), with permission from Elsevier)

total spine loss in a density-dependent manner, with animals dying at the highest microalgal concentrations.

In hepatopancreatic cells of *Penaeus vannamei*, PLTX extracts induced lipid peroxidation and protein carbonylation, resulting in cellular damage, as reported by Cen et al. (2019). Nevertheless, while no effect on the total antioxidant capacity of the hepatopancreas was noted, PTX increased the expression of immune-related genes (*crustin*, *toll*). Concurrently, the expression levels of *manganese superoxide dismutase* decreased gradually from 6 to 48 h and returned to normal levels after 72 h. These mechanisms may, at least in part, explain the tolerance of shrimps to PLTX.

Unlike the other fish species mentioned above, *Sparus aurata* does not appear to bioaccumulate PLTX-like toxins when fed contaminated mussels. This intriguing exception suggests species-specific differences in toxin uptake or metabolism and raises questions about the mechanisms behind this natural resilience. Giuliani et al. (2024) observed that seabreams fed with toxin-contaminated *M. galloprovincialis* started to reject the food after 6 days. Although no detectable toxin levels were found in muscle, liver, gills, and gastrointestinal tracts, the toxin-enriched diet induced alterations in lipid metabolism in the liver, displaying a decreased content of total lipids and fatty acids, along with overexpression of fatty acid biosynthetic genes, downregulation of β-oxidation genes, and modulation of several genes related to lipid transport and regulation. The authors attribute the lack of bioaccumulation to general mechanisms, such as the activation of biotransformation and detoxification processes, which are particularly effective in vertebrate marine organisms compared to invertebrates. This results in the production of different metabolites that can be more easily excreted.

3.7 Spirolides

Spirolides (SPXs, Fig. 3.19) are a family of marine phycotoxins found in shellfish that contain cyclic imines, a feature also found in other marine toxins, such as pinnatoxins, pteriatoxins, prorocentrolides, spiro-prorocentrimine, and gymnodimine (Munday et al. 2012). For the pharmacological profile, see Vilariño et al.

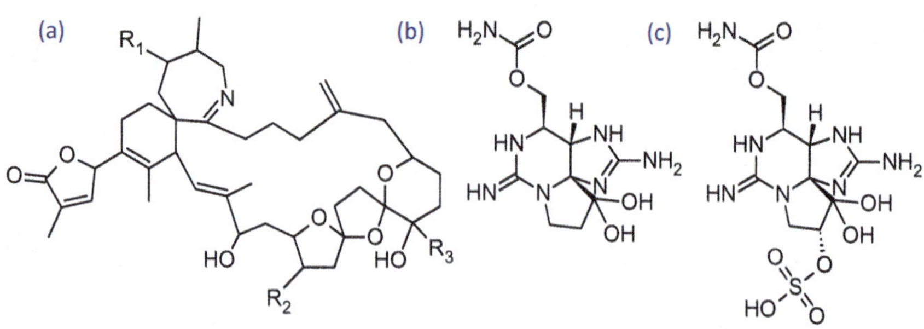

Fig. 3.19 Structure of spirolides (**a**), saxitoxin (**b**), and (**c**) gonyautoxin (GTX2)

(2015), and for the toxicological profile, see Molgó et al. (2016). Spirolides inhibit acetylcholine-induced calcium signaling and are likely irreversible antagonists of these receptors, as evidenced by the persistence of symptoms (Wandscheer et al. 2010). The primary producer of SPXs is the dinoflagellate *Alexandrium ostenfeldii*, which also produces saxitoxin. SPXs were first isolated and characterized in scallops and mussels harvested in Nova Scotia, Canada, in the early 1990s. Environmental conditions cause the SPX profile produced by *A. ostenfeldii* to vary significantly between geographical locations (EFSA 2010) and strains (Suikkanen et al. 2013; Van de Waal et al. 2015). These toxins accumulate in shellfish, crustaceans, and other aquatic organisms, such as fish. Since the discovery of SPX in 1995 (Hu et al. 1995), several analogues have been isolated (Nieva et al. 2020; Finch et al. 2024). SPXs have been found in algae and shellfish from a wide array of geographic locations, including the Netherlands, Ireland, Norway, France, Denmark, Spain, Greenland, Canada, the USA, Mexico, Argentina, Chile, and New Zealand (Amzil et al. 2007; Gerssen et al. 2010; Touzet et al. 2011; Tillmann et al. 2014; Van de Waal et al. 2015; Fabro et al. 2017; Paredes-Banda et al. 2018; Finch et al. 2024). These toxins have also been identified in the Mediterranean Sea and the brackish Baltic Sea (Suikkanen et al. 2013). This broad distribution underscores the global presence and ecological relevance of SPX-producing algae. Notably, isolates from Greenland did not contain PSP toxin or gymnodimines but were found to produce multiple spirolide congeners, highlighting regional variation in toxin profiles. In total, 12 different SPX analogs were detected, including 8 novel compounds that had not been previously identified (Tillmann et al. 2014).

Van de Waal et al. (2015) reported that *A. ostenfeldii* cells exhibit high allelopathic potency, with EC_{50} values based on lysis assays against the cryptophyte *Rhodomonas salina* ranging from 212 to 525 *A. ostenfeldii* cells mL^{-1}. Although SPX can accumulate in fish, trophic dilution is observed for the marine lipophilic phycotoxins (including SPX) analyzed in the above-mentioned Hong Kong study, with the low trophic magnification factor of 0.109 (Fig. 3.7) (Li et al. 2021). It remains unclear to what extent this study reflects the broader patterns of marine algal toxin bioaccumulation across diverse food webs. Further research is needed to determine the generalizability of these findings across different regions, species, and environmental contexts.

3.8 Saxitoxins

Saxitoxin (STX, Fig. 3.19) and its 57 analogues are powerful natural neurotoxic alkaloids, collectively known as paralytic shellfish toxins (PST). These compounds are structurally defined by a unique perhydropurine skeleton, which underlies their potent ability to bind to and block voltage-gated sodium channels in nerve cells, effectively halting nerve signal transmission and leading to paralysis. STX acts by reversibly inhibiting the passage of sodium ions through these channels, thereby preventing the propagation of action potentials in neurons (Fig. 3.1) (Durán-Riveroll and Cembella 2017). This inhibition results in muscle paralysis and, in severe cases, respiratory failure and death. For a toxicological profile, see Suarez-Isla (2016). Notably, although chronic exposure to sublethal concentrations of STX (1 μg L^{-1}) reduced antioxidant activity and immunity in zebrafish, it did not affect reproductive parameters (Haque et al. 2022). The authors found no gender differences in either phenotypic or biomolecular parameters.

Furthermore, PSP toxins are unique among neurotoxins because they are produced by organisms across both marine and freshwater environments, belonging to two entirely different kingdoms of life. In freshwater ecosystems, prokaryotic cyanobacteria are the primary source of PSP toxins, whereas in marine environments, the main producers are eukaryotic dinoflagellates. This duality underscores the evolutionary convergence in toxin production in ecologically and phylogenetically diverse organisms.

Filter-feeding bivalves, copepods, and fish contribute significantly to bioaccumulation through the food web (Doucette et al. 2005; Cusick and Sayler 2013). Often, herbivores reduce feed intake after ingesting toxic algae (Abdulhussain et al. 2024). Importantly, herbivory often increases the toxin levels in marine dinoflagellates (Yang et al. (2011), details below). Marine PSP toxin-producing dinoflagellates belong to the genera *Alexandrium* (formerly *Gonyaulax* and *Protogonyaulax*), *Gymnodinium*, and *Pyrodinium*, while freshwater PSP toxin production has been identified in several cyanobacterial genera, including *Anabaena*, *Raphidiopsis* (*Cylindrospermopsis*), *Aphanizomenon*, *Planktothrix*, and *Microseira (Lyngbya) wollei* (Lagos et al. 1999; Wiese et al. 2010; Hackett et al. 2013). The ecological role of STX is not well understood in either cyanobacteria or dinoflagellates. However, the evolution of STX production in these two ecologically distinct groups of organisms suggests that this toxin confers a benefit to producers that is not yet fully understood (Hackett et al. 2013).

Saxitoxin and its analogues can be divided into three categories. First, the carbamate compounds, which include STX, neo-saxitoxin, and gonyautoxins 1–4; second, the N-sulfocarbamoyl compounds, which include the C and B toxins; and finally, the decarbamoyl compounds, distinguished by the presence or absence of 1-N-hydroxyl, 11-hydroxy sulfate, and 21-N-sulfocarbamoyl substitutions, as well as epimerization at the C-11 position. Over the past few decades, at least 24 structurally related imidazoline guanidinium PSP derivatives have been identified and characterized from dinoflagellate species (Wang 2008).

Saxitoxins accumulate in a wide range of marine organisms through trophic transfer in the food web. While bivalves are considered the traditional vectors—concentrating toxins via suspension feeding on toxic dinoflagellates—STXs have also been detected in non-traditional vectors, such as crustaceans and gastropods (Fig. 3.20). STXs can move up the marine food web and ultimately affect top predators, including marine mammals and seabirds. These animals are exposed to the toxins through various pathways depending on their diet.

Marine mammals, such as sea otters and seabirds, that consume shellfish contaminated with STX can develop symptoms of paralytic shellfish poisoning (PSP), often leading to severe morbidity or mortality. Baleen whales can accumulate

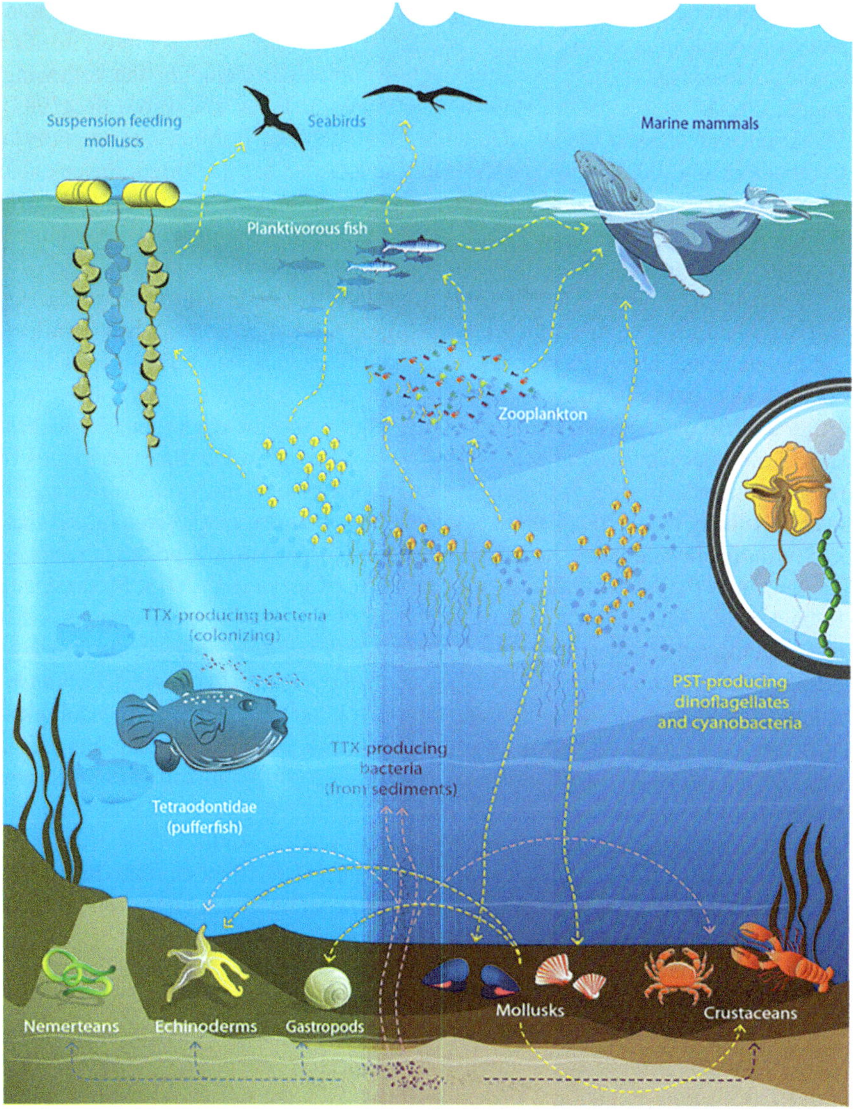

Fig. 3.20 Origin and fate of saxitoxins (STXs) and tetrodotoxins (TTXs) in the marine environment, indicating food web transfer vectors and distribution among species (from Cembella and Durán-Riveroll (2021), with permission from Elsevier)

significant amounts of STXs by filtering large volumes of krill and other zooplankton that have fed on toxic dinoflagellates. Carnivorous (toothed) whales, including dolphins and orcas, can acquire high toxin loads by consuming fish that have bioaccumulated STXs through their diet (Doucette et al. 2006; Cembella and Durán-Riveroll 2021).

The ingestion of PSP toxins via algal or zooplankton vectors has been demonstrated to result in mortality in adult and larval fish in field and laboratory feeding studies. This phenomenon has been well-documented, including the following, almost classic works: White (1977, 1980, 1981), Potts and Edwards (1987), Moestrup and Hansen (1988), Gosselin et al. (1989), Robineau et al. (1991), Cembella et al. (2002). However, the opposite has also been reported: the calanoid copepods *Eurytemora affinis* and *Acartia bifilosa*, from the northern Baltic Sea, do not graze on *Alexandrium ostenfeldii* due to rapid behavioral disturbance and incapacitation by the algal cells or their exudates (Sopanen et al. 2011).

A recent study shows that PSP toxin has a carry-over effect during reproduction and early ontogeny in the carnivorous gastropod *Chorus giganteus*, directly affecting the elemental bioenergetic constituents of the parents, and consequently influencing key reproductive parameters (fertility, egg size, hatching rate) (Andrade-Villagrán et al. 2024). Notably, high levels of eicosapentaenoic acid (EPA) and docosahexaenoic acid (DHA) were detected in the digestive glands of toxic individuals, tripling the levels quantified in non-toxic individuals of *C. giganteus*. It is likely that some of these essential fatty acids are synthesized by the snail itself. Marine invertebrates, including snails, have been shown to possess the necessary enzymes for the biosynthesis of ω-3 PUFAs (Kabeya et al. 2018; Monroig and Kabeya 2018). The increase in these important PUFAs in toxic snails suggests that the consumption of a diet containing PSP toxin may induce changes in the fatty acid profiles of *C. giganteus*, especially in the content of essential fatty acids (Andrade-Villagrán et al. 2022). These findings point to a potentially adaptive physiological response to PSP toxin exposure; however, the specific benefits and trade-offs of this biosynthetic adjustment remain unclear and warrant further investigation in future studies.

In another carnivorous snail (*Acanthina monodon*), Andrade-Villagrán et al. (2019) found that PSP toxin consumption had a negative effect on reproduction, as intoxicated adults had lower egg masses and delayed onset of oviposition. In addition, the development of embryos inside the capsules was delayed, and the number of hatched juveniles was lower; however, these few juveniles from intoxicated parents achieved higher growth rates during the next 6 months.

In the context of global warming, the Neotropical freshwater fish (*Rhamdia quelen*) was exposed to the crude extract of STX-producing *Raphidiopsis raciborskii*. Vicentini et al. (2024) demonstrated that the increase in temperature resulting from global warming intensifies the effects of STXs, with notable differences between the sexes. At 25 °C, the crude *Raphidiopsis* extract caused alterations in the gill antioxidant system in both sexes. In the liver, kidney, and gills, DNA damage increased in females at 30 °C and in males at both temperatures. More damage to macromolecules occurred in females at the highest temperature.

3.8.1 Biotransformation, Depuration, and Tolerance

In their review, Leal and Cristiano (2022) highlight that the uptake, transformation, and excretion of toxins by organisms are highly dependent on various factors, including the metabolic specificities of each species and the molecular structure of the toxins. The authors provide an overview of 45 bioconversion reactions in organisms (dinoflagellates, bivalves, and humans), noting that bivalves exhibit a higher number of identified reactions compared to dinoflagellates and humans. The main types of reactions are epimerization, hydrolysis, elimination and reductive cleavage, oxidation, sulfonylation, desulfonylation, hydroxylation, and glucuronidation, some of which are common across different organisms. All of these bioconversions directly impact toxicity.

STX biotransformation has been demonstrated in several aquatic species. For example, in *M. gigas* exposed to *Gymnodinium catenatum* (García-Lagunas et al. 2013) or *Alexandrium minutum* (Fabioux et al. 2015), in the copepod *Tigriopus japonicus* fed *A. minutum* or *G. catenatum* (Han et al. 2021a), and in *Artemia franciscana* also exposed to *G. catenatum* (Han et al. 2021b). Additionally, the four *A. franciscana hsp* genes (*hsp26*, *hsp40*, *hsp70*, and *hsp90*) are upregulated in response to *G. catenatum*, indicating that these genes are highly associated with cellular defense mechanisms (Han et al. 2021c). Other examples of biotransformation include the Chinese scallop (*Azumapecten farreri*) exposed to STX directly (Cao et al. 2018) or fed *Alexandrium* spp. (Lou et al. 2020, Kong et al. 2021, Wang et al. 2021a, b), *Mytilus galloprovincialis* and *Ostrea chilensis* fed *A. minutum* (Ding et al. 2017; Ortiz et al. 2019), Yesso scallops exposed to *Alexandrium* spp. (Liu et al. 2020, Lou et al. 2020, Kong et al. 2021, Wang et al. 2021a, b), *Perna viridis* on dietary *A. minutum* (Andres et al. 2019), *Oncorhynchus mykiss* exposed to extracts of *A. catenella* and *A. ostenfeldii* (Painefilú et al. 2020), and *Spisula solidissima* toxified with *A. catenella* (Bricelj et al. 2014). The latter authors highlight that toxin bioconversions may have a limited effect on the net detoxification rate, as some conversions can lead to increased toxicity. Notably, only a few studies have investigated the mechanism of MXR exporter detoxification. Notably, Xie et al. (2013) suggested that chitosan facilitates toxin depuration in *Magallana* (*Ostrea*) *rivularis*, although the mechanism remains to be elucidated.

Xu et al. (2021) highlight another possible detoxification pathway involving complement 1Q (C1Q), a vital recognition molecule in the immunological complement system that may also be involved in stress responses induced by endotoxins or exotoxins, possibly through detoxification processes. This biomarker, therefore, requires further validation.

Bivalves exhibit considerable variation in their ability to eliminate accumulated toxins and can be broadly categorized into two groups: fast and slow detoxifiers. Fast detoxifiers, such as *Mytilus edulis*, are capable of reducing toxin levels to within regulatory limits ($\leq$800 µg kg^{-1} of mollusk flesh (FAO and WHO 2008–2015, 2016), within just a few weeks, with daily losses of up to 15%. In contrast, slow detoxifiers like *Placopecten magellanicus* or *Haliotis midae* may retain toxins for months or even years, showing elimination rates of less than 3% per day (Pitcher et al. 2001). Toxin biotransformation, which can alter net toxicity, varies widely among species. Some bivalves, such as *Leukoma staminea* and *Spisula solidissima*, exhibit rapid enzymatic decarboxylation, while others, like *Mya arenaria* and *Mytilus edulis*, show limited toxin metabolism and are thus useful indicators of toxin source. Significant changes in toxin composition occur when algae are rich in low-potency N-sulfocarbamoyl toxins.

Comprehensive reviews of aquatic organisms in general, and copepods in particular, have been published by Landsberg (2002), Landsberg et al. (2005), Valério et al. (2010), and Turner (2014), highlighting varying effects of feeding on harmful algae. Copepods are capable of detecting toxins, including STX, through chemoreception. Teegarden (1999) reported that three tested marine copepods (*Acartia tonsa*, *Centropages hamatus*, and *Eurytemora herdmani*) were able to differentiate between toxic and non-toxic *Alexandrium* spp. cells, suggesting that selective behavior, rather than physiological effects, controls the grazing response of copepods exposed to toxic prey. However, *Acartia clausii* that ingested high levels of *Alexandrium* cells experienced decreased hatching success and nauplius production, as well as delayed nauplius development compared to those fed non-toxic diets (Frangoulos et al. 2000). Similar incapacitation occurred in *A. clausii* exposed to the toxic *Alexandrium minutum* (Dutz 1998).

Literature data of *Alexandrium–Acartia* interactions showed a substantial dispersion for the variables ingestion rate and egg production. Therefore, Felpeto et al. (2024) conducted a systematic quantitative review of this interaction at the genus level. *Alexandrium* and its toxins do not affect food intake and egg production. However, *Alexandrium* toxins have adverse effects on hatching (bi-phasic) and mortality. The results support the notion that chronically exposed copepod populations have improved physiological performance relative to naïve populations over a wide geographic range. Details of potential tolerance mechanisms are discussed in Chap. 5.

If not depurated, *Alexandrium* toxins trigger the defense mechanism in marine organisms. In the Chinese scallop *Chlamys farreri*, gene multiplication in the HSP70 family plays a significant role in coping with chemical stress. Hu et al. (2019) identified 65 *hsp70* genes, with a considerable expansion in the Hspa12 subfamily. Phylogenetic analysis indicates that several *hsp70* genes underwent mussel-specific gene duplication. This is a common strategy employed by animals in extremely stressful environments, enabling them to enhance their stress response mechanisms (Zhang et al. 2012).

3.8.2 Toxic Effects of Saxitoxin

3.8.2.1 Invertebrates

Yang et al. (2011) conducted the first transcriptomic study exploring how grazer presence influences toxicity in the protist *Alexandrium minutum*, specifically in response to the copepod *Centropages typicus*. The authors identified 14 genes whose expression was affected by the presence of copepods. The differentially expressed genes were compared after one day—before the onset of enhanced toxicity—and after 3 days, when cellular toxin levels in the induced samples were five times higher than in the control cultures. Comparison with previously identified differences between toxic and non-toxic strains revealed two sequences associated with PSP toxin content, one of which is a NAD:arginine ADP-ribosyltransferase domain. However, the precise role of this domain in toxin biosynthesis or regulation remains to be fully clarified.

In freshwater ecosystems, cladocerans such as *Daphnia* spp. are the primary vector for poisoning pelagic food webs. Nogueira et al. (2004) reported that *D. magna* challenged by diets containing *Cuspidothrix* (*Aphanizomenon*) *issatschenkoi* exhibits poorer growth and survival than even starved individuals. In contrast, Soares et al. (2009) reported conflicting findings, showing that a strain of *Raphidiopsis* (*Cylindrospermopsis*) *raciborskii*—which produced STXs and gonyautoxins rather than cylindrospermopsin—did not exhibit acute toxicity toward *Daphnia*.

When *Alexandrium* spp. vegetative cells are filtered by oysters and other mollusks, a portion of the ingested cells can survive passage through the digestive tract, remaining viable after egestion. However, in the eastern oyster *Crassostrea virginica* (and most likely other mollusks), significant changes in the toxin composition of the cells occur after intestinal passage. Smith et al. (2011) found that vegetative cells and pellicle cysts from feces had increased levels of STX and decreased levels of one gonyautoxin per cell compared to levels before gut passage. After passing through the intestine, pellicle cysts demonstrated better survival rates than vegetative cells, supporting the notion that pellicle cyst formation is an effective survival strategy under adverse conditions.

In addition to the potential mechanisms leading to toxin resistance described in the section above regarding okadaic acid, Bricelj et al. (2005) identified a sodium channel mutation that resulted in STX resistance in the soft-shell clam *Mya arenaria*. This mutation appears to have played a crucial role in the evolution of clam populations in regions with frequent blooms of STX-producing *Alexandrium* spp. Clams from the bloom regions exhibit 24–44% lower mortality when exposed to *Alexandrium* spp. than clams from non-HAB regions. This toxin resistance is due to a single glutamate-to-aspartate substitution in the outer pore loop region of the voltage-gated sodium channel domain II. Although glutamate and aspartate differ by only one methylene group, this substitution makes resistant clams approximately 1500 times less sensitive to STX than non-resistant clams (Fig. 3.21) (Bricelj et al. 2005).

STXs interfere with the apoptosis pathway. Rolland et al. (2014) described a feedback mechanism between pro-apoptotic (*bax*) and anti-apoptotic *bax inhibitor-1*) genes in the digestive gland of the oyster *M. gigas* exposed to the paralytic shellfish toxin producer *A. catenella* (Fig. 3.22). In the Yesso scallop, Zhu et al. (2021) identified and characterized genes encoding inhibitors of apoptosis proteins (IAP). After exposure to *A. catenella*, fourteen IAP genes showed significant responses in the hepatopancreas and kidney. In addition, the authors highlighted gene duplications that may explain the scallop's adaptation to the fluctuating and stressful marine

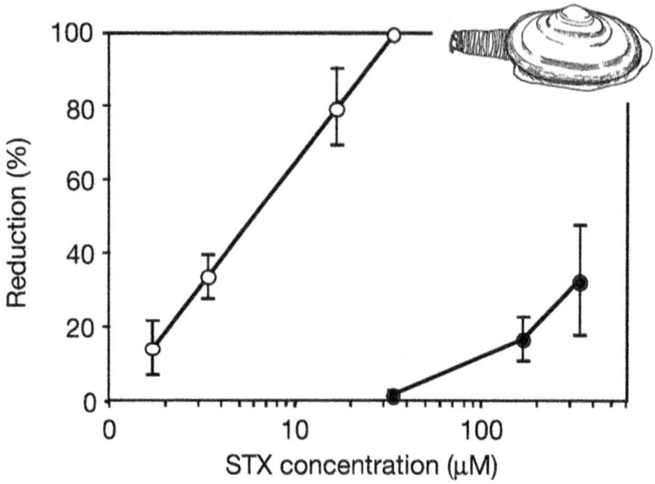

Fig. 3.21 Effect of STX on action potentials in nerves of *Mya arenaria*. Percentage decrease in action potentials (relative to pre-exposure conditions) measured in isolated nerves of sensitive (○) and resistant (●) clams exposed to increasing STX concentrations (from Bricelj et al. (2005) with permission from Springer Nature; image of living clam with siphon protruding, credit to Baker and Mann (1990))

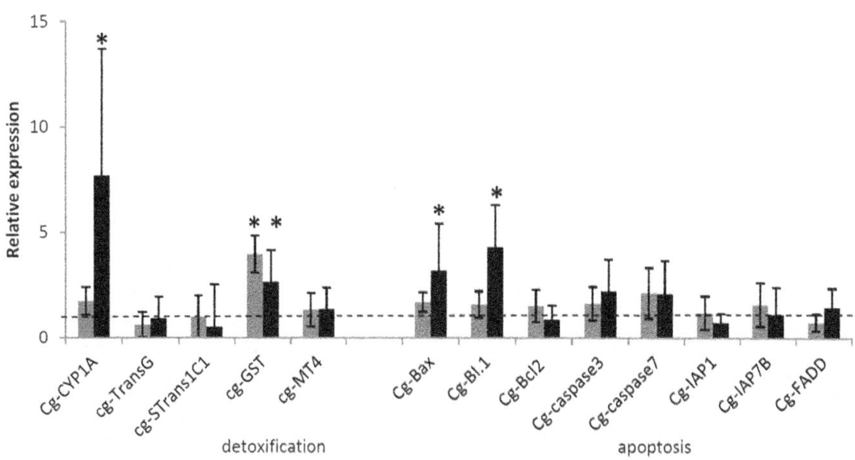

Fig. 3.22 The relative expression of 5 detoxification-related genes and 8 apoptotic-related genes (normalized to ribosomal protein F40) in the digestive gland of *Magallana gigas* exposed for 48 h to *Alexandrium tamarense* (grey) or to *Alexandrium catenella* (black) compared to the non--exposed oysters. The dashed line represents the mean value of the control group (non-exposed oysters) for each gene. Bars represent mean ± standard error ($N = 10$ animals). * $P < 0.01$ (from Rolland et al. (2014), credit to MDPI, Basel)

environment. However, the anti-apoptotic activity could eventually lead to tumor formation, which needs to be addressed experimentally.

Chan et al. (2021) confirmed a general proteomic stress response in the hepatopancreas of the bivalve *Meretrix meretrix* exposed to PSP toxin-producing dinoflagellate *Gymnodinium catenatum*. They reported significant changes in the expression of 15 proteins involved in crucial biological functions, including detoxification and antioxidant defense (increase in catalase and superoxide dismutase) after PSP toxin exposure.

It is worth noting that even farmed omnivorous and carnivorous shrimps are often exposed to various toxic substances during the farming process, which impacts their survival and quality. Pérez-Linares et al. (2008) evaluated the effects of PSP toxins from *G. catenatum* on the white leg shrimp *P. vannamei*, finding that mortality occurred at doses >1.1 μg g^{-1} STX. Lower doses induced paralysis of the pereiopods (walking legs), imbalance, and abdominal spasms in the animals. Target organs such as the heart and brain are severely damaged, including loss of cohesion and reduced cell density. As a result, PSP toxins can move up the marine food chain, as demonstrated in the Bay of Fundy, where a bloom of *A. catenella* caused significant mortalities of Atlantic salmon in aquaculture facilities (Sephton et al. 2007).

Similar to the transcriptional assessment of the effects of *Prorocentrum lima* (OkA producer) in the Pacific oyster as discussed above, Garcia-Lagunas et al. (2016) examined the transcriptional effects of *G. catenatum* in this bivalve. High transcription levels of the *p21* gene, a cyclin-dependent kinase inhibitor, and the *p53* gene, a tumor suppressor protein, were found in oysters exposed to dinoflagellate diets, indicating an activation of cell cycle checkpoints induced by DNA damage. Histologically, oysters exposed to the toxic dinoflagellate showed exfoliation, epithelial rupture, and inflammation, which correlated with changes in the expression level of *casp1*, a gene involved in the initial inflammatory response. Overall, the transcriptional response of the Pacific oyster to STX exposure closely mirrors that of OkA exposure, suggesting that both phycotoxins activate the same defense pathway.

An effective immune response is essential for growth, reproduction, and survival. In a comprehensive study, Dong et al. (2022) investigated the transcriptional and physiological responses to gain insight into the mechanisms regulating the pro-inflammatory to anti-inflammatory balance in two species of mussels, *Mytilus galloprovincialis* and *Argopecten irradians*, exposed to PSP toxins. Significant changes were observed in the mitogen-activated protein kinase (MAPK) and AMP-activated protein kinase (AMPK) signaling pathways, as well as in the specific regulatory pathway, phosphatidylinositol 3-kinase (PI3K)-protein kinase B (AKT). In *M. galloprovincialis*, the c-Jun N-terminal kinase (JNK) and p38 MAPK pathways were activated, which in turn activated the MAPK pathway. Furthermore, there was a significant expression of inhibitor genes, including *tnf*, *nf-κb*, and *ampk*, particularly during the slow metabolic phase. Consequently, the intense pro-inflammatory response and a corresponding anti-inflammatory response in *M. galloprovincialis* provided an effective mechanism for high filtration activity and low tissue damage after exposure to PSP toxins. However, *A. irradians* appears to focus on regulating

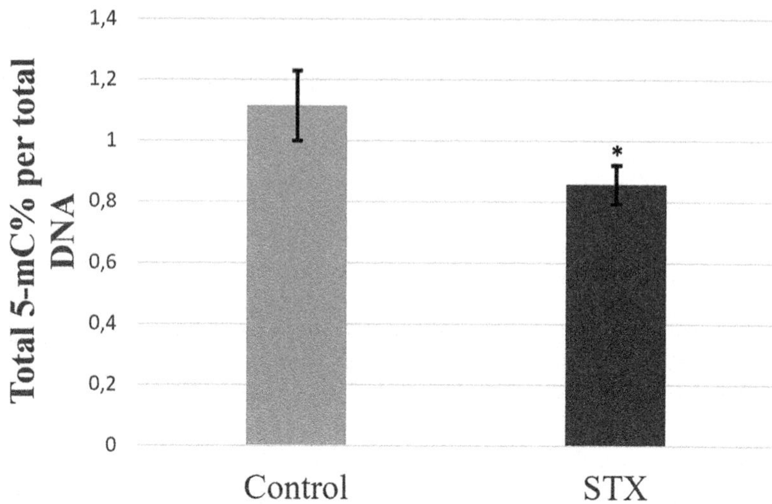

Fig. 3.23 Total 5-methylcytosine (5-mC) DNA methylation levels in *Daphnia magna* exposed to 30 μg L^{-1} of saxitoxin (STX, in dark grey) and control organisms (Control, in light grey). Error bars are SEM; *statistically significant differences (Independent samples t-test or Mann-Whitney U test, $P < 0.05$) between the exposed and control organisms (from Pinto et al. (2024), credit to Elsevier)

downstream cell metabolism and apoptosis, leading to immune imbalance, severe tissue damage, and high mortality.

Raising new concerns for environmental and public health safety, Pinto et al. (2024) revealed that even a brief, 24-hour exposure to STX at concentrations deemed safe for recreational freshwaters by the World Health Organization (30 μg L^{-1}) can have profound biological effects on *Daphnia magna.* This short-term pulse exposure significantly reduced feeding and heart rates, lowered total cholinesterase and catalase activity, and increased oxidative stress markers such as glutathione peroxidase and lipid peroxidation. Strikingly, DNA damage levels rose markedly in the exposed individuals (Fig. 3.23), accompanied by a substantial reduction in global DNA 5-methylcytosine (5-mC) levels. This study showed that even short-term exposure to STX has significant effects on critical molecular and cellular pathways, modulating swimming patterns in *D. magna* individuals at levels considered safe for humans by the WHO. The observed decrease in total 5-mC DNA methylation may be explained by changes in DNA methyltransferase activity and/or alterations in the methionine cycle, which affect the availability of methyl donors. However, an analysis of target genes is missing, which should be conducted next.

3.8.2.2 Fish Species

Saxitoxins don't just stop at their source; they travel up the food web. In a striking study by Jiang et al. (2007), researchers traced the transfer of these potent neurotoxins from the toxic dinoflagellate *A. catenella* to the small freshwater grazer *Moina mongolica*, and eventually into the tissues of red drum fish (*Sciaenops ocellatus*). Interestingly, with each step up the food chain, the toxin concentrations declined, suggesting that both the crustacean and fish can metabolize or transform the saxitoxins to some extent.

What happens to fish exposed to STXs through the food chain? Do the limited studies on trophic transfer suggest these phycotoxins are safely broken down before reaching top consumers? Actually, no. One of the first recorded instances of massive fish mortality was made in the fifteenth century by Valentim Fernandes in Arguin Bay, Mauritania. Paralytic shellfish toxicity was first associated with marine dinoflagellates, namely *A. catenella*, in the 1930s (Costa 2016). Later, Barrientos et al. (2019) reported STX-induced mortality in green turtles (*Chelonia mydas*) feeding on contaminated Caribbean sharpnose-puffer (*Canthigaster rostrata*). The likely STX producers were the dinoflagellates *Prorocentrum micans* and *Protoperidinium divergens*. These few studies and historical records indicate that SXTs are relatively non-degradable, can be transferred to higher trophic levels, and cause significant damage.

In addition to STX effects, Costa 2016 also provides a review of documented fish kills associated with PSP toxins. As of 2014, only fourteen events had been reported in the literature. These events occurred from 1977 to 2009 and were mainly located on the North and South American continents, with only one instance in Europe, specifically England, in 1968, when sand eels (*Ammodytes* spp.) served as both vectors and victims of PSP toxins. In Asia, there are two reports: one from Japan involving farmed yellowtail (*Seriola quinqueradiata*) and another from Taiwan regarding a mass mortality of milkfish (*Chanos chanos*). On the African continent, there were some reports from Lobito and Luanda Bay, both in Angola, where *Alexandrium* species may have been involved in fish kills (Costa (2016) and references therein). Since then, records of PSP toxin-associated fish kills have increased only slightly (Table 3.4). Focusing on the distribution of (harmful) PSP toxin-producing algal blooms, recent long-term surveys in the Philippines and Malaysia, China, Korea, Russia, and Japan, and along the Atlantic margin of Europe are provided by Yñiguez et al. (2021), Sakamoto et al. (2021), and Bresnan et al. (2021). Unfortunately, the effects on fish or invertebrate species are not reported.

3.8.3 Emerging PSP Toxins

The toxic dinoflagellate *Gymnodinium catenatum* produces a novel group of hydroxybenzoate PSP toxins called GC toxins (Fig. 3.24). Lin et al. (2024) studied the biokinetics of GC toxins in various tissues of the scallop *Argopecten irradians*. When exposed to *G. catenatum*, scallops retain most of the PSP toxins in the viscera, followed by the adductor muscle, gill, and mantle. The toxin load in the adductor muscle was surprisingly high and surpassed the regulatory safety threshold within 6 days of the experiment. Most of the GC toxins were enzymatically hydrolyzed to decarbamoyl toxins in scallops, and the viscera had higher enzymatic activity than the other tissues.

Table 3.4 Recent fish kills associated with PSP toxins

Affected fish species	Dinoflagellate species	Region	References
Acipenser brevirostrum	*Alexandrium* sp.	Northeastern USA coast	Fire et al. (2012)
Acipenser oxyrhynchus, *Osmerus mordax*, *Ammodytes* spp.	*Alexandrium tamarense*	St. Lawrence Estuary, Canada	Starr et al. (2017)
Anguilla reinhardtii	*Amphidinium carterae*	Coastal lagoon in Sydney, Australia	Murray et al. (2015)
Chelon labrosus	*Alexandrium minutum*	Coastal lagoon, S. Jorge Island, Azores	Santos et al. (2014)
Menidia menidia	*Gyrodinium jinhaense*	Avery Point, Groton, CT, USA.	Sprecher et al. (2021)
Salmo salar	*Alexandrium tamarense*	Aquaculture, Nova Scotia	Cembella et al. (2002)
	Alexandrium catenella	Chilean fjords	Mardones et al. (2015)
Sprattus fuegensis		Beagle Channel, South America	Cadaillon et al. (2024)
Grouper, snapper, sea bass	3 *Alexandrium* spp.	Aquacultures, Aman Island, Strait of Malacca	Razali et al. (2015)
Groupers, threadfin, golden trevally, rabbitfish	*Karenia mikimotoi*, *Karlodinium* cf. *australe*, *K.* cf. *veneficum*, *Takayama xiamenensis*	Johor Straits	Leong et al. (2015)
Not mentioned	*Alexandrium catenella*	Western Arctic Sea	Walsh et al. (2011)
Various cases, species not mentioned	*Alexandrium minutum*	Hunts Bay, Kingston Harbor, Jamaica	Ranston et al. (2007)

3.9 Domoic Acid

Domoic acid (DA, Fig. 3.26) is an unusual amino acid that shares structural similarities with kainic acid, proline, and the endogenous excitatory neurotransmitter glutamate. As a structural analog of glutamic acid, DA binds with exceptionally high affinity—up to one hundred times greater than L-glutamic acid—to specific ionotropic glutamate receptor subtypes in the brain (Cendes et al. 1995) (Costa et al. 2014). This intense and prolonged activation of receptors, particularly those involved in excitatory neurotransmission, disrupts normal neuronal signaling.

Exposure to domoic acid triggers excitotoxicity, leading to both cellular damage and structural brain injury (Ramsdell 2007; Grant et al. 2010). Research from experimental models and clinical case studies consistently shows that DA primarily targets the hippocampus, leading to seizures and impairments in cognitive functions. The neurobehavioral effects of DA are distinctive, involving both temporary and lasting memory deficits similar to those observed in human anterograde amnesia. Clinically, DA exposure manifests as amnesic shellfish poisoning (ASP), a

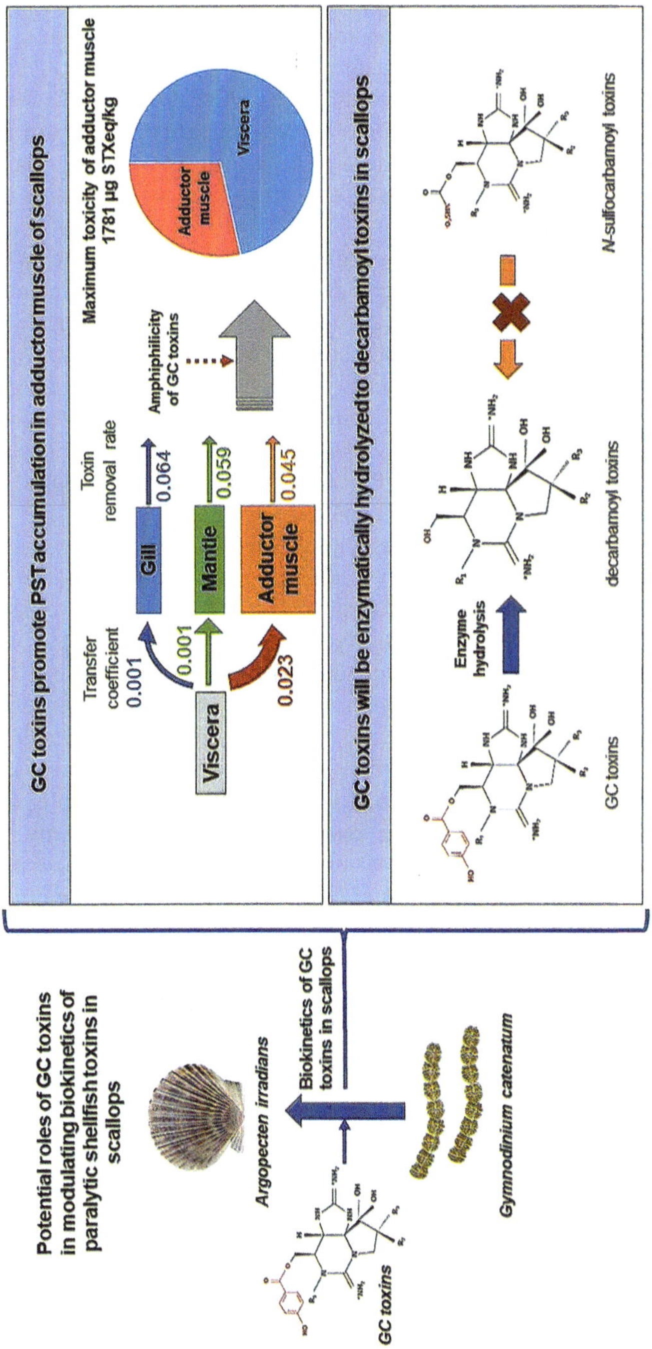

Fig. 3.24 Biokinetics of hydroxybenzoate paralytic shellfish toxins in the scallop *Argopecten irradians* (from Lin et al. (2024), with permission from Elsevier)

condition characterized by gastrointestinal symptoms, confusion, memory loss, and in severe cases, seizures or death.

DA exhibits a particularly strong binding affinity for kainate receptors, which are primarily involved in modulating excitatory transmission in the central nervous system. It binds to these receptors more potently than to AMPA (α-amino-3-hydroxy-5--methyl-4-isoxazole propionic acid) or NMDA (*N*-methyl-D-aspartate) receptors, although it does interact with those to a lesser extent as well. The selective and high-affinity binding of DA to kainate receptors contributes significantly to its neurotoxic potential.

The biosynthesis of DA was only recently elucidated. It is now known to be produced by certain species of the marine diatom genus *Pseudo-nitzschia*, with the domoic acid biosynthesis (*dab*) gene cluster identified as the core genetic element responsible for its production (Brunson et al. 2018). This discovery has advanced our understanding of the ecological and molecular basis of DA production in marine systems. For further details on the biosynthesis and physiological function of DA in diatoms, as well as its mechanisms of toxicity, readers are referred to Stonik and Stonik (2015) and Zabaglo et al. (2016). A comprehensive toxicological profile is provided in Tasker (2016).

Domoic acid is primarily produced by marine diatoms of the genus *Pseudo-nitzschia*, with 26 toxicogenic species (out of 52 total species) (Bates et al. 2018). In 2015, *Pseudo-nitzschia* formed the largest harmful algal bloom ever recorded in the Pacific Ocean, extending along nearly the entire coast from Alaska to Mexico and causing massive ecological and economic damage (McCabe et al. 2016). High DA concentrations are often observed during the colder seasons of the year (Horner et al. 1997; Kvrgić et al. 2022). In addition to diatoms, the red alga *Chondria armata* can produce DA, from which it was first isolated in 1957 and used in Japan as a remedy against intestinal parasites and as an insecticide (Zabaglo et al. 2016). Despite various hypotheses about the ecological role of DA in marine plankton, conflicting results prevent a satisfactory explanation of current phenomena.

Domoic acid (DA) can reach high concentrations in both invertebrate and vertebrate plankton feeders and may be passed up the food chain to top predators. Some species of *Pseudo-nitzschia* capable of producing DA have been associated with the deaths of southern right whale (*Eubalaena australis*) calves along Argentina's Península Valdés, Argentinia (D'Agostino et al. 2015, 2017). Interestingly, filter-feeding crustaceans, such as the sand crab (*Emerita analoga*) sometimes accumulate more DA than better-known vectors such as the mussel *Mytilus californianus* (Powell et al. 2002). Trainer et al. (2012) and Tasker (2021) provide comprehensive lists of organisms in which DA is internalized.

Although numerous invertebrate species are known to accumulate DA, documented cases of these organisms showing signs of DA-related toxicity remain rare Tasker (2021). One laboratory study reported adverse effects on larval development in the king scallop (*Pecten maximus*) (Liu et al. 2007). Another study suggests impaired innate immunity in the bay scallop (*Argopecten irradians*) (Chi et al. 2019). This impairment is indicated by the transcription of the *fibrinogen-related protein* (*frep*) gene (Fig. 3.25). FREP is a pattern recognition receptor in the innate

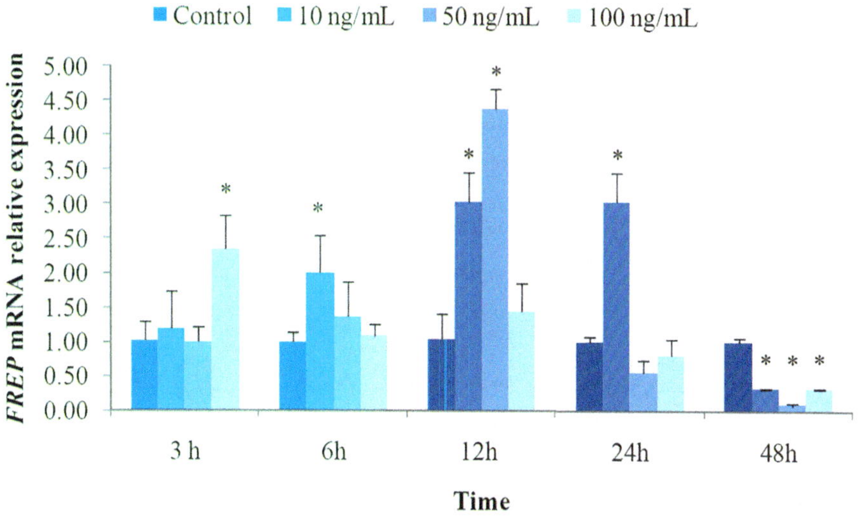

Fig. 3.25 Effects of domoic acid (DA) on non-specific immune responses of the bay scallop *Argopecten irradians* at different time intervals after exposure to DA at three concentrations (10, 50, and 100 ng mL^{-1}): Transcription of fibrinogen-related protein. Data represent mean ± SD ($N = 3$) at the same sampling time. * denotes significant differences between the treatments of each exposure time ($P < 0.05$) (from Chi et al. (2019), credit to MDPI, Basel)

immune response that scavenges xenobiotics (Zhang et al. 2009). *Frep* was expressed at 3–24 h post-exposure (hpe) to DA; however, it was inhibited in all treatment groups at 48 hpe. These results highlight that prolonged exposure to DA inhibits the ability of scallops to recognize and detoxify xenobiotic particles (Chi et al. 2019).

Whether the apparent insensitivity to DA in certain species reflects differences in the structure of the kainate receptor subunits or an inability to distribute DA from the digestive glands remains unclear. Mauriz and Blanco (2010) speculated about a lack of transport proteins in the digestive gland of the king scallop. However, a definitive explanation for the lack of toxicity in invertebrate species is still lacking (Tasker 2021).

The mode of DA intoxication is thought to arise from its structural similarity to glutamic acid, a neurotransmitter (Clayden et al. 2005). To understand the mechanisms of neurotoxic injury associated with DA exposure in fish, Lefebvre et al. (2009) examined the transcriptome of whole brains from zebrafish (*Danio rerio*) after intracoelomic injection of DA at both symptomatic (1.2 μg DA g^{-1}) and asymptomatic (0.47 μg DA g^{-1}) doses. Interestingly, the study indicated that the regulation of genes involved in various biochemical pathways is affected at DA doses below those causing observable behavioral symptoms. It is noteworthy that several genes involved in critical subcellular processes, including immune function, RNA processing, ion transport, metabolism, and signal transduction, were found to be downregulated following low-level DA exposures. These exposures did not result in observable neurobehavioral damage, suggesting the potential for a neurological risk

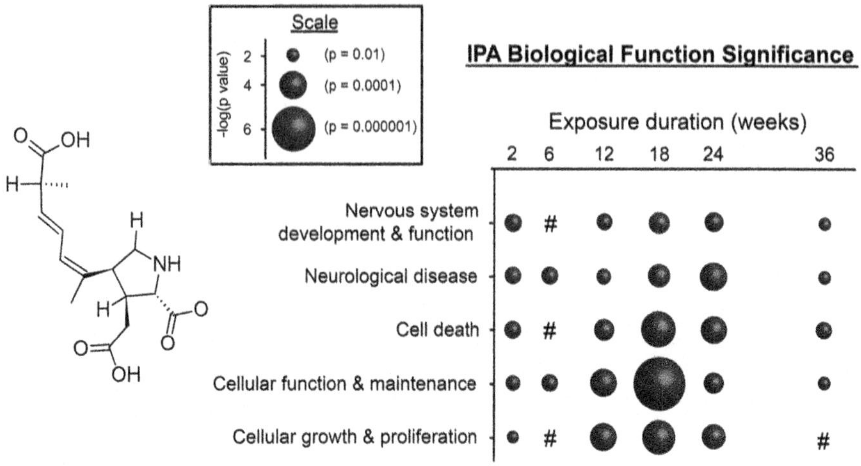

Fig. 3.26 Structure of domoic acid. Significantly altered biological functions in zebrafish (from Ingenuity Pathway Analysis, IPA) associated with cellular and neurological health, function and development. Bubble diameter corresponds to −log(P value) for functional categories and time points possessing ≥3 genes; # indicates not significant and/or < 3 genes within the category. Scale provides reference for relative bubble diameter (from Hiolski et al. (2014), with permission from Elsevier)

even with asymptomatic exposures. One specific gene, the zebrafish homolog of the human tumor suppressor gene *ndrg4*, which is associated with memory loss, was significantly downregulated in the zebrafish model.

In a subsequent study, Hiolski et al. (2014) reported that chronic low-level DA exposure altered gene transcription and impaired mitochondrial function in the central nervous system. The IPA Bio Functions are shown in Fig. 3.26. Notably, the IPA biofunctions were significantly altered by DA exposure at five to six of the six total exposure time points. Additionally, low-level (asymptomatic) exposure to seafood toxins impaired mitochondrial functions. Specifically, oxidative stress-related mitochondrial dysfunction appears to play a role in chronic low-level exposure, similar to findings from previous acute exposure studies. Furthermore, Hong et al. (2015) highlighted additional aspects of embryonic developmental toxicity. DA exposure not only disrupted normal heart development but also altered the expression of genes and calcium ion channels associated with heart development.

Several studies have highlighted the differences in strategies used by oysters and mussels to cope with domoic acid (DA) exposure. Mafra et al. (2010) compared *Crassostrea virginica* (the Eastern oyster) and *Mytilus edulis* (the blue mussel). They found that oysters accumulated significantly less DA per unit body weight when exposed to toxic *Pseudo-nitzschia multiseries* cells. Interestingly, smaller oysters accumulated more DA than larger ones, likely due to higher uptake rates, yet even the largest oysters never exceeded regulatory toxin thresholds. In contrast, the relationship between body size and DA accumulation in Blue mussels was less consistent, resulting in greater variability in weight-specific DA concentrations among larger individuals. DA was cleared more rapidly from the viscera than from other tissues in both species, suggesting redistribution of the toxin to other tissues rather than excretion. This internal transfer appeared more pronounced in oysters. Moreover, although small oysters eliminated DA faster than larger ones, their excretion rates were still consistently lower than those of similarly sized mussels. Thus, the reduced DA accumulation in oysters compared to mussels cannot be attributed to faster detoxification but likely reflects species-specific differences in uptake and internal processing.

In another oyster (*M. gigas*) and another mussel (*M. californianus*), DA triggered different responses. Jones et al. (1995) documented specific differences among these bivalves in their ability to tolerate and avoid exposure to *Pseudo-nitzschia*. The oyster rapidly accumulated significant concentrations of DA in soft tissue, resulting in a generalized stress response characterized by shell closure, hemolymph acidosis, and acute transient hypoxia. In contrast, the mussel increased its ventilatory flow, resulting in a mild, uncompensated respiratory alkalosis.

In *Mytilus galloprovincialis*, Pazos et al. (2017) identified a total of 1158 differentially expressed unigenes, comprising 686 upregulated and 472 downregulated ones. Several membrane transporters belonging to the SLC family (solute carriers) were overexpressed in exposed mussels. This overexpression is of great interest since some of these transporters may be involved in the absorption and/or excretion of DA. Some of the other enriched gene families (sulfotransferases, aldo/keto reductases, and carboxylesterases) may be involved in detoxification processes, oxidative stress response, and immunological processes.

DA moves through marine food webs largely via bioaccumulation in invertebrates, but planktivorous fish also play a key role in transferring the toxin to higher trophic levels. Among these, the northern anchovy (*Engraulis mordax*) has received the most attention, repeatedly identified as a major vector in DA-related poisonings of marine mammals and fish-eating birds (Tasker (2021) and references therein).

A striking study from Monterey Bay, California, revealed that blooms of the DA-producing diatom *Pseudo-nitzschia* have led to the death of numerous marine mammals and seabirds over the past decade. Key links in this toxic chain are Euphausiids (krill), vital members of the local zooplankton grazer community. As a primary food source for squid, baleen whales, and many seabirds, krill serve as potent vectors for transferring DA to higher trophic levels, thereby amplifying its impact throughout the ecosystem.

Bargu et al. (2002) reported toxin analyses and gut content examinations of krill *Euphausia pacifica* collected from Monterey Bay in 2000. The presence of *Pseudo-nitzschia australis* frustules in the digestive tract of *E. pacifica* confirmed that this toxic diatom species is an important part of their diet, implicating this alga as a source of DA.

Besides krill, anchovies and copepods are also important vectors for the uptake of DA into the food chain. For example, DA-related deaths of birds (e.g., pelicans, cormorants), sea lions, and minke whales off the coast of California, USA, are associated with ingestion of contaminated planktivorous fish, such as the northern anchovy (*Engraulis mordax*), which had fed directly on the toxic diatoms (Lefebvre et al. 2002; Fire et al. 2010).

Comparatively few studies have analyzed the grazing behavior of copepods or their ability to accumulate and retain DA when exposed to *Pseudo-nitzschia*. Nevertheless, the limited available research indicates that copepods show no discernible difference in grazing rates when exposed to either toxic or non-toxic *Pseudo-nitzschia* spp. Additionally, retention of DA by copepods after removal of toxic *Pseudo-nitzschia* cells has been demonstrated. These findings, although preliminary, suggest that several copepod taxa may serve as potential vectors for DA transfer in marine food webs (Leandro et al. 2010). These authors specifically documented the role of *Calanus finmarchicus* in the trophic transfer of DA, which often dominates North Atlantic zooplankton assemblages and constitutes a major component of the diet of the critically endangered North Atlantic right whale (*Eubalaena glacialis*). Grazing and toxin accumulation/depuration experiments showed that *C. finmarchicus* consumes DA-producing *Pseudo-nitzschia multiseries*. After ingestion, the copepod accumulated DA and retained it for up to 48 h post-exposure. These results underscore the potential of *C. finmarchicus* to facilitate DA trophic transfer in marine food webs where toxic *Pseudo-nitzschia* are present (Leandro et al. 2010). Similarly, reproductive problems in the southern right whale (*Eubalaena australis*) have been linked to DA transfer by copepods (D'Agostino et al. 2017). This study highlights the need for further research into the transfer of phycotoxins from mothers to calves. However, the specific health effects of DA on adult whales and their offspring remain unknown.

Despite the well-documented neurotoxicity and mass mortality observed in vertebrates that consume DA-contaminated fish, there is little evidence that the fish vectors themselves suffer from neurotoxic effects. One hypothesis suggests that fish may not absorb DA from the digestive tract, making them insensitive to dietary DA exposure. To test this hypothesis, Lefebvre et al. (2007) conducted oral gavage experiments on juvenile coho salmon (*Oncorhynchus kisutch*), administering ecologically relevant doses of DA and tracking uptake, excretion, and tissue distribution over time. Surprisingly, DA was detected in the plasma within 15 min post-exposure, as well as in various organs, including the liver, heart, spleen, kidney, muscle, brain, and bile. Nevertheless, the fish showed no signs of excitotoxicity. These findings indicate that while DA is clearly absorbed from the gut and distributed systemically, fish do not exhibit neurotoxic effects even at the highest naturally occurring oral exposure levels. The pattern of DA update and elimination indicates that most of the absorbed toxin was efficiently excreted through the kidneys and bile, thereby preventing the accumulation of toxic levels in sensitive neural tissues. Notably, within just 1 h of injection, over 20% of the administered DA was found sequestered in the bile, providing strong evidence for the role of biliary excretion in clearing the toxin from the bloodstream.

A subsequent literature review on DA and fish behavior reported a variety of effects, including alterations in metabolic activity in brain, liver, and gill tissues. Collectively, these studies shed light on the potential mechanisms of DA-induced toxicity in vertebrates but do little to address the implications for field-exposed fish populations. Lefebvre et al. (2012) concluded that fish are not significantly impacted by domoic acid during typical toxigenic *Pseudo-nitzschia* blooms in nature.

Almost simultaneously, Roy et al. (2013) reviewed contrasting aspects of toxin transfer in marine food webs. They highlighted that toxins produced by phytoplankton can be transferred across multiple trophic levels. Analysis of North Atlantic right whale (*Eubalaena glacialis*) feces from the Bay of Fundy revealed the presence of two types of toxins: paralytic shellfish toxins produced by *Alexandrium* species and domoic acid (DA) produced by *Pseudo-nitzschia* (Doucette et al. 2012), with copepods, specifically *Calanus finmarchicus*, as vectors (Leandro et al. 2009). More than 70% of the fecal samples contained PSP toxins, while 25–30% contained DA. Although the direct adverse effects of these toxins on *E. glacialis* were not measured, previous research has speculated about the adverse physiological consequences for cetaceans consuming toxic copepods. For instance, Durbin et al. (2002) postulated that the toxins could affect respiratory function, feeding behavior, and ultimately the reproductive status of the whale population.

Overall, similar to marine invertebrates, fish appear to be largely resistant to DA, although the research topic has received little attention beyond a few isolated experimental studies. While oral exposure typically results in minimal neurotoxic symptoms, studies have shown that when DA bypasses the gastrointestinal tract—such as through injection or gavage—mild neurotoxic effects may occur. This suggests that certain neuronal populations in fish may be susceptible to DA under specific exposure routes. However, the rapid elimination of DA from the bloodstream, particularly through renal and biliary excretion pathways, combined with limited absorption from the digestive tract, likely underpins the apparent resilience of fish to dietary DA exposure and might explain the different susceptibility between fish and vertebrates to DA (Tasker 2021).

A yet completely neglected factor in the different susceptibility to DA may lie in fundamental differences in brain structure and function across vertebrates. The distinctive characteristics of the fish blood-brain barrier (BBB)—which regulates ion transport and permeability differently than in mammals—might play a critical role in limiting DA's neurotoxic effects in fish (Bernstein and Streicher 1965; Dunton et al. 2021). Adding another layer of complexity, fish exhibit unique glutamate receptor compositions; for example, zebrafish possess duplicated sets of eight AMPA receptor subunits (gria1–4a and b), unlike the four subunits found in mammals (Gria1–4) (Hoppmann et al. 2008). Whether these receptor variants differ in their affinity or sensitivity to DA remains unknown, but could be pivotal. Moreover, while the fish dorsolateral pallium (Dl) serves as a functional homolog to the mammalian hippocampus (Selim and Alva 2024)—the primary target of DA—a lower neuronal density and differing receptor organization in this region may further reduce vulnerability. These anatomical and molecular distinctions may contribute to, or even be the main reason for, the observed resilience to DA exposure in fish. This hypothesis underscores the urgent need for targeted research to unravel these evolutionary adaptations within aquatic ecosystems.

3.10 Oxylipins

Oxylipins are a diverse and functionally important class of lipid-derived signaling molecules formed through the enzymatic oxidation of polyunsaturated fatty acids (PUFAs). These PUFAs are first liberated from membrane phospholipids by phospholipase A2 (PLA2) and then undergo stereospecific oxidation catalyzed by cyclooxygenases (COX), lipoxygenases (LOX), or cytochrome P450 monooxygenases (CYPs) (Mosblech et al. 2009; Barquissau et al. 2017). As membrane-permeable and transient bioactive compounds, oxylipins play a crucial role by modulating inflammation, immune responses, and cellular homeostasis (Cambiaggi et al. 2023). These molecules are widely distributed across aerobic organisms, including plants, animals, and fungi. In plants, oxylipins—most notably the jasmonates—are central regulators of growth, defense signaling, and environmental stress responses (Wasternack and Feussner 2018). In animals, the major oxylipins are the eicosanoids, which are formed from C_{20} PUFAs and exhibit potent, often opposing, biological effects (Steinberg 2022).

In their review, Ruocco et al. (2020) summarized that certain oxylipins, particularly polyunsaturated aldehydes (PUAs), are toxic compounds produced by diatoms that negatively impact the reproductive success of several marine invertebrate consumers. By acting as chemical deterrents against grazers, they affect gamete viability, fertilization, embryogenesis, and larval fitness (Fig. 3.27). The majority of

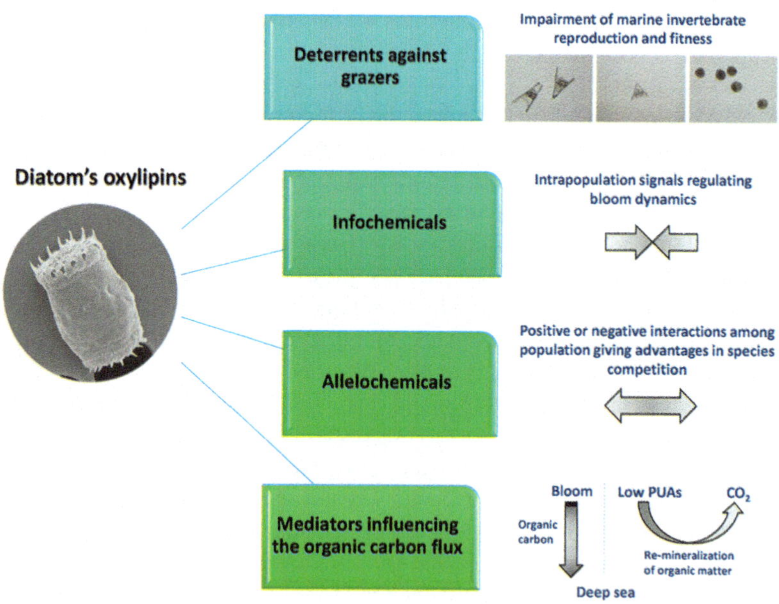

Fig. 3.27 Oxylipins in aquatic environments can act as deterrents against grazers, info-chemicals, allelochemicals, and mediators that influence carbon recycling (from Ruocco et al. (2020), credit to MDPI, Basel)

recent studies have focused on the effects of oxylipins on sea urchins and copepods. Given the ecological significance of diatom blooms in the marine environment, expanding these studies to include other marine invertebrates would provide valuable insights into the mechanisms of oxylipin responses in marine organisms and their ecological implications.

In addition to their synthesis during wound activation, these oxygenated fatty acid derivatives can also be actively produced by intact diatom cells through mechanisms that are still unknown. These small chemical mediators play a crucial role in regulating cell-to-cell communication within the same species and between different species, thereby influencing the structure and composition of phytoplankton communities (Fig. 3.27).

Ruocco et al. (2020) provided a comprehensive overview of studies examining the adverse effects of oxylipins, focusing primarily on experiments involving pure, commercially available PUAs, including 2*E*,4*E*-heptadienal, 2*E*,4*E*-octadienal, 2*E*,4*E*,7*Z*-octatrienal, 2*E*,4*E*-decadienal, and 2*E*,4*E*,7*Z*-decatrienal. In addition, the review covered investigations using four oxygenated fatty acids characterized by hydroxy functional groups. All of these studies focused on the reproduction of sea urchins, tunicates, and copepods, which are common diatom feeders in both benthic and planktonic environments. The major novelty of these studies was the use of genomic approaches to elucidate the effects of oxylipins on gene expression patterns in these organisms. Comparative analyses of the results suggest that some common pathways are activated in response to grazing on oxylipin-producing diatoms by these marine invertebrates from different phyla, as shown in Fig. 3.28.

3.11 Emerging Toxins

New toxic marine algae continue to be identified, expanding our understanding of these organisms. The ongoing discovery of new toxic marine algae and the complex compounds they produce not only broadens our knowledge of marine biotoxins but also raises new concerns regarding their ecological roles and potential impacts on human and animal health.

Among these emerging toxins, pinnatoxins have attracted significant attention. These are a class of potent cyclic imine toxins known for their ability to inhibit both neuronal and muscle-type nicotinic acetylcholine receptors, resulting in neurotoxic effects (Fig. 3.29 depicts the structure of pinnatoxin G). The chemistry of pinnatoxins is detailed in (Mabe and Zakarian 2015), while the toxicological profile of pinnatoxins is comprehensively reviewed by Molgó et al. (2016).

Pinnatoxins were first isolated from bivalve mollusks of the genus *Pinna*, hence their name (Uemura et al. 1995), and the causative organism was later identified as the benthic dinoflagellate *Vulcanodinium rugosum* (Nézan and Chomérat 2011; Rhodes et al. 2011).

In addition to pinnatoxins, *V. rugosum* also produces portimines, another group of cyclic imines that display high cytotoxicity in mammalian cells. Studies, such as that by Selwood et al. (2013), have shown that portimines activate caspase

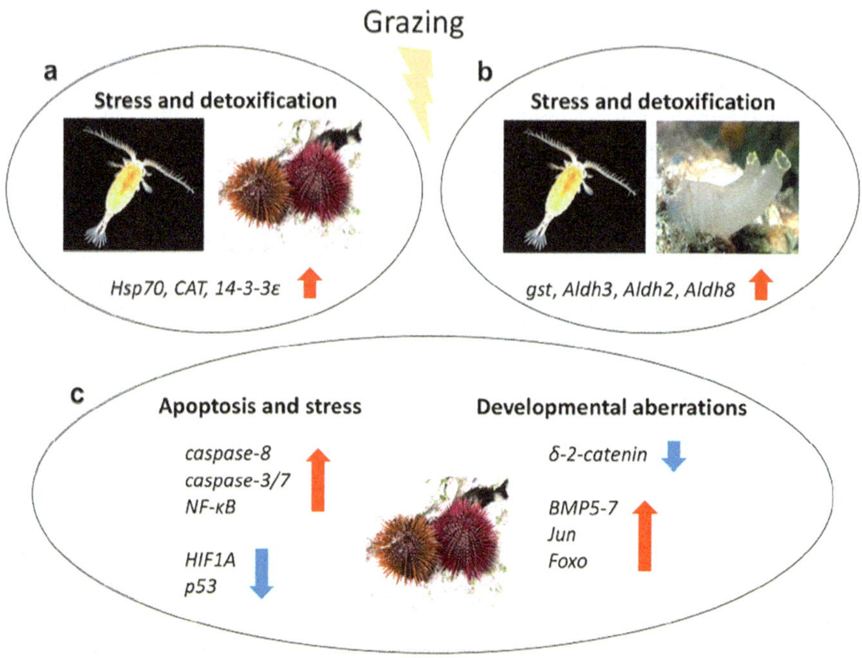

Fig. 3.28 Molecular response of marine invertebrates to diatom-derived oxylipins; potential common molecular pathways between copepods vs. sea urchins (**a**) and copepods vs. tunicates (**b**), together with some of the most strongly affected genes in sea urchins (**c**). Red arrows indicate gene upregulation; blue arrows indicate gene downregulation (from Ruocco et al. (2020), credit to MDPI, Basel, with photos of copepods, sea urchins, and tunicates from https://www.marinespecies.org/. Genes and their function, see Abbreviations and Glossary

pathways, indicating their pro-apoptotic activity. The dual production of both pinnatoxins and portimines by a single organism underlines the complex and potentially synergistic nature of emerging marine toxins. Both pinnatoxins and portimines are lipophilic, accumulate in filter-feeding organisms such as bivalves (Otero et al. 2020; Norambuena and Mardones 2023), and can enter the human food chain, posing significant risks to consumers (Arnich et al. 2020).

In a seminal study, Bouquet et al. (2022) demonstrated that juvenile mullets (*Chelon* (*Liza*) *ramada*) can accumulate pinnatoxins and portimines from feeding live *V. rugosum* (Fig. 3.29). While these fish did not show signs of toxicity, the discovery of viable cysts in their feces after 3 days of feeding suggests that fish may serve as both vectors for toxin bioaccumulation and agents of algal dispersal. The germination of these cysts within 8–22 h in enriched seawater culture medium indicates a possible mechanism for the expansion of harmful algal populations in coastal ecosystems. This raises ecological concerns about the role of fish in facilitating the persistence and spread of toxic dinoflagellates, particularly in shallow and nutrient-enriched lagoon systems.

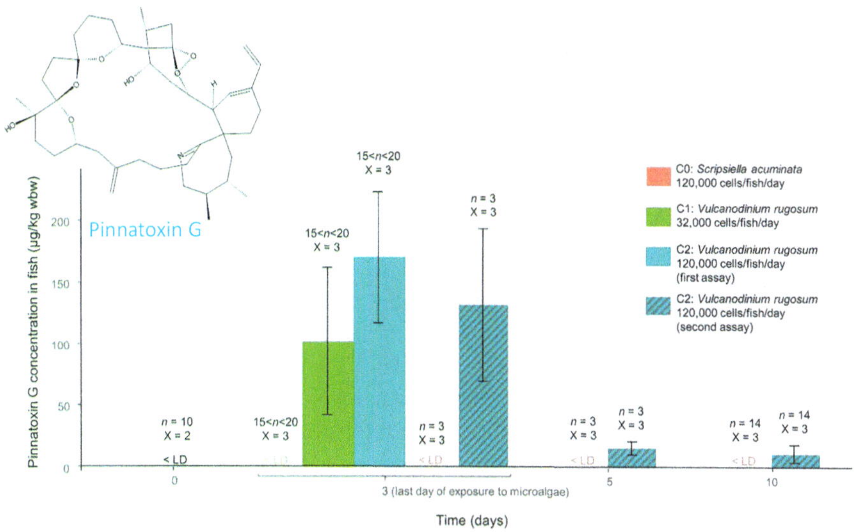

Fig. 3.29 Bar chart of mean pinnatoxin G concentrations found in fish before the experiment (day 0), after 3 days of exposure, and 2 days and 7 days after the end of the exposure. *n*: number of individuals per replicate. X: number of replicates. Error bars represent ± SD from triplicate mean values (from Bouquet et al. (2022), credit to MDPI, Basel)

Further research is needed to determine the response of other fish species that share similar feeding habits with *C. ramada* or consume *Vulcanodinium*-contaminated prey, as well as to assess whether long-term exposure to these toxins affects life-history traits, including gut microbiota composition.

In one such system, the Ingril Lagoon in France, recognized as a hotspot for *V. rugosum* blooms, Hort et al. (2023) found persistent contamination of *Mytilus galloprovincialis* with pinnatoxin A and pinnatoxin G throughout the year. A notable finding was the identification of several fatty acid esters of pinnatoxin-G, with palmitoyl-conjugates being predominant. Although concentrations of these esters were low, their potential significance lies in their bioavailability. Fatty acid esters may act as latent toxin reservoirs, releasing active pinnatoxins during digestion via enzymatic hydrolysis, potentially increasing the risk of intoxication. The authors caution that the in vivo toxicity of these metabolites should not be underestimated, as the fatty acid esters of marine biotoxins may be hydrolyzed by lipases and other enzymes during digestion, potentially releasing free toxin forms into the gastrointestinal tract of consumers.

These findings point to a critical knowledge gap in our understanding of toxin metabolism in both invertebrate and vertebrate vectors. The bioavailability, persistence, and toxicokinetics of such esterified toxins in consumers remain largely unexplored. Moreover, the long-term effects of chronic exposure to low levels of cyclic imines—whether through direct algal ingestion or via trophic transfer—are unknown. As *Vulcanodinium rugosum* and other toxin-producing marine algae continue to expand their range—possibly facilitated by climate change, eutrophication, and human-mediated dispersal—the monitoring and assessment of their toxins becomes increasingly important. Given that cyclic imines are not yet regulated in many regions, including parts of the EU, early detection methods and risk assessments must be developed to protect seafood safety and ecosystem health. Multidisciplinary efforts combining field monitoring, molecular biology, toxicology, and oceanography will be essential to anticipate and mitigate the risks associated with these potent marine biotoxins.

3.12 *Pfiesteria* Toxin Controversy

The toxicogenicity of *Pfiesteria* dinoflagellates has been the focus of an intense debate. Are these organisms true "killer organisms" as claimed by Shimizu (2003)? Or was the hunt for *Pfiesteria* akin to the "cold fusion" of biology, as suggested by Kaiser (2002)—a high-profile scientific claim that initially garnered attention but was later met with skepticism due to a lack of reproducible evidence? The controversy surrounding *Pfiesteria*, particularly surrounding the claims of being a super-killer made by North Carolina State University marine biologist JoAnn Burkholder and her coworkers (Burkholder et al. 1992), was the hottest topic in the field of harmful microalgae and microalgal toxins in recent decades. The organism was portrayed as a true monster with an unusual 24-stage life cycle that enabled it to attack fish and humans in a manner reminiscent of an extraterrestrial invader. The organism was reported to release an ultra-potent neurotoxin, purportedly harmful to both humans and fish. The toxin was reported to be so potent that even people in proximity to dead fish or the cultures of the organism became seriously ill, necessitating that research had to be conducted in a P3 safety level laboratory (Shimizu 2003). The scientific accounts of the controversy are well summarized in an editorial in "Science" (Kaiser 2002).

Central to the debate is the toxicogenicity of the organism. Drgon et al. (2005) conducted a small-scale study in which in vitro exposure of fish larvae to clonal *P. piscicida* showed that fish mortality was directly proportional to the density of dinospore cells, a fundamental principle in (eco)toxicology. Dinospores clustered around the mouth, eyes, and opercula, suggesting that the toxic effects derive from direct interactions with skin, gill epithelia, or mucosal surfaces of the fish.

Furthermore, toxic strains of *Pseudopfiesteria* (*Pfiesteria*) *shumwayae* have been implicated in fish kills, although its ability to secrete an exotoxin lethal to its prey remains controversial (Miller and Belas 2003). A study published in 2002 reiterated that this dinoflagellate could kill fish through direct contact and by feeding on their skin through micropredation (Vogelbein et al. 2002). However, skepticism intensified when Berry et al. (2002) reported that *P. shumwayae* lacks the polyketide synthetase genes responsible for producing dinoflagellate toxins, especially ichthyotoxins. While compounds other than polyketides could potentially be responsible for toxicity, the evidence was inconclusive. In a later study, a new complex of hydrophilic, metal-containing *Pfiesteria* toxins was identified (Moeller et al. 2007). This water-soluble *Pf*Tx complex consists of ligated copper compounds, with toxicity attributed to the formation of carbon-sulfur metal-based radicals. These toxin congeners are highly unstable, maintaining activity for only 2–5 days in purified form, with degradation accelerated by white light, pH changes, and heat. Overall, toxicity levels also appear to vary depending on the strains and assays used in the laboratory studies (Burkholder et al. 2005; Burkholder and Marshall 2012).

Given that none of the opponents of *Pfiesteria*'s purported toxicity claimed that fish mortality was primarily due to feeding interactions, it appears that this contentious and highly engaging scientific debate may now be drawing to a close.

3.13 Grazing, Nutrient Imbalances, and Xenobiotics Increase Promote Growth and Toxin Production

As with the cyanotoxins in the previous chapter, it is essential to highlight an often-ignored issue: Grazing herbivores and exposure to xenobiotics can promote growth and enhance toxin production in harmful marine algae, such as dinoflagellates and diatoms. Notably, phosphorus or nitrogen deficiency does not significantly alter toxin production, as shown by the expression of *Alexandrium minutum* genes encoding homologs of proteins involved in saxitoxin biosynthesis (Akbar et al. 2023).

3.13.1 Grazing

The phenomenon of toxin increase through grazing is well understood, as evidenced by the production of saxitoxin and domoic acid.

3.13.1.1 Paralytic Shellfish Toxins with Emphasis on Saxitoxin

The natural function of PSP toxins has been hypothesized to be a chemical defense mechanism against grazers. Evidence for this was provided by Selander et al. (2006), who showed that the presence of copepod-related waterborne cues from *Acartia tonsa* was sufficient to induce the production of PSP toxins in the HAB-forming dinoflagellate *Alexandrium minutum*. When exposed to these cues, *A. minutum* produced up to 2.5 times more toxins than controls not exposed to such cues, thereby becoming more resistant to further grazing by copepods. Interestingly, the ingestion of non-toxic alternative prey was not affected by the presence of induced *A. minutum*, suggesting that the dinoflagellate's ability to detect and respond to the presence of grazers significantly contributes to HAB formation in the marine environments (Heuschele and Selander 2014). Further research by Selander et al. (2015) identified the signaling molecules responsible for this response: copepodamides, which are polar lipids that link taurine via an amide to an isoprenoid fatty acid conjugate of variable composition (Fig. 3.31c). New copepodamides continue to be identified in ongoing studies (Grebner et al. 2019).

In a study in which *Acartia hudsonica* grazed on *Alexandrium catenella*, Griffin et al. (2019) unexpectedly found that interspecific alarm cues increased toxin production, but conspecific cues did not. The presence of grazers dramatically induced

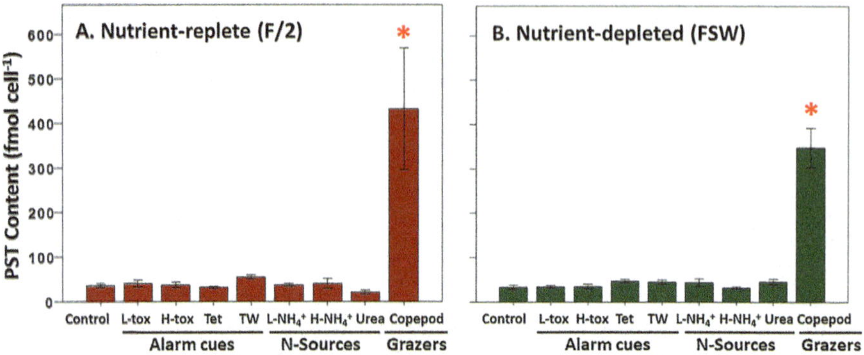

Fig. 3.30 Summary of cell PSP toxin content of *Alexandrium catenella* (strain BF-5) in F/2 medium (**a**) or filtered seawater (**b**). Grazer = copepod *Acartia hudsonica*. Control (filtered seawater); **H-tox** high toxigenic *A. catenella* (strain BF-5); **L-tox** low toxigenic *A. catenella* (strain GTCN-16); **Tet** *Tetraselmis* sp.; **L-NH$_4^+$** 0.083 µM; **H-NH$_4^+$** 50 µM; urea = 50 µM. * = Higher PSP toxin production based on the SNK multiple comparisons test ($P < 0.001$). Values are mean ± SD (from Griffin et al. (2019), with permission from Elsevier)

PSP toxin production in *A. catenella*, regardless of nutrient conditions, and this effect was an order of magnitude greater than any of the other variables tested (Fig. 3.30). These results support the hypothesis that grazer-induced PSP toxin induction is not an artifact of nitrogenous excretion products leading to increased nutrient availability but is likely a direct response that has evolved as a defense mechanism against grazing.

Different copepods release distinct mixtures of copepodamines A-F, and *Alexandrium minutum* responds to pico- to nanomolar concentrations of these signals in a species-specific response. While some copepods induced several-fold increases in PSP toxin production, others induced only weak responses, similar to those of *A. catenella* above. For instance, the copepod *Centropages typicus* is the most potent inducer, likely due to its high proportion of the more active copepodamides A–C (Fig. 3.31). In contrast, weaker inducers have approximately three times lower proportions of these potent copepodamines, leading to lower toxin production in *A. minutum*. This research suggests that the copepodamide fingerprint—essentially, the unique combination of copepodamines produced by different copepod species—plays a crucial role in determining the level of toxin production in *A. minutum* and, consequently, the intensity and impact of HAB events in marine ecosystems.

In general, marine copepodamides induce not only defensive traits such as toxin production, but also bioluminescence (Fig. 3.32), and colony size plasticity in various marine phytoplankton algae. However, copepodamides also occur in freshwater, where their role remains unknown. Arnoldt et al. (2024) reported the consistent presence of copepodamides in copepods from six Swedish freshwater lakes. While the composition of these freshwater copepodamines overlaps significantly with that of their marine counterparts, distinct differences exist. Marine copepods typically contain both subgroups of copepodamides, namely copepodamides (CA) and dihydro-copepodamides (dhCA) (Fig. 3.31c), whereas freshwater copepods are

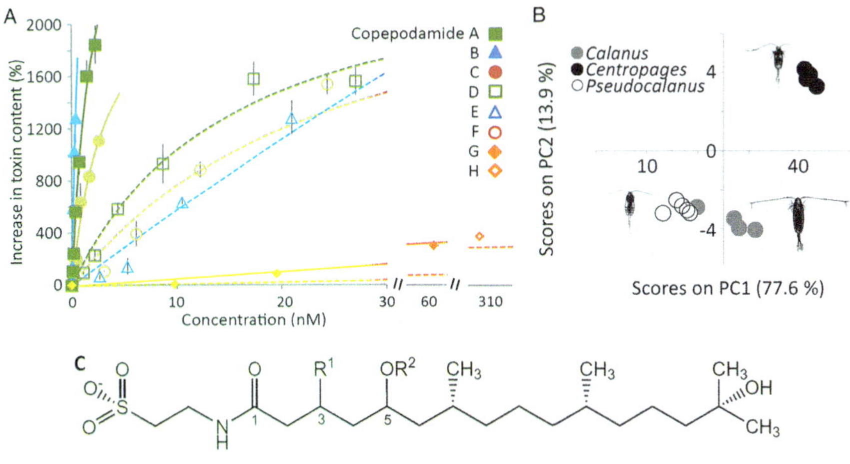

Fig. 3.31 (**a**) Dose–response curves showing induction of toxicity following exposure of *Alexandrium* cells to individual, pure copepodamides A–H. Each symbol represents the mean of three independent replicates ± SE. Curves represent nonlinear least-square fits to the Michaelis–Menten equation. At doses near 2000 nM, copepodamide H was even less active than at 60–300 nM (symbols represent mean values ± SE). (**b**) PCA representation of copepodamides A–F in whole body extracts of common marine copepods *Calanus* sp., *Centropages typicus*, and *Pseudocalanus* sp (from Selander et al. (2015), credit to the National Academy of Sciences of the USA). (**c**) General structure of copepodamides. Two main subgroups exist, determined by the presence of methylene (copepodamide/CA) or methyl (dihydro-copepodamide/dhCA) at R^1. The blend is species-specific, but the fatty acid side chain (at position R^2) changes with diet. Copepodamides are named by the acyl group (Grebner et al. 2019) followed by the scaffold name, e.g., 22:6 dihydro-copepodamide for a dhCA scaffold with a docosahexaenoic acyl group in position R^2 (from Arnoldt et al. (2024), credit to Springer Nature)

dominated by dhCAs. The consistency of copepodamide profiles across different taxonomic groups, sampling sites, and time points suggests the existence of species-specific copepodamide signatures.

The presence of copepodamides in limnic copepods also warrants further research into their potential role as predator alarm cues in freshwater ecosystems.

Applying a functional genomic approach to *Alexandrium tamarense*, Wohlrab et al. (2010) proposed that the regulation of serine/threonine kinase signaling pathways plays a significant role in translating external stimuli, such as copepodamides, into different intracellular cascades and networks. These findings suggest that *A. tamarense* can detect potential predatory copepods and respond by increasing its toxin production, even in a species-specific manner.

Because dinoflagellates are large and slow-growing, they are more susceptible to top-down control by copepods than faster-growing phytoplankton. A second defense mechanism against copepod grazing involves short flashes of blue light (bioluminescence) produced by dinoflagellates. Lindström et al. (2017) demonstrated that strains of the dinoflagellates *Lingulodinium polyedra* and *Alexandrium tamarense*

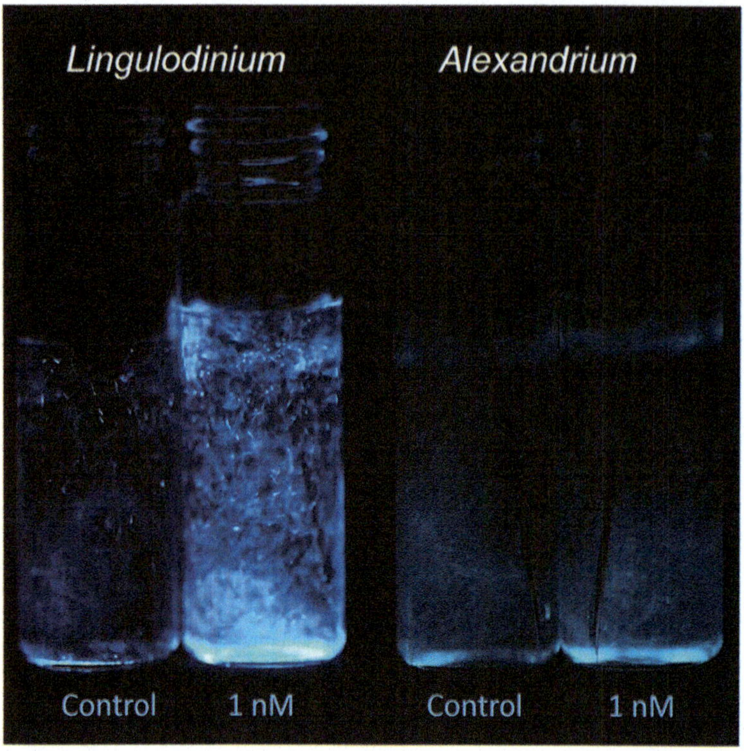

Fig. 3.32 Image showing total luminescence from *Lingulodinium polyedra* and *Alexandrium tamarense*. Cells are exposed to 1 nM copepodamides (right vial) or no copepodamides (control, left vial). Culture samples (5 mL) are simultaneously acidified by the addition of acetic acid (1 M, 1 mL). *L. polyedra* cells produce visibly more light after pre-exposure to copepodamides as compared to the control and to *Alexandrium* (from Lindström et al. (2017), credit to Springer Nature)

can increase their total bioluminescence capacity in response to copepodamides (Fig. 3.32).

However, these defense mechanisms do come with costs. Park and Dam (2021) and Park et al. (2023) demonstrated that reduced growth, evidenced by decreased expression of cell growth-related genes, is a direct fitness cost associated with grazer-induced toxin production in *A. catenella*. A similar effect was observed in *A. minutum*, which produced gonyautoxins when exposed to copepodamides (Fig. 3.33) (Brown et al. 2022). A similar effect was observed with A*lexandrium minutum*, which produces gonyautoxins, a group of potent neurotoxins belonging to the saxitoxin family, when exposed to copepodamides (Fig. 3.33) (Brown et al. 2022).

3.13.1.2 Domoic Acid

Several diatoms also defend themselves against copepod grazing by increasing their production of domoic acid (DA). To investigate this response, Harðardóttir et al. (2015) exposed *Pseudo-nitzschia seriata* to the copepods *Calanus hyperboreus* and

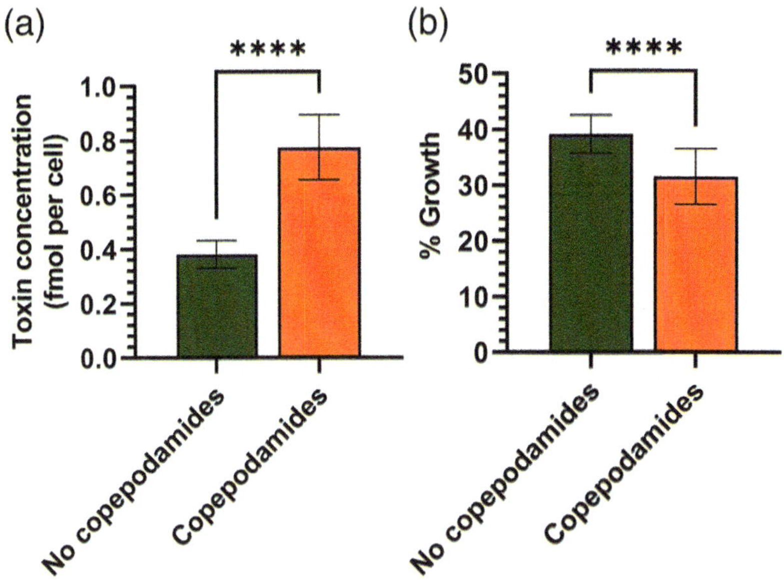

Fig. 3.33 Exposure to copepodamides induces toxin production and reduces growth in *Alexandrium minutum*. Bars represent the mean ± SD of (**a**) cellular toxin concentration of gonyautoxins 1–4 and (**b**) percent growth of *A. minutum* when exposed to a solvent control (no copepodamides, $N = 20$) or 1 nM copepodamides (copepodamides, $N = 20$). Data were analyzed using an unpaired Welch's t-test for the toxin analysis and an unpaired t-test for the growth analysis. Asterisks indicate level of statistical significance; ****$P < 0.0001$ (from Brown et al. (2022), with permission from Wiley & Sons)

C. finmarchicus, both directly and indirectly (using a membrane barrier). Remarkably, DA concentrations in *P. seriata* rose significantly in both scenarios, suggesting that the diatoms can detect grazer presence through chemical cues and rapidly escalate their toxin production. These findings raise compelling questions about the ecological role of DA and the sophistication of chemical communication in planktonic interactions—topics explored in greater detail in Chap. 4.

Lundholm et al. (2018) found that the induction of DA production in diatoms is grazer and diatom-specific. Among the diatoms exposed to herbivorous copepod grazing, only two out of six *Pseudo-nitzschia* species, and none of *Nitzschia* or *Fragilariopsis* strains, responded by producing DA. This suggests that not all *Pseudo-nitzschia* species or strains are capable of DA production and that different diatom species may have evolved varying strategies to cope with grazing pressure. Additionally, a negative correlation between zooplankton growth rate and cellular DA content indicates an allocation cost associated with toxin production. Long-term grazing experiments demonstrated higher mortality rates in grazers fed with toxic diatoms, supporting the hypothesis that DA production is an induced defense mechanism.

Further research by Zhang et al. (2021) using *Artemia* nauplii as grazers revealed that DA content exerts adverse effects on the grazers. Specifically, a negative correlation between cellular DA content in *Pseudo-nitzschia multiseries* and the clearance and/or ingestion rates of *Artemia* nauplii indicates that DA may reduce nauplii grazing efficiency. Such an interaction could potentially decrease grazing pressure on toxic bloom species, thereby prolonging the duration of HAB.

The Arctic copepods *Calanus hyperboreus* and *C. glacialis* showed reduced escape responses after feeding on a DA-producing *Pseudo-nitzschia*, increasing their own susceptibility to predation. This, in turn, reduces grazing pressure on toxic diatoms, consequently facilitating the transfer of DA to higher trophic levels of the food web (Harðardóttir et al. 2018).

However, questions remain regarding the overall efficiency of these defense mechanisms against grazing. For example, although *Calanus finmarchicus* and *C. glacialis* accumulate DA in their body tissue, their egg production and hatching success remain unaffected (Miesner et al. 2016). This suggests that the long-term population numbers of these herbivorous predators are not affected by DA toxicity.

As demonstrated with saxitoxin and domoic acid, copepodamides possess a remarkable capacity to broadly induce defensive traits at the base of the marine food web. These chemically mediated, non-consumptive effects trigger the production of toxic secondary metabolites and promote the development of smaller, more resilient phenotypes in phytoplankton, thereby contributing to the onset and persistence of harmful algal blooms (Selander et al. 2019). Despite growing evidence of their ecological significance, the evolutionary reasons why copepods synthesize and release copepodamides into the surrounding water remain enigmatic. One compelling hypothesis suggests that copepodamides may serve physiological functions within the copepod digestive system, possibly acting as emulsifiers—similar to taurine-containing lipids in crab gastric fluids or bile salts in vertebrates (Grebner et al. 2019). Whether this internal role coincidentally results in powerful ecological signaling or whether copepodamides have evolved dual functions remains an open and fascinating question for future research.

3.13.2 Stoichiometric Nutrient Imbalance

In a comprehensive meta-analysis, Brandenburg et al. (2020) demonstrated that bloom toxicity is not only species-specific but also strongly modulated by environmental nutrient ratios, particularly nitrogen (N) to phosphorus (P) availability. This relationship is partly due to the biochemical demands of toxin synthesis, as many algal toxins are rich in nitrogen (N) and their production varies under nitrogen (N)- or phosphorus (P)- limited conditions. Another critical factor influencing bloom toxicity is the qualitative composition of toxin analogs produced. Multiple structural variants (analogs) of a single toxin can exist, differing in their amino acid sequences or functional group substitutions, which can significantly affect their toxicity (Shimizu 1986; Vale 2010; McLellan and Manderville 2017). Depending on the toxin group, more or less harmful analogs can be produced in response to N availability. For example, *Microcystis* spp. produce a range of microcystin (MC) variants. Under N-replete conditions, there is a notable increase in the production of

MC-RR—a less toxic analog compared to the more harmful, N-leaner MC-LR (Rinehart et al. 1994; Dai et al. 2008; Gobler et al. 2016). Similarly, *Alexandrium tamarense* adjusts its paralytic shellfish toxin profile in response to nutrient availability. Under P-limited conditions, this species produces higher proportions of the more toxic gonyautoxins rather than the less potent N-sulfocarbamoyl derivatives C1 and C2 (Anderson et al. 1990). Field studies have confirmed this pattern: Montoya et al. (2010) observed that both the cellular toxin content and analog composition of *A. tamarense* populations in southern South America were regulated by N availability, with a marked reduction in the amounts and proportions of GTX1 and GTX4 under N limitation.

3.13.3 Xenobiotic Exposure

Low-to-medium concentrations of glyphosate, the most important herbicide in history (Duke and Powles 2008), stimulate the growth of the toxic dinoflagellate *Prorocentrum cordatum* in seawater, while high concentrations exhibit algicidal activity (Fig. 3.34). This stimulatory effect is reflected in enhanced photosynthetic efficiency and increased enzymatic activity of exposed cells. *P. cordatum* is capable of utilizing glyphosate as an alternative phosphorus source when inorganic phosphorus is scarce, yet it remains sensitive to glyphosate even in phosphorus-replete conditions (Shoman et al. 2024). Similarly, *Prorocentrum micans,* a closely related species, also showed growth stimulation in response to glyphosate exposure (Solomonova et al. 2024). This hormetic effect may be due to the switch of phototrophic microorganisms to mixotrophic nutrition with glyphosate as the phosphorus source. An increase in glyphosate concentrations in coastal waters, primarily due to agricultural runoff, could lead to shifts in phytoplankton community composition, potentially favoring the proliferation of harmful or bloom-forming species such as *P. cordatum*. Such changes may disrupt natural biocenoses, alter trophic dynamics, and increase the frequency or severity of harmful algal blooms (Shoman et al. 2024).

The switch may be one mechanism explaining the observed growth stimulation. However, Cedergreen and Olesen (2010) and Nascentes et al. (2018) have shown in barley, sugarcane, and eucalyptus that photosynthetic performance, measured as CO_2 fixation, is also increased by low-dose glyphosate exposure through an as-yet-undiscovered mechanism. However, studies in obligate phototrophic plants such as barley, sugarcane, and eucalyptus have shown that low-dose glyphosate exposure can increase photosynthetic performance, measured as CO_2 fixation, through an as-yet-unidentified mechanism (Cedergreen and Olesen 2010; Nascentes et al. 2018). Notably, this stimulatory effect is not restricted to glyphosate. Other phosphorus-free herbicides, such as the ACCase inhibitor clodinafop and synthetic antiauxins, have also been shown to elicit hormetic responses in plants (Belz and Duke 2017). These findings suggest that mechanisms beyond phosphorus metabolism—possibly involving stress-induced overcompensation or altered hormonal signaling—may underlie the hormetic effects observed in phototrophic organisms.

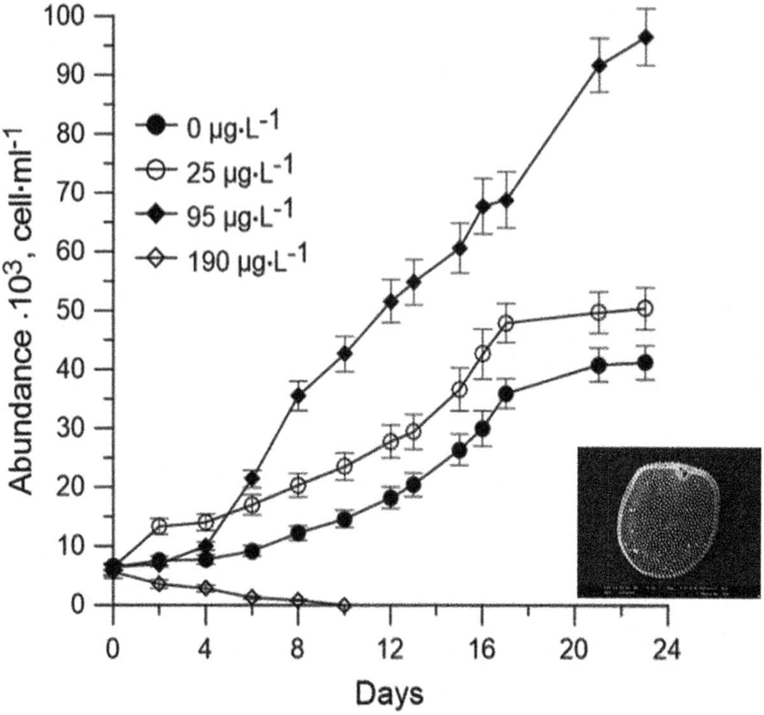

Fig. 3.34 Hormetic effect of glyphosate on the growth of *Prorocentrum cordatum* cultivated in seawater. Error bars indicate standard deviation ($N = 3$) (from Shoman et al. (2024) with permission from Elsevier); SEM image of *P. cordatum* from the Adriatic (from Monti-Birkenmeier et al. (2019), credit to Inter-Research Science Publisher)

Understanding these alternative pathways is critical, especially in marine systems where glyphosate and other agrochemicals increasingly co-occur. If similar hormetic responses occur broadly in phytoplankton, even in the absence of phosphorus limitation, this could contribute to unexpected shifts in bloom dynamics and ecosystem functioning under sublethal herbicide exposure. Further investigation into the cellular targets and metabolic effects in marine algae is essential to predict the ecological consequences of chronic low-dose contamination.

In the harmful alga, *Prymnesium parvum*, Dabney and Patiño (2018) observed growth stimulation by glyphosate and glyphosate-based herbicides at environmentally realistic low doses. The mechanisms do not appear to be related to the associated small increases in total phosphorus from glyphosate degradation.

In a subsequent study, Chávez Montes et al. (2024) investigated the transcriptomic responses of *P. parvum* to increasing concentrations of glyphosate (Fig. 3.35). Their findings reveal a clear dose-dependent shift in the molecular response. At the lowest concentration tested (0.1 mg L^{-1}), the ratio of up- to downregulated KEGG Orthologues (KOs) was 17.4, indicating a pronounced activation of gene expression. However, as glyphosate concentrations increased to 2 and 4 mg L^{-1}, this ratio

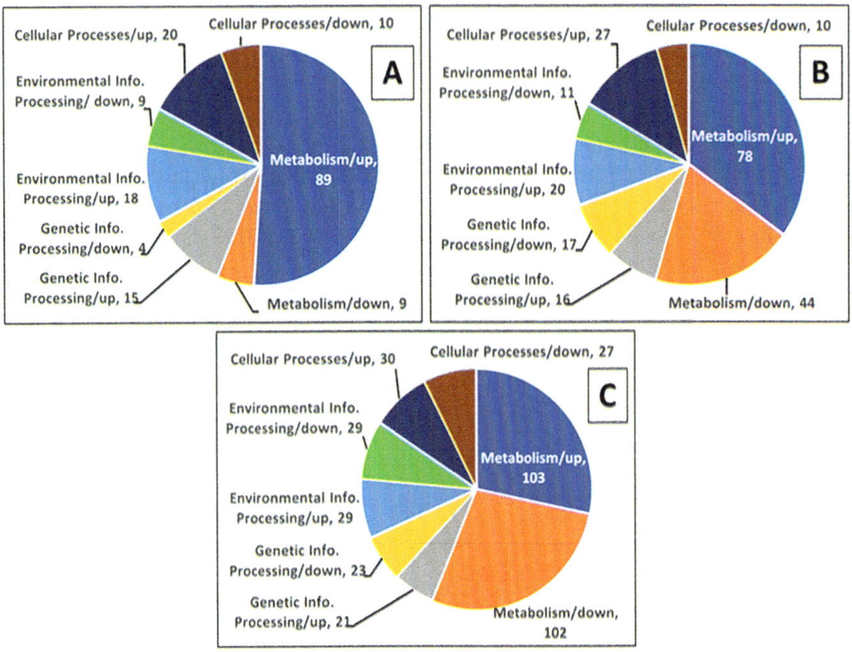

Fig. 3.35 Pie chart showing the number of Kyoto Encyclopedia of Genes and Genomes pathways upregulated (up) or downregulated (down) within various pathway categories following a 9-day exposure of *Prymnesium parvum* to glyphosate. (**a**), 0.1 mg L^{-1}; (**b**), 2 mg L^{-1}; (**c**) 4 mg L^{-1}. The number next to each broad category name is the number of upregulated or downregulated pathways. Notice how the proportion of downregulated pathways within some of the categories increases as glyphosate concentration increases, especially for metabolism (from Chávez Montes et al. (2024), with permission from Elsevier)

dropped markedly to 6.0 and 1.4, respectively, suggesting a progressive suppression or exhaustion of transcriptional activation. The disproportionate change in differentially regulated pathways was observed in most pathway categories. It was particularly notable for Metabolism (Fig. 3.35), where the ratio of up- to downregulated pathways decreased from 9.9 to 1.8 to 1.0 as glyphosate increased from 0.1 to 2 to 4 mg L^{-1}. These findings suggest that while low concentrations of glyphosate may trigger broad metabolic activation—consistent with a hormetic response—higher concentrations begin to overwhelm the regulatory capacity of the algae, potentially tipping the balance toward stress or toxicity.

Overall, *P. parvum* appears not only to tolerate environmental exposure to glyphosate but may actually benefit from it, as suggested by increased growth under low concentrations of the herbicide. This observation highlights the potential role of glyphosate as an environmental driver shaping the distribution and intensity of *P. parvum* blooms, both historically and in present-day aquatic systems. Understanding this relationship may prove critical for anticipating bloom dynamics under changing agricultural runoff patterns.

While enhanced growth under glyphosate exposure has been observed in *P. parvum* and other potentially harmful marine algae, it remains to be confirmed whether an increase in toxin production accompanies such hormetic stimulation. This remains a critical gap in our knowledge and a priority for future studies. By contrast, in freshwater systems, substantial evidence indicates that sublethal concentrations of pollutants—such as herbicides, heavy metals, and pharmaceuticals—not only stimulate the growth of harmful cyanobacteria but also increase the production and accumulation of their toxins (Agathokleous et al. 2022). If a similar pattern holds for marine species like *P. parvum*, the implications for ecosystem health and human exposure risk could be significant, particularly in coastal areas receiving agricultural runoff. Hence, a better mechanistic understanding of how glyphosate and other contaminants modulate both growth and toxicity in marine microalgae is urgently needed.

3.14 Concluding Remarks

Hormesis is a two-phased dose-response relationship to an environmental agent whereby low-dose amounts have a beneficial effect and high-dose amounts are either inhibitory to function or toxic (Calabrese 2014; Costantini and Borremans 2019; Mahalakshmi et al. 2022). Hormesis is considered an adaptive response of an organism to stress and may even involve non-identical pathways at low and high doses, as shown by transcriptomic and metabolomic analyses (Celorio-Mancera et al. 2011, 2012; Menzel et al. 2012; Yi et al. 2018; Zhang et al. 2025).

As a demonstration of the hormetic principle, we refer to two examples:

- Treatment of the coccal green alga *Coelastrella vacuolata* (formerly *Scenedesmus vacuolatus*) with triclosan, an antibacterial and antifungal agent (Larras et al. 2020). Growth exhibited an increasing sigmoidal signal of inhibition along the concentration gradient while photosynthesis exhibited hormesis at the lowest concentration (Fig. 3.36), with promotion occurring at low exposure concentrations.
- Treatment of the predatory sting bug *Podisus distinctus* with too low a dose of the powerful insecticide permethrin. The result is an increase in reproduction rather than the elimination of the pest (Fig. 3.37) (Guedes et al. 2009).

The phenomenon of hormesis is well-documented and accepted in laboratory studies of harmful algae exposed to herbicides, as well as in greenhouse and field studies of crops and medicinal plants exposed to glyphosate (Ferenc 2001; Belz and Duke 2017) and other herbicides (Mathiassen et al. 2021). It has also occasionally been used commercially: the use of low doses of glyphosate and other herbicides as ripeners in sugarcane to enhance sugar production, the use of herbicide safeners to induce tolerance in crops to a follow-up herbicide treatment, and the use of low doses of several auxins and antiauxins as bioregulators for yield improvement (Belz and Duke 2017) with references therein). The principle of hormesis applies to all

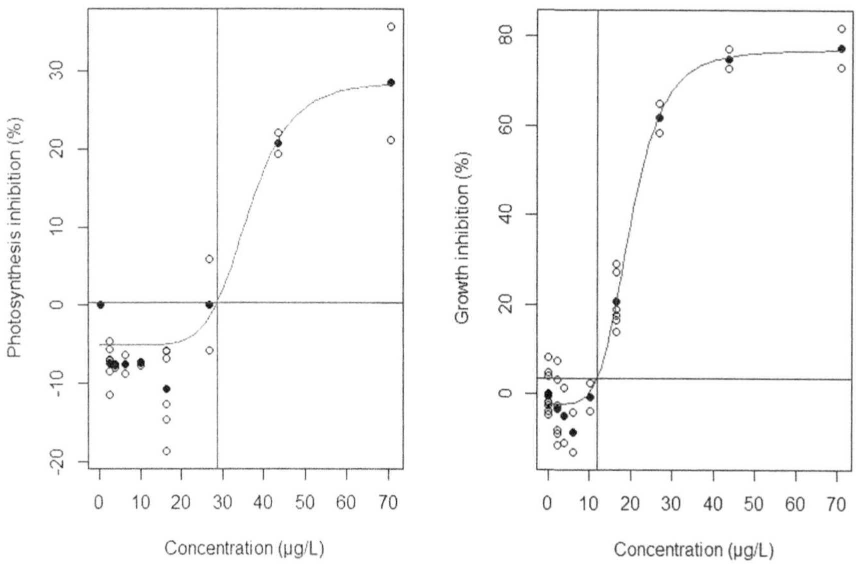

Fig. 3.36 Concentration-response curves of photosynthesis and growth inhibition of *Coelastrella* (*Scenedesmus*) *vacuolatus* after exposure to a concentration gradient of triclosan for 14 h fitted with a Hill model. Black dots represent the mean, while white dots represent replicate values for each concentration. Blue lines represent the abscissae and ordinates of the BMD-1SD value. (From Larras et al. (2020) credit to Elsevier)

ionizing radiation as well as inorganic and organic xenobiotic chemicals, including plant allelochemicals (Ma et al. 2019) and toxins (Calabrese 2005).

This also applies to animals. Several papers have been published on xenobiotics, epigenetics, and transgenerational inheritance in insects. For example, Ayyanath et al. (2015) demonstrated that hormetic stimulation of reproduction in aphids can occur after exposure to low doses of an insect growth regulator that interferes with juvenile hormone function. In addition, Ayyanath et al. (2014) reported imidacloprid-induced hormesis in the form of stimulated reproduction in the green peach aphid (*Myzus persicae*) exposed to this neonicotinoid insecticide. At hormetic concentrations, downregulation of *hsp60* was followed by upregulation of this gene in the following generation. Similarly, the expression of dispersal and developmental genes varied with concentration, life stage and generation. These results indicate that reproductive hormesis in *M. persicae* is accompanied by a complex transgenerational pattern of up- and downregulation of genes, likely reflecting trade-offs in gene expression and related physiological processes during the phenotypic dose-response. Furthermore, global DNA methylation in second-generation *M. persicae* occurred at higher doses than in first-generation aphids, suggesting that heritable adaptability to low doses of the stressor may have occurred.

Later, Brevik et al. (2021) examined the effects of sublethal exposure to the same insecticide on DNA methylation in the Colorado potato beetle (*Leptinotarsa decemlineata*). Exposure to insecticide resulted in a decrease in global DNA methylation

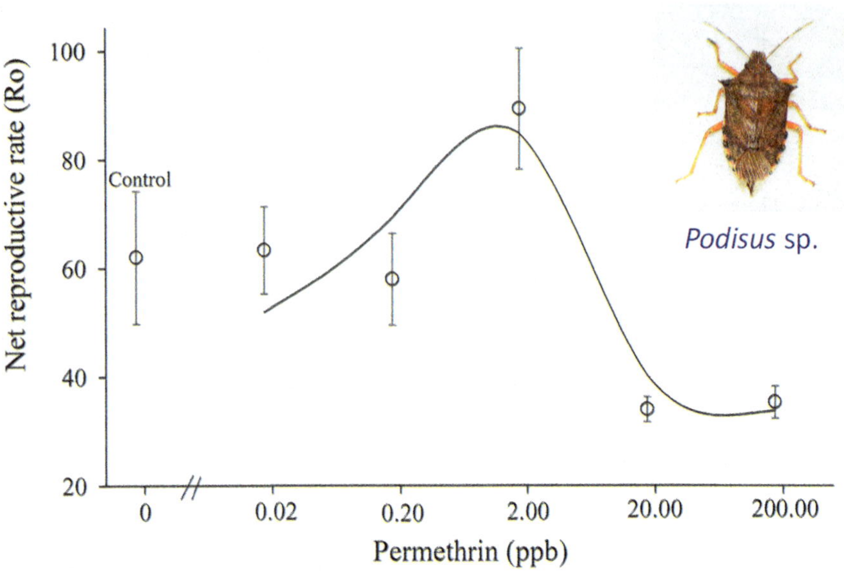

Fig. 3.37 Effect of sublethal doses of the insecticide permethrin on the net reproductive rate of the predatory stink bug *Podisus distinctus*. Each symbol represents the mean of three replicates and the vertical bars indicate the standard error of the means (from Guedes et al. (2009) with permission from the Oxford University Press). *Podisus* sp., image credit to Douglas Hooper via iNaturalist under https://creativecommons.org/licenses/by/4.0/)

for both the parental and F2 generations. Many of the sites of methylation changes are located within genes associated with insecticide resistance, such as cytochrome P450s, or within transposable elements. Exposure to sublethal doses of this insecticide resulted in heritable changes in DNA methylation in this agricultural herbivore, characterized by reduced global methylation and altered methylation patterns (Fig. 3.38). Overall, pesticides may drive the evolution of resistance via epigenetic processes. Moreover, insect pests primed by pesticides may become more tolerant to other stressors, thereby further enhancing their ability to adapt to stressful environments (Brevik et al. 2018).

If insecticide exposure alters DNA methylation of transposable elements, they may be more likely to generate mutations in affected insects, and these mutations may be associated with resistance. For example, studies on the fruit fly (*Drosophila melanogaster*) have shown that repeated insertions of transposable elements within stress response genes are associated with increased stress tolerance (Merenciano et al. 2016). In addition, changes in DNA methylation patterns, including reduced global DNA methylation as found in the potato beetle, increase epigenetic diversity (Vogt 2023), which in turn allows the organism to respond more flexibly to environmental challenges than members of the naïve population (Liebl et al. 2013; Hawes et al. 2019; Ellesse Lauer et al. 2024).

The mechanisms, including the role of epigenetics, identified for synthetic xenobiotics in insects are likely to occur with natural xenobiotics, such as phycotoxins,

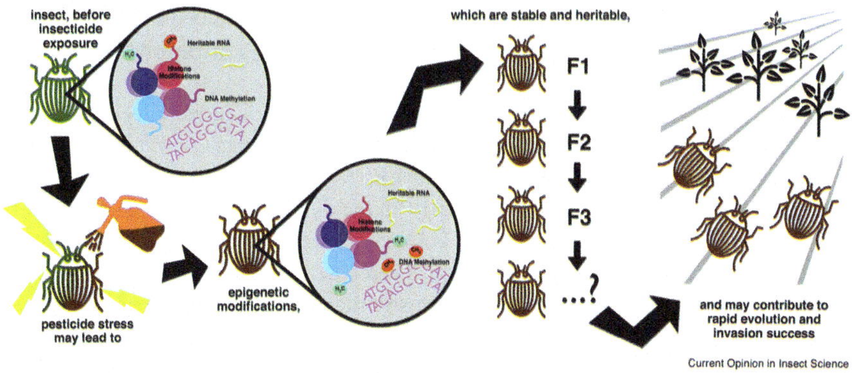

Fig. 3.38 How exposure to a stressor may lead to heritable epigenetic changes (here, DNA hypomethylation) that could lead to stress-resistant phenotype in an invasive agricultural insect pest (from Brevik et al. (2018), with permission from Elsevier)

and their predators, leading to occasional, trained, and probably heritable tolerance in marine animals (→Chap. 5). Unlike mammals, lower vertebrates (such as fish) and invertebrates do not develop specific immunity when challenged with pathogens. However, these animals may benefit from trained immunity, a form of innate immune memory in which immune cells "remember" past encounters with pathogens, allowing them to mount a stronger defense against subsequent infections. This is achieved through the reprogramming of immune cells. Recent studies show that allochthonous methyl donors can also train innate immunity. This has been proven with dietary supplements, such as β-glucans (Petit and Wiegertjes 2016; Roy et al. 2020), peptidoglycans (Lee and Kim 2025), mannan-oligosaccharides (MOS) (Wang et al. 2024), poly(I:C) (Lafont et al. 2017, 2020), or a combination of MOS with β-glucan or with inulin (Lu et al. 2019) and likely also with humic substances (Lieke et al. 2021, Zhao et al. 2025). It can be hypothesized here that low (hormetic) concentrations of several phycotoxins serve as methyl donors when metabolized. In light of the aforementioned hormetic response of algae and invertebrates to synthetic xenobiotics, this hypothesis warrants experimental testing.

However, these mechanisms have not yet been thoroughly studied in relation to phycotoxins. Instead, most of the existing studies in this area are based on the superficial assumption that toxins are only toxic and do not, or only inadequately, cover the low-exposure, hormetic range in terms of exposure dose. As a result, the response of aquatic animals is incompletely covered, and an essential range of possible responses of exposed animals is not examined at all. Essential knowledge is being withheld.

While the incidence of poisoning and mortality from marine toxins in aquatic animals has been extensively monitored and studied, certain mechanistic aspects remain poorly understood. These include, in addition to the role of epigenetics, the roles of gut microbiota, both fundamental and prophylactic. There is a lack of studies demonstrating the modulating effect of gut microbiota during exposure to marine phycotoxins, leaving unanswered the question of how appropriate nutrition can reduce the potential toxic effects or even lead to toxin tolerance in farmed aquatic animals. Therefore, we need to draw on knowledge from freshwater cyanotoxin studies.

Prophylactic nutrition aims to prevent, delay, and/or reduce the effects of disease or toxicity. For example, in zebrafish, a probiotic *Lactobacillus rhamnosus* strain has been shown to ameliorate microcystin-LR (MC-LR) neurotoxicity through the gut-brain axis (Luan et al. 2024). Similarly, in the Pacific white giant shrimp, a probiotic *L. plantarum* strain increased both growth and survival rates in MC-LR stressed individuals (Xu et al. 2023). In these cases, MC-LR stress disrupted gut microbiota, while the use of *L. plantarum* significantly increased the abundance of immune- and metabolically related bacteria and inhibited the growth of pathogenic bacteria, thereby maintaining gut microbial balance and gut health. These promising results should serve as a model for similar marine studies.

From a human perspective, the United Nations Office for Disaster Risk Reduction states that there is no antidote for any of the marine algal toxins, and supportive care remains the primary treatment approach.[4] Prevention, technical treatments, such as coagulation or adsorption, which have been shown to make a significant contribution, but have not been applied in real-world conditions (Miranda et al. 2017; Oh et al. 2023; Chambers et al. 2024), and particularly the avoidance of harmful algal blooms, have been and will continue to be the preferred strategies for the marine aquaculture industry to prevent economic loss of valuable farmed aquatic animals. Nevertheless, probiotic intervention in the form of prophylactic and therapeutic nutrition should be adapted to aquaculture and further developed using Artificial Intelligence. The ultimate goal is to modulate the gut microbiota to improve health outcomes and stress tolerance as practiced in various clinical settings (D'Urso and Broccolo 2024).

References

Abdulhussain AH, Cook KB, Bresnan E, Lacaze JP, Mayor DJ (2024) The influence of the toxin-producing dinoflagellate, *Alexandrium catenella*, on feeding, reproduction and toxin retention in *Calanus helgolandicus*. Harmful Algae 132. https://doi.org/10.1016/j.hal.2023.102564

Absalon D, Matysik M, Woźnica A, Janczewska N (2023) Detection of changes in the hydrobiological parameters of the Oder River during the ecological disaster in July 2022 based on multi-parameter probe tests and remote sensing methods. Ecol Indic 148:110103. https://doi.org/10.1016/j.ecolind.2023.110103

Accoroni S, Totti C (2016) The toxic benthic dinoflagellates of the genus *Ostreopsis* in temperate areas: A review. Adv Oceanogr Limnol 7(1):1–15. https://doi.org/10.4081/aiol.2016.5591

Accoroni S, Ubaldi M, Bacchiocchi S, Neri F, Siracusa M, Buonomo MG, Campanelli A, Totti C (2022) Palytoxin-analogues accumulation in natural mussel banks during an *Ostreopsis* cf. *ovata* bloom. J Mar Sci Eng 10(10). https://doi.org/10.3390/jmse10101402

Adolf JE, Krupatkina D, Bachvaroff T, Place AR (2007) Karlotoxin mediates grazing by *Oxyrrhis marina* on strains of *Karlodinium veneficum*. Harmful Algae 6(3):400–412. https://doi.org/10.1016/j.hal.2006.12.003

[4] https://www.who.int/news-room/fact-sheets/detail/natural-toxins-in-food, accessed 2024-07-16

Agathokleous E, Peñuelas J, Azevedo RA, Rillig MC, Sun H, Calabrese EJ (2022) Low levels of contaminants stimulate harmful algal organisms and enrich their toxins. Environ Sci Technol 56(17):11991–12002. https://doi.org/10.1021/acs.est.2c02763

Ajani P, Brett S, Krogh M, Scanes P, Webster G, Armand L (2013) The risk of harmful algal blooms (HABs) in the oyster-growing estuaries of New South Wales, Australia. Environ Monit Assess 185(6):5295–5316. https://doi.org/10.1007/s10661-012-2946-9

Ajuzie CC (2008) Toxic *Prorocentrum lima* induces abnormal behaviour in juvenile sea bass. J Appl Phycol 20(1):19–27. https://doi.org/10.1007/s10811-007-9176-5

Akbar MA, Yusof NYM, Usup G, Ahmad A, Baharum SN, Bunawan H (2023) Nutrient deficiencies impact on the cellular and metabolic responses of saxitoxin producing *Alexandrium minutum*: a transcriptomic perspective. Mar Drugs 21(9). https://doi.org/10.3390/md21090497

Aligizaki K, Katikou P, Nikolaidis G, Panou A (2008) First episode of shellfish contamination by palytoxin-like compounds from *Ostreopsis* species (Aegean Sea, Greece). Toxicon 51(3):418–427. https://doi.org/10.1016/j.toxicon.2007.10.016

Amaya O, Quintanilla R, Stacy BA, Bottein MYD, Flewelling L, Hardy R, Dueñas C, Ruiz G (2018) Large-scale sea turtle mortality events in El Salvador attributed to paralytic shellfish toxin-producing algae blooms. Front Marine Sci 9(Nov). https://doi.org/10.3389/fmars.2018.00411

Amzil Z, Sibat M, Royer F, Masson N, Abadie E (2007) Report on the first detection of pectenotoxin-2, spirolide-a and their derivatives in French shellfish. Mar Drugs 5(4):168–179. https://doi.org/10.3390/md504168

Amzil Z, Sibat M, Chomerat N, Grossel H, Marco-Miralles F, Lemee R, Nezan E, Sechet V (2012) Ovatoxin-a and palytoxin accumulation in seafood in relation to *Ostreopsis* cf. *ovata* blooms on the French Mediterranean coast. Mar Drugs 10(2):477–496. https://doi.org/10.3390/md10020477

Anderson DM, Kulis DM, Sullivan JJ, Hall S (1990) Toxin composition variations in one isolate of the dinoflagellate *Alexandrium fundyense*. Toxicon 28(8):885–893. https://doi.org/10.1016/0041-0101(90)90018-3

Anderson DM, Cembella AD, Hallegraeff GM (2012) Progress in understanding harmful algal blooms: Paradigm shifts and new technologies for research, monitoring, and management. Annu Rev Mar Sci 4:143–176. https://doi.org/10.1146/annurev-marine-120308-081121

Andrade-Villagrán PV, Navarro JM, Aliste S, Chaparro OR, Ortíz A (2019) Trophic transfer of paralytic shellfish toxin (PST): Physiological and reproductive effects in the carnivorous gastropod *Acanthina monodon* (Pallas, 1774). Aquat Toxicol 212:37–46. https://doi.org/10.1016/j.aquatox.2019.04.017

Andrade-Villagrán PV, Agüero MJ, Navarro JM, Urzúa Á (2022) The paralytic shellfish toxin effect on bioenergetic constituents of the fishery resource *Chorus giganteus* (Gastropoda: Muricidae). Mar Environ Res 180. https://doi.org/10.1016/j.marenvres.2022.105735

Andrade-Villagrán PV, Navarro JM, Villanueva PA, Polanco Y, Urzúa Á (2024) Ingestion of paralytic shellfish toxins in a carnivorous gastropod (*Chorus giganteus*): effects on their elemental composition and reproductive traits. Mar Environ Res 195:106366. https://doi.org/10.1016/j.marenvres.2024.106366

Andres JK, Yñiguez AT, Maister JM, Turner AD, Olano DEB, Mendoza J, Salvador-Reyes L, Azanza RV (2019) Paralytic shellfish toxin uptake, assimilation, depuration, and transformation in the southeast asian green-lipped mussel (*Perna viridis*). Toxins 11(8). https://doi.org/10.3390/toxins11080468

Araoz R, Servent D, Molgó J, Iorga BI, Fruchart-Gaillard C, Benoit E, Gu Z, Stivala C, Zakarian A (2011) Total synthesis of pinnatoxins A and G and revision of the mode of action of pinnatoxin A. J Am Chem Soc 133(27):10499–10511. https://doi.org/10.1021/ja201254c

Armitage JM, Gobas FAPC (2007) A terrestrial food-chain bioaccumulation model for POPs. Environ Sci Technol 41(11):4019–4025. https://doi.org/10.1021/es0700597

Arnich N, Abadie E, Delcourt N, Fessard V, Fremy JM, Hort V, Lagrange E, Maignien T, Molgó J, Peyrat MB, Vernoux JP, Mattei C (2020) Health risk assessment related to pinnatoxins in French shellfish. Toxicon 180:1–10. https://doi.org/10.1016/j.toxicon.2020.03.007

Arnoldt S, Pourdanandeh M, Spikkeland I, Andersson MX, Selander E (2024) Mass spectroscopy reveals compositional differences in copepodamides from limnic and marine copepods. Sci Rep 14(1):3147. https://doi.org/10.1038/s41598-024-53247-1

Ayyanath MM, Cutler GC, Scott-Dupree CD, Prithiviraj B, Kandasamy S, Prithiviraj K (2014) Gene expression during imidacloprid-induced hormesis in green peach aphid. Dose-Response 12(3):480–497. https://doi.org/10.2203/dose-response.13-057.Cutler

Ayyanath MM, Scott-Dupree CD, Cutler GC (2015) Effect of low doses of precocene on reproduction and gene expression in green peach aphid. Chemosphere 128:245–251. https://doi.org/10.1016/j.chemosphere.2015.01.061

Baden, D. G. (1983). Marine food-borne dinoflagellate toxins. Int Rev Cytol G. H. Bourne, J. F. Danielli and K. W. Jeon, 82: 99–150

Bagnis R, Kuberski T, Laugier S (1979) Clinical observations on 3,009 cases of ciguatera (fish poisoning) in the South Pacific. Am J Trop Med Hyg 28(6):1067–1073. https://doi.org/10.4269/ajtmh.1979.28.1067

Bagnis R, Chanteau S, Chungue E, Hurtel JM, Yasumoto T, Inoue A (1980) Origins of ciguatera fish poisoning: a new dinoflagellate, *Gambierdiscus toxicus* Adachi and Fukuyo, definitively involved as a causal agent. Toxicon 18(2):199–208. https://doi.org/10.1016/0041-0101(80)90074-4

Baker PK, Mann R (1990) Habitat requirements for the softshell clam, *Mya arenaria* in the Chesapeake Bay, Special scientific report no. 125. Virginia Institute of Marine Science, William & Mary. https://doi.org/10.25773/v5-jc3f-tk19

Bargu S, Powell CL, Coale SL, Busman M, Doucette GJ, Silver MW (2002) Krill: a potential vector for domoic acid in marine food webs. Mar Ecol Prog Ser 237:209–216. https://doi.org/10.3354/meps237209

Barkoh A, Smith DG, Schlechte JW (2003) An effective minimum concentration of un-ionized ammonia nitrogen for controlling *Prymnesium parvum*. N Am J Aquac 65(3):220–225. https://doi.org/10.1577/C02-046

Barkoh A, Smith DG, Schlechte JW, Paret JM (2004) Ammonia tolerance by sunshine bass fry: Implication for use of ammonium sulfate to control *Prymnesium parvum*. N Am J Aquac 66(4):305–311. https://doi.org/10.1577/a03-063.1

Barquissau V, Ghandour RA, Ailhaud G, Klingenspor M, Langin D, Amri E-Z, Pisani DF (2017) Control of adipogenesis by oxylipins, GPCRs and PPARs. Biochimie 136:3–11. https://doi.org/10.1016/j.biochi.2016.12.012

Barrientos RG, Hernández-Mora G, Alegre F, Field T, Flewelling L, McGrath S, Deeds J, Chacón YS, Rojas Arrieta K, Vargas EC, Artavia KB, Stacy BA (2019) Saxitoxin poisoning in green turtles (*Chelonia mydas*) linked to scavenging on mass mortality of Caribbean sharpnose puffer fish (*Canthigaster rostrata* – Tetraodontidae). Front Veterin Sci 6. https://doi.org/10.3389/fvets.2019.00466

Bates SS, Hubbard KA, Lundholm N, Montresor M, Leaw CP (2018) Pseudo-nitzschia, Nitzschia, and domoic acid: New research since 2011. Harmful Algae 79:3–43. https://doi.org/10.1016/j.hal.2018.06.001

Beasley V, Carmichael W, Haschek WM, Colegrove KM, Solter P (2023) Phycotoxins. In: Haschek WM, Rousseaux CG, Wallig MA et al (eds) Haschek and Rousseaux's handbook of toxicologic pathology, Environmental toxicologic pathology and major toxicant classes, vol 3. Academic, pp 305–391

Belz RG, Duke SO (2017) Herbicide-mediated hormesis. ACS Symp Ser 1249:135–148. https://doi.org/10.1021/bk-2017-1249.ch010

Bennett CT, Robertson A (2021) Depuration kinetics and growth dilution of Caribbean ciguatoxin in the omnivore *Lagodon rhomboides*: Implications for trophic transfer and ciguatera risk. Toxins 13. https://doi.org/10.3390/toxins13110774

Bernstein JJ, Streicher E (1965) The blood-brain barrier of fish. Exp Neurol 11(4):464–473. https://doi.org/10.1016/0014-4886(65)90060-9

Berry JP, Reece KS, Rein KS, Baden DG, Haas LW, Ribeiro WL, Shields JD, Snyder RV, Vogelbein WK, Gawley RE (2002) Are Pfiesteria species toxicogenic? Evidence against production of ichthyotoxins by *Pfiesteria shumwayae*. Proc Natl Acad Sci USA 99(17):10970–10975. https://doi.org/10.1073/pnas.172221699

Bian Y, Feng XS, Zhang Y, Du C, Wen YQ (2024) Marine toxins in environment: Recent updates on depuration techniques. Ecotoxicol Environ Saf 284. https://doi.org/10.1016/j.ecoenv.2024.116990

Binzer SB, Svenssen DK, Daugbjerg N, Alves-de-Souza C, Pinto E, Hansen PJ, Larsen TO, Varga E (2019) A-, B- and C-type prymnesins are clade specific compounds and chemotaxonomic markers in *Prymnesium parvum*. Harmful Algae 81:10–17. https://doi.org/10.1016/j.hal.2018.11.010

Biré R, Trotereau S, Lemée R, Delpont C, Chabot B, Aumond Y, Krys S (2013) Occurrence of palytoxins in marine organisms from different trophic levels of the French Mediterranean coast harvested in 2009. Harmful Algae 28:10–22. https://doi.org/10.1016/j.hal.2013.04.007

Botana A, López VR (2015) Chemistry and analysis of PSP toxins. In: Botana LM, Alfonso A (eds) Phycotoxins: Chemistry and Biochemistry. John Wiley & Sons, Ltd, Chichester, West Sussex, pp 69–84

Bouquet A, Perdrau MA, Laabir M, Foucault E, Chomérat N, Rolland JL, Abadie E (2022) *Liza ramada* juveniles after exposure to the toxic dinoflagellate *Vulcanodinium rugosum*: Effects on fish viability, tissue contamination and microalgae survival after gut passage. Toxins 14(6). https://doi.org/10.3390/toxins14060401

Brandenburg K, Siebers L, Keuskamp J, Jephcott TG, Van de Waal DB (2020) Effects of nutrient limitation on the synthesis of N-rich phytoplankton toxins: A meta-analysis. Toxins 12(4). https://doi.org/10.3390/toxins12040221

Bresnan E, Arévalo F, Belin C, Branco MAC, Cembella AD, Clarke D, Correa J, Davidson K, Dhanji-Rapkova M, Lozano RF, Fernández-Tejedor M, Guðfinnsson H, Carbonell DJ, Laza-Martinez A, Lemoine M, Lewis AM, Menéndez LM, Maskrey BH, McKinney A, Pazos Y, Revilla M, Siano R, Silva A, Swan S, Turner AD, Schweibold L, Provoost P, Enevoldsen H (2021) Diversity and regional distribution of harmful algal events along the Atlantic margin of Europe. Harmful Algae 102. https://doi.org/10.1016/j.hal.2021.101976

Brevik K, Lindström L, McKay SD, Chen YH (2018) Transgenerational effects of insecticides — implications for rapid pest evolution in agroecosystems. Curr Opin Insect Sci 26:34–40. https://doi.org/10.1016/j.cois.2017.12.007

Brevik K, Bueno EM, McKay S, Schoville SD, Chen YH (2021) Insecticide exposure affects intergenerational patterns of DNA methylation in the Colorado potato beetle, *Leptinotarsa decemlineata*. Evol Appl 14(3):746–757. https://doi.org/10.1111/eva.13153

Bricelj VM, Connell L, Konoki K, MacQuarrie SP, Scheuer T, Catterall WA, Trainer VL (2005) Sodium channel mutation leading to saxitoxin resistance in clams increases risk of PSP. Nature 434(7034):763–767. https://doi.org/10.1038/nature03415

Bricelj VM, Cembella AD, Laby D (2014) Temperature effects on kinetics of paralytic shellfish toxin elimination in Atlantic surfclams, *Spisula solidissima*. Deep-Sea Res Part II Top Stud Oceanogr 103:308–317. https://doi.org/10.1016/j.dsr2.2013.05.014

Brooks BW, Grover JP, Roelke DL (2011) *Prymnesium parvum*: An emerging threat to inland waters. Environ Toxicol Chem 30(9):1955–1964. https://doi.org/10.1002/etc.613

Brovedani V, Pelin M, D'Orlando E, Poli M (2016) Brevetoxins: toxicological profile. In: Gopalakrishnakone P, Haddad V, Tubaro A, Kim E, Kem WR (eds) Marine and freshwater toxins. Springer, Dordrecht, pp 113–127

Brown ER, Moore SG, Gaul DA, Kubanek J (2022) Predator cues target signaling pathways in toxic algal metabolome. Limnol Oceanogr 67(6):1227–1237. https://doi.org/10.1002/lno.12069

Brownlee EF, Sellner SG, Sellner KG, Nonogaki H, Adolf JE, Bachvaroff TR, Place AR (2008) Responses of *Crassostrea virginica* (Gmelin) and *C. ariakensis* (Fujita) to bloom-forming phytoplankton including ichthyotoxic *Karlodinium veneficum* (Ballantine). J Shellfish Res 27(3):581–591. https://doi.org/10.2983/0730-8000(2008)27[581:ROCVGA]2.0.CO;2

Brunson JK, McKinnie SMK, Chekan JR, McCrow JP, Miles ZD, Bertrand EM, Bielinski VA, Luhavaya H, Oborník M, Smith GJ, Hutchins DA, Allen AE, Moore BS (2018) Biosynthesis of the neurotoxin domoic acid in a bloom-forming diatom. Science 361(6409):1356–1358. https://doi.org/10.1126/science.aau0382

Burkholder JM, Marshall HG (2012) Toxigenic Pfiesteria species – updates on biology, ecology, toxins, and impacts. Harmful Algae 14:196–230. https://doi.org/10.1016/j.hal.2011.10.022

Burkholder JM, Noga EJ, Hobbs CH, Glasgow HB Jr, Smith SA (1992) New 'phantom' dinoflagellate is the causative agent of major estuarine fish kills. Nature 358(6385):407–410. https://doi.org/10.1038/358407a0

Burkholder JM, Gordon AS, Moeller PD, Law JM, Coyne KJ, Lewitus AJ, Ramsdel JS, Marshall HG, Deamer NJ, Cary SC, Kempton JW, Morton SL, Rublee PA (2005) Demonstration of toxicity to fish and to mammalian cells by *Pfiesteria* species: Comparison of assay methods and strains. Proc Natl Acad Sci USA 102(9):3471–3476. https://doi.org/10.1073/pnas.0500168102

Cadaillon AM, Mattera B, Albizzi A, Montoya N, Maldonado S, Raya Rey A, Riccialdelli L, Almandoz GO, Schloss IR (2024) Multispecies mass mortality in the Beagle Channel associated with paralytic shellfish toxins. Harmful Algae 132. https://doi.org/10.1016/j.hal.2024.102581

Cai P, He S, Zhou C, Place AR, Haq S, Ding L, Chen H, Jiang Y, Guo C, Xu Y, Zhang J, Yan X (2016) Two new karlotoxins found in *Karlodinium veneficum* (strain GM2) from the East China Sea. Harmful Algae 58:66–73. https://doi.org/10.1016/j.hal.2016.08.001

Calabrese EJ (2005) Paradigm lost, paradigm found: The re-emergence of hormesis as a fundamental dose response model in the toxicological sciences. Environ Pollut 138(3):378–411. https://doi.org/10.1016/j.envpol.2004.10.001

Calabrese EJ (2014) Hormesis: a fundamental concept in biology. Microb Cell 1(5):145–149. https://doi.org/10.15698/mic2014.05.145

Cambiaggi L, Chakravarty A, Noureddine N, Hersberger M (2023) The role of α-linolenic acid and its oxylipins in human cardiovascular diseases. Int J Mol Sci 24(7). https://doi.org/10.3390/ijms24076110

Cao R, Wang D, Wei Q, Wang Q, Yang D, Liu H, Dong Z, Zhang X, Zhang Q, Zhao J (2018) Integrative biomarker assessment of the influence of saxitoxin on marine bivalves: A comparative study of the two bivalve species oysters, *Crassostrea gigas*, and scallops, *Chlamys farreri*. Front Physiol 9(Aug). https://doi.org/10.3389/fphys.2018.01173

Carella F, Sardo A, Mangoni O, Di Cioccio D, Urciuolo G, De Vico G, Zingone A (2015) Quantitative histopathology of the Mediterranean mussel (*Mytilus galloprovincialis* L.) exposed to the harmful dinoflagellate *Ostreopsis* cf. *ovata*. J Invertebr Pathol 127:130–140. https://doi.org/10.1016/j.jip.2015.03.001

Cedergreen N, Olesen CF (2010) Can glyphosate stimulate photosynthesis? Pestic Biochem Physiol 96(3):140–148. https://doi.org/10.1016/j.pestbp.2009.11.002

Celorio-Mancera M d LP, Ahn SJ, Vogel H, Heckel DG (2011) Transcriptional responses underlying the hormetic and detrimental effects of the plant secondary metabolite gossypol on the generalist herbivore *Helicoverpa armigera*. BMC Genomics 12:575. https://doi.org/10.1186/1471-2164-12-575

Celorio-Mancera M d LP, Heckel DG, Vogel H (2012) Transcriptional analysis of physiological pathways in a generalist herbivore: Responses to different host plants and plant structures by the cotton bollworm, *Helicoverpa armigera*. Entomol Exp Appl 144(1):123–133. https://doi.org/10.1111/j.1570-7458.2012.01249.x

Cembella AD, Durán-Riveroll LM (2021) Marine guanidinium neurotoxins: Biogenic origins and interactions, biosynthesis and pharmacology. In: Novelli A, Fernández-Sánchez M-T, Aschner M, Costa LG (eds) Advances in neurotoxicology, vol 6. Academic, pp 1–47

Cembella AD, Quilliam MA, Lewis NI, Bauder AG, Dell'Aversano C, Thomas K, Jellett J, Cusack RR (2002) The toxigenic marine dinoflagellate *Alexandrium tamarense* as the probable cause of mortality of caged salmon in Nova Scotia. Harmful Algae 1(3):313–325. https://doi.org/10.1016/S1568-9883(02)00048-3

Cen J, Cui L, Duan Y, Zhang H, Lin Y, Zheng J, Lu S (2019) Effects of palytoxins extracted from *Ostreopsis ovata* on the oxidative stress and immune responses in Pacific white shrimp *Litopenaeus vannamei*. Fish Shellf Immunol 95:670–678. https://doi.org/10.1016/j.fsi.2019.11.001

Cendes F, Andermann F, Carpenter S, Zatorre RJ, Cashman NR (1995) Temporal lobe epilepsy caused by domoic acid intoxication: Evidence for glutamate receptor–mediated excitotoxicity in humans. Ann Neurol 37(1):123–126. https://doi.org/10.1002/ana.410370125

Chambers C, Grimes S, Fire S, Reza MT (2024) Influence of biochar on the removal of microcystin-LR and saxitoxin from aqueous solutions. Sci Rep 14(1). https://doi.org/10.1038/s41598-024-61802-z

Chan KK, Tam NFY, Ng C, Kwok CSN, Xu SJL, Sze ETP, Lee FWF (2021) Proteome response of *Meretrix* bivalves hepatopancreas exposed to paralytic shellfish toxins producing dinoflagellate *Gymnodinium catenatum*. J Mar Sci Eng 9(9). https://doi.org/10.3390/jmse9091039

Chávez Montes RA, Mary MA, Rashel RH, Fokar M, Herrera-Estrella L, Lopez-Arredondo D, Patiño R (2024) Hormetic and transcriptomic responses of the toxic alga *Prymnesium parvum* to glyphosate. Sci Total Environ 954:176451. https://doi.org/10.1016/j.scitotenv.2024.176451

Cheng J, Xun X, Kong Y, Wang S, Yang Z, Li Y, Kong D, Wang S, Zhang L, Hu X, Bao Z (2016) *Hsp70* gene expansions in the scallop *Patinopecten yessoensis* and their expression regulation after exposure to the toxic dinoflagellate *Alexandrium catenella*. Fish Shellfish Immunol 58:266–273. https://doi.org/10.1016/j.fsi.2016.09.009

Chi C, Giri SS, Jun JW, Kim HJ, Yun S, Kim SG, Park SC (2016) Marine toxin okadaic acid affects the immune function of bay scallop (*Argopecten irradians*). Molecules 21(9). https://doi.org/10.3390/molecules21091108

Chi C, Giri SS, Jun JW, Kim HJ, Kim SW, Yun S, Park SC (2017) Effects of algal toxin okadaic acid on the non-specific immune and antioxidant response of bay scallop (*Argopecten irradians*). Fish Shellf Immunol 65:111–117. https://doi.org/10.1016/j.fsi.2017.03.031

Chi C, Giri SS, Jun JW, Kim SW, Kim HJ, Kang JW, Park SC (2018) Detoxification- and immune-related transcriptomic analysis of gills from bay scallops (*Argopecten irradians*) in response to algal toxin okadaic acid. Toxins 10(8). https://doi.org/10.3390/toxins10080308

Chi C, Zhang C, Liu J, Zheng X (2019) Effects of marine toxin domoic acid on innate immune responses in Bay Scallop *Argopecten irradians*. J Mar Sci Eng 7(11). https://doi.org/10.3390/jmse7110407

Chinain M, Gatti CMI, Darius HT, Quod JP, Tester PA (2021) Ciguatera poisonings: a global review of occurrences and trends. Harmful Algae 102. https://doi.org/10.1016/j.hal.2020.101873

Ciminiello P, Dell'Aversano C, Forino M (2015) Chemistry of palytoxin and its analogues. In: Botana LM, Alfonso A (eds) Phycotoxins: chemistry and biochemistry. Wiley, Chichester, pp 85–111

Clausing RJ, Ben Gharbia H, Sdiri K, Sibat M, Rañada-Mestizo ML, Lavenu L, Hess P, Chinain M, Bottein MYD (2024) Tissue distribution and metabolization of ciguatoxins in an herbivorous fish following experimental dietary exposure to *Gambierdiscus polynesiensis*. Mar Drugs 22(1). https://doi.org/10.3390/md22010014

Clayden J, Read B, Hebditch KR (2005) Chemistry of domoic acid, isodomoic acids, and their analogues. Tetrahedron 61(24):5713–5724. https://doi.org/10.1016/j.tet.2005.04.003

Conte FS (1984) Economic impact of paralytic shellfish poison on the oyster industry in the Pacific United States. Aquaculture 39(1):331–343. https://doi.org/10.1016/0044-8486(84)90275-8

Corriere M, Baptista M, Paula JR, Repolho T, Rosa R, Costa PR, Soliño L (2020) Impaired fish swimming performance following dietary exposure to the marine phycotoxin okadaic acid. Toxicon 179:53–59. https://doi.org/10.1016/j.toxicon.2020.02.022

Costa PR (2016) Impact and effects of paralytic shellfish poisoning toxins derived from harmful algal blooms to marine fish. Fish Fish 17(1):226–248. https://doi.org/10.1111/faf.12105

Costa LG, Giordano G, Aschner M (2014) Domoic acid. In: Aminoff MJ, Daroff RB (eds) Encyclopedia of the neurological sciences. Academic, Oxford, pp 1016–1017

Costa PR, Estévez P, Soliño L, Castro D, Rodrigues SM, Timoteo V, Leao-Martins JM, Santos C, Gouveia N, Diogène J, Gago-Martínez A (2021) An update on ciguatoxins and ctx-like toxicity in fish from different trophic levels of the Selvagens Islands (NE Atlantic, Madeira, Portugal). Toxins 13(8). https://doi.org/10.3390/toxins13080580

Costantini D, Borremans B (2019) The linear no-threshold model is less realistic than threshold or hormesis-based models: an evolutionary perspective. Chem Biol Interact 301:26–33. https://doi.org/10.1016/j.cbi.2018.10.007

Cusick K, Sayler G (2013) An overview on the marine neurotoxin, saxitoxin: genetics, molecular targets, methods of detection and ecological functions. Mar Drugs 11(4):991–1018. https://doi.org/10.3390/md11040991

D'Agostino VC, Hoffmeyer MS, Almandoz GO, Sastre V, Degrati M (2015) Potentially toxic *Pseudo-nitzschia* species in plankton and fecal samples of *Eubalaena australis* from Península Valdés calving ground, Argentina. J Sea Res 106:39–43. https://doi.org/10.1016/j.seares.2015.09.004

D'Agostino VC, Degrati M, Sastre V, Santinelli N, Krock B, Krohn T, Dans SL, Hoffmeyer MS (2017) Domoic acid in a marine pelagic food web: Exposure of southern right whales *Eubalaena australis* to domoic acid on the Península Valdés calving ground, Argentina. Harmful Algae 68:248–257. https://doi.org/10.1016/j.hal.2017.09.001

D'Urso F, Broccolo F (2024) Applications of Artificial Intelligence in microbiome analysis and probiotic interventions—an overview and perspective based on the current state of the art. Appl Sci 14(19). https://doi.org/10.3390/app14198627

Dabney BL, Patiño R (2018) Low-dose stimulation of growth of the harmful alga, *Prymnesium parvum*, by glyphosate and glyphosate-based herbicides. Harmful Algae 80:130–139. https://doi.org/10.1016/j.hal.2018.11.004

Dai R, Liu H, Qu J, Zhao X, Ru J, Hou Y (2008) Relationship of energy charge and toxin content of *Microcystis aeruginosa* in nitrogen-limited or phosphorous-limited cultures. Toxicon 51(4):649–658. https://doi.org/10.1016/j.toxicon.2007.11.021

Darius HT, Ponton D, Revel T, Cruchet P, Ung A, Tchou Fouc M, Chinain M (2007) Ciguatera risk assessment in two toxic sites of French Polynesia using the receptor-binding assay. Toxicon 50(5):612–626. https://doi.org/10.1016/j.toxicon.2007.05.007

Davis SL, Roelke DL, Brooks BW, Lundgren VM, Withrow F, Scott WC (2015) Rotifer-*Prymnesium parvum* interactions: role of lake bloom history on rotifer adaptation to toxins produced by *P. parvum*. Aquat Microb Ecol 75(1):55–68. https://doi.org/10.3354/ame01748

Deeds JR, Hoesch RE, Place AR, Kao JPY (2015) The cytotoxic mechanism of karlotoxin 2 (KmTx 2) from *Karlodinium veneficum* (Dinophyceae). Aquat Toxicol 159:148–155. https://doi.org/10.1016/j.aquatox.2014.11.028

Delcourt N, Lagrange E, Abadie E, Fessard V, Frémy JM, Vernoux JP, Peyrat MB, Maignien T, Arnich N, Molgó J, Mattei C (2019) Pinnatoxins' deleterious effects on cholinergic networks: From experimental models to human health. Mar Drugs 17(7). https://doi.org/10.3390/md17070425

Dietrich W, Hesse KJ (1990) Local fish kill in a pond at the German North Sea coast associated with a mass development of *Prymnesium* sp. Meeresforschung-Reports Mar Res 33(1):104–106

Ding L, Qiu J, Li A (2017) Proposed biotransformation pathways for new metabolites of Paralytic Shellfish Toxins based on field and experimental mussel samples. J Agric Food Chem 65(27):5494–5502. https://doi.org/10.1021/acs.jafc.7b02101

Dolah FMV (2000) Marine algal toxins: origins, health effects, and their increased occurrence. Environ Health Perspect 108(suppl 1):133–141. https://doi.org/10.1289/ehp.00108s1133

Domínguez-Pérez D, Martins JC, Almeida D, Costa PR, Vasconcelos V, Campos A (2021) Transcriptomic profile of the cockle *Cerastoderma edule* exposed to seasonal diarrhetic shellfish toxin contamination. Toxins 13(11). https://doi.org/10.3390/toxins13110784

Dong C, Wu H, Zheng G, Peng J, Guo M, Tan Z (2022) Transcriptome analysis reveals MAPK/AMPK as a key regulator of the inflammatory response in PST detoxification in *Mytilus galloprovincialis* and *Argopecten irradians*. Toxins 14(8). https://doi.org/10.3390/toxins14080516

Dou M, Jiao Y-h, Zheng J-w, Zhang G, Li H-y, Liu J-s, Yang W-d (2020) *De novo* transcriptome analysis of the mussel *Perna viridis* after exposure to the toxic dinoflagellate *Prorocentrum lima*. Ecotoxicol Environ Saf 192:110265. https://doi.org/10.1016/j.ecoenv.2020.110265

Doucette GJ, Turner JT, Powell CL, Keafer BA, Anderson DM (2005) ECOHAB-Gulf of Maine. Trophic accumulation of PSP toxins in zooplankton during *Alexandrium fundyense* blooms in

Casco Bay, Gulf of Maine, April–June 1998. I. Toxin levels in *A. fundyense* and zooplankton size fractions. Deep-Sea Res Part II Top Stud Oceanogr 52(19–21):2764–2783. https://doi.org/10.1016/j.dsr2.2005.06.031

Doucette GJ, Cembella AD, Martin JL, Michaud J, Cole TVN, Rolland RM (2006) Paralytic shellfish poisoning (PSP) toxins in North Atlantic right whales *Eubalaena glacialis* and their zooplankton prey in the Bay of Fundy, Canada. Mar Ecol Prog Ser 306:303–313. https://doi.org/10.3354/meps306303

Doucette GJ, Mikulski CM, King KL, Roth PB, Wang Z, Leandro LF, DeGrasse SL, White KD, De Biase D, Gillett RM, Rolland RM (2012) Endangered North Atlantic right whales (*Eubalaena glacialis*) experience repeated, concurrent exposure to multiple environmental neurotoxins produced by marine algae. Environ Res 112:67–76. https://doi.org/10.1016/j.envres.2011.09.010

Drgon T, Saito K, Gillevet PM, Sikaroodi M, Whitaker B, Krupatkina DN, Argemi F, Vasta GR (2005) Characterization of ichthyocidal activity of *Pfiesteria piscicida*: Dependence on the dinospore cell density. Appl Environ Microbiol 71(1):519–529. https://doi.org/10.1128/AEM.71.1.519-529.2005

Duke SO, Powles SB (2008) Glyphosate: a once-in-a-century herbicide. Pest Manag Sci 64(4):319–325. https://doi.org/10.1002/ps.1518

Dunton AD, Göpel T, Ho DH, Burggren W (2021) Form and function of the vertebrate and invertebrate blood-brain barriers. Int J Mol Sci 22(22). https://doi.org/10.3390/ijms222212111

Durán-Riveroll LM, Cembella AD (2017) Guanidinium toxins and their interactions with voltage-gated sodium ion channels. Mar Drugs 15(10). https://doi.org/10.3390/md15100303

Durbin E, Teegarden G, Campbell R, Cembella A, Baumgartner MF, Mate BR (2002) North Atlantic right whales, *Eubalaena glacialis*, exposed to paralytic shellfish poisoning (PSP) toxins via a zooplankton vector, *Calanus finmarchicus*. Harmful Algae 1(3):243–251. https://doi.org/10.1016/S1568-9883(02)00046-X

Dutz J (1998) Repression of fecundity in the neritic copepod *Acartia clausi* exposed to the toxic dinoflagellate *Alexandrium lusitanicum*: relationship between feeding and egg production. Mar Ecol Prog Ser 175:97–107. https://doi.org/10.3354/meps175097

EFSA (2010) Scientific Opinion on marine biotoxins in shellfish—Cyclic imines (spirolides, gymnodimines, pinnatoxins and pteriatoxins). EFSA J 8(6):39. https://doi.org/10.2903/j.efsa.2010.1628

Ellesse Lauer M, Kodak H, Albayrak T, Lima MR, Ray D, Simpson-Wade E, Tevs DR, Sheldon EL, Martin LB, Schrey AW (2024) Introduced house sparrows (*Passer domesticus*) have greater variation in DNA methylation than native house sparrows. J Hered 115(1):11–18. https://doi.org/10.1093/jhered/esad067

Escoffier N, Gaudin J, Mezhoud K, Huet H, Chateau-Joubert S, Turquet J, Crespeau F, Edery M (2007) Toxicity to medaka fish embryo development of okadaic acid and crude extracts of *Prorocentrum* dinoflagellates. Toxicon 49(8):1182–1192. https://doi.org/10.1016/j.toxicon.2007.02.008

Fabioux C, Sulistiyani Y, Haberkorn H, Hégaret H, Amzil Z, Soudant P (2015) Exposure to toxic *Alexandrium minutum* activates the detoxifying and antioxidant systems in gills of the oyster *Crassostrea gigas*. Harmful Algae 48:55–62. https://doi.org/10.1016/j.hal.2015.07.003

Fabro E, Almandoz GO, Ferrario M, John U, Tillmann U, Toebe K, Krock B, Cembella A (2017) Morphological, molecular, and toxin analysis of field populations of *Alexandrium* genus from the Argentine Sea. J Phycol 53(6):1206–1222. https://doi.org/10.1111/jpy.12574

FAO and WHO (2008–2015) Standard for live and raw Bivalve Molluscs. F. WHO. Rome, FAO & WHO. CXS 292–2008

FAO and WHO (2016) Toxicity equivalence factors for marine biotoxins associated with bivalve molluscs, Technical paper. FAO & WHO

Farabegoli F, Blanco L, Rodríguez LP, Vieites JM, Cabado AG (2018) Phycotoxins in marine shellfish: Origin, occurrence and effects on humans. Mar Drugs 16(6). https://doi.org/10.3390/md16060188

Fauquier DA, Flewelling LJ, Maucher J, Manire CA, Socha V, Kinsel MJ, Stacy BA, Henry M, Gannon J, Ramsdell JS, Landsberg JH (2013a) Brevetoxin in blood, biological fluids, and tissues of sea turtles naturally exposed to *Karenia brevis* blooms in central west Florida. J Zoo Wildl Med 44(2):364–375. https://doi.org/10.1638/2012-0164r.1

Fauquier DA, Flewelling LJ, Maucher JM, Keller M, Kinsel MJ, Johnson CK, Henry M, Gannon JG, Ramsdell JS, Landsberg JH (2013b) Brevetoxicosis in seabirds naturally exposed to Karenia brevis blooms along the central west coast of Florida. J Wildl Dis 49(2):246–260. https://doi.org/10.7589/2011-09-270

Faustino LS, Queiroga FR, Hégaret H, Marques-Santos LF, Neves RAF, Nascimento S, da Silva PM (2021) Effects of the toxic dinoflagellates *Prorocentrum lima* and *Ostreopsis* cf. *ovata* on immune responses of cultured oysters *Crassostrea gasar*. Aquat Toxicol 236. https://doi.org/10.1016/j.aquatox.2021.105846

Felpeto AB, Rivera MF, Vasconcelos VM (2024) The effects of the toxic dinoflagellate *Alexandrium* on feeding, reproduction and mortality of the copepod *Acartia*: A systematic review employing weighted linear models. Harmful Algae 137. https://doi.org/10.1016/j.hal.2024.102659

Ferenc SA (ed) (2001) Impacts of low dose, high potency herbicides on nontarget and unintended plant species. SETAC

Finch SC, Harwood DT, Boundy MJ, Selwood AI (2024) A review of cyclic imines in shellfish: worldwide occurrence, toxicity and assessment of the risk to consumers. Mar Drugs 22(3). https://doi.org/10.3390/md22030129

Fire SE, Wang Z, Berman M, Langlois GW, Morton SL, Sekula-Wood E, Benitez-Nelson CR (2010) Trophic transfer of the harmful algal toxin domoic acid as a cause of death in a minke whale (*Balaenoptera acutorostrata*) stranding in southern California. Aquat Mamm 36(4):342–350. https://doi.org/10.1578/AM.36.4.2010.342

Fire SE, Pruden J, Couture D, Wang Z, Bottein MYD, Haynes BL, Knott T, Bouchard D, Lichtenwalner A, Wippelhauser G (2012) Saxitoxin exposure in an endangered fish: Association of a shortnose sturgeon mortality event with a harmful algal bloom. Mar Ecol Prog Ser 460:145–153. https://doi.org/10.3354/meps09768

Fire SE, Flewelling LJ, Stolen M, Durden WN, De Wit M, Spellman AC, Wang Z (2015) Brevetoxin-associated mass mortality event of bottlenose dolphins and manatees along the east coast of Florida, USA. Mar Ecol Prog Ser 526:241–251. https://doi.org/10.3354/meps11225

Flaherty KE, Landsberg JH (2011) Effects of a persistent red tide (*Karenia brevis*) bloom on community structure and species-specific relative abundance of nekton in a Gulf of Mexico Estuary. Estuar Coasts 34(2):417–439. https://doi.org/10.1007/s12237-010-9350-x

Fleming LE, Kirkpatrick B, Backer LC, Walsh CJ, Nierenberg K, Clark J, Reich A, Hollenbeck J, Benson J, Cheng YS, Naar J, Pierce R, Bourdelais AJ, Abraham WM, Kirkpatrick G, Zaias J, Wanner A, Mendes E, Shalat S, Hoagland P, Stephan W, Bean J, Watkins S, Clarke T, Byrne M, Baden DG (2011) Review of Florida red tide and human health effects. Harmful Algae 10(2):224–233. https://doi.org/10.1016/j.hal.2010.08.006

Flewelling LJ, Naar JP, Abbott JP, Baden DG, Barros NB, Bossart GD, Bottein MY, Hammond DG, Haubold EM, Heil CA, Henry MS, Jacocks HM, Leighfield TA, Pierce RH, Pitchford TD, Rommel SA, Scott PS, Steidinger KA, Truby EW, Van Dolah FM, Landsberg JH (2005) Brevetoxicosis: red tides and marine mammal mortalities. Nature 435(7043):755–756. https://doi.org/10.1038/nature435755a

Florez-Barros F, Prado-Alvarez M, Mendez J, Fernandez-Tajes J (2011) Evaluation of genotoxicity in gills and hemolymph of clam *Ruditapes decussatus* fed with the toxic dinoflagellate *Prorocentrum lima*. J Toxicol Environ Health Part A Curr Issues 74(15–16):971–979. https://doi.org/10.1080/15287394.2011.582025

Frangoulos M, Guisande C, Maneiro I, Riveiro I, Franco J (2000) Short-term and long-term effects of the toxic dinoflagellate *Alexandrium minutum* on the copepod *Acartia clausi*. Mar Ecol Prog Ser 203:161–169. https://doi.org/10.3354/meps203161

Friedman MA, Fleming LE, Fernandez M, Bienfang P, Schrank K, Dickey R, Bottein MY, Backer L, Ayyar R, Weisman R, Watkins S, Granade R, Reich A (2008) Ciguatera fish poisoning: treatment, prevention and management. Mar Drugs 6(3):456–479. https://doi.org/10.3390/md20080022

Fukui M, Murata M, Inoue A, Gawel M, Yasumoto T (1987) Occurrence of palytoxin in the trigger fish *Melichtys vidua*. Toxicon 25(10):1121–1124. https://doi.org/10.1016/0041-0101(87)90269-8
Galimany E, Sunila I, Hegaret H, Ramon M, Wikfors GH (2008) Pathology and immune response of the blue mussel (*Mytilus edulis* L.) after an exposure to the harmful dinoflagellate *Prorocentrum minimum*. Harmful Algae 7(5):630–638. https://doi.org/10.1016/j.hal.2008.01.001
García-Lagunas N, Romero-Geraldo R, Hernández-Saavedra NY (2013) Genomics study of the exposure effect of *Gymnodinium catenatum*, a Paralyzing Toxin Producer, on *Crassostrea gigas*' defense system and detoxification fenes. PLoS One 8(9):e72323. https://doi.org/10.1371/journal.pone.0072323
Garcia-Lagunas N, Romero-Geraldo RD, Hernandez-Saavedra NY (2016) Changes in gene expression and histological injuries as a result of exposure of *Crassostrea gigas* to the toxic dinoflagellate *Gymnodinium catenatum*. J Molluscan Stud 82:193–200. https://doi.org/10.1093/mollus/eyv053
Genitsaris S, Kormas KA, Moustaka-Gouni M (2009) Microscopic eukaryotes living in a dying lake (Lake Koronia, Greece). FEMS Microbiol Ecol 69(1):75–83. https://doi.org/10.1111/j.1574-6941.2009.00686.x
Gerssen A, Pol-Hofstad IE, Poelman M, Mulder PPJ, van den Top HJ, Dde Boer J (2010) Marine toxins: chemistry, toxicity, occurrence and detection, with special reference to the dutch situation. Toxins 2(4):878–904. https://doi.org/10.3390/toxins2040878
Giuliani ME, Bacchiocchi S, Accoroni S, Siracusa M, Campacci D, Notarstefano V, Mezzelani M, Piersanti A, Totti C, Benedetti M, Regoli F, Gorbi S (2024) Subcellular effects and lipid metabolism alterations in the gilthead seabream *Sparus aurata* fed on ovatoxins-contaminated mussels. Chemosphere 352. https://doi.org/10.1016/j.chemosphere.2024.141413
Gleibs S, Mebs D, Werding B (1995) Studies on the origin and distribution of palytoxin in a Caribbean coral reef. Toxicon 33(11):1531–1537. https://doi.org/10.1016/0041-0101(95)00079-2
Gobler CJ, Burkholder JM, Davis TW, Harke MJ, Johengen T, Stow CA, Van de Waal DB (2016) The dual role of nitrogen supply in controlling the growth and toxicity of cyanobacterial blooms. Harmful Algae 54:87–97. https://doi.org/10.1016/j.hal.2016.01.010
Gong Y, Zhang K, Geng N, Wu M, Yi X, Liu R, Challis JK, Codling G, Xu EG, Giesy JP (2021) Molecular mechanisms of zooplanktonic toxicity in the okadaic acid-producing dinoflagellate *Prorocentrum lima*. Environ Pollut 279. https://doi.org/10.1016/j.envpol.2021.116942
Gonzalez-Romero R, Suarez-Ulloa V, Rodriguez-Casariego J, Garcia-Souto D, Diaz G, Smith A, Pasantes JJ, Rand G, Eirin-Lopez JM (2017) Effects of Florida Red Tides on histone variant expression and DNA methylation in the Eastern oyster *Crassostrea virginica*. Aquat Toxicol 186:196–204. https://doi.org/10.1016/j.aquatox.2017.03.006
Gorbi S, Bocchetti R, Binelli A, Bacchiocchi S, Orletti R, Nanetti L, Raffaelli F, Vignini A, Accoroni S, Totti C, Regoli F (2012) Biological effects of palytoxin-like compounds from *Ostreopsis* cf. *ovata*: a multibiomarkers approach with mussels *Mytilus galloprovincialis*. Chemosphere 89(5):623–632. https://doi.org/10.1016/j.chemosphere.2012.05.064
Gordon N, Colorni A (2008) *Prymnesium parvum*, an ichthyotoxic alga in an ornamental fish farm in southern Israel. Isr J Aquacult Bamidgeh 60(1):5–8
Gosselin S, Fortier L, Gagné JA (1989) Vulnerability of marine fish larvae to the toxic dinoflagellate *Protogonyaulax tamarensis*. Mar Ecol Prog Ser 57:1–10. https://doi.org/10.3354/MEPS057001
Grant KS, Burbacher TM, Faustman EM, Grattan L (2010) Domoic acid: Neurobehavioral consequences of exposure to a prevalent marine biotoxin. Neurotoxicol Teratol 32(2):132–141. https://doi.org/10.1016/j.ntt.2009.09.005
Grebner W, Berglund EC, Berggren F, Eklund J, Harðadóttir S, Andersson MX, Selander E (2019) Induction of defensive traits in marine plankton—new copepodamide structures. Limnol Oceanogr 64(2):820–831. https://doi.org/10.1002/lno.11077
Griffin JE, Park G, Dam HG (2019) Relative importance of nitrogen sources, algal alarm cues and grazer exposure to toxin production of the marine dinoflagellate *Alexandrium catenella*. Harmful Algae 84:181–187. https://doi.org/10.1016/j.hal.2019.04.006

Guedes RNC, Magalhães LC, Cosme LV (2009) Stimulatory sublethal response of a generalist predator to permethrin: Hormesis, hormoligosis, or homeostatic regulation? J Econ Entomol 102(1):170–176. https://doi.org/10.1603/029.102.0124

Guo M, Harrison PJ, Taylor FJR (1996) Fish kills related to *Prymnesium parvum* N. Carter (Haptophyta) in the People's Republic of China. J Appl Phycol 8(2):111–117. https://doi.org/10.1007/BF02186313

Habermann E (1989) Palytoxin acts through Na^{+},K^{+}-ATPase. Toxicon 27(11):1171–1187. https://doi.org/10.1016/0041-0101(89)90026-3

Habibi N, Uddin S, Bottein M-YD, Faizuddin M (2021) Ciguatera in the Indian Ocean with special insights on the Arabian Sea and adjacent gulf and seas: a review. Toxins 13. https://doi.org/10.3390/toxins13080525

Hackett JD, Wisecaver JH, Brosnahan ML, Kulis DM, Anderson DM, Bhattacharya D, Plumley FG, Erdner DL (2013) Evolution of saxitoxin synthesis in cyanobacteria and dinoflagellates. Mol Biol Evol 30(1):70–78. https://doi.org/10.1093/molbev/mss142

Hallegraeff GM (1993) A review of harmful algal blooms and their apparent global increase. Phycologia 32(2):79–99. https://doi.org/10.2216/i0031-8884-32-2-79.1

Hamamoto Y, Tachibana K, Holland PT, Shi F, Beuzenberg V, Itoh Y, Satake M (2012) Brevisulcenal-F: a polycyclic ether toxin associated with massive fish-kills in New Zealand. J Am Chem Soc 134(10):4963–4968. https://doi.org/10.1021/ja212116q

Hambright KD, Zamor RM, Easton JD, Glenn KL, Remmel EJ, Easton AC (2010) Temporal and spatial variability of an invasive toxigenic protist in a North American subtropical reservoir. Harmful Algae 9(6):568–577. https://doi.org/10.1016/j.hal.2010.04.006

Han J, Park JS, Lee J, Park Y, Shin HH, Choi YU, Lee KW (2021a) Effects of toxic dinoflagellate *Alexandrium minutum* on the expression of detoxification-related genes in the marine copepod *Tigriopus japonicus*. J Exp Mar Biol Ecol 540. https://doi.org/10.1016/j.jembe.2021.151563

Han J, Park JS, Park Y, Lee J, Shin HH, Lee KW (2021b) Effects of paralytic shellfish poisoning toxin-producing dinoflagellate *Gymnodinium catenatum* on the marine copepod *Tigriopus japonicus*. Mar Pollut Bull 163. https://doi.org/10.1016/j.marpolbul.2020.111937

Han J, Park Y, Shin HH, Shin AY, Kang HM, Lee J, Choi YU, Lee KW (2021c) Effects of dinoflagellate *Gymnodinium catenatum* on swimming behavior and expression of heat shock protein (hsp) genes in the brine shrimp *Artemia franciscana*. Harmful Algae 110. https://doi.org/10.1016/j.hal.2021.102146

Haque MN, Nam SE, Han YS, Park HS, Rhee JS (2022) Chronic exposure to sublethal concentrations of saxitoxin reduces antioxidant activity and immunity in zebrafish but does not affect reproductive parameters. Aquat Toxicol 243. https://doi.org/10.1016/j.aquatox.2021.106070

Harðardóttir S, Pančić M, Tammilehto A, Krock B, Møller EF, Nielsen TG, Lundholm N (2015) Dangerous relations in the Arctic marine food web: Interactions between toxin producing *Pseudo-nitzschia* diatoms and *Calanus* copepodites. Mar Drugs 13(6):3809–3835. https://doi.org/10.3390/md13063809

Harðardóttir S, Krock B, Wohlrab S, John U, Nielsen TG, Lundholm N (2018) Can domoic acid affect escape response in copepods? Harmful Algae 79:50–52. https://doi.org/10.1016/j.hal.2018.08.009

Hashimoto Y, Fusetani N, Kimura S (1969) Aluterin: a toxin of filefish, *Alutera scripta*, probably originating from a zoantharian, *Palythoa tuberculosa*. Nippon Suisan Gakkaishi 35(11):1086–1093. https://doi.org/10.2331/suisan.35.1086

Hawes NA, Amadoru A, Tremblay LA, Pochon X, Dunphy B, Fidler AE, Smith KF (2019) Epigenetic patterns associated with an ascidian invasion: a comparison of closely related clades in their native and introduced ranges. Sci Rep 9(1). https://doi.org/10.1038/s41598-019-49813-7

He ZB, Duan GF, Liang CY, Li HY, Liu JS, Yang WD (2019) Up-regulation of Nrf2-dependent antioxidant defenses in *Perna viridis* after exposed to *Prorocentrum lima*. Fish Shellf Immunol 90:173–179. https://doi.org/10.1016/j.fsi.2019.05.003

Heuschele J, Selander E (2014) The chemical ecology of copepods. J Plankton Res 36(4):895–913. https://doi.org/10.1093/plankt/fbu025

Hidalgo J, Liberona JL, Molgó J, Jaimovich E (2002) Pacific ciguatoxin-1b effect over Na^+ and K^+ currents, inositol 1,4,5-triphosphate content and intracellular Ca^{2+} signals in cultured rat myotubes. Br J Pharmacol 137(7):1055–1062. https://doi.org/10.1038/sj.bjp.0704980

Hiolski EM, Kendrick PS, Frame ER, Myers MS, Bammler TK, Beyer RP, Farin FM, Wilkerson H-w, Smith DR, Marcinek DJ, Lefebvre KA (2014) Chronic low-level domoic acid exposure alters gene transcription and impairs mitochondrial function in the CNS. Aquat Toxicol 155:151–159. https://doi.org/10.1016/j.aquatox.2014.06.006

Holdway PA, Watson RA, Moss B (1978) Aspects of the ecology of *Prymnesium parvum* (Haptophyta) and water chemistry in the Norfolk Broads, England. Freshw Biol 8(4):295–311. https://doi.org/10.1111/j.1365-2427.1978.tb01451.x

Hong Z, Zhang Y, Zuo Z, Zhu R, Gao Y (2015) Influences of domoic acid exposure on cardiac development and the expression of cardiovascular relative genes in zebrafish (*Danio rerio*) embryos. J Biochem Mol Toxicol 29(6):254–260. https://doi.org/10.1002/jbt.21692

Hoppmann V, Wu JJ, Søviknes AM, Helvik JV, Becker TS (2008) Expression of the eight AMPA receptor subunit genes in the developing central nervous system and sensory organs of zebrafish. Dev Dyn 237(3):788–799. https://doi.org/10.1002/dvdy.21447

Horner RA, Garrison DL, Plumley FG (1997) Harmful algal blooms and red tide problems on the U.S. west coast. Limnol Oceanogr 42(5 II):1076–1088

Hort V, Bastardo-Fernández I, Nicolas M (2023) Exploration of *Vulcanodinium rugosum* toxins and their metabolism products in mussels from the Ingril Lagoon hotspot in France. Mar Drugs 21(8):429. https://doi.org/10.3390/md21080429

Hu T, Curtis JM, Oshima Y, Quilliam MA, Walter JA, Watson-Wright WM, Wright JLC (1995) Spirolides B and D, two novel macrocycles isolated from the digestive glands of shellfish. J Chem Soc Chem Commun 20:2159–2161. https://doi.org/10.1039/C39950002159

Hu B, Li M, Yu X, Xun X, Lu W, Li X, Li Y, Lou J, Wang S, Zhang L, Cheng J, Hu X, Bao Z (2019) Diverse expression regulation of Hsp70 genes in scallops after exposure to toxic *Alexandrium* dinoflagellates. Chemosphere 234:62–69. https://doi.org/10.1016/j.chemosphere.2019.06.034

Huang L, Wang J, Chen W-C, Li H-Y, Liu J-S, Tao J, Yang W-D (2014) P-glycoprotein expression in *Perna viridis* after exposure to *Prorocentrum lima*, a dinoflagellate producing DSP toxins. Fish Shellfish Immunol 39(2):254–262. https://doi.org/10.1016/j.fsi.2014.04.020

Huang L, Liu S-L, Zheng J-W, Li H-Y, Liu J-S, Yang W-D (2015a) *P*-glycoprotein and its inducible expression in three bivalve species after exposure to *Prorocentrum lima*. Aquat Toxicol 169:123–132. https://doi.org/10.1016/j.aquatox.2015.10.012

Huang L, Zou Y, Weng HW, Li HY, Liu JS, Yang WD (2015b) Proteomic profile in *Perna viridis* after exposed to *Prorocentrum lima*, a dinoflagellate producing DSP toxins. Environ Pollut 196:350–357. https://doi.org/10.1016/j.envpol.2014.10.019

Huang JH, Jiao YH, Li L, Li DW, Li HY, Yang WD (2021) Small RNA analysis of *Perna viridis* after exposure to *Prorocentrum lima*, a DSP toxins-producing dinoflagellate. Aquat Toxicol 239. https://doi.org/10.1016/j.aquatox.2021.105950

Imai I, Yamaguchi M (2012) Life cycle, physiology, ecology and red tide occurrences of the fish-killing raphidophyte *Chattonella*. Harmful Algae 14:46–70. https://doi.org/10.1016/j.hal.2011.10.014

Jeong HJ, Lim AS, Jang SH, Yih WH, Kang NS, Lee SY, Yoo YD, Kim HS (2012) First report of the epiphytic dinoflagellate *Gambierdiscus caribaeus* in the temperate waters off Jeju Island, Korea: Morphology and molecular characterization. J Eukaryot Microbiol 59(6):637–650. https://doi.org/10.1111/j.1550-7408.2012.00645.x

Jiang TJ, Wang DZ, Niu T, Xu YX (2007) Trophic transfer of paralytic shellfish toxins from the cladoceran (*Moina mongolica*) to larvae of the fish (*Sciaenops ocellatus*). Toxicon 50(5):639–645. https://doi.org/10.1016/j.toxicon.2007.05.008

Johnsen TM, Eikrem W, Olseng CD, Tollefsen KE, Bjerknes V (2010) *Prymnesium parvum*: The Norwegian experience. J Am Water Resour Assoc 46(1):6–13. https://doi.org/10.1111/j.1752-1688.2009.00386.x

Jones TO, Whyte JNC, Townsend LD, Ginther NG, Iwama GK (1995) Effects of domoic acid on haemolymph pH, PCO_2 and PO_2 in the Pacific oyster, *Crassostrea gigas* and the California mussel, *Mytilus californianus*. Aquat Toxicol 31(1):43–55. https://doi.org/10.1016/0166-445X(94)00057-W

Kaartvedt S, Johnsen TM, Aksnes DL, Lie U, Svendsen H (1991) Occurrence of the toxic phytoflagellate *Prymnesium parvum* and associated fish mortality in a Norwegian fjord system. Can J Fish Aquat Sci 48(12):2316–2323. https://doi.org/10.1139/f91-272

Kabeya N, Fonseca MM, Ferrier DEK, Navarro JC, Bay LK, Francis DS, Tocher DR, Castro LFC, Monroig Ó (2018) Genes for de novo biosynthesis of omega-3 polyunsaturated fatty acids are widespread in animals. Sci Adv 4(5). https://doi.org/10.1126/sciadv.aar6849

Kaiser J (2002) The science of *Pfiesteria*: Elusive, subtle, and toxic. Science 298(5592):346–349. https://doi.org/10.1126/science.298.5592.346

Katikou P, Gokbulut C, Kosker AR, Campàs M, Ozogul F (2022) An updated review of tetrodotoxin and its peculiarities. Mar Drugs 20(1). https://doi.org/10.3390/md20010047

Khan S, Arakawa O, Onoue Y (1997) Neurotoxins in a toxic red tide of *Heterosigma akashiwo* (Raphidophyceae) in Kagoshima Bay, Japan. Aquac Res 28(1):9–14. https://doi.org/10.1111/j.1365-2109.1997.tb01309.x

Kirkpatrick B, Fleming LE, Squicciarini D, Backer LC, Clark R, Abraham W, Benson J, Cheng YS, Johnson D, Pierce R, Zaias J, Bossart GD, Baden DG (2004) Literature review of Florida red tide: Implications for human health effects. Harmful Algae 3(2):99–115. https://doi.org/10.1016/j.hal.2003.08.005

Kitchen SA, Bourdelais AJ, Taylor AR (2018) Interaction of a dinoflagellate neurotoxin with voltage-activated ion channels in a marine diatom. PeerJ 2018(4). https://doi.org/10.7717/peerj.4533

Kodama AM, Hokama Y, Yasumoto T, Fukui M, Jo Manea S, Sutherland N (1989) Clinical and laboratory findings implicating palytoxin as cause of ciguatera poisoning due to *Decapterus macrosoma* (mackerel). Toxicon 27(9):1051–1053. https://doi.org/10.1016/0041-0101(89)90156-6

Kong L, Liu P, Li M, Wang H, Shi J, Hu J, Li Y, Hu X (2021) Transcriptional responses of flavin-containing monooxygenase genes in scallops exposed to PST-producing dinoflagellates implying their involvements in detoxification. Front Mar Sci 8:1229. https://doi.org/10.3389/fmars.2021.732000

Kozlowsky-Suzuki B, Carlsson P, Rühl A, Granéli E (2006) Food selectivity and grazing impact on toxic *Dinophysis* spp. by copepods feeding on natural plankton assemblages. Harmful Algae 5(1):57–68. https://doi.org/10.1016/j.hal.2005.05.002

Kozlowsky-Suzuki B, Koski M, Hallberg E, Wallén R, Carlsson P (2009) Glutathione transferase activity and oocyte development in copepods exposed to toxic phytoplankton. Harmful Algae 8(3):395–406. https://doi.org/10.1016/j.hal.2008.08.025

Kubanek J, Snell TW, Pirkle C (2007) Chemical defense of the red tide dinoflagellate *Karenia brevis* against rotifer grazing. Limnol Oceanogr 52(3):1026–1035. https://doi.org/10.4319/lo.2007.52.3.1026

Kudela RM, Ryan JP, Blakely MD, Lane JQ, Peterson TD (2008) Linking the physiology and ecology of *Cochlodinium* to better understand harmful algal bloom events: A comparative approach. Harmful Algae 7(3):278–292. https://doi.org/10.1016/j.hal.2007.12.016

Kuhl H, Strassert JFH, Čertnerová D, Varga E, Kreuz E, Lamatsch DK, Wuertz S, Köhler J, Monaghan MT, Stöck M (2024) The haplotype-resolved *Prymnesium parvum* (type B) microalga genome reveals the genetic basis of its fish-killing toxins. Curr Biol 34(16):3698–3706.e3694. https://doi.org/10.1016/j.cub.2024.06.033

Kvrgić K, Lešić T, Džafić N, Pleadin J (2022) Occurrence and seasonal monitoring of domoic acid in three shellfish species from the Northern Adriatic Sea. Toxins 14(1). https://doi.org/10.3390/toxins14010033

Lafont M, Petton B, Vergnes A, Pauletto M, Segarra A, Gourbal B, Montagnani C (2017) Long-lasting antiviral innate immune priming in the Lophotrochozoan Pacific oyster, *Crassostrea gigas*. Sci Rep 7(1):13143. https://doi.org/10.1038/s41598-017-13564-0

Lafont M, Vergnes A, Vidal-Dupiol J, d. Lorgeril J, Gueguen Y, Haffner P, Petton B, Chaparro C, Barrachina C, Destoumieux-Garzon D, Mitta G, Gourbal B, Montagnani C (2020) A sustained immune response supports long-term antiviral immune priming in the Pacific oyster, *Crassostrea gigas*. MBio 11(2):e02777–02719. https://doi.org/10.1128/mbio.02777-19

Lagos N, Onodera H, Zagatto PA, Andrinolo D o, Azevedo SMFQ, Oshima Y (1999) The first evidence of paralytic shellfish toxins in the freshwater cyanobacterium *Cylindrospermopsis raciborskii*, isolated from Brazil. Toxicon 37(10):1359–1373. https://doi.org/10.1016/S0041-0101(99)00080-X

Landsberg JH (2002) The effects of harmful algal blooms on aquatic organisms. Rev Fish Sci 10(2):113–390. https://doi.org/10.1080/20026491051695

Landsberg J, Van Dolah F, Doucette G (2005) Marine and estuarine harmful algal blooms: Impacts on human and animal health. Oceans Health Pathog Mar Environ:165–215

Landsberg JH, Flewelling LJ, Naar J (2009) *Karenia brevis* red tides, brevetoxins in the food web, and impacts on natural resources: Decadal advancements. Harmful Algae 8(4):598–607. https://doi.org/10.1016/j.hal.2008.11.010

Larras F, Billoir E, Scholz S, Tarkka M, Wubet T, Delignette-Muller ML, Schmitt-Jansen M (2020) A multi-omics concentration-response framework uncovers novel understanding of triclosan effects in the chlorophyte *Scenedesmus vacuolatus*. J Hazard Mater 397. https://doi.org/10.1016/j.jhazmat.2020.122727

Lasley-Rasher RS, Nagel K, Angra A, Yen J (2016) Intoxicated copepods: Ingesting toxic phytoplankton leads to risky behaviour. Proc R Soc B Biol Sci 283(1829). https://doi.org/10.1098/rspb.2016.0176

Le Du J, Tovar-Ramírez D, Núñez-Vázquez EJ (2017) Embryotoxic effects of dissolved okadaic acid on the development of longfin yellowtail *Seriola rivoliana*. Aquat Toxicol 190:210–216. https://doi.org/10.1016/j.aquatox.2017.07.012

Leal JF, Cristiano MLS (2022) Marine paralytic shellfish toxins: chemical properties, mode of action, newer analogues, and structure–toxicity relationship. Nat Prod Rep 39(1):33–57. https://doi.org/10.1039/d1np00009h

Leandro LF, Rolland RM, Roth PB, Lundholm N, Wang Z, Doucette GJ (2009) Exposure of the North Atlantic right whale *Eubalaena glacialis* to the marine algal biotoxin, domoic acid. Mar Ecol Prog Ser 398:287–303. https://doi.org/10.3354/meps08321

Leandro LF, Teegarden GJ, Roth PB, Wang ZH, Doucette GJ (2010) The copepod *Calanus finmarchicus*: a potential vector for trophic transfer of the marine algal biotoxin, domoic acid. J Exp Mar Biol Ecol 382(2):88–95. https://doi.org/10.1016/j.jembe.2009.11.002

Ledreux A, Brand H, Chinain M, Bottein MYD, Ramsdell JS (2014) Dynamics of ciguatoxins from *Gambierdiscus polynesiensis* in the benthic herbivore *Mugil cephalus*: Trophic transfer implications. Harmful Algae 39:165–174. https://doi.org/10.1016/j.hal.2014.07.009

Lee KM, Kim KH (2025) Long-term and non-specific immune memory induced by muramyl dipeptide and its protective effect against *Vibrio anguillarum* in rainbow trout (*Oncorhynchus mykiss*). Develop Comparat Immunol 167:105386. https://doi.org/10.1016/j.dci.2025.105386

Lee T, Fong F, Ho K-C, Lee F (2016) The mechanism of diarrhetic shellfish poisoning toxin production in *Prorocentrum* spp.: physiological and molecular perspectives. Toxins 8(10):272. https://doi.org/10.3390/toxins8100272

Lefebvre KA, Bargu S, Kieckhefer T, Silver MW (2002) From sanddabs to blue whales: The pervasiveness of domoic acid. Toxicon 40(7):971–977. https://doi.org/10.1016/S0041-0101(02)00093-4

Lefebvre KA, Noren DP, Schultz IR, Bogard SM, Wilson J, Eberhart B-TL (2007) Uptake, tissue distribution and excretion of domoic acid after oral exposure in coho salmon (*Oncorhynchus kisutch*). Aquat Toxicol 81(3):266–274. https://doi.org/10.1016/j.aquatox.2006.12.009

Lefebvre KA, Tilton SC, Bammler TK, Beyer RP, Srinouanprachan S, Stapleton PL, Farin FM, Gallagher EP (2009) Gene expression profiles in zebrafish brain after acute exposure to domoic acid at symptomatic and asymptomatic doses. Toxicol Sci 107(1):65–77. https://doi.org/10.1093/toxsci/kfn207

Lefebvre KA, Frame ER, Kendrick PS (2012) Domoic acid and fish behavior: a review. Harmful Algae 13:126–130. https://doi.org/10.1016/j.hal.2011.09.011

Leite ID, Sdiri K, Taylor A, Viallon J, Gharbia HB, Mafra Júnior LL, Swarzenski P, Oberhaensli F, Darius HT, Chinain M, Bottein M-YD (2021a) Experimental evidence of ciguatoxin accumulation and depuration in carnivorous lionfish. Toxins 13. https://doi.org/10.3390/toxins13080564

Leite IDP, Sandrini-Neto L, Squella FL, Alves TP, Schramm MA, Calado SLDM, Silva de Assis HC, Mafra LL (2021b) Toxin accumulation, detoxification and oxidative stress in bivalve (*Anomalocardia flexuosa*) exposed to the dinoflagellate *Prorocentrum lima*. Aquat Toxicol 232. https://doi.org/10.1016/j.aquatox.2020.105738

Lenz F (1933) Untersuchungen zur Limnologie von Strandseen. Verh Int Ver Limnol 6:166–177. https://doi.org/10.1080/03680770.1933.11898529

Leong SCY, Lim LP, Chew SM, Kok JWK, Teo SLM (2015) Three new records of dinoflagellates in Singapore's coastal waters, with observations on environmental conditions associated with microalgal growth in the Johor Straits. Raffles Bull Zool 31:24–36

Li Y, Yu J, Sun T, Liu C, Sun Y, Wang Y (2018) Using the marine Rotifer Brachionus plicatilis as an endpoint to evaluate whether ROS-dependent hemolytic toxicity is involved in the allelopathy induced by Karenia mikimotoi. Toxins 10(11). https://doi.org/10.3390/toxins10110439

Li J, Ruan Y, Mak YL, Zhang X, Lam JCW, Leung KMY, Lam PKS (2021) Occurrence and trophodynamics of marine lipophilic phycotoxins in a subtropical marine food web. Environ Sci Technol 55(13):8829–8838. https://doi.org/10.1021/acs.est.1c01812

Li L, Chen S, Xu SY, Li DW, Li HY, Yang WD (2022) Toxicity and underlying mechanism of the toxic dinoflagellate *Gambierdiscus caribaeus* to the fish *Oryzias melastigma*. Ecotoxicol Environ Saf 247. https://doi.org/10.1016/j.ecoenv.2022.114223

Li Q, Mahmudiono T, Mohammadi H, Nematollahi A, Hoseinvandtabar S, Mehri F, Hasanzadeh V, Limam I, Fakhri Y, Thai VN (2023) Concentration ciguatoxins in fillet of fish: A global systematic review and meta-analysis. Heliyon 9(8). https://doi.org/10.1016/j.heliyon.2023.e18500

Liebert F, Deerns WM (1920) Onderzoek naar de oorsaak van een Vissterfte in den polder Workumer-Nieuwland, nabij Workum. Verb Rijksinst Visscherijonderz 1:81–93

Liebl AL, Schrey AW, Richards CL, Martin LB (2013) Patterns of DNA methylation throughout a range expansion of an introduced songbird. Integr Comp Biol 53(2):351–358. https://doi.org/10.1093/icb/ict007

Lieke T, Steinberg CEW, Bittmann S, Behrens S, Hoseinifar SH, Meinelt T, Knopf K, Kloas W (2021) Fulvic acid accelerates hatching and stimulates antioxidative protection and the innate immune response in zebrafish larvae. Sci Total Environ 796. https://doi.org/10.1016/j.scitotenv.2021.148780

Lin JN, Yan T, Zhang QC, Wang YF, Liu Q, Zhou MJ (2014) *In situ* detrimental impacts of *Prorocentrum donghaiense* blooms on zooplankton in the East China Sea. Mar Pollut Bull 88(1–2):302–310. https://doi.org/10.1016/j.marpolbul.2014.08.026

Lin ZR, Geng HX, Yu RC (2024) Potential roles of hydroxybenzoate paralytic shellfish toxins in modulating toxin biokinetics in scallops. J Hazard Mater 469. https://doi.org/10.1016/j.jhazmat.2024.133896

Lindholm T, Öhman P, Kurki-Helasmo K, Kincaid B, Meriluoto J (1999) Toxic algae and fish mortality in a brackish-water lake in Åland, SW Finland. Hydrobiologia 397:109–120. https://doi.org/10.1023/a:1003667728458

Lindström J, Grebner W, Rigby K, Selander E (2017) Effects of predator lipids on dinoflagellate defence mechanisms – increased bioluminescence capacity. Sci Rep 7(1):13104. https://doi.org/10.1038/s41598-017-13293-4

Liu H, Kelly MS, Campbell DA, Dong SL, Zhu JX, Wang SF (2007) Exposure to domoic acid affects larval development of king scallop *Pecten maximus* (Linnaeus, 1758). Aquat Toxicol 81(2):152–158. https://doi.org/10.1016/j.aquatox.2006.11.012

Liu Y, Kong FZ, Xun XG, Dai L, Geng HX, Hu XL, Yu RC, Bao ZM, Zhou MJ (2020) Biokinetics and biotransformation of paralytic shellfish toxins in different tissues of Yesso scallops, *Patinopecten yessoensis*. Chemosphere 261. https://doi.org/10.1016/j.chemosphere.2020.128063

Long M, Krock B, Castrec J, Tillmann U (2021) Unknown extracellular and bioactive metabolites of the genus *Alexandrium*: A review of overlooked toxins. Toxins 13(12). https://doi.org/10.3390/toxins13120905

Lou J, Cheng J, Xun X, Li X, Li M, Zhang X, Li T, Bao Z, Hu X (2020) Glutathione *S*-transferase genes in scallops and their diverse expression patterns after exposure to PST-producing dinoflagellates. Mar Life Sci Technol 2(3):252–261. https://doi.org/10.1007/s42995-020-00050-2

Louzao MC, Espiña B, Cagide E, Ares IR, Alfonso A, Vieytes MR, Botana LM (2010) Cytotoxic effect of palytoxin on mussel. Toxicon 56(5):842–847. https://doi.org/10.1016/j.toxicon.2010.02.027

Louzao MC, Ares IR, Cagide E, Espiña B, Vilariño N, Alfonso A, Vieytes MR, Botana LM (2011) Palytoxins and cytoskeleton: an overview. Toxicon 57(3):460–469. https://doi.org/10.1016/j.toxicon.2010.09.017

Louzao MC, Fernández DA, Abal P, Fraga M, Vilariño N, Vieytes MR, Botana LM (2015a) Diarrhetic effect of okadaic acid could be related with its neuronal action: changes in neuropeptide Y. Toxicol Lett 237(2):151–160. https://doi.org/10.1016/j.toxlet.2015.06.004

Louzao MC, Fraga M, Vilariño N (2015b) Pharmacology of palytoxins and ostreocins. In: Botana LM, Alfonso A (eds) Phycotoxins: chemistry and biochemistry, pp 113–135

Lu J, Bu X, Xiao S, Lin Z, Wang X, Jia Y, Wang X, Qin JG, Chen L (2019) Effect of single and combined immunostimulants on growth, anti-oxidation activity, non-specific immunity and resistance to *Aeromonas hydrophila* in Chinese mitten crab (*Eriocheir sinensis*). Fish Shellf Immunol 93:732–742. https://doi.org/10.1016/j.fsi.2019.08.027

Luan N, Zuo J, Niu Q, Yan W, Hung T-C, Liu H, Wu Q, Wang G, Deng P, Ma X, Qin J, Li G (2024) Probiotic *Lactobacillus rhamnosus* alleviates the neurotoxicity of microcystin-LR in zebrafish (*Danio rerio*) through the gut-brain axis. Sci Total Environ 908:168058. https://doi.org/10.1016/j.scitotenv.2023.168058

Lundgren VM, Roelke DL, Brooks BW, Graneli E, Davis SL, Baty T, Scott WC (2015) *Prymnesium parvum* invasion success into coastal bays of the Gulf of Mexico: Galveston Bay case study. Harmful Algae 43:31–45. https://doi.org/10.1016/j.hal.2014.12.008

Lundholm N, Krock B, John U, Skov J, Cheng J, Pančić M, Wohlrab S, Rigby K, Nielsen TG, Selander E, Harðardóttir S (2018) Induction of domoic acid production in diatoms—types of grazers and diatoms are important. Harmful Algae 79:64–73. https://doi.org/10.1016/j.hal.2018.06.005

Ma K, Tang Q, Liang P, Xia J, Zhang B, Gao X (2019) Toxicity and sublethal effects of two plant allelochemicals on the demographical traits of cotton aphid, *Aphis gossypii* Glover (Hemiptera: Aphididae). PLoS One 14(11). https://doi.org/10.1371/journal.pone.0221646

Mabe P, Zakarian A (2015) Chemistry of pinnatoxins. In: Botana LM, Alfonso A (eds) Phycotoxins: chemistry and biochemistry. Wiley, Chichester, pp 49–68

Mafra LL, Bricelj VM, Fennel K (2010) Domoic acid uptake and elimination kinetics in oysters and mussels in relation to body size and anatomical distribution of toxin. Aquat Toxicol 100(1):17–29. https://doi.org/10.1016/j.aquatox.2010.07.002

Mafra LL, Lopes D, Bonilauri VC, Uchida H, Suzuki T (2015) Persistent contamination of octopuses and mussels with lipophilic shellfish toxins during spring *Dinophysis* blooms in a subtropical estuary. Mar Drugs 13(6):3920–3935. https://doi.org/10.3390/md13063920

Mahalakshmi R, Priyanga J, Bhakta-Guha D, Guha G (2022) Hormetic effect of low doses of rapamycin triggers anti-aging cascades in WRL-68 cells by modulating an mTOR-mitochondria cross-talk. Mol Biol Rep 49(1):463–476. https://doi.org/10.1007/s11033-021-06898-6

Mahnir VM, Kozlovskaya EP, Kalinovsky AI (1992) Sea anemone *Radianthus macrodactylus*–a new source of palytoxin. Toxicon 30(11):1449–1456. https://doi.org/10.1016/0041-0101(92)90520-f

Mak YL, Wai TC, Murphy MB, Chan WH, Wu JJ, Lam JCW, Chan LL, Lam PKS (2013) Pacific ciguatoxins in food web components of coral reef systems in the Republic of Kiribati. Environ Sci Technol 47(24):14070–14079. https://doi.org/10.1021/es403175d

Mak YL, Li J, Liu CN, Cheng SH, Lam PKS, Cheng J, Chan LL (2017) Physiological and behavioural impacts of Pacific ciguatoxin-1 (P-CTX-1) on marine medaka (*Oryzias melastigma*). J Hazard Mater 321:782–790. https://doi.org/10.1016/j.jhazmat.2016.09.066

Manfrin C, Dreos R, Battistella S, Beran A, Gerdol M, Varotto L, Lanfranchi G, Venier P, Pallavicini A (2010) Mediterranean mussel gene expression profile induced by okadaic acid exposure. Environ Sci Technol 44(21):8276–8283. https://doi.org/10.1021/es102213f

Manivasagan P, Kim SK (2015) An overview of harmful algal blooms on marine organisms. In: Handbook of marine microalgae: biotechnology advances. S K Kim, pp 517–526

Manning SR, La Claire JW (2010) Prymnesins: toxic metabolites of the golden alga, *Prymnesium parvum* Carter (Haptophyta). Mar Drugs 8(3):678–704. https://doi.org/10.3390/md8030678

Mardones JI, Dorantes-Aranda JJ, Nichols PD, Hallegraeff GM (2015) Fish gill damage by the dinoflagellate *Alexandrium catenella* from Chilean fjords: synergistic action of ROS and PUFA. Harmful Algae 49:40–49. https://doi.org/10.1016/j.hal.2015.09.001

Mardones J, Holland D, Anderson L, Le Bihan V, Gianella F, Clément A, Davidson K, Sakamoto S, Yoshida T, Trainer V (2020) Estimating and mitigating the economic costs of harmful algal blooms on commercial and recreational shellfish harvesters. GlobalHAB. In: Trainer VL (ed) Evaluating, reducing and mitigating the cost of harmful algal blooms: a compendium of case studies, PICES scientific report no. 59. Marine Science Organization (PICES), Sidney, pp 66–83

Martínez-Escauriaza R, Lozano V, Pérez-Parallé ML, Blanco J, Sánchez JL, Pazos AJ (2021) Expression analyses of genes related to multixenobiotic resistance in *Mytilus galloprovincialis* after exposure to okadaic acid-producing *Dinophysis acuminata*. Toxins 13(9). https://doi.org/10.3390/toxins13090614

Martins JC, Domínguez-Pérez D, Azevedo C, Braga AC, Costa PR, Osório H, Vasconcelos V, Campos A (2020) Molecular responses of mussel *Mytilus galloprovincialis* associated to accumulation and depuration of marine biotoxins okadaic acid and dinophysistoxin-1 revealed by shotgun proteomics. Front Mar Sci 7. https://doi.org/10.3389/fmars.2020.589822

Mathiassen SK, Boutin C, Strandberg B, Carpenter D, Damgaard C (2021) Effects of low doses of herbicides on different endpoints in the life cycle of nontarget terrestrial plants. Environ Toxicol Chem 40(5):1389–1404. https://doi.org/10.1002/etc.4992

Matus Hernández MA, Hernández Saavedra NY (2021) Effect of different species of *Prorocentrum* genus on the Japanese oyster *Crassostrea gigas* proteomic profile. Toxins 13(7). https://doi.org/10.3390/toxins13070504

Mauriz A, Blanco J (2010) Distribution and linkage of domoic acid (amnesic shellfish poisoning toxins) in subcellular fractions of the digestive gland of the scallop *Pecten maximus*. Toxicon 55(2):606–611. https://doi.org/10.1016/j.toxicon.2009.10.017

McCabe RM, Hickey BM, Kudela RM, Lefebvre KA, Adams NG, Bill BD, Gulland FMD, Thomson RE, Cochlan WP, Trainer VL (2016) An unprecedented coastwide toxic algal bloom linked to anomalous ocean conditions. Geophys Res Lett 43(19):10,366-310,376. https://doi.org/10.1002/2016GL070023

McCall JR, Sausman KT, Keeler DM, Brown AP, Turrise SL (2022) Immune modulating brevetoxins: Monocyte cytotoxicity, apoptosis, and activation of M1/M2 response elements is dependent on reactive groups. Mar Drugs 20(4). https://doi.org/10.3390/md20040233

McFarland K, Jean F, Thébault J, Volety AK (2016) Potential impacts of blooms of the toxic dinoflagellate *Karenia brevis* on the growth, survival and juvenile recruitment of the non-native green mussel *Perna viridis* in southeastern United States. Toxicon 109:94–102. https://doi.org/10.1016/j.toxicon.2015.11.017

McGowan S (2016) Algal blooms. In: Biological and environmental hazards, risks, and disasters. Elsevier, pp 5–43

McLellan NL, Manderville RA (2017) Toxic mechanisms of microcystins in mammals. Toxicol Res 6(4):391–405. https://doi.org/10.1039/c7tx00043j

Mebs D (1998) Occurrence and sequestration of toxins in food chains. Toxicon 36(11):1519–1522. https://doi.org/10.1016/S0041-0101(98)00143-3

Mello DF, De Oliveira ES, Vieira RC, Simoes E, Trevisan R, Dafre AL, Barracco MA (2012) Cellular and transcriptional responses of *Crassostrea gigas* hemocytes exposed *in vitro* to brevetoxin (PbTx-2). Mar Drugs 10(3):583–597. https://doi.org/10.3390/md10030583

Menzel R, Menzel S, Swain SC, Pietsch K, Tiedt S, Witczak J, Stürzenbaum SR, Steinberg CEW (2012) The nematode *Caenorhabditis elegans*, stress and aging: identifying the complex interplay of genetic pathways following the treatment with humic substances. Front Genet 3. https://doi.org/10.3389/fgene.2012.00050

Merenciano M, Ullastres A, de Cara MAR, Barrón MG, González J (2016) Multiple independent retroelement insertions in the promoter of a stress response gene have variable molecular and functional effects in *Drosophila*. PLoS Genet 12(8):e1006249. https://doi.org/10.1371/journal.pgen.1006249

Messieh SN, El-Sabh MI (1990) The marine killers: Dinoflagellates in estuarine and coastal waters. Nat Hazards 3(1):69–86. https://doi.org/10.1007/BF00144975

Meyer KA, O'Neil JM, Hitchcock GL, Heil CA (2014) Microbial production along the West Florida Shelf: responses of bacteria and viruses to the presence and phase of *Karenia brevis* blooms. Harmful Algae 38(C):110–118. https://doi.org/10.1016/j.hal.2014.04.015

Meyer L, Capper A, Carter S, Simpfendorfer C (2016) An investigation into ciguatoxin bioaccumulation in sharks. Toxicon 119:234–243. https://doi.org/10.1016/j.toxicon.2016.06.007

Miesner AK, Lundholm N, Krock B, Nielsen TG (2016) The effect of *Pseudo-nitzschia seriata* on grazing and fecundity of *Calanus finmarchicus* and *Calanus glacialis*. J Plankton Res 38(3):564–574. https://doi.org/10.1093/plankt/fbw015

Miller TR, Belas R (2003) *Pfiesteria piscicida*, *P. shumwayae*, and other *Pfiesteria*-like dinoflagellates. Res Microbiol 154(2):85–90. https://doi.org/10.1016/S0923-2508(03)00027-5

Miranda M, Noyma N, Pacheco FS, de Magalhães L, Pinto E, Santos S, Soares MFA, Huszar VL, Lürling M, Marinho MM (2017) The efficiency of combined coagulant and ballast to remove harmful cyanobacterial blooms in a tropical shallow system. Harmful Algae 65:27–39. https://doi.org/10.1016/j.hal.2017.04.007

Moeller PDR, Beauchesne KR, Huncik KM, Davis WC, Christopher SJ, Riggs-Gelasco P, Gelasco AK (2007) Metal complexes and free radical toxins produced by *Pfiesteria piscicida*. Environ Sci Technol 41(4):1166–1172. https://doi.org/10.1021/es0617993

Moestrup Ø, Hansen PJ (1988) On the occurrence of the potentially toxic dinoflagellates *Alexandrium tamarense* (= *Gonyaulax excavata*) and *A. ostenfeldii* in Danish and faroese waters. Ophelia 28(3):195–213. https://doi.org/10.1080/00785326.1988.10430813

Molgó J, Benoit E, Aráoz R, Zakarian A, Iorga BI (2016) Spirolides and cyclic imines: toxicological profile. In: Gopalakrishnakone P, Haddad V, Tubaro A, Kim E, Kem WR (eds) Marine and freshwater toxins. Springer, Dordrecht, pp 193–217

Monroig Ó, Kabeya N (2018) Desaturases and elongases involved in polyunsaturated fatty acid biosynthesis in aquatic invertebrates: a comprehensive review. Fish Sci 84(6):911–928. https://doi.org/10.1007/s12562-018-1254-x

Monti-Birkenmeier M, Berden Zrimec M, Drinovec L, Beran A, Zrimec A, Cataletto B, Fonda Umani S (2019) Influence of salinity on growth and cell volume in three strains of *Prorocentrum cordatum* (Dinophyceae). Aquat Biol 28:1–12. https://doi.org/10.3354/ab00704

Montoya NG, Fulco VK, Carignan MO, Carreto JI (2010) Toxin variability in cultured and natural populations of *Alexandrium tamarense* from southern South America – evidences of diversity and environmental regulation. Toxicon 56(8):1408–1418. https://doi.org/10.1016/j.toxicon.2010.08.006

Moore RE, Bartolini G (1981) Structure of palytoxin. J Am Chem Soc 103(9):2491–2494. https://doi.org/10.1021/ja00399a093

Mora D, Schlüsener M, Fischer H, Kleinteich J, Schulz M, Ternes T, Thiel J, Wick A, Krenek S (2024) From genes to toxins: profiling *Prymnesium parvum* during a riverine harmful algal bloom. Harmful Algae:102644. https://doi.org/10.1016/j.hal.2024.102644

Mosblech A, Feussner I, Heilmann I (2009) Oxylipins: structurally diverse metabolites from fatty acid oxidation. Plant Physiol Biochem 47(6):511–517. https://doi.org/10.1016/j.plaphy.2008.12.011

Mu C, Ren X, Li J (2023) Immune response of the ridgetail white prawn *Exopalaemon carinicauda* after exposure to the dinoflagellate *Prorocentrum minimum*. J Ocean Univ China 22(3):821–830. https://doi.org/10.1007/s11802-023-5336-8

Munday R, Quilliam MA, LeBlanc P, Lewis N, Gallant P, Sperker SA, Ewart HS, MacKinnon SL (2012) Investigations into the toxicology of spirolides, a group of marine phycotoxins. Toxins 4(1):1–14. https://doi.org/10.3390/toxins4010001

Murray SA, Kohli GS, Farrell H, Spiers ZB, Place AR, Dorantes-Aranda JJ, Ruszczyk J (2015) A fish kill associated with a bloom of *Amphidinium carterae* in a coastal lagoon in Sydney, Australia. Harmful Algae 49:19–28. https://doi.org/10.1016/j.hal.2015.08.003

Naar JP, Flewelling LJ, Lenzi A, Abbott JP, Granholm A, Jacocks HM, Gannon D, Henry M, Pierce R, Baden DG, Wolny J, Landsberg JH (2007) Brevetoxins, like ciguatoxins, are potent ichthyotoxic neurotoxins that accumulate in fish. Toxicon 50(5):707–723. https://doi.org/10.1016/j.toxicon.2007.06.005

Nascentes RF, Carbonari CA, Simões PS, Brunelli MC, Velini ED, Duke SO (2018) Low doses of glyphosate enhance growth, CO_2 assimilation, stomatal conductance and transpiration in sugarcane and eucalyptus. Pest Manag Sci 74(5):1197–1205. https://doi.org/10.1002/ps.4606

Neves RAF, Fernandes T, Santos LN d, Nascimento SM (2017) Toxicity of benthic dinoflagellates on grazing, behavior and survival of the brine shrimp *Artemia salina*. PLoS One 12(4). https://doi.org/10.1371/journal.pone.0175168

Neves RAF, Santiago TC, Carvalho WF, Silva E d S, da Silva PM, Nascimento SM (2019) Impacts of the toxic benthic dinoflagellate *Prorocentrum lima* on the brown mussel *Perna perna*: Shell-valve closure response, immunology, and histopathology. Mar Environ Res 146:35–45. https://doi.org/10.1016/j.marenvres.2019.03.006

Nézan E, Chomérat N (2011) *Vulcanodinium rugosum* gen. nov., sp. nov. (Dinophyceae): a new marine dinoflagellate from the French Mediterranean coast (in French). Cryptogam Algol 32(1):3–18, 16. https://doi.org/10.7872/crya.v32.iss1.2011.003

Nicholson GM, Lewis RJ (2006) Ciguatoxins: cyclic polyether modulators of voltage-gated ion channel function. Mar Drugs 4(3):82–118. https://doi.org/10.3390/md403082

Nielsen P, Krock B, Hansen PJ, Vismann B (2020) Effects of the DSP-toxic dinoflagellate *Dinophysis acuta* on clearance and respiration rate of the blue mussel, *Mytilus edulis*. PLoS One 15(3):e0230176. https://doi.org/10.1371/journal.pone.0230176

Nieva JA, Tebben J, Tillmann U, Wohlrab S, Krock B (2020) Mass spectrometry-based characterization of new spirolides from *Alexandrium ostenfeldii* (Dinophyceae). Mar Drugs 18(10). https://doi.org/10.3390/md18100505

Nishimura T, Sato S, Tawong W, Sakanari H, Uehara K, Shah MMR, Suda S, Yasumoto T, Taira Y, Yamaguchi H, Adachi M (2013) Genetic diversity and distribution of the ciguatera-causing dinoflagellate *Gambierdiscus* spp. (Dinophyceae) in coastal areas of Japan. PLoS One 8(4). https://doi.org/10.1371/journal.pone.0060882

Nogueira ICG, Pereira P, Dias E, Pflugmacher S, Wiegand C, Franca S, Vasconcelos VM (2004) Accumulation of paralytic shellfish toxins (PST) from the cyanobacterium *Aphanizomenon issatschenkoi* by the cladoceran *Daphnia magna*. Toxicon 44(7):773–780. https://doi.org/10.1016/j.toxicon.2004.08.006

Norambuena L, Mardones JI (2023) Emerging phycotoxins in the Chilean coast: First localized detection of the neurotoxic cyclic imine Pinnatoxin-G in shellfish banks. Mar Pollut Bull 190. https://doi.org/10.1016/j.marpolbul.2023.114878

Oh J-W, Pushparaj SS, Muthu M, Gopal J (2023) Review of harmful algal blooms (HABs) causing marine fish kills: Toxicity and mitigation. Plants 12. https://doi.org/10.3390/plants12233936

Oikonomou A, Katsiapi M, Karayanni H, Moustaka-Gouni M, Kormas KA (2012) Plankton microorganisms coinciding with two consecutive mass fish kills in a newly reconstructed lake. Sci World J 2012. https://doi.org/10.1100/2012/504135

Ortiz A, Navarro JM, Pizarro G, Villanueva PA, Segura CJ (2019) Accumulation and biotransformation dynamics of the neurotoxic complex, saxitoxin, in different life stages of *Ostrea chilensis*. Mar Environ Res 144:240–245. https://doi.org/10.1016/j.marenvres.2019.01.014

Otero P, Vale C, Boente-Juncal A, Costas C, Louzao MC, Botana LM (2020) Detection of cyclic imine toxins in dietary supplements of green lipped mussels (*Perna canaliculus*) and in shellfish *Mytilus chilensis*. Toxins 12(10). https://doi.org/10.3390/toxins12100613

Otterstrøm CV, Steeman Nielsen E (1940) Two cases of extensive mortality in fishes caused by the flagellate *Prymnesium parvum*, Carter. Rep Danish Biol Stat 44:1–24

Painefilú JC, Bianchi VA, Krock B, De Anna JS, Kristoff G, Luquet CM (2020) Effects of paralytic shellfish toxins on the middle intestine of *Oncorhynchus mykiss*: Glutathione metabolism, oxidative status, lysosomal function and ATP-binding cassette class C (ABCC) proteins activity. Ecotoxicol Environ Saf 204. https://doi.org/10.1016/j.ecoenv.2020.111069

Paredes-Banda P, García-Mendoza E, Ponce-Rivas E, Blanco J, Almazán-Becerril A, Galindo-Sánchez C, Cembella A (2018) Association of the toxigenic dinoflagellate *Alexandrium ostenfeldii* with spirolide accumulation in cultured mussels (*Mytilus galloprovincialis*) from Northwest Mexico. Front Marine Sci 5. https://doi.org/10.3389/fmars.2018.00491

Park G, Dam HG (2021) Cell-growth gene expression reveals a direct fitness cost of grazer-induced toxin production in red tide dinoflagellate prey. Proc R Soc B Biol Sci 288(1944). https://doi.org/10.1098/rspb.2020.2480

Park G, Norton L, Avery D, Dam HG (2023) Grazers modify the dinoflagellate relationship between toxin production and cell growth. Harmful Algae 126. https://doi.org/10.1016/j.hal.2023.102439

Parnas I (1963) The toxicity of *Prymnesium parvum*. Isr J Zool 12(1–4):15–23. https://doi.org/10.1080/00212210.1963.10688166

Patin NV, Brown E, Chebli G, Garfield C, Kubanek J, Stewart FJ (2020) Microbial and chemical dynamics of a toxic dinoflagellate bloom. PeerJ 8. https://doi.org/10.7717/peerj.9493

Pazos AJ, Ventoso P, Martínez-Escauriaza R, Pérez-Parallé ML, Blanco J, Triviño JC, Sánchez JL (2017) Transcriptional response after exposure to domoic acid-producing *Pseudo-nitzschia* in the digestive gland of the mussel *Mytilus galloprovincialis*. Toxicon 140:60–71. https://doi.org/10.1016/j.toxicon.2017.10.002

Pelin M, Sosa S, Tubaro A (2016) Palytoxins: toxicological profile. In: Gopalakrishnakone P, Haddad V, Tubaro A, Kim E, Kem WR (eds) Marine and freshwater toxins. Springer Netherlands, Dordrecht, pp 129–145

Peng JN, Place AR, Yoshida W, Anklin C, Hamann MT (2010) Structure and absolute configuration of karlotoxin-2, an ichthyotoxin from the marine dinoflagellate *Karlodinium veneficum*. J Am Chem Soc 132(10):3277. https://doi.org/10.1021/ja9091853

Pérez-Linares J, Ochoa JL, Gago-Martínez A (2008) Effect of PSP toxins in white leg shrimp *Litopenaeus vannamei* Boone, 1931. J Food Sci 73(4):T69–T73. https://doi.org/10.1111/j.1750-3841.2008.00710.x

Petit J, Wiegertjes GF (2016) Long-lived effects of administering β-glucans: Indications for trained immunity in fish. Dev Comp Immunol 64:93–102. https://doi.org/10.1016/j.dci.2016.03.003

Pinto A, Macário IPE, Marques SM, Lourenço J, Domingues I, Botelho MJ, Asselman J, Pereira P, Pereira JL (2024) A short-term exposure to saxitoxin triggers a multitude of deleterious effects in *Daphnia magna* at levels deemed safe for human health. Sci Total Environ 951:175431. https://doi.org/10.1016/j.scitotenv.2024.175431

Pires E, de Melo Tarouco F, Alves TP, da Rosa CE, da Cunha Lana P, Mafra LL (2021) Antioxidant responses and okadaic acid accumulation in *Laeonereis acuta* (Annelida) exposed to the harmful dinoflagellate *Prorocentrum* cf. *lima*. Toxicon 203:104–114. https://doi.org/10.1016/j.toxicon.2021.10.004

Pitcher GC, Franco JM, Doucette GJ, Powell CL, Mouton A (2001) Paralytic shellfish poisoning in the abalone *Haliotis midae* on the West Coast of South Africa. J Shellfish Res 20(2):895–904

Place AR, Bowers HA, Bachvaroff TR, Adolf JE, Deeds JR, Sheng J (2012) *Karlodinium veneficum*—the little dinoflagellate with a big bite. Harmful Algae 14:179–195. https://doi.org/10.1016/j.hal.2011.10.021

Poli MA, Musser SM, Dickey RW, Eilers PP, Hall S (2000) Neurotoxic shellfish poisoning and brevetoxin metabolites: a case study from Florida. Toxicon 38(7):981–993. https://doi.org/10.1016/s0041-0101(99)00191-9

Pope EK, Bigelow LJ, Bernard PB (2024) Glutamatergic neurotransmission and toxicity: domoic acid and kainic acid (glutamic acid analogs). In: Woodward KN, Marrs T (eds) Neurotransmitters and toxicology, vol 48. Royal Society of Chemistry, p 0

Potts GW, Edwards JM (1987) The impact of a *Gyrodinium aureolum* bloom on inshore young fish populations. J Mar Biol Assoc U K 67(2):293–297. https://doi.org/10.1017/S0025315400026618

Pouil S, Clausing RJ, Metian M, Bustamante P, Dechraoui Bottein MY (2018) The role of marine biotoxins on the trophic transfer of Mn and Zn in fish. Aquat Toxicol 198:198–205. https://doi.org/10.1016/j.aquatox.2018.03.004

Poulin RX, Hogan S, Poulson-Ellestad KL, Brown E, Fernández FM, Kubanek J (2018) *Karenia brevis* allelopathy compromises the lipidome, membrane integrity, and photosynthesis of competitors. Sci Rep 8(1). https://doi.org/10.1038/s41598-018-27845-9

Poulson KL, Sieg RD, Prince EK, Kubanek J (2010) Allelopathic compounds of a red tide dinoflagellate have species-specific and context-dependent impacts on phytoplankton. Mar Ecol Prog Ser 416:69–78. https://doi.org/10.3354/meps08788

Poulson-Ellestad K, McMillan E, Montoya JP, Kubanek J (2014a) Are offshore phytoplankton susceptible to *Karenia brevis* allelopathy? J Plankton Res 36(5):1344–1356. https://doi.org/10.1093/plankt/fbu064

Poulson-Ellestad KL, Jones CM, Roy J, Viant MR, Fernandez FM, Kubanek J, Nunn BL (2014b) Metabolomics and proteomics reveal impacts of chemically mediated competition on marine plankton. Proc Natl Acad Sci USA 111(24):9009–9014. https://doi.org/10.1073/pnas.1402130111

Powell CL, Ferdin ME, Busman M, Kvitek RG, Doucette GJ (2002) Development of a protocol for determination of domoic acid in the sand crab (*Emerita analoga*): a possible new indicator species. Toxicon 40(5):485–492. https://doi.org/10.1016/S0041-0101(01)00236-7

Prego-Faraldo MV, Martínez L, Méndez J (2018) RNA-seq analysis for assessing the early response to DSP toxins in *Mytilus galloprovincialis* digestive gland and gill. Toxins 10(10). https://doi.org/10.3390/toxins10100417

Ramos-Sosa MJ, Hernández López V, Sanchez-Henao A, García-Álvarez N, Díaz-Marrero AR, Rosario I, Real F, Fernández JJ (2022) Metabolic hepatic disorders caused by ciguatoxins in goldfish (*Carassius auratus*). Animals 12(24). https://doi.org/10.3390/ani12243500

Ramos-Sosa MJ, García-Álvarez N, Sanchez-Henao A, Padilla D, Sergent FS, Gago-Martínez A, Diogène J, Caballero MJ, Fernández A, Real F (2023) Ciguatoxin-like toxicity distribution in flesh of amberjack (*Seriola* spp.) and dusky grouper (*Epinephelus marginatus*). Environ Res 228. https://doi.org/10.1016/j.envres.2023.115869

Ramsdell JS (2007) The molecular and integrative basis to domoic acid toxicity. In: Botana LM (ed) Phycotoxins: Chemistry and Biochemistry. Wiley-Blackwell, Ames, IA, USA, pp 223–250

Ranston ER, Webber DF, Larsen J (2007) The first description of the potentially toxic dinoflagellate, *Alexandrium minutum* in Hunts Bay, Kingston Harbour, Jamaica. Harmful Algae 6(1):29–47. https://doi.org/10.1016/j.hal.2006.05.006

Rasmussen SA, Andersen AJC, Andersen NG, Nielsen KF, Hansen PJ, Larsen TO (2016a) Chemical diversity, origin, and analysis of phycotoxins. J Nat Prod 79(3):662–673. https://doi.org/10.1021/acs.jnatprod.5b01066

Rasmussen SA, Meier S, Andersen NG, Blossom HE, Duus JG, Nielsen KF, Hansen PJ, Larsen TO (2016b) Chemodiversity of ladder-frame prymnesin polyethers in *Prymnesium parvum*. J Nat Prod 79(9):2250–2256. https://doi.org/10.1021/acs.jnatprod.6b00345

Ray JL, Althammer J, Skaar KS, Simonelli P, Larsen A, Stoecker D, Sazhin A, Ijaz UZ, Quince C, Nejstgaard JC, Frischer M, Pohnert G, Troedsson C (2016) Metabarcoding and metabolome analyses of copepod grazing reveal feeding preference and linkage to metabolite classes in dynamic microbial plankton communities. Mol Ecol 25(21):5585–5602. https://doi.org/10.1111/mec.13844

Razali RM, Leaw CP, Lim HC, Nyanti L, Ishak I, Lim PT (2015) Harmful microalgae assemblage in the aquaculture area of Aman Island, Northern Strait of Malacca. Malaysian J Sci 34(1):20–32. https://doi.org/10.22452/mjs.vol34no1.3

Reguera B, Riobó P, Rodríguez F, Díaz P, Pizarro G, Paz B, Franco J, Blanco J (2014) *Dinophysis* toxins: causative organisms, distribution and fate in shellfish. Mar Drugs 12(1):394–461. https://doi.org/10.3390/md12010394

Reich K, Aschner M (1947) Mass development and control of the phytoflagellate *Prymnesium parvum* in fish ponds in Palestine. Palestine J Bot 4:14–23

Rhodes K, Hubbs C (1992) Recovery of Pecos River fishes from a red tide fish kill. Southwest Nat 37(2):178–187. https://doi.org/10.2307/3671666

Rhodes L, Smith K, Selwood A, McNabb P, Munday R, Suda S, Molenaar S, Hallegraeff G (2011) Dinoflagellate *Vulcanodinium rugosum* identified as the causative organism of pinnatoxins in Australia, New Zealand and Japan. Phycologia 50(6):624–628. https://doi.org/10.2216/11-19.1

Rinehart KL, Namikoshi M, Choi BW (1994) Structure and biosynthesis of toxins from blue-green algae (Cyanobacteria). J Appl Phycol 6(2):159–176. https://doi.org/10.1007/BF02186070

Robineau B, Gagné JA, Fortier L, Cembella AD (1991) Potential impact of a toxic dinoflagellate (*Alexandrium excavatum*) bloom on survival of fish and crustacean larvae. Mar Biol 108(2):293–301. https://doi.org/10.1007/BF01344344

Roelke DL, Errera RM, Kiesling R, Brooks BW, Grover JP, Schwierzke L, Urena-Boeck F, Baker J, Pinckney JL (2007) Effects of nutrient enrichment on *Prymnesium parvum* population dynamics and toxicity: results from field experiments, Lake Possum Kingdom, USA. Aquat Microb Ecol 46(2):125–140. https://doi.org/10.3354/ame046125

Roelke DL, Grover JP, Brooks BW, Glass J, Buzan D, Southard GM, Fries L, Gable GM, Schwierzke-Wade L, Byrd M, Nelson J (2011) A decade of fish-killing *Prymnesium parvum* blooms in Texas: roles of inflow and salinity. J Plankton Res 33(2):243–253. https://doi.org/10.1093/plankt/fbq079

Roelke DL, Barkoh A, Brooks BW, Grover JP, Hambright KD, LaClaire J II, Moeller PR, Patino R (2016) A chronicle of a killer alga in the west: ecology, assessment, and management of *Prymnesium parvum* blooms. Hydrobiologia 764(1):29–50. https://doi.org/10.1007/s10750-015-2273-6

Rolland JL, Medhioub W, Vergnes A, Abi-khalil C, Savar V, Abadie E, Masseret E, Amzil Z, Laabir M (2014) A feedback mechanism to control apoptosis occurs in the digestive gland of the oyster *Crassostrea gigas* exposed to the paralytic shellfish toxins producer *Alexandrium catenella*. Mar Drugs 12(9):5035–5054. https://doi.org/10.3390/md12095035

Rolton A, Soudant P, Vignier J, Pierce R, Henry M, Shumway SE, Bricelj VM, Volety AK (2015) Susceptibility of gametes and embryos of the eastern oyster, *Crassostrea virginica*, to *Karenia brevis* and its toxins. Toxicon 99:6–15. https://doi.org/10.1016/j.toxicon.2015.03.002

Rolton A, Vignier J, Volety AK, Pierce RH, Henry M, Shumway SE, Bricelj VM, Hegaret H, Soudant P (2016) Effects of field and laboratory exposure to the toxic dinoflagellate *Karenia brevis* on the reproduction of the eastern oyster, *Crassostrea virginica*, and subsequent development of offspring. Harmful Algae 57:13–26. https://doi.org/10.1016/j.hal.2016.04.011

Romero-Geraldo R d J, Hernández-Saavedra NY (2014) Stress gene expression in *Crassostrea gigas* (Thunberg, 1793) in response to experimental exposure to the toxic dinoflagellate *Prorocentrum lima* (Ehrenberg) Dodge, 1975. Aquac Res 45(9):1512–1522. https://doi.org/10.1111/are.12100

Romero-Geraldo RDJ, Garciá-Lagunas N, Hernańdez-Saavedra NY (2014) Effects of *in vitro* exposure to diarrheic toxin producer *Prorocentrum lima* on gene expressions related to cell cycle regulation and immune response in *Crassostrea gigas*. PLoS One 9(5):e97181. https://doi.org/10.1371/journal.pone.0097181

Romero-Geraldo RD, Garcia-Lagunas N, Hernandez-Saavedra NY (2016) *Crassostrea gigas* exposure to the dinoflagellate *Prorocentrum lima*: Histological and gene expression effects on the digestive gland. Mar Environ Res 120:93–102. https://doi.org/10.1016/j.marenvres.2016.07.011

Rossini GP, Bigiani A (2011) Palytoxin action on the Na^+, K^+-ATPase and the disruption of ion equilibria in biological systems. Toxicon 57(3):429–439. https://doi.org/10.1016/j.toxicon.2010.09.011

Roué M, Darius HT, Picot S, Ung A, Viallon J, Gaertner-Mazouni N, Sibat M, Amzil Z, Chinain M (2016) Evidence of the bioaccumulation of ciguatoxins in giant clams (*Tridacna maxima*) exposed to *Gambierdiscus* spp. cells. Harmful Algae 57:78–87. https://doi.org/10.1016/j.hal.2016.05.007

Roy JS, Poulson-Ellestad KL, Sieg RD, Poulin RX, Kubanek J (2013) Chemical ecology of the marine plankton. Nat Prod Rep 30(11):1364–1379. https://doi.org/10.1039/c3np70056a

Roy S, Bossier P, Norouzitallab P, Vanrompay D (2020) Trained immunity and perspectives for shrimp aquaculture. Rev Aquac 12(4):2351–2370. https://doi.org/10.1111/raq.12438
Ruocco N, Albarano L, Esposito R, Zupo V, Costantini M, Ianora A (2020) Multiple roles of diatom-derived oxylipins within marine environments and their potential biotechnological applications. Mar Drugs 18(7). https://doi.org/10.3390/md18070342
Sakamoto S, Lim WA, Lu D, Dai X, Orlova T, Iwataki M (2021) Harmful algal blooms and associated fisheries damage in East Asia: current status and trends in China, Japan, Korea and Russia. Harmful Algae 102. https://doi.org/10.1016/j.hal.2020.101787
Salierno JD, Kane AS (2019) Brevetoxin and saxitoxin alter shoaling dynamics in mummichog, *Fundulus heteroclitus*. J Exp Mar Biol Ecol 520. https://doi.org/10.1016/j.jembe.2019.151210
Sanchez-Henao A, García-álvarez N, Padilla D, Ramos-Sosa M, Sergent FS, Fernández A, Estévez P, Gago-Martínez A, Diogène J, Real F (2021) Accumulation of c-ctx1 in muscle tissue of goldfish (*Carassius auratus*) by dietary experience. Animals 11(1):1–13. https://doi.org/10.3390/ani11010242
Santos M, Costa PR, Porteiro FM, Moita MT (2014) First report of a massive bloom of *Alexandrium minutum* (Dinophyceae) in middle North Atlantic: A coastal lagoon in S. Jorge Island, Azores. Toxicon 90(1):265–268. https://doi.org/10.1016/j.toxicon.2014.08.065
Sardo A, Rossi R, Soprano V, Ciminiello P, Fattorusso E, Cirino P, Zingone A (2021) The dual impact of *Ostreopsis* cf. *ovata* on *Mytilus galloprovincialis* and *Paracentrotus lividus*: toxin accumulation and pathological aspects. Mediterr Mar Sci 22(1):59–72. https://doi.org/10.12681/mms.24160
Sasaki M, Ebine M, Takagi H, Takakura H, Shida T, Satake M, Oshima Y, Igarashi T, Yasumoto T (2004) Synthesis of the CDE/FG ring models of prymnesins: Reassignment of the relative configuration of the E/F ring juncture. Org Lett 6(9):1501–1504. https://doi.org/10.1021/ol049569l
Selander E, Thor P, Toth G, Pavia H (2006) Copepods induce paralytic shellfish toxin production in marine dinoflagellates. Proc Royal Soc B-Biol Sci 273(1594):1673–1680. https://doi.org/10.1098/rspb.2006.3502
Selander E, Kubanek J, Hamberg M, Andersson MX, Cervin G, Pavia H (2015) Predator lipids induce paralytic shellfish toxins in bloom-forming algae. Proc Natl Acad Sci USA 112(20):6395–6400. https://doi.org/10.1073/pnas.1420154112
Selander E, Berglund EC, Engström P, Berggren F, Eklund J, Harðardóttir S, Lundholm N, Grebner W, Andersson MX (2019) Copepods drive large-scale trait-mediated effects in marine plankton. Sci Adv 5(2):eaat5096. https://doi.org/10.1126/sciadv.aat5096
Selim KA, Alva V (2024) PII-like signaling proteins: a new paradigm in orchestrating cellular homeostasis. Curr Opin Microbiol 79:102453. https://doi.org/10.1016/j.mib.2024.102453
Selwood AI, Wilkins AL, Munday R, Shi F, Rhodes LL, Holland PT (2013) Portimine: a bioactive metabolite from the benthic dinoflagellate *Vulcanodinium rugosum*. Tetrahedron Lett 54(35):4705–4707. https://doi.org/10.1016/j.tetlet.2013.06.098
Sephton DH, Haya K, Martin JL, LeGresley MM, Page FH (2007) Paralytic shellfish toxins in zooplankton, mussels, lobsters and caged Atlantic salmon, *Salmo salar*, during a bloom of *Alexandrium fundyense* off Grand Manan Island, in the Bay of Fundy. Harmful Algae 6(5):745–758. https://doi.org/10.1016/j.hal.2007.03.002
Shilo M, Shilo M (1953) Conditions which determine the efficiency of ammonium sulphate in the control of *Prymnesium parvum* in fish breeding ponds. Appl Microbiol 1(6):330–333
Shimizu Y (1986) Toxigenesis and biosynthesis of saxitoxin analogues. Pure Appl Chem 58(2):257–262. https://doi.org/10.1351/pac198658020257
Shimizu Y (2003) Microalgal metabolites. Curr Opin Microbiol 6(3):236–243. https://doi.org/10.1016/S1369-5274(03)00064-X
Shoman N, Solomonova E, Akimov A (2024) Growth stimulation of toxic dinoflagellate *Prorocentrum cordatum* with herbicide glyphosate. J Exp Mar Biol Ecol 577. https://doi.org/10.1016/j.jembe.2024.152017
Silva M, Rodriguez I, Barreiro A, Kaufmann M, Neto AI, Hassouani M, Sabour B, Alfonso A, Botana LM, Vasconcelos V (2015) First report of ciguatoxins in two starfish species:

Ophidiaster ophidianus and *Marthasterias glacialis*. Toxins 7(9):3740–3757. https://doi.org/10.3390/toxins7093740
Sipler RE, McGuinness LR, Kirkpatrick GJ, Kerkhof LJ, Schofield OM (2014) Bacteriocidal effects of brevetoxin on natural microbial communities. Harmful Algae 38(C):101–109. https://doi.org/10.1016/j.hal.2014.04.009
Smith BC, Persson A, Selander E, Wikfors GH, Alix J (2011) Toxin profile change in vegetative cells and pellicle cysts of *Alexandrium fundyense* after gut passage in the eastern oyster *Crassostrea virginica*. Aquat Biol 13(2):193–201. https://doi.org/10.3354/ab00362
Soares MCS, Lürling M, Panosso R, Huszar V (2009) Effects of the cyanobacterium *Cylindrospermopsis raciborskii* on feeding and life-history characteristics of the grazer *Daphnia magna*. Ecotoxicol Environ Saf 72(4):1183–1189. https://doi.org/10.1016/j.ecoenv.2008.09.004
Solomonova E, Shoman N, Akimov A (2024) Differential responses of diatom and dinophyte microalgae to the presence of glyphosate herbicide in the medium. Reg Stud Mar Sci 73. https://doi.org/10.1016/j.rsma.2024.103514
Sopanen S, Setälä O, Piiparinen J, Erler K, Kremp A (2011) The toxic dinoflagellate *Alexandrium ostenfeldii* promotes incapacitation of the calanoid copepods *Eurytemora affinis* and *Acartia bifilosa* from the northern Baltic Sea. J Plankton Res 33(10):1564–1573. https://doi.org/10.1093/plankt/fbr052
Sosa S, Tubaro A (2016) Okadaic acid and other diarrheic toxins: Toxicological profile. In: Gopalakrishnakone P, Haddad V, Tubaro A, Kim E, Kem WR (eds) Marine and Freshwater Toxins. Springer Netherlands, Dordrecht, pp 147–168
Sosa S, Pelin M, Cavion F, Hervé F, Hess P, Tubaro A (2020) Acute oral toxicity of pinnatoxin G in mice. Toxins 12(2). https://doi.org/10.3390/toxins12020087
Souid G, Souayed N, Haouas Z, Maaroufi K (2018) Does the phycotoxin okadaic acid cause oxidative stress damages and histological alterations to seabream (*Sparus aurata*)? Toxicon 144:55–60. https://doi.org/10.1016/j.toxicon.2018.02.001
Southard GM, Fries LT, Barkoh A (2010) *Prymnesium parvum*: the Texas experience. J Am Water Resour Assoc 46(1):14–23. https://doi.org/10.1111/j.1752-1688.2009.00387.x
Sprecher BN, Zhang H, Park G, Lin S (2021) Isolation from a fish kill and transcriptomic characterization of *Gyrodinium jinhaense* off Long Island Sound. Harmful Algae 110. https://doi.org/10.1016/j.hal.2021.102136
Starr M, Lair S, Michaud S, Scarratt M, Quilliam M, Lefaivre D, Robert M, Wotherspoon A, Michaud R, Ménard N, Sauvé G, Lessard S, Béland P, Measures L (2017) Multispecies mass mortality of marine fauna linked to a toxic dinoflagellate bloom. PLoS One 12(5). https://doi.org/10.1371/journal.pone.0176299
Steinberg CEW (2012) Stress ecology–environmental stress as ecological driving force and key player in evolution. Springer, Dordrecht
Steinberg CEW (2022) Essential fatty acids—'*Fueling versus controlling*'. In: Aquatic animal nutrition: organic macro- and micro-nutrients. Springer International Publishing, Cham, pp 673–721
Stoecker DK, Adolf JE, Place AR, Glibert PM, Meritt DW (2008) Effects of the dinoflagellates *Karlodinium veneficum* and *Prorocentrum minimum* on early life history stages of the eastern oyster (*Crassostrea virginica*). Mar Biol 154(1):81–90. https://doi.org/10.1007/s00227-007-0901-z
Stonik VS, Stonik I (2015) Low-molecular-weight metabolites from diatoms: Structures, biological roles and biosynthesis. Mar Drugs 13(6):3672–3709. https://doi.org/10.3390/md13063672
Suarez-Isla BA (2016) Saxitoxin and other paralytic toxins: toxicological profile. In: Gopalakrishnakone P, Haddad V, Tubaro A, Kim E, Kem WR (eds) Marine and freshwater toxins. Springer Netherlands, Dordrecht, pp 23–41
Sud P, Su MK, Greller HA, Majlesi N, Gupta A (2013) Case series: Inhaled coral vapor—toxicity in a tank. J Med Toxicol 9(3):282–286. https://doi.org/10.1007/s13181-013-0307-x

Suikkanen S, Kremp A, Hautala H, Krock B (2013) Paralytic shellfish toxins or spirolides? The role of environmental and genetic factors in toxin production of the *Alexandrium ostenfeldii* complex. Harmful Algae 26:52–59. https://doi.org/10.1016/j.hal.2013.04.001

Summerson HC, Peterson CH (1990) Recruitment failure of the bay scallop, *Argopecten irradians concentricus*, during the first red tide, *Ptychodiscus brevis*, outbreak recorded in North Carolina. Estuaries 13(3):322–331. https://doi.org/10.2307/1351923

Swift AEB, Swift TR (1993) Ciguatera. Clin Toxicol 31(1):1–29. https://doi.org/10.3109/15563659309000371

Takeuchi T, Koyanagi R, Gyoja F, Kanda M, Hisata K, Fujie M, Goto H, Yamasaki S, Nagai K, Morino Y, Miyamoto H, Endo K, Endo H, Nagasawa H, Kinoshita S, Asakawa S, Watabe S, Satoh N, Kawashima T (2016) Bivalve-specific gene expansion in the pearl oyster genome: implications of adaptation to a sessile lifestyle. Zool Letters 2:3. https://doi.org/10.1186/s40851-016-0039-2

Tang H, Vanderploeg HA, Johengen TH, Liebig JR (2014) Quagga mussel (*Dreissena rostriformis bugensis*) selective feeding of phytoplankton in Saginaw Bay. J Great Lakes Res 40(S1):83–94. https://doi.org/10.1016/j.jglr.2013.11.011

Tang XW, Yu R, Fan MH, Yang Z, Liao Z, Yang ZX, Xie CY, Xuan YK, Wang JX, Zhang XL, Yan XJ (2022) Physiological and transcriptome analysis of *Mytilus coruscus* in response to *Prorocentrum lima* and microplastics. Front Mar Sci 9. https://doi.org/10.3389/fmars.2022.1087667

Taniyama S, Mahmud Y, Tanu MB, Takatani T, Arakawa O, Noguchi T (2001) Delayed haemolytic activity by the freshwater puffer *Tetraodon* sp. toxin. Toxicon 39(5):725–727. https://doi.org/10.1016/S0041-0101(00)00197-5

Taniyama S, Mahmud Y, Terada M, Takatani T, Arakawa O, Noguchi T (2002) Occurrence of a food poisoning incident by palytoxin from a serranid *Epinephelus* sp. in Japan. J Nat Toxins 11(4):277–282

Taniyama S, Arakawa O, Terada M, Nishio S, Takatani T, Mahmud Y, Noguchi T (2003) *Ostreopsis* sp., a possible origin of palytoxin (PTX) in parrotfish *Scarus ovifrons*. Toxicon 42(1):29–33. https://doi.org/10.1016/s0041-0101(03)00097-7

Tanyag BE, Perelonia KBS, Cambia FD, Montojo UM (2021) Screening of ciguatoxins in The Philippines by animal assay: symptoms, levels, and distribution in fish tissue. Philippine J Fisher 28(1):77–85. https://doi.org/10.31398/tpjf/28.1.2020A0015

Tasker RA (2016) Domoic acid and other amnesic toxins: toxicological profile. In: Gopalakrishnakone P, Haddad V, Tubaro A, Kim E, Kem WR (eds) Marine and freshwater toxins. Springer Netherlands, Dordrecht, pp 93–112

Tasker RA (2021) Domoic acid. In: Novelli A, Fernández-Sánchez M-T, Aschner M, Costa LG (eds) Advances in neurotoxicology, vol 6. Academic, pp 49–88

Teegarden GJ (1999) Copepod grazing selection and particle discrimination on the basis of PSP toxin content. Mar Ecol Prog Ser 181:163–176. https://doi.org/10.3354/meps181163

Ternon E, Pavaux AS, Marro S, Thomas OP, Lemée R (2018) Allelopathic interactions between the benthic toxic dinoflagellate *Ostreopsis* cf. *ovata* and a co-occurring diatom. Harmful Algae 75:35–44. https://doi.org/10.1016/j.hal.2018.04.003

Ternon E, Glukhov E, Trytten E, Lemée R, Gerwick WH (2022) On the hunt for new toxin families produced by a Mediterranean strain of the benthic Dinoflagellate *Ostreopsis* cf. *ovata*. Toxins 14(4). https://doi.org/10.3390/toxins14040234

Tillmann U, Kremp A, Tahvanainen P, Krock B (2014) Characterization of spirolide producing *Alexandrium ostenfeldii* (Dinophyceae) from the western Arctic. Harmful Algae 39:259–270. https://doi.org/10.1016/j.hal.2014.08.008

Touzet N, Lacaze JP, Maher M, Turrell E, Raine R (2011) Summer dynamics of *Alexandrium ostenfeldii* (Dinophyceae) and spirolide toxins in Cork Harbour, Ireland. Mar Ecol Prog Ser 425:21–33. https://doi.org/10.3354/meps08983

Trainer VL, Bates SS, Lundholm N, Thessen AE, Cochlan WP, Adams NG, Trick CG (2012) *Pseudo-nitzschia* physiological ecology, phylogeny, toxicity, monitoring and impacts on ecosystem health. Harmful Algae 14:271–300. https://doi.org/10.1016/j.hal.2011.10.025

Turner JT (2014) Planktonic marine copepods and harmful algae. Harmful Algae 32:81–93. https://doi.org/10.1016/j.hal.2013.12.001

Uemura D, Chou T, Haino T, Nagatsu A, Fukuzawa S, Zheng S-z, Chen H-s (1995) Pinnatoxin A: a toxic amphoteric macrocycle from the Okinawan bivalve *Pinna muricata*. J Am Chem Soc 117(3):1155–1156. https://doi.org/10.1021/ja00108a043

Uliano-Silva M, Americo JA, Brindeiro R, Dondero F, Prosdocimi F, de Freitas Rebelo M (2014) Gene discovery through transcriptome sequencing for the invasive mussel *Limnoperna fortunei*. PLoS One 9(7):e102973. https://doi.org/10.1371/journal.pone.0102973

Ulitzur S (1973) The amphiphatic nature of *Prymnesium parvum* hemolysin. BBA-Biomembranes 298(3):673–679. https://doi.org/10.1016/0005-2736(73)90083-7

Valdiglesias V, Prego-Faraldo MV, Pasaro E, Mendez J, Laffon B (2013) Okadaic acid: More than a diarrheic toxin. Mar Drugs 11(11):4328–4349. https://doi.org/10.3390/md11114328

Vale P (2010) New saxitoxin analogues in the marine environment: developments in toxin chemistry, detection and biotransformation during the 2000s. Phytochem Rev 9(4):525–535. https://doi.org/10.1007/s11101-010-9196-7

Vale P, Rangel I, Silva B, Coelho P, Vilar A (2009) Atypical profiles of paralytic shellfish poisoning toxins in shellfish from Luanda and Mussulo bays, Angola. Toxicon 53(1):176–183. https://doi.org/10.1016/j.toxicon.2008.10.029

Vale C, Antelo Á, Martín V (2015) Pharmacology of ciguatoxins. In: Botana LM, Alfonso A (eds) Phycotoxins: chemistry and biochemistry. John Wiley & Sons, Ltd, Chichester, West Sussex, pp 23–48

Valenti TW Jr, James SV, Lahousse MJ, Schug KA, Roelke DL, Grover JP, Brooks BW (2010) A mechanistic explanation for pH-dependent ambient aquatic toxicity of *Prymnesium parvum* Carter. Toxicon 55(5):990–998. https://doi.org/10.1016/j.toxicon.2009.09.014

Valério E, Chaves S, Tenreiro R (2010) Diversity and impact of prokaryotic toxins on aquatic environments: a review. Toxins 2(10):2359–2410. https://doi.org/10.3390/toxins2102359

Van de Waal DB, Tillmann U, Martens H, Krock B, van Scheppingen Y, John U (2015) Characterization of multiple isolates from an *Alexandrium ostenfeldii* bloom in The Netherlands. Harmful Algae 49:94–104. https://doi.org/10.1016/j.hal.2015.08.002

Vanderploeg HA, Liebig JR, Carmichael WW, Agy MA, Johengen TH, Fahnenstiel GL, Nalepa TF (2001) Zebra mussel (*Dreissena polymorpha*) selective filtration promoted toxic *Microcystis* blooms in Saginaw Bay (Lake Huron) and Lake Erie. Can J Fish Aquat Sci 58(6):1208–1221. https://doi.org/10.1139/cjfas-58-6-1208

VanLandeghem MM, Farooqi M, Farquhar B, Patino R (2013) Impacts of golden alga *Prymnesium parvum* on fish populations in reservoirs of the Upper Colorado River and Brazos River Basins, Texas. Trans Am Fish Soc 142(3):581–595. https://doi.org/10.1080/00028487.2012.754786

Vasas G, M-Hamvas M, Borics G, Gonda S, Máthé C, Jámbrik K, Nagy ZL (2012) Occurrence of toxic *Prymnesium parvum* blooms with high protease activity is related to fish mortality in Hungarian ponds. Harmful Algae 17:102–110. https://doi.org/10.1016/j.hal.2012.03.007

Vicentini M, Calado SLM, Pessati JBK, Perussolo MC, Lirola JR, Marcondes FR, Nascimento ND, Beghetto CL, Vilar MCP, Mela M, Coral LAA, Magalhães VF, Prodocimo V, Cestari MM, de Assis HCS (2024) Temperature rise and its influence on the toxic effects caused by cyanotoxins in a neotropical catfish. Environ Pollut 353:124166. https://doi.org/10.1016/j.envpol.2024.124166

Vilariño N, Ferreiro SF, Crespo A, Gil J (2015) Pharmacology of the cyclic imines. In: Botana LM, Alfonso A (eds) Phycotoxins: chemistry and biochemistry. Wiley, Chichester, pp 343–360

Vogelbein WK, Lovko VJ, Shields JD, Reece KS, Mason PL, Haas LW, Walker CC (2002) *Pfiesteria shumwayae* kills fish by micropredation not exotoxin secretion. Nature 418(6901):967–970. https://doi.org/10.1038/nature01008

Vogt G (2023) Environmental adaptation of genetically uniform organisms with the help of epigenetic mechanisms—an insightful perspective on ecoepigenetics. Epigenomes 7(1). https://doi.org/10.3390/epigenomes7010001

Waggett RJ, Hardison DR, Tester PA (2012) Toxicity and nutritional inadequacy of *Karenia brevis*: Synergistic mechanisms disrupt top-down grazer control. Mar Ecol Prog Ser 444:15–30. https://doi.org/10.3354/meps09401

Wagstaff BA, Pratscher J, Rivera PPL, Hems ES, Brooks E, Rejzek M, Todd JD, Murrell JC, Field RA (2021) Assessing the toxicity and mitigating the impact of harmful *Prymnesium* blooms in eutrophic waters of the Norfolk Broads. Environ Sci Technol 55(24):16538–16551. https://doi.org/10.1021/acs.est.1c04742

Walsh JJ, Dieterle DA, Chen FR, Lenes JM, Maslowski W, Cassano JJ, Whitledge TE, Stockwell D, Flint M, Sukhanova IN, Christensen J (2011) Trophic cascades and future harmful algal blooms within ice-free Arctic Seas north of Bering Strait: a simulation analysis. Prog Oceanogr 91(3):312–343. https://doi.org/10.1016/j.pocean.2011.02.001

Wandscheer CB, Vilariño N, Espiña B, Louzao MC, Botana LM (2010) Human muscarinic acetylcholine receptors are a target of the marine toxin 13-desmethyl C spirolide. Chem Res Toxicol 23(11):1753–1761. https://doi.org/10.1021/tx100210a

Wang DZ (2008) Neurotoxins from marine dinoflagellates: a brief review. Mar Drugs 6(2):349–371. https://doi.org/10.3390/md20080016

Wang H, Liu S, Xun X, Li M, Lou J, Zhang Y, Shi J, Hu J, Bao Z, Hu X (2021a) Toxin- and species-dependent regulation of ATP-binding cassette (ABC) transporters in scallops after exposure to paralytic shellfish toxin-producing dinoflagellates. Aquat Toxicol 230. https://doi.org/10.1016/j.aquatox.2020.105697

Wang X, Wang D, Zhang T, Zhang Q, Zhao J (2021b) iTRAQ-based quantitative proteomic analysis reveals toxicity mechanisms in *Chlamys farreri* exposed to okadaic acid. Front Mar Sci 8. https://doi.org/10.3389/fmars.2021.792050

Wang DZ, Xin YH, Wang MH (2022) *Gambierdiscus* and its associated toxins: a minireview. Toxins 14(7). https://doi.org/10.3390/toxins14070485

Wang Z, Hu J, Mei Z, Zhang Y, Liu Q, Yang D (2024) Mannan-oligosaccharide induces trained immunity activation and alleviates pathological liver injury in turbot (*Scophthalmus maximus*). Aquaculture 578. https://doi.org/10.1016/j.aquaculture.2023.740097

Wasternack C, Feussner I (2018) The oxylipin pathways: Biochemistry and function. Annu Rev Plant Biol 69(69):363–386. https://doi.org/10.1146/annurev-arplant-042817-040440

Watkins SM, Reich A, Fleming LE, Hammond R (2008) Neurotoxic shellfish poisoning. Mar Drugs 6(3):431–455. https://doi.org/10.3390/md6030431

Wattenberg EV (2011) Modulation of protein kinase signaling cascades by palytoxin. Toxicon 57(3):440–448. https://doi.org/10.1016/j.toxicon.2010.11.003

Wei XM, Lu MY, Duan GF, Li HY, Liu JS, Yang WD (2019) Responses of CYP450 in the mussel *Perna viridis* after short-term exposure to the DSP toxins-producing dinoflagellate *Prorocentrum lima*. Ecotoxicol Environ Saf 176:178–185. https://doi.org/10.1016/j.ecoenv.2019.03.073

White AW (1977) Dinoflagellate toxins as probable cause of an Atlantic herring (*Clupea harengus harengus*) kill, and pteropods as apparent vector. J Fish Res Board Can 34(12):2421–2424. https://doi.org/10.1139/f77-328

White AW (1980) Recurrence of kills of Atlantic herring (*Clupea harengus harengus*) caused by dinoflagellate toxins transferred through herbivorous zooplankton. Can J Fish Aquat Sci 37(12):2262–2265. https://doi.org/10.1139/f80-271

White AW (1981) Marine zooplankton can accumulate and retain dinoflagellate toxins and cause fish kills. Limnol Oceanogr 26:103–109

Wiederhold J, Buchinger S, Duester L, Fischer H, Hahn J, Helms M, Hermes N, Jewell K, Kleinteich J, Krenek S, Löffler D, Mora D, Rademacher S, Schlüsener M, Schütze K, Wahrendorf D-S, Wick A, Ternes T (2023) Investigations on fish mass kill in the Odra river in August 2022. Koblenz, BfG Bundesanstalt für Gewässerkunde. https://doi.org/10.5675/BfG-2143_EN

Wiese M, D'Agostino PM, Mihali TK, Moffitt MC, Neilan BA (2010) Neurotoxic alkaloids: Saxitoxin and its analogs. Mar Drugs 8(7):2185–2211. https://doi.org/10.3390/md8072185

Wiles JS, Vick JA, Christensen MK (1974) Toxicological evaluation of palytoxin in several animal species. Toxicon 12(4):427–433. https://doi.org/10.1016/0041-0101(74)90011-7

Willén T, Mattsson R (1997) Water-blooming and toxin-producing cyanobacteria in Swedish fresh and brackish waters, 1981–1995. Hydrobiologia 353(1):181–192. https://doi.org/10.1023/a:1003047019422

Winter A, Tosteson TR (1992) Ciguatera toxins in the food chain revealed by stable isotopes. Bulletin de la Societe de pathologie exotique (1990) 85(5 Pt 2):510–513

Wohlrab S, Iversen MH, John U (2010) A molecular and co-evolutionary context for grazer induced toxin production in *Alexandrium tamarense*. PLoS One 5(11). https://doi.org/10.1371/journal.pone.0015039

Wu CH (2009) Palytoxin: Membrane mechanisms of action. Toxicon 54(8):1183–1189. https://doi.org/10.1016/j.toxicon.2009.02.030

Xie W, Liu X, Yang X, Zhang C, Bian Z (2013) Accumulation and depuration of paralytic shellfish poisoning toxins in the oyster *Ostrea rivularis* Gould—Chitosan facilitates the toxin depuration. Food Control 30(2):446–452. https://doi.org/10.1016/j.foodcont.2012.07.035

Xu L, Ding X, Wang T, Mou S, Sun H, Hou T (2019) Voltage-gated sodium channels: structures, functions, and molecular modeling. Drug Discov Today 24(7):1389–1397. https://doi.org/10.1016/j.drudis.2019.05.014

Xu K, Wang Y, Lian S, Hu N, Chen X, Dai X, Zhang L, Wang S, Hu J, Hu X, Bao Z (2021) Expansion of *C1Q* genes in Zhikong scallop and their expression profiling after exposure to the toxic dinoflagellates. Front Mar Sci 8. https://doi.org/10.3389/fmars.2021.640425

Xu W, Hu X, Li H, Tian X, Ouyang Z, Du Y, Chen J (2023) Effects of *Lactobacillus plantarum* Ep-M17 on growth, immunity and intestinal microbiota of *Penaeus vannamei* under microcystin-LR stress. Aquat Toxicol 265. https://doi.org/10.1016/j.aquatox.2023.106763

Yan M, Leung PTY, Ip JCH, Cheng JP, Wu JJ, Gu JR, Lam PKS (2017) Developmental toxicity and molecular responses of marine medaka (*Oryzias melastigma*) embryos to ciguatoxin P-CTX-1 exposure. Aquat Toxicol 185:149–159. https://doi.org/10.1016/j.aquatox.2017.02.006

Yan M, Mak MYL, Cheng J, Li J, Gu JR, Leung PTY, Lam PKS (2020) Effects of dietary exposure to ciguatoxin P-CTX-1 on the reproductive performance in marine medaka (*Oryzias melastigma*). Mar Pollut Bull 152. https://doi.org/10.1016/j.marpolbul.2019.110837

Yang I, Selander E, Pavia H, John U (2011) Grazer-induced toxin formation in dinoflagellates: a transcriptomic model study. Eur J Phycol 46(1):66–73. https://doi.org/10.1080/09670262.2011.552194

Yang Z, Luo Q, Liang Y, Mazumder A (2016) Processes and pathways of ciguatoxin in aquatic food webs and fish poisoning of seafood consumers. Environ Rev 24(2):144–150. https://doi.org/10.1139/er-2015-0054

Yasumoto T, Yasumura D, Ohizumi Y, Takahashi M, Alcala AC, Alcala LC (1986) Palytoxin in two species of xanthid crab from The Philippines. Agric Biol Chem 50(1):163–167. https://doi.org/10.1080/00021369.1986.10867344

Ye MH, Li DW, Cai QD, Jiao YH, Liu Y, Li HY, Yang WD (2022) Toxic responses of different shellfish species after exposure to *Prorocentrum lima*, a DSP toxins producing dinoflagellate. Toxins 14(7). https://doi.org/10.3390/toxins14070461

Yi Y, Dou G, Yu Z, He H, Wang C, Li L, Zhou J, Liu D, Shi J, Li G, Pang L, Yang N, Huang Q, Qi H (2018) Z-ligustilide exerted hormetic effect on growth and detoxification enzymes of *Spodoptera litura* larvae. Evid Based Complement Alternat Med 2018. https://doi.org/10.1155/2018/7104513

Yñiguez AT, Lim PT, Leaw CP, Jipanin SJ, Iwataki M, Benico G, Azanza RV (2021) Over 30 years of HABs in The Philippines and Malaysia: what have we learned? Harmful Algae 102. https://doi.org/10.1016/j.hal.2020.101776

Yuan K-K, Chen Z-M, Liu Y-X, Li H-Y, Yang W-D (2023) Possible role of docosahexaenoic acid in response to diarrhetic shellfish toxins in the mussel *Perna viridis*. Mar Drugs 21(3). https://doi.org/10.3390/md21030155

Zabaglo K, Chrapusta E, Bober B, Kaminski A, Adamski M, Bialczyk J (2016) Environmental roles and biological activity of domoic acid: A review. Algal Res 13:94–101. https://doi.org/10.1016/j.algal.2015.11.020

Zhang H, Wang L, Song L, Song X, Wang B, Mu C, Zhang Y (2009) A fibrinogen-related protein from bay scallop *Argopecten irradians* involved in innate immunity as pattern recognition receptor. Fish Shellfish Immunol 26(1):56–64. https://doi.org/10.1016/j.fsi.2008.07.019

Zhang G, Fang X, Guo X, Li L, Luo R, Xu F, Yang P, Zhang L, Wang X, Qi H, Xiong Z, Que H, Xie Y, Holland PWH, Paps J, Zhu Y, Wu F, Chen Y, Wang J, Peng C, Meng J, Yang L, Liu J, Wen B, Zhang N, Huang Z, Zhu Q, Feng Y, Mount A, Hedgecock D, Xu Z, Liu Y, Domazet-Loso T, Du Y, Sun X, Zhang S, Liu B, Cheng P, Jiang X, Li J, Fan D, Wang W, Fu W, Wang T, Wang B, Zhang J, Peng Z, Li Y, Li N, Wang J, Chen M, He Y, Tan F, Song X, Zheng Q, Huang R, Yang H, Du X, Chen L, Yang M, Gaffney PM, Wang S, Luo L, She Z, Ming Y, Huang W, Zhang S, Huang B, Zhang Y, Qu T, Ni P, Miao G, Wang J, Wang Q, Steinberg CEW, Wang H, Li N, Qian L, Zhang G, Li Y, Yang H, Liu X, Wang J, Yin Y, Wang J (2012) The oyster genome reveals stress adaptation and complexity of shell formation. Nature 490(7418):49–54. https://doi.org/10.1038/nature11413

Zhang NS, Li HY, Liu JS, Yang WD (2014) Gene expression profiles in zebrafish (*Danio rerio*) liver after acute exposure to okadaic acid. Environ Toxicol Pharmacol 37(2):791–802. https://doi.org/10.1016/j.etap.2014.02.005

Zhang S, Zheng T, Lundholm N, Huang X, Jiang X, Li A, Li Y (2021) Chemical and morphological defenses of *Pseudo-nitzschia* multiseries in response to zooplankton grazing. Harmful Algae 104. https://doi.org/10.1016/j.hal.2021.102033

Zhang C, Ma Y, Steinberg CEW, Cai H, He X, Duan W, Chen F, Pan B (2025) Contrasting physiological and metabolomic responses of ryegrass to low and high doses of sewage sludge biochars. Preprint. https://doi.org/10.2139/ssrn.5049595

Zhao J, Yuan Q, Lei X, Lieke T, Liu Y, et al. (2025) The double-edged environmental effect of dissolved organic matter in global climate change. Environmental and Biogeochemical Processes 1:e009. https://doi.org/10.48130/ebp-0025-0009

Zhu X, Zhang F, Lian S, Wang Y, Hu N, Chen X, Dai X, Hu X, Wang S, Bao Z (2021) *IAP*s gene expansion in the scallop *Patinopecten yessoensis* and their expression profiles after exposure to the toxic dinoflagellate. Front Physiol 12. https://doi.org/10.3389/fphys.2021.633301

Zhu J, Lee WH, Wu J, Zhou S, Yip KC, Liu X, Kirata T, Chan LL (2022) The occurrence, distribution, and toxicity of high-risk ciguatera fish species (grouper and snapper) in Kiritimati Island and Marakei Island of the Republic of Kiribati. Toxins 14(3). https://doi.org/10.3390/toxins14030208

Zhu J, Li J, Wu J, Liu X, Lin Y, Deng H, Qin X, Wong MH, Chan LL (2024) The prevalence of marine lipophilic phycotoxins causes potential risks in a tropical small Island developing state. Environ Sci Technol. https://doi.org/10.1021/acs.est.4c00512

Zlotnick BA, Hintz S, Park DL, Auerbach PS (1995) Ciguatera poisoning after ingestion of imported jellyfish: diagnostic application of serum immunoassay. Wildern Environ Med 6(3):288–294. https://doi.org/10.1580/1080-6032(1995)006[0288:CPAIOI]2.3.CO;2

Zou Y, Wei X-M, Weng H-W, Li H-Y, Liu J-S, Yang W-D (2015) Expression profile of eight glutathione S-transferase genes in *Crassostrea ariakensis* after exposure to DSP toxins producing dinoflagellate *Prorocentrum lima*. Toxicon 105:45–55. https://doi.org/10.1016/j.toxicon.2015.08.018

The Functions of Phycotoxins

4

Many microbes shape their world by releasing substances that modify their surroundings—a cooperative behavior seen across nature. This phenomenon, known as quorum sensing, is a cell-to-cell communication mechanism used by microorganisms to detect and respond to changes in population density. It involves the production, release, and detection of chemical signaling molecules called autoinducers. When the concentration of these molecules reaches a threshold—indicating a critical population size—microbes collectively alter their gene expression. This allows coordinated behaviors such as biofilm formation, virulence, or bioluminescence. Toxic algal blooms are no exception, coordinating their toxin production via such substances. Far from being random outbreaks, they are increasingly understood as the product of sophisticated, population-level strategies, where billions of cells act in concert. From this perspective, toxigenesis isn't a biological accident—it's an evolutionary tool.

According to this view, the primary ecological benefits responsible for the evolution of toxigenesis are considered to be cooperative elimination of competitors (=allelopathy, Adolf et al. (2006)) and predators ("grazer killing," Adolf et al. (2007); Skovgaard and Hansen (2003)). Kim et al. (2016) reported that slow-growing *Alexandrium pohangense* may out-compete other toxic *Alexandrium* species by lowering its mortality rate due to predation. For *A. pohangense*, killing potential heterotrophic protist predators by producing and excreting lytic or lytic-related substances may be a compensation for its relatively slow growth rate in the competition among harmful algal bloom (HAB) organisms. But what exactly are the substances responsible for these effects? And how do we begin to unravel their chemical identity and ecological function? This chapter takes us deeper into the molecular arsenal of toxic algae, focusing on how these organisms weaponize their metabolism to shape food webs.

Already three decades ago, Paerl and Millie criticized the view of phycotoxin (cyanotoxin) research. "Toxicity," a problematic ecophysiological trait of certain cyanobacteria, is largely an anthropomorphic term determined by negative impacts

T. Lieke, C. E. W. Steinberg, *The Ecology of Phycotoxins*,
https://doi.org/10.1007/978-3-032-18380-4_4

of various secondary metabolites (e.g., alkaloids, peptides, organo-phosphate esters) on mammalian health. Most target organisms are neither natural enemies nor consumers of cyanobacteria. This dichotomy has hindered our understanding of the ecophysiological rationales for toxin production. The physical and chemical factors forcing cyanobacterial bloom initiation and persistence appear to act as controls for metabolite production in a complex and variable manner. Toxin production may simply be a mechanism for shunting excess metabolites during periods of environmental stress and may be metabolically linked to the production of other secondary metabolites, including those causing taste and odor problems. Toxins may also help establish mutually beneficial microbial consortia by chemotactically attracting and supporting microorganisms involved in nutrient cycling and the production of growth factors. Conversely, antagonistic organisms may be repelled by toxins. Evaluations concerning the production of toxins and the roles these and other secondary metabolites may play should consider the chemical and biotic interactions cyanobacteria experience during growth and bloom formation (Paerl and Millie, 1996).

Well said—and remarkably ahead of its time! Let's look at the progress since then. In the decades that followed, research has made significant strides, answering some key questions—yet many others remain unresolved. Cyanobacteria in general and *Microcystis* in particular are the best-studied and understood phycotoxin producers—thereby emerging as model systems for unraveling the ecological and biological roles of phycotoxins. As a result, much of our current understanding of toxin function is rooted in studies of cyanobacteria. In the following section, we will explore what is now known about the *primary functions* of cyanotoxins, grounded in evolutionary and ecological contexts. Broader *secondary effects*, including unintended consequences and interactions across taxa, will be considered more broadly to paint a complete picture of these complex metabolites.

4.1 Role of Cyanotoxins with Focus on Microcystins

Cyanobacterial blooms are known to pose a significant threat to ecosystem function, water quality, and the health of animals and humans. Yet this cannot be their original role, as cyanobacteria appeared on Earth billions of years before any of the organisms now affected by their toxins even evolved (Stanier and Cohen-Bazire 1977). A recent review of the biological function of microcystins (MCs) (Wei et al., 2024) has clarified major aspects of the original function of these phycotoxins, which had previously been the subject of speculation.

Cyanobacterial blooms typically consist of a mixture of MC-producing (toxic) and non-MC-producing (non-toxic) genotypes that either possess or lack biosynthetic genes for MCs (Kurmayer et al. 2014; Dick et al., 2021). These genes are organized in a gene cluster known as the *mcy* gene cluster, typically spanning approximately 55 kb in *Microcystis aeruginosa* and other cyanobacteria capable of producing microcystin. The cluster encodes a set of nonribosomal peptide synthetases (NRPSs) and polyketide synthases (PKSs), along with several tailoring enzymes, regulatory elements, and transport proteins.

The core biosynthetic genes for microcystin production include *mcyA*, *mcyB*, and *mcyC*, which encode large, multifunctional enzymes that combine NRPS and PKS domains. These enzymes are responsible for assembling the peptide backbone of microcystin. Additionally, *mcyD*, *mcyE*, and *mcyG* encode PKS modules that are involved in incorporating and modifying fatty acid- and amino acid-derived units during microcystin synthesis. Furthermore, the gene cluster contains genes for enzymes involved in tailoring and modifying, transport, and resistance, as well as regulation of expression in response to the environmental conditions of the specific MC.

The toxic strains show competitive advantages under unfavorable conditions (Briand et al., 2008). In particular, Dziallas and Grossart (2011) demonstrated that increasing reactive oxygen species (ROS) and water temperature select for toxic *Microcystis* sp. In support of this statement, multiyear studies in Lake Taihu, China, found that toxic *Microcystis* dominated from August to October, indicating that temperature was the main trigger for the transformation between toxic and non-toxic strains (Hu et al., 2016; Shan et al., 2019; Zuo et al., 2021). The latter authors showed that three *Phenylobacterium* strains, a genus of gram-negative, aerobic bacteria, known for their ability to degrade complex organic compounds, found during *Microcystis* blooms in Lake Taihu, promoted the dominance of the toxic MC-strains in laboratory experiments when cocultured with the non-toxic mutant. It is possible that *Phenylobacterium* directly uses MCs as a carbon or nitrogen source. These results suggest that toxic cyanobacteria may perform better than non-toxic ones under these oxidative stress conditions. In addition, simulation experiments conducted by Feng et al. (2019) found that toxic *Microcystis* survived more easily than non-toxic *Microcystis* in sediments. These selected examples may suffice to demonstrate that MC-producing strains adapt better to a changing environment than non-toxic strains and maintain dominance (Fig. 4.1).

Under normal conditions, most MCs remain sequestered within cyanobacterial cells and are not actively released into the surrounding environment. However, under adverse environmental conditions, a significant increase in both MC production and the expression of *mcy* biosynthetic genes has been consistently observed in toxin-producing strains. This suggests that intracellular MCs play vital roles in cellular stress responses rather than functioning primarily as external toxins. Specifically, MCs protect cyanobacterial cells from photooxidative damage and oxidative stress, particularly under iron-limiting conditions. They can act as ROS scavengers when cells are exposed to intermediate levels of ultraviolet A (UVA)[1] radiation. Moreover, MCs facilitate the survival of *Microcystis* populations by

[1] Zhao et al. (2025) reported that eutrophicated high-altitude lakes, such as those from the Yunnan Plateau (China), exhibit a shift in cyanobacterial blooms toward the winter period when strong UV irradiation is lower than in summer.

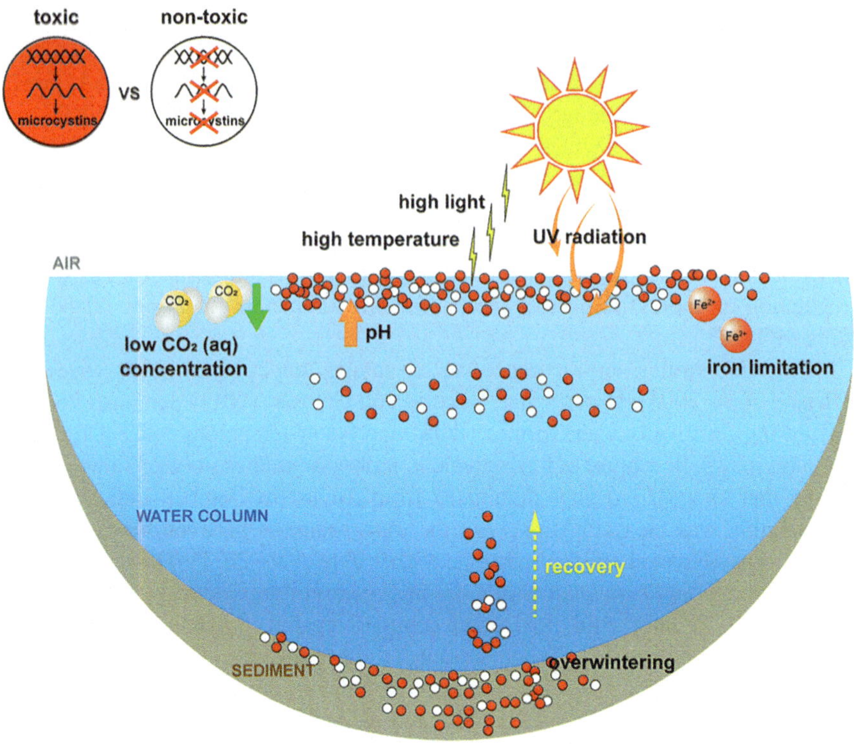

Fig. 4.1 Schematic representation of circumstances in which microcystin-producing (toxic) cyanobacteria dominate in competition with non-microcystin-producing (non-toxic) cyanobacteria. Red circles symbolize toxic cells, and white circles represent non-toxic cells. Toxic cells are more resistant to high light, high temperature, ultraviolet (UV) radiation, iron limitation, and CO_2 depletion than non-toxic cells. Toxic cells survive more easily in sediments and recover earlier and more readily into the water column from sediments in spring than non-toxic cells. Overall, MCs enhance the plasticity of toxic cyanobacteria to changing environments (from Wei et al. (2024), with permission from Elsevier)

supporting adaptation to benthic (sediment) environments, aiding in overwintering and persistence during unfavorable seasons.

Furthermore, intracellular MCs function in carbon/nitrogen metabolism, as well as antioxidation, and MC biosynthesis is closely linked to these processes in *Microcystis* cells (Ginn et al., 2010; Kuniyoshi et al., 2011). Field studies have shown that nitrogen speciation and inorganic carbon availability might be important drivers of toxic *Microcystis* population dynamics, and the absence of NH_4^+ may trigger MC production (Beversdorf et al., 2015). More physiological details can be found in Wei et al. (2024).

Extracellular MCs function in cell-to-cell communication and colony formation (Fig. 4.2). A signaling role for MCs outside *Microcystis* cells was first proposed in the study by Schatz et al. (2007). Their results showed that both the lysis of toxic *Microcystis* cells and the external addition of pure microcystin-LR (MC-LR) led to

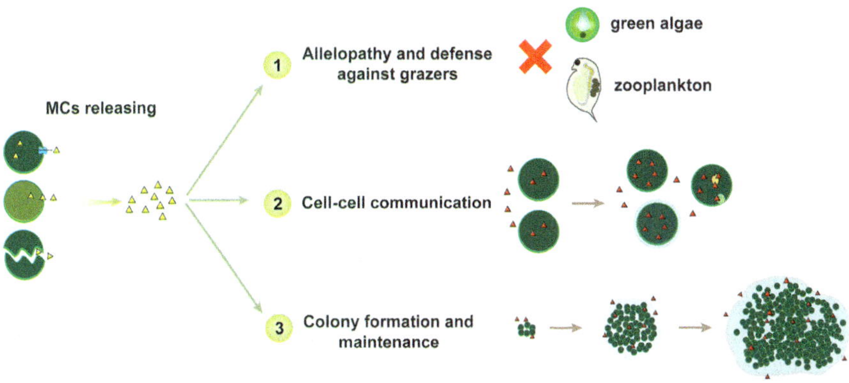

Fig. 4.2 Schematic representation of the ecological significance of released microcystins (MCs). MCs are released from *Microcystis* cells into the water via putative ABC transporter export, cell aging, or cell lysis. Released MCs may serve an allelochemical function by inhibiting the growth of other photoautotrophs, such as green algae, and/or defending against grazers, such as zooplankton. For the producers, released MCs serve as signaling molecules, inducing MC production, enhancing intracellular MC binding to their protein targets, and promoting the production and secretion of extracellular polysaccharides in the remaining intact cells, which enhances cellular defense capacity and fitness in their corresponding ecological niche. MCs at environmentally relevant concentrations are involved in promoting and maintaining *Microcystis* cell aggregates (from Wei et al. (2024), with permission from Elsevier)

a significant increase in McyB abundance (MC biosynthesis) and intracellular MC content in the remaining intact cells. This suggests that the extracellular release of MCs may act as a regulatory signal, influencing *mcy* gene expression and promoting MC production within *Microcystis* cells.

This response could enhance the toxic defense capability and fitness of *Microcystis* in its respective ecological niche. During the response, MCs could serve as signaling molecules in intraspecies communication. Similarly, Yeung et al. (2016) studied the different responses of the toxic *Microcystis* strain PCC 7806 to iron limitation and iron repletion and found that extracellular MCs were only observed in iron-limited cultures. Notably, cells in iron-limited cultures remained intact, suggesting a potential MC export upon exposure to iron-limited conditions, likely involved in the upregulation of transporter proteins that were simultaneously discovered in iron-limited cultures. Considering the higher total MC levels in iron--limited cultures than in iron-rich cultures and the protective role of intracellular MCs against ROS upon exposure to iron limitation, it was proposed that extracellular MCs may serve as signaling molecules that function to stimulate intracellular MC production (from Wei et al. (2024), excerpted with permission from Elsevier and appropriately referenced).

Beyond their role as signaling molecules, growing evidence suggests that cyanotoxins may also function as physiological enhancers. In freshwater systems, nutrient availability—particularly phosphorus (P)—is a key limiting factor for the proliferation of phytoplankton. Intriguingly, several studies have shown that MC production increases as P becomes scarce, suggesting a potential adaptive role for these compounds under nutrient-stressed conditions (Oh et al., 2000).

In more detail, it has been shown that when P is limited, autotrophs take up inorganic P from the environment through the secretion of alkaline phosphatases (ALPase) (Raven, 2010). A recent study has suggested that the presence of cylindrospermopsin (CYN) causes the secretion of ALPase by other phytoplankton, thereby increasing the amount of inorganic P available to the CYN producer and allowing it to outcompete other species in environments with limited inorganic P (Bar-Yosef et al., 2010). The study showed that adding spent media from *Umezakia (Aphanizomenon) ovalisporum*—rich in CYN or purified CYN itself to various phytoplankton cultures, including *Chlamydomonas reinhardtii*, induced the upregulation of genes typically activated under inorganic P limitation. Remarkably, this response occurred despite the presence of high inorganic P levels in the medium and was accompanied by increased extracellular alkaline phosphatase (ALPase) activity (Bar-Yosef et al., 2010). These findings were further supported by field data from Lake Kinneret, Israel, where a strong correlation was observed between the abundance of cylindrospermopsin-producing *U. ovalisporum* and ALPase activity across various phytoplankton species—though notably not in *Umezakia* itself (Bar-Yosef et al., 2010). Similarly, an earlier study also reported increased ALPase activity during blooms of *U. ovalisporum* (Hadas et al., 1999). These results, as summarized by Holland and Kinnear (2013), point to the potential role of *Umezakia* in manipulating community-level P metabolism.

Microcystins also play a role in iron scavenging (Utkilen and Gjølme, 1995; Lyck et al., 1996; Alexova et al., 2011). At circumneutral pH, iron availability is generally low, particularly in freshwater systems prone to toxic cyanobacterial blooms, which typically have a pH near neutral or slightly alkaline. In such environments, ferric iron (Fe^{3+}) is the dominant form, but it is poorly soluble and thus largely biologically inaccessible. Notably, MC-producing cyanobacterial strains have been shown to possess more efficient iron uptake systems than their non-MC--producing counterparts (Utkilen and Gjølme, 1995; Lyck et al., 1996). It has been proposed that microcystins act as intracellular Fe chelators, binding and inactivating free iron—most likely in the ferrous (Fe^{2+}) state, which is more reactive and potentially harmful under cellular conditions. This chelation could serve as a protective mechanism against oxidative stress, while also helping to regulate iron homeostasis within the cell.

It has also been suggested that MC production is regulated by the amount of free Fe in the cell (Utkilen and Gjølme, 1995). This has recently been demonstrated using iron-deficient media, increasing transcription of the *mcyD* gene responsible for MC production and a corresponding increase in MC levels (Sevilla et al., 2008). However, Alexova et al. (2011) found that transcription of some MC-producing genes, such as *mcyA* and *mcyH*, did not increase with a decrease in Fe, whereas MC concentrations did. The authors suggested that toxin production appears to give MC-producing cyanobacteria an advantage in the early stages of exposure to severe Fe stress and may protect the cell from reactive oxygen species-induced damage (from Holland and Kinnear (2013), credit to MDPI and appropriately referenced).

4.1.1 Autointoxication

Cyanobacterial bloom dynamics hinge on the delicate balance between population growth and loss. While extensive research has focused on the environmental factors and physiological traits that promote or limit cyanobacterial growth—such as light, nutrients, and temperature—the processes that contribute to population decline have received far less attention. Yet, paradoxically, blooms are often observed to collapse quite suddenly and dramatically (Harris et al., 2024), suggesting that loss mechanisms play a critical, albeit understudied, role in bloom dynamics. Some hypotheses have been proposed to explain the mechanisms behind the sudden decline in cyanobacterial biomass. These include a range of biological, chemical, and physical processes, such as:

- Sensitivity to internal oxidative stress under strong light and UV-B irradiation (Leunert et al., 2013; Zhang et al., 2013; Rastogi et al., 2014)
- Sensitivity to external H_2O_2 generated by illuminated dissolved organic matter (Leunert et al., 2014)
- Cyanophages (Yoshida et al., 2006; McKindles et al., 2020)
- Fungal parasites (Gerphagnon et al., 2015)
- Predatory bacteria (Mayali and Azam, 2004)
- Bacteria with cyanobactericidal activity, such as *Paucibacter aquatile* (Le et al., 2022)
- Lack of promotion of cyanoHABs by *Phenylobacterium* strains or comparable α-*Proteobacteria* (Zuo et al., 2021)
- Sedimentation of the population due to an increase in carbohydrates at lower temperatures (Visser et al., 1995; Welker et al., 2007).

Nevertheless, the decline process in cyanoHABs remains poorly understood (Harris et al., 2024).

Recently, Sha et al. (2025) introduced a compelling new hypothesis to the conversation—one that adds a fresh perspective and warrants closer attention. The authors demonstrated that *Microcystis* produces autotoxic substances during its growth, which accumulate in the surrounding medium and trigger the degradation of cellular chlorophyll. This self-poisoning mechanism likely plays a natural role in bloom decline, adding a new layer of complexity to the life cycle of *Microcystis* blooms. The potential autotoxic components are primarily small, hydrophobic bases, and their accumulation appears to be influenced by environmental stressors such as nutrient limitation and temperature fluctuations (Fig. 4.3).

In addition to these primary functions of phycotoxins, which are better understood for microcystins than for other phycotoxins, microcystins also perform several secondary functions, as shown in the top row (Fig. 4.2), which we will now consider as a whole.

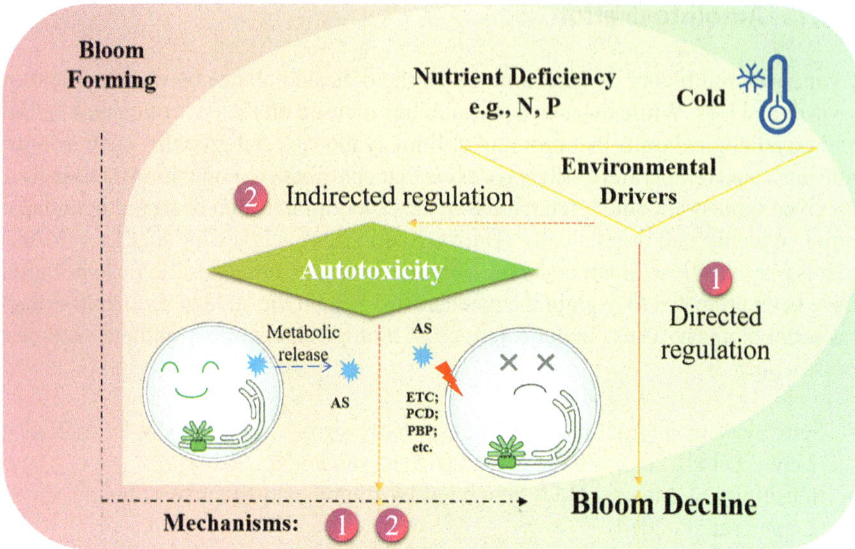

Fig. 4.3 Possible pathways of autotoxic DOM-mediated temperature and nutrient-driven *Microcystis* blooms decline (*AS* autotoxic substance, *ETC* electron transfer chain, *PCD* programmed cell death, *PBP* physiological and biochemical processes). (From Sha et al. (2025), with permission from Elsevier)

4.2 Role of Marine Phycotoxins

In her review, Pistocchi (2016) explored the ecophysiological roles of marine phycotoxins. Much like MCs, intracellular marine phycotoxins participate in carbon and nitrogen metabolism and act as antioxidants. Their biosynthesis is intricately tied to these metabolic pathways (Brandenburg et al., 2020).

A consistent pattern emerges from multiple studies:

1. Phosphorus limitation leads to higher toxin content per cell compared with nutrient-replete or nitrogen-limited conditions.
2. Cells produce more toxins when ammonium serves as the N source rather than nitrate.

In a meta-analysis, Brandenburg et al. (2020) confirmed that the levels of N-rich toxins increase with P limitation and warned that P limitation, achieved through water quality management, may cause the accumulation of cellular toxins, leading to unexpected increases in bloom toxicity.

The ecological advantage of increased cytotoxicity under P limitation can be explained by either an increased nitrogen storage capacity or an attempt to acquire a higher competitive ability for species with low Ks for P, and/or by the possibility of redirecting predators toward non-toxic species. Regarding the last aspect, it has nevertheless been observed that the production of STXs does not appear to be a universal defense mechanism against copepods (Granéli and Flynn, 2006; Hardison et al., 2013; Pistocchi, 2016).

Specifically, Hardison et al. (2013) demonstrated that in *Karenia brevis,* brevetoxin levels increase under both N- and P-limitation. However, the increase per cell under P-limitation is roughly double that seen under N-limitation, mainly due to differences in cell size. Consequently, for an equal concentration of cells, a P-limited bloom could be about twice as toxic as an N-limited bloom experiencing the same degree of growth rate restriction. The benthic dinoflagellate *Prorocentrum lima* shows a comparable pattern, with okadaic acid and the closely related dinophysistoxin-1 increasing under both N- and P-limitation, and particularly elevated under P-limitation (Vanucci et al., 2010).

A similar response was observed in *Protoceratium reticulatum,* where reducing the P content of the medium to one-tenth of its normal level triggered a tenfold increase in yessotoxin (YTX) production (Röder et al., 2012).

In the diatom *Pseudo-nitzschia multiseries*, the production of domoic acid (DA) at steady state is positively correlated with the N:P ratios, suggesting that P deficiency promotes DA production (Pan et al., 1996; Hagström et al., 2011). In a different experimental setup, Auro and Cochlan (2013) tested only nitrogen utilization and toxin production by two diatoms of the *P. pseudodelicatissima* complex (*P. cuspidata, P. fryxelliana*). The authors found that these two morphologically similar species can readily utilize both organic and inorganic N sources. Notably, nitrate-supported *P. cuspidata* cultures grown under low light conditions are the most toxic cells identified during the laboratory study. This finding may explain the elevated concentrations of dissolved and particulate DA observed at subsurface depths for natural assemblages of *P. cuspidata* in the NO_3^--rich waters off the coast of Washington State (Trainer et al., 2009a, b).

Field studies continue to show correlations between the presence of *Pseudo-nitzschia* and various environmental parameters (Husson et al., 2016; Thorel et al., 2017; Pednekar et al., 2018), as well as a correlation between *Pseudo-nitzschia* toxicity and limiting concentrations of both silicate and phosphate (Thorel et al., 2017). A mechanistic model was developed to explore abiotic factors that control the production of DA by *Pseudo-nitzschia* (Terseleer et al., 2013). Model simulations showed that excess light, in combination with Si or P limitation, favors DA production, provided that N is sufficient; this supports previous laboratory studies and is consistent with DA being a secondary metabolite (reviewed in Lelong et al. (2012)). Despite the model not yet implementing the effects of inorganic vs. organic N, trace metals, salinity, temperature, pH, bacteria, or grazers, the simulations were generally realistic (from Bates et al. (2018), credit to Elsevier, appropriately referenced).

In contrast, a strain of *Pseudo-nitzschia australis* showed no increase in DA production under P or Si limitation (Cochlan et al., 2023), hinting that the role of (macro)nutrient limitation in toxin regulation is far from straightforward. This complexity is underscored by observations that oceanic *Pseudo-nitzschia turgidula* not only produce and accumulate DA, but also retain this capacity even after Fe enrichment in high-nitrate, low-chlorophyll regions (Trick et al., 2010).

Marine phycotoxins, much like MCs, may double as clever nutrient scavengers. In *Protoceratium reticulatum*, growth depends on both selenium and iron—but intriguingly, only selenium scarcity causes a surge in yessotoxin (YTX) production, for reasons still unknown (Mitrovic et al., 2004). Even more striking, He et al. (2010) found that iron limitation in *Alexandrium tamarense* triggered a dramatic rise in paralytic shellfish poisoning toxins (saxitoxin and its derivatives), suggesting a provocative idea: these potent molecules may not only defend but also actively help the cell hunt for missing nutrients.

In many toxigenic marine algae, higher ambient salinity often coincides with elevated phycotoxin levels. This consistent pattern has led to the intriguing hypothesis that phycotoxins may function as osmolytes. Osmolytes—low-molecular-weight, uncharged, or zwitterionic molecules—are well-known allies for organisms coping with high or fluctuating salt conditions. By accumulating in the cytoplasm without disrupting essential macromolecules, they help cells maintain osmotic balance in challenging environments. They raise intracellular osmotic pressure in response to high environmental salt concentrations, yet do not interfere with protein function, and more importantly, enhance protein stability (Arakawa et al., 2006).

In the diatom *Pseudo-nitzschia multiseries*, low salinity and low toxicity were correlated, with DA increasing at salinities of 30 and 40 psu (Doucette et al., 2008). Regarding YTXs, the increase in cellular toxicity in *Protoceratium reticulatum* suggests that toxin accumulation may be an indirect response to increased cytoplasmic osmotic pressure (Guerrini et al., 2007). However, strain-specific responses have been observed, as YTX production was negatively affected by salinity in the range of 20–34 psu in a strain from Spain (Paz et al., 2006).

The Adriatic benthic dinoflagellate *Ostreopsis* cf. *ovata* grew best at salinities of 36 and 40 psu, with the highest content of palytoxin-like compounds per cell measured at salinity 32 psu (Pezzolesi et al., 2012).

Taken together, these findings underscore the urgent need to look beyond the traditional paradigm of single-nutrient limitation in regulating algal communities and their toxin production. While historical management strategies have often targeted reductions of individual nutrients—most commonly P or N—they may inadvertently promote harmful algal blooms that are richer in cellular toxins. Singular nutrient reductions can disrupt the delicate stoichiometric balance in aquatic ecosystems, leading to elevated N:P ratios or secondary limitations that shift metabolic pathways toward increased toxin biosynthesis. Several studies demonstrate that N and P co-limitation is common in aquatic systems, with excessive supply or reduction of one nutrient able to push communities toward greater toxin output (Carrick et al., 2023).

Furthermore, the intricate interplay between nutrient availability, light, salinity, temperature, and trace metals underscores that toxin production cannot be attributed to any single limiting factor in isolation (Carrick et al., 2023). Toxic responses reflect dynamic physiological adaptations to multiple, often fluctuating, environmental stressors, including the presence of organic matter and variations in micronutrient status. For instance, changes in dissolved organic carbon, trace metals, or temperature can modify nutrient limitation regimes, algal assemblage composition, and toxin profiles. Management approaches that overlook these interdependencies risk not only shifting the composition of phytoplankton communities but also generating conditions conducive to more toxic and persistent blooms, with direct implications for food safety, ecosystem health, and economic interests (Anderson et al., 2012; Wells et al., 2015; Xie et al., 2016; Griffith and Gobler, 2020).

Therefore, future water quality management and mitigation strategies must adopt a more holistic, ecosystem-based framework. This includes monitoring nutrient ratios, their sources (organic vs. inorganic), and their interactions with other abiotic and biotic factors such as light, temperature, and trace elements (Morón-López, 2021; Li et al., 2023). Only through such integrative perspectives can we move toward more effective control of phycotoxin-producing algae, minimizing unintended consequences and safeguarding both aquatic environments and their associated human uses.

In summary, the ecological context and multifactorial nature of toxin regulation in marine algae demand a paradigm shift: from simplistic nutrient reduction policies to adaptive management informed by a comprehensive understanding of phytoplankton physiology, community interactions, and ecosystem processes. This remains a vital frontier for both scientific research and practical policy-making in the management of coastal and marine systems (Glibert and Burkholder, 2018).

4.3 Elimination of Competitors

Competition within aquatic ecosystems extends beyond the direct acquisition of nutrients, occupation of space, or interception of light. Harmful bloom-forming flagellates and other phytoplankton taxa have evolved a suite of strategies to secure ecological advantage, especially when faced with lower nutrient uptake affinity compared to dominant groups, such as diatoms (Smayda, 1997). Among these adaptations, the production of bioactive secondary metabolites—allelochemicals—enables certain species to inhibit the growth or survival of potential competitors. Smayda (1997) identified four primary adaptations compensating for their ecological disadvantage due to low nutrient affinity: migratory nutrient retrieval, mixotrophy, allelochemical-enhanced competition, and allelopathic antipredator defenses. Such chemically mediated interference, or allelopathy, provides a competitive edge by selectively suppressing rival phytoplankton, including raphidophytes, other dinoflagellates, and diverse algal taxa (Glibert et al., 2012; Li et al., 2019) (see below for details).

In his recent review, Tan (2023) provides a valuable summary of allelochemicals in competition for space and resources, which is cited here to demonstrate the importance of this chemical warfare. In marine reef habitats, space is a precious and limited resource—especially on hard substrates within the photic zone, where enough sunlight penetrates the water to fuel photosynthesis. Due to the intense crowding of sessile benthic organisms in this zone, many have evolved allelopathic strategies to secure limited space and vital resources. Many allelochemicals used during competition are potential cytotoxic compounds, causing tissue necrosis and cell damage to prevent or arrest the growth of nearby competitors (Singh and Thakur, 2016). For instance, soft corals produce structurally diverse terpenoids with antifouling and cytotoxic properties (Coll, 1992). The soft coral *Sarcophyton glaucum* was investigated for its natural product variation as well as its competitive interactions with the scleractinian coral *Acropora robusta*. In field experiments, high concentrations of the diterpene sarcophytoxide were found in female soft coral colonies that were placed in contact with *A. robusta* (Fleury et al., 2006).

Sponges, as common benthic invertebrates, rank among the top spatial competitors in reef ecosystems. Their success stems not only from their ability to exploit a wide variety of nutritional sources but also from their arsenal of diverse chemical defenses that protect against predators, microbial invaders, and competing sessile organisms. Notably, sponge-derived bioactive allelochemicals have been linked to the rising abundance of animals on reefs in Zanzibar, highlighting their powerful ecological role in shaping reef communities (Helber et al., 2018). Of the ten dominant sponges found on reefs in Zanzibar, 80% of the sponge-derived crude extracts had antimicrobial activity. In contrast, seven sponge species, including *Callyspongia* sp. and *Scopalina hapalia*, exhibited cytotoxic properties based on the brine shrimp toxicity assay. Specific sponge-derived allelochemicals have also been identified to confer a competitive edge in interactions with other marine invertebrates. In a study involving the excavating sponge *Cliona tenuis*, clionapyrrolidine A was found to be toxic to scleractinian corals, killing coral tissue upon contact and lowering the photosynthetic potential of coral zooxanthellae (Chaves-Fonnegra et al., 2008). The steroidal b-sitosterol, isolated from another sponge species such as *Cinachyrella* cf. *cavernos*a, was shown to cause bleaching of the aggressive zoanthids *Zoanthus sansibaricus*, possibly by affecting the zoanthid's symbiotic zooxanthellae (Singh and Thakur, 2021).

Macroalgae are chemically rich marine plants known to produce inhibitory lipid-soluble allelochemicals used to compete with other benthic organisms, particularly cnidarians, for space in reef systems. These allelochemicals are usually produced by specialized cells and localized onto the macroalgal surfaces. As such, contact is generally required for these molecules to exert an allelopathic effect on potential competitors (Longo and Hay, 2017). Such allelopathic interactions may suppress/limit coral reef recovery due to the proliferation of macroalgae in constant contact with remaining corals. The chemical structures of allelochemicals have been reported in several macroalgal species, using traditional extraction and bioassay-guided purification methods. Several terpenoids, including loliolide derivatives and acetylated diterpenes, isolated from two macroalgae, *Galaxaura filamentosa* and *Chlorodesmis fastigiata*, respectively, were identified as potent allelochemicals against the corals *Montipora digitata*, *Acropora millepora*, and *Pocillopora damicornis* (Rasher et al., 2011). These hydrophobic allelochemicals could exert their allelopathic effects by impacting coral photosystems, leading to coral bleaching and death through their transfer from macroalgal surfaces. In another in situ study on several brown algal species belonging to *Lobophora* sp., three macroalgal C_{21} polyunsaturated alcohols, lobophorenols, were found to cause bleaching in some coral species, particularly *Acropora muricata*, upon direct contact (Vieira et al., 2016). In addition, an a-pyrone macrolide allelochemical, neurymenolide A, was found on all surfaces of the red macroalga *Phacelocarpus neurymenioides*, especially on the basal portions of the algal blades (Andras et al., 2012). This molecule was found to induce bleaching when the macroalga came into contact with the natural population of *Porites rus*. Previously isolated from another red macroalga—*Neurymenia fraxinifolia* (from Fiji), neurymenolide A is found to possess antibacterial and cytotoxic properties (Motuhi et al., 2019) (from Tan (2023) with credit to MDPI, Basel; appropriate references added).

A range of allelochemicals is also used by bloom-forming marine and freshwater microbes, such as cyanobacteria and dinoflagellates, to compete with sympatric macrophytes, algae, or other microbes. Many studies support the notion that microcystins play an important physiological role in producing strains even within the same species. For instance, a spontaneous *Microcystis* mutant, strain MG-J, bearing a deletion in *mcy* and unable to produce microcystins, was isolated from a toxic wild-type strain MG-K (Schatz et al., 2005). Laboratory and field experiments comparing the wild type with this spontaneous mutant indicated an advantage of the toxic strain over its non-toxic successor.

With respect to cyanobacteria, allelopathy is considered one of the factors promoting and maintaining massive cyanobacterial and algal blooms in freshwater, brackish, and marine ecosystems around the world (Figueredo et al., 2007; Żak and Kosakowska, 2015; Allen et al., 2018; Felpeto et al., 2018; Śliwińska-Wilczewska et al., 2018). Consequently, the number of mainly laboratory-based studies investigating the allelopathic effects of various cyanobacteria and microalgae has been steadily increasing in recent years (Dias et al., 2017; Pichierri et al., 2017; Wang et al., 2017; Allen et al., 2018; Felpeto et al., 2018). While much of this work focuses on controlled experimental systems, a smaller but growing body of research is beginning to document the adverse impacts of algal toxins—or toxic algae themselves—on bacteria and phytoplankton, shedding light on their potential role in shaping microbial and planktonic community dynamics in natural environments.

Śliwińska-Wilczewska and Latała (2018) measured the allelopathic activity of a picoplanktonic *Synechococcus* sp. on the growth, chlorophyll fluorescence, and photosynthetic performance of the green algae *Chlorella vulgaris* and *Oocystis submarina* and the diatoms *Bacillaria paxillifera* and *Skeletonema marinoi* by adding cell-free filtrate obtained from cultures growing in nutrient-sufficient, nitrogen-deficient, or phosphorus-deficient culture medium. The authors found that the allelopathic activity of *Synechococcus* sp. was influenced by nutrient availability: the higher the nutrient availability, the higher the production of allelochemicals.

Karenia's allelochemical arsenal against competitors has been well documented over decades of research. However, unraveling the specific mechanisms behind its competitive edge is complicated by the fact that many *Karenia* species produce multiple toxins simultaneously. This chemical cocktail means that the observed effects in both laboratory and field settings cannot always be attributed to a single compound. Instead, they may result from the combined effects of several bioactive molecules.

Kubanek et al. (2005) and Prince et al. (2008a, 2008b) found that *K. brevis* produces allelochemical compounds that suppress the growth of most other algae species tested. Prince et al. (2010) concluded that these compounds are not brevetoxin, but a variety of unstable, polar organics. Poulson et al. (2010) found that *K. brevis* produces multiple compounds other than brevetoxin that are allelopathic to other phytoplankton species. Various studies have shown that *K. mikimotoi* produces lipophilic toxins, with tissue culture experiments revealing toxic effects of a variety of polyunsaturated fatty acids (PUFAs) (Fossat et al., 1999; Sola et al., 1999). Gentien et al. (2007) further demonstrated that these PUFAs exert strong allelopathic effects on diatoms and, intriguingly, also display a degree of autotoxicity—suggesting that the chemical weapons of *K. mikimotoi* may, under certain conditions, turn against itself. Chang (2011) reported that *K. concordia, K. brevisulcata*, and *K. mikimotoi* produce potent lipophilic allelochemicals capable of inhibiting the growth of other species of algae. These compounds were particularly effective against cryptophytes, raphidophytes, prasinophytes, and diatoms, while showing comparatively lower activity against other dinoflagellates. Their mode of action appears to involve the disruption of cell membranes, likely through interference with ion pumps and osmotic regulation. Complementing these findings, Chang et al. (2008) demonstrated that *K. concordia* populations in New Zealand exhibited both hemolytic and cytotoxic activity, underscoring the multifaceted and aggressive chemical arsenal employed by these species in competitive interactions.

Mooney et al. (2007, 2011) showed that *K. brevis, K. mikimotoi, K. papilionacea*, and *K. umbella* all produce unusual lipids, sterols, and PUFAs and suggested that most species in the family Kareniaceae produce PUFAs and sterols that are ichthyotoxic. The mode of toxicity of the sterols is unknown, but it is thought that they may also possess allelochemical properties. The toxicity of the PUFAs can be enhanced by lipid peroxidation in the presence of reactive oxygen species (Mooney et al., 2007). Zou et al. (2010) demonstrated that compounds produced by *K. mikimotoi* are hemolytic; however, direct cell contact is necessary to induce mortality. Most of these bioactive compounds remain only partially characterized, leaving their ecological roles shrouded in uncertainty—but intriguingly, at least one has been fully described, offering a rare glimpse into the chemical weaponry of *Karenia*. Gymnodimine is a spirocyclic imine toxin produced by *K. selliformis* (Miles et al., 2003; Munday et al., 2004; Mountfort et al., 2006; Molgó et al., 2007). It acts as a neuromuscular blocking agent by activating calcium receptors and disrupting nicotinic acetylcholine receptors in muscles (Molgó et al., 2007), ultimately leading to widespread mortality in both invertebrates and vertebrates. Whether gymnodimine also serves an allelochemical role remains unresolved. Evidence suggests that this toxin is not confined to localized events but is instead widespread across marine food webs, having been detected in shellfish from New Zealand, Tunisia, Canada, and other regions worldwide (Seki et al., 1995; Munday et al., 2004; Molgó et al., 2007). In addition to gymnodimine, *K. mikimotoi* has been shown to produce gymnocin A and B (Satake et al., 2002, 2005), though these compounds appear to be only weakly toxic to fish (Silke et al., 2005) (from Brand et al. (2012), with permission from Elsevier and appropriately referenced).

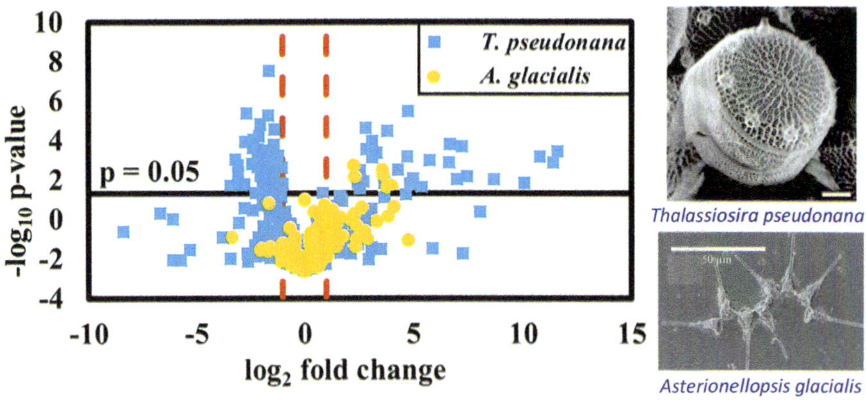

Fig. 4.4 Volcano plot summarizes the differences in the lipidomes of *T. pseudonana* (blue squares) and *A. glacialis* (yellow circles) when exposed vs. not exposed to *Karenia brevis* allelopathy. The relative abundances of 80 metabolites were significantly different ($P < 0.05$ after Bonferroni correction) in *T. pseudonana* upon exposure to *K. brevis* allelopathy. Red lines indicate $\log_2$ fold difference of ±1. Six metabolites with concentrations that were significantly different in *A. glacialis* when exposed to *K. brevis* were also significantly different in concentration when *T. pseudonana* was exposed to *K. brevis* (from Poulin et al. (2018), credit to Springer Nature). Image of *T. pseudonana* from Kröger and Poulsen (2006) with permission from Springer Nature and of *A. glacialis* credit to Sunesen and Sar (2007) and Revista Chilena de Historia Natural. Scale bars: *T.p.* 1 μm, *A.g.* 50 μm

Using NMR spectroscopy and metabolomics, Poulin et al. (2018) uncovered striking insights into the allelopathic strategies of *Karenia brevis*. When two nutrient competitors—the diatoms *Asterionellopsis glacialis* and *Thalassiosira pseudonana*—were exposed to its exuded compounds, both exhibited profound shifts in their lipidomes (Fig. 4.4). Yet their responses were far from equal. *A. glacialis* displayed metabolic resilience, maintaining much of its normal physiology despite the chemical assault. In contrast, *T. pseudonana* displayed severe disruptions in lipid synthesis, cell membrane integrity, and photosynthetic processes. Membrane-associated lipids were significantly suppressed, causing even living cells to become permeable and unstable. These findings reveal that *K. brevis* allelopathy specifically undermines lipid biosynthesis, with cascading effects on multiple physiological pathways. The exuded compounds can significantly alter the physiology of competitors, giving *K. brevis* an edge over sensitive species.

In OkA, there is some, but not clear, evidence of allelopathy (Windust et al., 1997; Sugg and Van Dolah, 1999). Although the exact mechanism by which OkA impairs growth in microalgae is not yet fully understood, studies using the green alga *Dunaliella tertiolecta* (Perreault et al., 2012) point toward two major routes: independent and dependent on photosynthesis. Accordingly, when a culture of *D. tertiolecta* is exposed to low OkA concentrations, a decrease in cell density is observed in both dark and light conditions, with toxicity being greater in the latter case. Additionally, exposure to OkA in light conditions results in a reduction of the photosynthetic electron transport rate that may lead to photooxidative stress and damage of the photosynthetic apparatus, increasing the observed effect of OkA on algal culture cell density (from Prego-Faraldo et al. (2013) references added).

Kitchen et al. (2018) reported a relatively weak mechanism of *Karenia* toxins in the diatom *Trieres (Odontella) chinensis*. Unlike in other animals, exposure to brevetoxins (PbTx) did not induce the typical hyperexcitability in membrane potential. Instead, the evoked voltage-activated currents underlying the action potential of the diatom were only partially inhibited, accompanied by a shift in ion selectivity that favored Ca^{2+} influx. These results suggest that the voltage-gated channels and cell signaling of *T. chinensis* are comparatively insensitive to PbTx at environmentally relevant concentrations.

Some recent examples of marine chemical warfare between primary producers are listed in Table 4.1.

The diatom-derived oxylipins can inhibit other diatoms, as in the case of the well-studied 2-trans, 4-trans-decadienal (DD), which leads to growth inhibition and cell death with programmed cell death characteristics in diatoms (Casotti et al., 2005; Vardi et al., 2006). Diatoms respond to DD with a Ca^{2+} burst within seconds to minutes and NO (Vardi et al., 2006; Gallina et al., 2014; Sabharwal et al., 2017)

Table 4.1 Selected reports of marine allelopathic interactions

Compartment	Allelochemical producer, chemical	Observed effect in	References
Plankton	2-trans, 4-trans-decadienal (DD) from diatoms	*Thalassiosira weissflogii* programmed cell death (Fig. 4.5)	Casotti et al. (2005)
	Skeletonema costatum	*Karenia mikimotoi*, membrane damage	Guo et al. (2022)
	Cylindrotheca closterium	*Prorocentrum obtusidens*, lipid peroxidation	Xu et al. (2020b)
	Karenia mikimotoi ↔ *Prorocentrum obtusidens* growth reduction		Zheng et al. (2023)
	Chattonella marina var. *marina*, *Gymnodinium impudicum*	*Gymnodinium catenatum*, growth reduction	Fernández--Herrera et al. (2022)
	Alexandrium minutum	*Chaetoceros muelleri* membrane disruption, photosynthesis inhibition	Long et al. (2021)
	A. catenella, Alexandrolide	*Skeletonema costatum*, growth inhibition	Satake et al. (2019)
	Phaeodactylum tricornutum	*P. obtusidens*, growth reduction	Zhang et al. (2022)
	Dinophysistoxin-1, presumably from *Prorocentrum* sp. or *Dinophysis* sp.	*Isochrysis galbana*, reduced photosynthesis and growth	Han et al. (2023)
Benthos	*Ostreopsis* cf. *ovata*, ovatoxins	*Licmophora paradoxa*, *Navicula arenaria*, *Prorocentrum lima*, *Coolia monotis*, weak growth inhibition	Ternon et al. (2022)

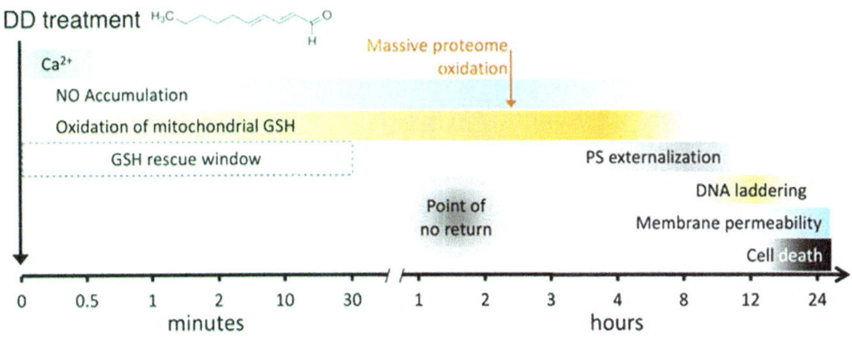

Fig. 4.5 Programmed cell death signaling cascade in response to 2-trans, 4-trans-decadie (DD) treatment in diatoms. Following exposure to lethal doses of DD, *P. tricornutum* responds with a cascade of signaling events involving the secondary messengers Ca^{2+}, NO, and ROS, which eventually lead to cell death. This cascade initiates with a short Ca^{2+} pulse, ending within seconds of exposure to DD, and by NO accumulation and the oxidation of the mitochondrial GSH pool, both of which have much longer durations. Massive proteome oxidation was measured at a single time point of ~2 h (Graff van Creveld, 2018). During the first ~30 min after the DD addition, external application of GSH can prevent later cell death (Graff van Creveld et al., 2015). However, after about an hour, cells can no longer be rescued in that manner, exposing the "point of no return," after which cells are already committed to cell death or have accumulated damage beyond repair (Graff van Creveld et al., 2015). At later time points, there is externalization of phosphatidylserine, and the DNA is being fragmented. Phosphatidylserine (PS) externalization and DNA laddering were monitored in *P. tricornutum* in response to H_2O_2. DNA laddering was also detected in *Thalassiosira weissflogii* in response to DD (Casotti et al., 2005). When cells are dying, their membranes become permeable in an irreversible manner, marking the cell's death. The intensity of each colored box represents roughly the timing and intensity of the parameter (from Graff van Creveld et al. (2022) with permission from Springer Nature)

and ROS accumulation for a few hours (Gallina et al., 2014; Graff van Creveld et al., 2015) (Figure 4.5). Although datasets of transcriptional responses to DD are available (Maheswari et al., 2010; Islam et al., 2020), specific DD-responsive genes remain to be discovered. A chemo-proteomic approach to identify cellular targets of DD revealed several putative proteins that exhibit potential DD binding and are involved in the Calvin cycle or the pentose phosphate pathway (Wolfram et al., 2015).

When tested in complex communities rather than in single-species laboratory cultures, a more nuanced picture of allelopathy emerges. Although *K. brevis* shows strong allelopathic effects against multiple competitors under controlled conditions, a microcosm experiment with natural plankton assemblages revealed that extracellular extracts from two *K. brevis* strains had no impact on specific taxa, while paradoxically stimulating the growth of some diatoms.

This indicates that in a species-rich ecological community under oligotrophic conditions, the relative importance of *K. brevis* allelopathy may not be as high as most laboratory studies predict (Poulson et al., 2010). Later, this finding is confirmed in offshore phytoplankton, where reactions to *Karenia* allelopathy range from strong sensitivity to resistance (Poulson-Ellestad et al., 2014). Recently, Shang et al. (2021) demonstrated that *Protogonyaulax* (*Alexandrium*) *leei* from Malaysia exhibits variable allelopathic effects to 13 microalgal species, with growth inhibition rates ranging from 0% to 100%. The adverse allelopathic effects of *P. leei* on microalgae include loss of flagella, thereby reducing motility, damage to the cell chain structure, deformation in cell morphology, and eventually cell lysis. These adverse effects are due to toxins that are neither PSTs nor spirolides.

The functional role of DA in diatom populations remains largely unknown, resulting in conflicting ecological evidence for DA. Lundholm et al. (2005) showed that the addition of DA-producing *P. multiseries* to cultures of other phytoplankton species (*Haptolina* (*Chrysochromulina*) *ericina*, *Heterocapsa triquetra*, *Eutreptiella gymnastica*, *Rhodomonas marina*) has no allelopathic effects. Likewise, Prince et al. (2013) demonstrated that the addition of DA has little effect on the growth of the diatom *Skeletonema marinoi* in monocultures. However, when *S. marinoi* is cocultured with *Pseudo-nitzschia delicatissima*, DA caused an increase in *P. delicatissima* cell number and a decrease in *S. marinoi* cell number. These results are observed only under iron-replete conditions and do not recur in iron-deficient conditions. This provides evidence that DA can improve the competitive ability of *Pseudo-nitzschia* species for iron. Studies have also indicated the role of DA as a binding ligand for trace metals. DA released into the water column is an effective chelator for Cu(II) and Fe(III), which increases the bioavailability of iron and decreases the toxicity of copper (Rue and Bruland, 2001; Trick et al., 2010).[2]

These few studies indicate that the "Elimination of Competitors" hypothesis does not generally apply to algal communities under allelopathic stress, but rather that species-specific and developmental stage-dependent responses occur in natural communities. This hypothesis explains many, but not all, of the findings to date. Natural assemblages of offshore phytoplankton dominated by diatoms are largely resistant to *K. brevis* allelopathy, even experiencing slight stimulation of growth from exposure to *K. brevis* exudates (Poulson-Ellestad et al., 2014). And even the reverse case can be observed: the exometabolome of the diatom *Licmophora paradoxa*, which cohabits with the toxic benthic dinoflagellate *Ostreopsis* cf. *ovata*, impairs the growth and photochemistry of the dinoflagellate (Ternon et al., 2018). In fact, both microalgae are capable of waging chemical warfare against each other via the exometabolome, indicated by reduced in vivo chlorophyll fluorescence (Fig. 4.6). The authors hypothesize that the accumulation of phycotoxins in the mucus yields deleterious effects.

Bacteria and eukaryotic algae compete intensely for inorganic nutrients, particularly N and P. Due to their high surface-to-volume ratio, bacteria are generally more efficient at nutrient uptake and therefore outcompete eukaryotic phytoplankton in oligotrophic waters (Stolte and Riegman, 1995; Grover et al., 2012; Kerimoglu et al., 2012). This competitive edge enables bacteria to dominate nutrient cycling and shape community structure under conditions of nutrient scarcity. However, size-dependent advantages in nutrient acquisition do not always hold true. Drakare (2002) demonstrated that, despite their small size, picophytoplankton are not superior competitors compared to heterotrophic bacteria, particularly in terms of

[2] Excerpted from Zabaglo et al. (2016), with permission from Elsevier.

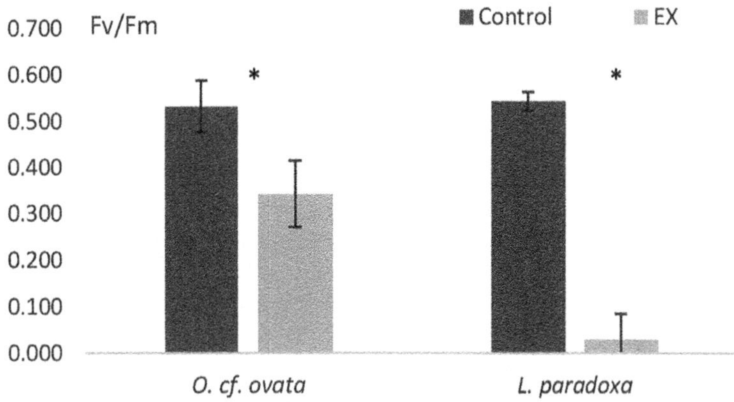

Fig. 4.6 Effect of the exometabolome of *Licmophora paradoxa* on the in vivo fluorescence (F_v/F_m) of *Ostreopsis* cf. *ovata* (left) and *O.* cf. *ovata* on *L. paradoxa* (right). Controls were performed with sterile media. Statistical significance was assessed using a Wilcoxon–Mann–Whitney test (* $P < 0.01$). Error bars correspond to the standard error between samples ($N = 4$) (from Ternon et al. (2018), with permission from Elsevier)

phosphorus uptake. In fact, like large phytoplankton, they are unable to successfully compete with bacteria under P limitation, underscoring the unique efficiency of bacteria in low-nutrient environments.

As one of the first studies, Sipler et al. (2014) demonstrated that brevetoxin has the potential to adversely impact some species of bacteria as well as bacterial communities as a whole. *Alphaproteobacteria* are the least affected by PbTx-2 and are the most abundant species overall, as they persisted throughout all PbTx-2 addition experiments, and increased in dominance in natural blooms with increasing concentrations of *K. brevis*. *Cyanobacteria* and *Actinobacterium*-like species are the most adversely impacted by PbTx-2. Through brevetoxin (PbTx-2), as many as 37% of bacterial species can be lost when it is introduced into less tolerant, previously unexposed systems. Simultaneously, Meyer et al. (2014) reported that bacterial abundance and production are significantly decreased, and protein synthesis becomes unbalanced in relation to nucleic acid synthesis during the maintenance phase bloom of *K. brevis*. Whether *K. brevis* gains a competitive edge for nutrients deserves further study.

Recently, Patin et al. (2020) investigated the microbial communities and metabolomes associated with a bloom of the toxic dinoflagellate *K. brevis* off the west coast of Florida in June 2018. The microbial communities and intracellular metabolite pools differed based on both bacterial lifestyle and bloom level, indicating a complex role for blooms in reshaping microbial processes. Network analysis identified *K. brevis* as an ecological hub in the planktonic ecosystem, with significant connections to diverse microbial taxa. These included four flavobacteria and one sequence variant unidentified past the domain level, suggesting uncharacterized diversity in phytoplankton-associated microbial communities. Additionally, intracellular metabolomic analyses associate high *K. brevis* levels with higher levels of aromatic compounds and lipids. The ecophysiological feedback to *K. brevis*, however, is not considered.

In their comprehensive review of toxic picocyanobacteria, Śliwińska-Wilczewska et al. (2018) compiled a broad spectrum of interactions with heterotrophic and phototrophic microorganisms as well as planktonic and benthic animals. Their synthesis highlights both inhibitory and facilitative effects: for example, one *Synechococcus* strain suppressed the growth of several cyanobacteria and diatoms, yet paradoxically promoted the growth of *Aphanizomenon flosaquae*. Other strains belonging to the genera *Synechococcus*, *Synechocystis*, and *Cyanobium* have been shown to exert detrimental effects on brine shrimp, sea urchins, and bivalves. Particularly striking is that the report reveals that exposure to one of the two tested *Synechococcus* strains altered the behavior of black perch (*Embiotoca jacksoni*), with the fish exhibiting a strong preference for the dark zone of a light—dark test—an established marker of elevated anxiety. Such findings underscore that even minute picocyanobacteria can exert profound and unexpected impacts across multiple trophic levels, extending from microbes to vertebrates.

Babica et al. (2007) reported that microcystins are unlikely to inhibit the proliferation of planktonic photoautotrophs under environmentally relevant conditions. These findings challenge the notion that microcystins play a direct allelopathic role in shaping phytoplankton communities—at least with respect to growth inhibition. Instead, their ecological function may lie elsewhere, for example, in mediating interactions with grazers, protecting against oxidative stress, or serving as intracellular regulators rather than as external allelochemicals. This highlights the need to look beyond simple growth effects when considering the potential ecological roles of microcystins.

Nevertheless, the catalog of microcystin (MC) effects on freshwater microbiota is expanding rapidly.

Yet, caution is warranted, as many experimental studies have applied concentrations far above those typically encountered in natural environments. A selection of environmentally representative investigations is summarized in Table 4.2. What emerges is that the initial hypothesis—that MCs generally suppress microbial growth and diversity—does not hold universally and requires refinement. In some systems, such as free-living bacteria in Lake Mochou and Lake Taihu, or in sediments of Dianchi Lake, cyanobacterial blooms are indeed associated with decreased bacterial diversity. In contrast, other studies report the opposite pattern: blooms can enhance microbial richness and community complexity, as observed for free-living bacteria in Lake Mochou and the Zhoucun drinking water reservoir. These contrasting outcomes highlight the context-dependent nature of MC effects, pointing to a more nuanced interplay between cyanobacterial toxins and microbial assemblages than previously assumed.

In Barberton Reservoir, the mean bacterial diversity (α-diversity) remains unchanged during the cyanobacterial bloom. Notably, Zhang et al. (2020) reported that cyanobacterial blooms are a crucial driver of antibiotic resistance gene diffusion and enrichment in freshwater. However, biomagnification in the food chain

Table 4.2 Selected studies of microcystin/*Microcystis* impact on microbial communities

Lake, lake type, or lake compartment	Cyanotoxin source	Observed effect	References
Anaerobic sediments	1 mg L^{-1} MC-LR under anaerobic conditions. MC degradation increases	*Exiguobacterium*↓ *and Acinetobacter*↓ → *Prosthecobacter*↑, *Dechloromonas*↑, *Agrobacterium*↑, *Dechloromonas*↑, *Pseudomonas*↑, *Hydrogenophaga*↑, *Agrobacterium*↑	Ding et al. (2020)
Barberton Reservoir, Ohio, USA	Bloom of *Planktolyngbya*, *Aphanizomenon*, *Pseudanabaena*, *Cylindrospermopsis*	Bacterial α-diversity↔	Wang et al. (2020)
Coastal lagoons, Brazil	MC-LR, 20 mg L^{-1}	*Actinobacteria*↓, *Proteobacteria*↑, MC-LR degraders (*Methylophilaceae*)↑	Santos et al. (2022)
Eutrophic lake sediment (Dianchi Lake, Kunming, China)	Not specified cyanobacterial bloom, most likely *Microcystis*	Bacterial diversity↓, archaeal diversity↓	Yang et al. (2021)
Eutrophicated urban lakes in Mexico City	*Microcystis* spp.	*Rhodobacter*↑, *Pseudomonas*↑, *Sphingomonas*↑, *Porphyrobacter*↑	Pineda--Mendoza et al. (2020)
Lake Erie, Lake Agawam	*Microcystis*	*Microcystis*-attached: *Gemmatimonadetes*↑, *Burkholderiaceae*↑, *Rhizobiales*↑, *Cytophagales*↑, *Actinobacteria*↓ distinct from the free-living microbiome	Jankowiak and Gobler (2020)
Lake Nanhu, China	Blooms of *Dolichospermum* (*Anabaena*) *circinale* and *Microcystis flos-aquae*	*Bacteroidetes*↑	Han et al. (2021)
Lake Taihu, China	*Microcystis*	Bacterial α-diversity↓	Su et al. (2017)
	Microcystis, *Synechococcus*	*Actinomycetales*↑, *Pirellulaceae*↑, *Sphingobacteriaceae*↑	Zhu et al. (2019)
Lake Taihu, China, enclosures	*Microcystis*	Sediment bacterial diversity↓	Gu et al. (2020)
Lake Waitawa, Wellington, New Zealand	Breaking down scum of *Microcystis*, *Dolichospermum*	*Aeromonas*↑, *Caulobacter*↑, *Brevundimonas*↑ (*Proteobacteria*)	Steiner et al. (2017)
Recreational lakes in South Germany	*Microcystis* nondominant, MC biosynthesis gene (*mcyB*)↑ → MC↑	*mcyB*↑ → *Bacteroidetes* fluctuating, *Actinobacteria*↔	Scherer et al. (2017)

(continued)

Table 4.2 (continued)

Lake, lake type, or lake compartment	Cyanotoxin source	Observed effect	References
Urban Lake Mochou, Nanjing, China	*Synechococcus*, *Cyanobium*	Diversity and richness of free-living bacteria↑, diversity and richness of particle-attached bacteria↓	Xu et al. (2020a)
Urban lakes in Nagpur city, India	*Microcystis*, *Anabaena*, *Trichodesmium*, *Microcoleus*, *Nodularia*	Catabolic profile↑, resistome profile↑, ecosystem services↓	Pal et al. (2021)
Utah Lake, USA	Succession of *Cyanobium*, *Aphanizomenon*, *Dolichospermum*, *Planktothrix*, *Microcystis aeruginosa*, *M. panniformis*	*Actinobacteria*↑, *Bacteroidetes*↑, *Alphaproteobacteria*↑	Li et al. (2020)
Western Lake Erie	*Microcystis* sp., *Synechococcus* sp.	*Actinobacteria*↑, *Alphaproteobacteria*(↑), *Betaproteobacteria*↑, *Bacteroidetes*↑	Berry et al. (2017)
Zhoucun drinking water reservoir, China	Bloom of *Synechococcus*, *Microcystis*, *Prochlorothrix*	Bacterial diversity↑, *Acinetobacter* sp.↑, *Limnobacter* sp.↑, *Synechococcus* sp.↑, *Roseomonas* sp.↑	Zhang et al. (2018)

↑increase/support; ↓ reduction/inhibition; ↔ unchanged

does not occur frequently or seems unlikely, but MCs simply accumulate (Kotak et al., 1996; Ibelings et al., 2005; Kim et al., 2019).

Emerging mechanistic insights are beginning to refine the hypotheses outlined above. Zhang et al. (2019) showed that *M. aeruginosa* can accelerate its growth via interactions with specific co-occurring microorganisms and the accumulation of several intermediates in energy metabolism-related pathways. Furthermore, *M. aeruginosa* can decrease the carbohydrate metabolism of co-occurring aquatic microorganisms, thereby disrupting microbial activities. This also has a negative effect on the richness and diversity of both bacterial and eukaryotic microbial communities. This process is further accelerated by MC, which is capable of decreasing the biomass of total phytoplankton in aquatic microcosms.

Furthermore, the ability of *M. aeruginosa* to compete with a coccal green alga (*Chlorella sorokiniana*) is determined by genomic differences beyond genes involved in microcystin generation and indicates an important role of host-associated bacteria in mediating phytoplankton interspecies interactions (Schmidt et al., 2020).

It has been hypothesized that allelopathy can prevent competitive exclusion and promote phytoplankton diversity in aquatic ecosystems, where numerous species coexist on a limited number of resources. Felpeto et al. (2018) presented experimental evidence to support this hypothesis by demonstrating that allelopathy promotes

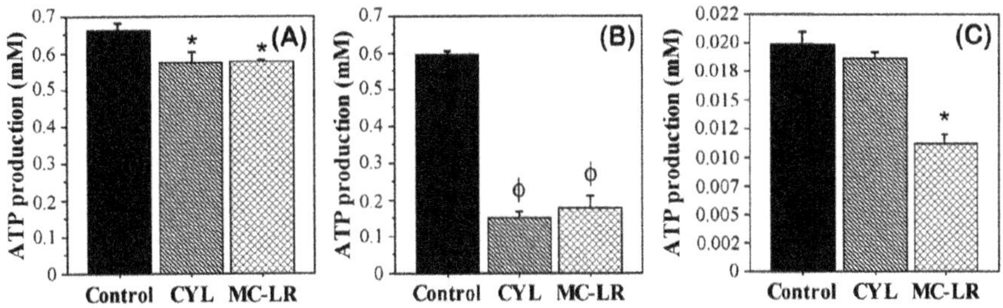

Fig. 4.7 Effects of microcystin-LR (MC-LR) and cylindrospermopsin (CYL) on the production of adenosine triphosphate (ATP) in *Chlamydomonas asymmetrica* (**a**), *Dunaliella salina* (**b**), and *Scenedesmus bijugus* var. *obtusiusculus* (**c**) after exposure for 15 h. Control: non-treated cell cultures. Results are presented as means ± SD. $*P < 0.05$, $\phi\ P < 0.001$ (from Teneva et al. (2023), credit to MDPI, Basel)

the coexistence of two phytoplankton species, *Ankistrodesmus falcatus* and *Oscillatoria* sp., that compete for a single limiting nutrient. And vice versa: even *Microcystis*, namely the tropical water-bloom forming *M. panniformis*,[3] may be outcompeted by another cyanobacterium, *Chrysosporum* (*Aphanizomenon*) *ovalisporum* (Zhang et al., 2017).

Recently, Teneva et al. (2023) confirmed the allelopathic potential of the cyanotoxins microcystin-LR and cylindrospermopsin on green algae (*Chlamydomonas asymmetrica*, *Dunaliella salina*, and *Scenedesmus bijugus* var. *obtusiusculus*) by detecting time-dependent inhibitory effects on the growth and motility of the green algae exposed to cyanotoxins. Changes in cell shape, granulation of the cytoplasm, and loss of flagella also occur. Additionally, cyanotoxins adversely affect the photosynthetic parameters of green algae and inhibit adenosine triphosphate (ATP) synthase (Fig. 4.7).

4.4 Elimination of Predators

Cyanobacterial and algal blooms are not just passive collections of cells drifting in water—they are battlegrounds where survival is a daily struggle. Blooms face threats from all directions: macropredators such as fish and invertebrates, micropredators like zooplankton, and fierce competition among neighboring algal species. While many studies have shown that the chemical arsenal of bloom-forming algae can act as an effective feeding deterrent, discouraging large generalist consumers (Liu et al., 2024), some algal defenses go beyond mere avoidance. In certain cases, these chemical weapons are so potent that they actively incapacitate or even eliminate predators, turning the tables in this microscopic arms race.

[3] In the past, they have been designated "*Microcystis aeruginosa*", "*Microcystis flos-aquae*," or "*Microcystis* cf. *lamelliformis*" in the floristic and hydrobiological literature (Komárek et al., 2002).

As many zooplankton grazers can discriminate between toxic and non-toxic prey, the capability of toxin production adds another advantage to inter- and intra-specific competition: the grazing pressure is redirected to non-toxic phytoplankton species and, within one species, to non-toxic strains (Guisande et al., 2002). This creates a dynamic feedback loop in which unicellular phytoplankton sense the presence of zooplankton and ramp up their defensive strategies, thereby shaping both predator behavior and community composition. However, when grazing is reduced simply because the grazer is harmed by the toxin, competing algal species benefit just as much, meaning the toxin-producing phytoplankton gains no clear competitive advantage. This remains an unresolved question, highlighting a key difference between terrestrial plants and phytoplankton: a tree can afford to sacrifice a few toxic leaves to deter herbivores, but a single-celled phytoplankton cannot (Pančić and Kiørboe, 2018).

After evaluating predator–prey interactions between the ciliate *Blepharisma americanum* and toxic (*Microcystis* spp.) and non-toxic (*Chlorella vulgaris*, *Microcystis* sp.) photosynthetic microbes in laboratory cultures, Chapman et al. (2019) concluded that the ubiquitous nature of protozoa as photosynthetic microbivores and their ability to digest only non-toxic cells indicate that, under certain conditions, feeding pressure favors the development of toxic HABs and alters microbial community dynamics.

4.4.1 Oxylipins[4]

Oxylipins are a diverse family of oxygenated natural products derived from fatty acids through pathways that involve at least one dioxygen-dependent oxidation step (Wasternack, 2007). Among them are well-known compounds, including polyunsaturated aldehydes (PUAs), eicosanoids, prostaglandins, and leukotrienes (Steinberg, 2022a, b). These small, polar lipid mediators originate from polyunsaturated fatty acids (PUFAs), including both ω-3 and ω-6 fatty acids (Watrous et al., 2019), and can be generated through both enzymatic and non-enzymatic oxidation processes (Camunas-Alberca et al., 2023).

When diatom populations are exposed to grazing, aging, or nutrient stress, cells can rapidly induce the biosynthesis of oxylipins, many of which are toxic to various organisms, including multicellular grazers and other algae (Pohnert and Boland, 2002; Leflaive and Ten-Hage, 2009). For instance, PUAs inhibit the growth of the toxic benthic dinoflagellate *Ostreopsis* cf. *ovata* (Pichierri et al., 2016).

Oxylipins elicit different cellular responses depending on their concentration, their recipient, and the context. It is suggested that they are involved in mediating biotic interactions across trophic levels: from interactions with other diatoms during allelopathy to grazing and interactions with bacteria.

[4]Adapted mainly from Graff van Creveld et al. (2022) with permission from Springer Nature.

Recent studies have shown that oxylipins repel grazers or reduce grazing rates in microzooplankton herbivory on diatoms (Franzè et al., 2018; Johnson et al., 2020). Oxylipins are not toxic to all grazers, and it has been suggested that during coevolution, some grazers have evolved resilience by using specific oxylipins to detect prey (Maibam et al., 2014, 2015). In addition, oxylipins have been suggested to play a role in mediating resistance to algicidal bacteria (Meyer et al., 2018; Meyer and Pohnert, 2019).

Little information is available on the effects of PUA on the pelagic food web. Therefore, Franzè et al. (2018) conducted a series of experiments in Chesapeake Bay and coastal Atlantic waters using dissolved PUA, natural assemblages of phytoplankton and microzooplankton, and the copepod *Acartia tonsa*. The results show that PUA primarily acts as a deterrent to microzooplankton herbivory on diatoms, while enhancing herbivory on picophytoplankton. This switch should favor PUA-producing diatoms by simultaneously reducing direct grazing losses and competition. In addition, PUA stimulated copepod predation on microzooplankton, particularly on ciliates (Fig. 4.8). Together, these effects have the potential to disrupt the growth-grazing balance in the pelagic food web and create an opportunity for bloom development.

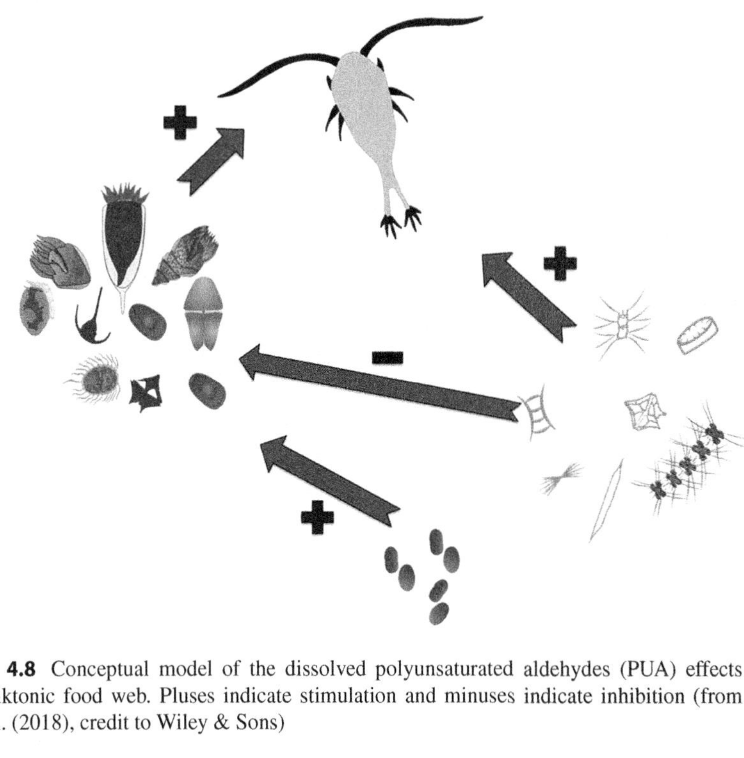

Fig. 4.8 Conceptual model of the dissolved polyunsaturated aldehydes (PUA) effects on the planktonic food web. Pluses indicate stimulation and minuses indicate inhibition (from Franzè et al. (2018), credit to Wiley & Sons)

4.4.2 Domoic Acid

Lundholm et al. (2018) found that induction of DA production in diatoms is grazer- and diatom-specific. Furthermore, bacteria are thought to play a role in the production or degradation of DA, as axenic cultures—those free of any associated microorganisms—generally produce less DA than xenic cultures, which maintain the natural microbial community alongside the algae (Kotaki et al., 2000; Lelong et al., 2012; Sison-Mangus et al., 2014). The specific roles of bacterial taxa include their potential involvement in both the production and degradation of DA, which influences bloom toxicity and persistence.

Among the diatoms exposed to grazing herbivorous copepods, only two of six species of *Pseudo-nitzschia* and none of the *Nitzschia* or *Fragilariopsis* strains respond by producing DA, indicating that not all *Pseudo-nitzschia* species/strains can produce DA, and that different diatom species might have different strategies for coping with grazing pressure. Zooplankton growth rate is negatively correlated to cellular DA content, indicating an allocation cost associated with toxin production. Long-term grazing experiments reveal higher mortality rates among grazers fed toxic diatoms, supporting the hypothesis that DA production serves as an induced defense mechanism.

Studying *Artemia* nauplii as grazers, Zhang et al. (2021) show that DA content exerts adverse effects on the grazers. A negative correlation between cellular DA content in *Pseudo-nitzschia multiseries* and clearance and/or ingestion rates of *Artemia* nauplii indicates that DA may harm the grazing of the nauplii. Such interaction might result in a decrease in grazing pressure on toxic blooming species, and hence, potentially a prolonged bloom.

As a sublethal defensive strategy, DA-producing *Pseudo-nitzschia* can impair predator behavior: the Arctic copepods *Calanus hyperboreus* and *C. glacialis* show a reduced escape response after feeding on these diatoms. By diminishing the escape response of copepods, DA-producing *Pseudo-nitzschia* indirectly increases copepod mortality through predation. This not only reduces grazing pressure on the toxic diatoms but also facilitates the transfer of domoic acid (DA) further up the food web, amplifying its ecological impact (Harðardóttir et al., 2018).

With respect to overall defense efficiency against grazing, however, several questions remain unanswered. The copepods *Calanus finmarchicus* and *C. glacialis* accumulate DA in their body tissue; yet, egg production and hatching success remain unaffected (Miesner et al., 2016).

4.4.3 Saxitoxin, Prymnesin & Co

Grazer killing appears to be a density-dependent phenomenon: when the density of certain toxic dinoflagellates is low, they are readily consumed by zooplankton; however, once a critical threshold is reached, the balance shifts—and the dinoflagellates can turn the tables, harming or even killing their grazers.

This perilous game has recently been displayed by Berge et al. (2012), showing that *Karlodinium armiger* (but not the related species *K. veneficum*) immobilizes the common marine copepod *Acartia tonsa* in a density-dependent manner and collectively ingests the grazer to promote its own growth rate. Copepods were able to graze and reproduce on a diet of *K. armiger* at concentrations below 1000 cells mL^{-1}. However, once densities exceeded 3500 cells mL^{-1}, the mixotrophic dinoflagellates turned the tables—immobilizing, killing, and feeding on the copepods, effectively inverting the food chain. Most importantly, *K. armiger* ingests different types of metazooplankton organisms, demonstrating that it may attack and feed on several different types of metazooplankton organisms, including larvae of commercially important bivalves and finfish.

Another perceived role of toxins is to serve as a predation instrument. For instance, Remmel and Hambright (2012) showed that *Prymnesium* attaches to zooplankton and fish, resulting in mortality. In contrast, exposure of these organisms to *Prymnesium* across a permeable membrane did not cause mortality. The authors also demonstrated that *Prymnesium* exotoxins are released independently of contact toxicity only in response to experimental procedures or natural causes of stress. The results are consistent with the idea that toxins have evolved for release during cell-to-cell contact in support of heterotrophy. The evolution of toxin-assisted micropredation would be consistent with mechanisms of natural selection favoring individual fitness as opposed to broadcast allelopathy, from which the benefits are more dispersed.

Using high-speed holographic microscopy, Sheng et al. (2010) measured the swimming behavior of several toxic and non-toxic strains of *Karlodinium veneficum* and their cryptomonad prey. They showed that toxic strains caused prey immobilization at rates that are consistent with the potency and dosage of the karlotoxins. Even prey cells that continued swimming slowed down after exposure to toxic predators.

Consistent with social evolution theory, structured habitats play a role in maintaining these traits in microbial model systems by localizing the benefits and separating strains that invest in these products from "cheater" strains that benefit without paying the cost. It is thus surprising that many unicellular, well-mixed microalgal populations invest in extracellular toxins that confer ecological benefits upon the entire population, for example, by eliminating nutrient competitors (allelopathy). Driscoll et al. (2013) tested the hypotheses that microalgal exotoxins are (1) exploitable public goods that benefit all cells, regardless of investment, or (2) non-exploitable private goods involved in cell-level functions. The authors tested these hypotheses using high-toxicity (TOX+) and low-toxicity (TOX−) strains of the damaging, mixotrophic microalga *Prymnesium parvum*, as well as two common competitors: green algae and diatoms. TOX+ strains gain a clear advantage in dense populations when competing with green algae, which serve both as competitors and as prey for *P. parvum*. This dual role provides TOX+ with a relative fitness edge over coexisting TOX strains. In contrast, when the competitors are non-prey species such as diatoms, TOX strains increase in frequency—even though they still benefit from the diatom-suppressing activity of TOX+. Such dynamics suggest that toxin production can act as an evolutionarily unstable and ecologically destructive public good, arising from traits that are adaptive at lower levels but have unforeseen consequences in novel environments.

Based on these and other experiences, Driscoll et al. (2016) drew more general conclusions. The transition from relatively subtle, cell-level "private" functions of toxicity to conspicuous, community-scale "public good" benefits frequently depends on threshold densities of toxic cells. The dinoflagellate *Karlodinium armiger* and haptophyte *Prymnesium parvum* can be safely consumed by animal and microbial predators at low densities; however, beyond threshold concentrations, the trophic interaction reverses, and the microalgae kill and feed upon their predators. Although these direct interactions may benefit the individual cells involved, there are nevertheless broader beneficial impacts not only for conspecifics but also for all (resistant) potential prey of the killed predators. A similar transition may occur simply via changes in the abundances of different predators. For example, individual colonies of toxic cyanobacteria benefit through toxin production when predators are capable of avoiding or rejecting toxic colonies; however, their negative effects on indiscriminate predators following ingestion can only benefit all remaining potential prey, including non-toxic populations.

Dense, highly toxic populations may also harm heterospecific populations that (directly or indirectly) benefit toxic lineages. Susceptible prey populations provide direct benefits to individual toxic predators, but are unlikely to persist at appreciable densities during ecosystem-disruptive blooms of their predators (Adolf et al., 2008). Similarly, even highly selective predators may suffer at extreme toxic population frequencies or densities (Vanderploeg et al., 2009). Loss of susceptible prey and selective predators would remove agents of selection that favor toxin production at the cellular level for both offensive and defensive functions, respectively.

If phagotrophy is a sufficient pathway for acquiring energy and nutrients, the question arises, why do mixotrophic microalgae maintain their photosynthetic capability? Or, conversely, if phototrophy is sufficient, why do many organisms combine both pathways? What is the benefit of combining both strategies? Mitra et al. (2014) attempted to find a synthesis by proposing an interesting new carbon flow paradigm for aquatic ecosystems that sees the bulk of the base of the new model food web supported by protist plankton communities that are mixotrophic—combining phototrophy and phagotrophy within a single cell. The photoautotrophic eukaryotic plankton and their heterotrophic microzooplankton grazers dominate only during the developmental phase of the spring bloom in temperate systems. With their flexible nutrition, mixotrophic protists are extremely well-prepared for and adapted to fluctuating nutrient and/or feed sources. With their flexible nutrition, mixotrophic protists dominate not only in more mature systems (e.g., temperate summer, established eutrophic as well as oligotrophic systems) but also in the carbon-limited aquatic system, such as very acidic mining lakes (Nixdorf et al., 2001). Figure 4.9

Fig. 4.9 (continued) nano-flagellates (MNFs), conforming to the current understanding of protist physiology. The MNFs are capable of eating bacteria (red-lined black arrow), and hence deriving nutrients for the support of their growth, and of their primary production, directly rather than (for ANFs in panel A) via the activities of HNFs + μZ. Dashed arrows indicate functions contributing to nutrient pools (blue for inorganic, brown for organic). Heavy black arrows indicate predatory links (from Mitra et al. (2014), credit to the European Geosciences Union)

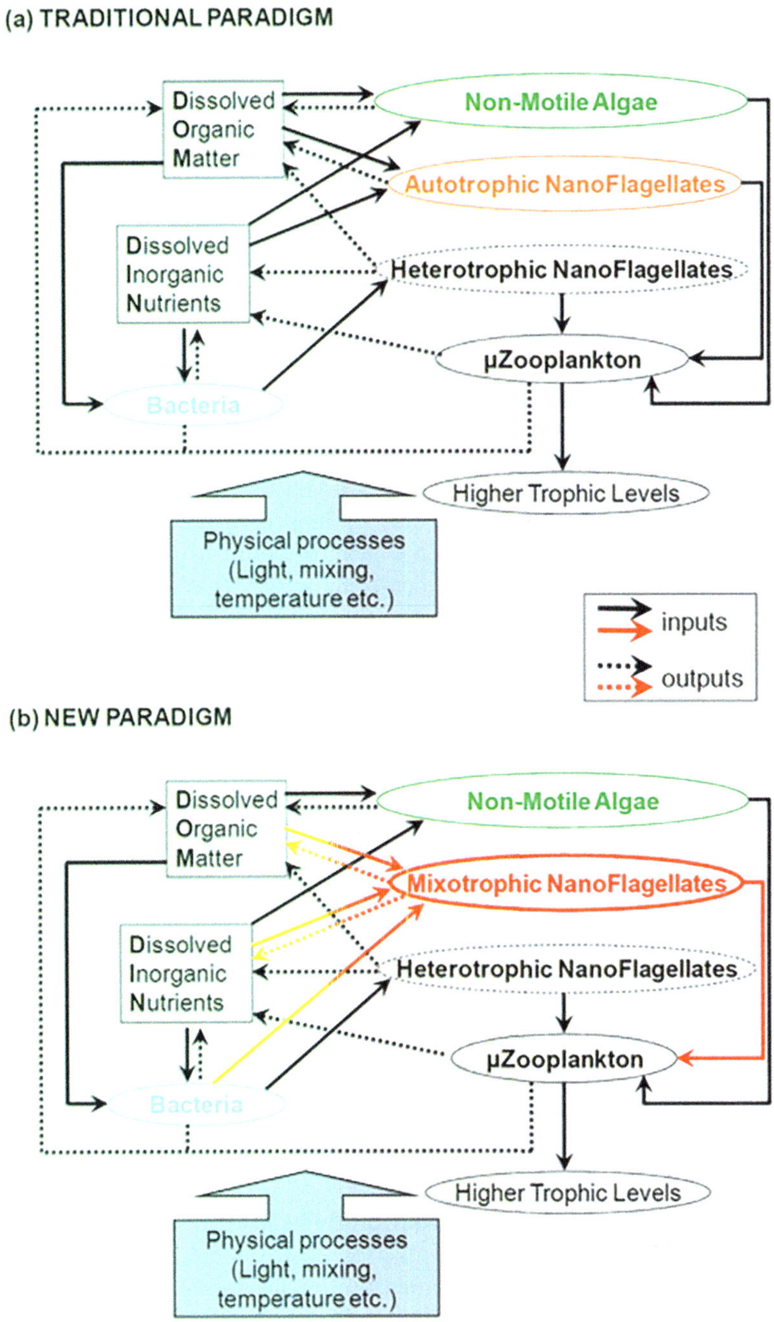

Fig. 4.9 Schematic representation of the two alternate modeled food web structures. In the traditional paradigm (**a**), the physiology of the photoautotrophic flagellates (ANFs) is similar to that of the nonmotile microalgae (NMA); neither is mixotrophic. Inorganic nutrients essential for primary production are regenerated through bacteria, heterotrophic nanoflagellates (HNFs), and phagotrophic microzooplankton (μZ). In the new paradigm (**b**), the ANFs are replaced with mixotrophic

presents the comparison of food webs based on the traditional paradigm and the new paradigm. In addition to the trophic relationships depicted in this figure, mixotrophic protists are able to use zooplankton as a food source (see below). The significance of mixotrophy increases further with the notion that certain phycotoxins, such as MCs, do not provide anti-herbivore defense against mixotrophic flagellates (Wilken et al., 2010). Ingestion rates of mixotrophic *Ochromonas* sp. are even slightly higher on the toxic *Microcystis aeruginosa* than on the non-toxic strain. Although even heterotrophic grazers can develop resistance to toxins if in contact with toxic algae (Zou et al., 2015; Huang et al., 2015), this protection is less secure than the direct digestion by *Ochromonas* sp. without any apparent adverse effect. Overall, the mixotrophy-dominated food web structure differs fundamentally in its flow of energy and nutrients, with a shortened and potentially more efficient chain from nutrient regeneration to primary production. Furthermore, mixotrophy provides a direct conduit for supporting primary production from bacterial production.

Notably, the fluctuating trophic relationships become even more complex because herbivory increases the toxin content of potentially harmful algae. This happens in *Alexandrium minutum* when being grazed by copepods (Yang et al., 2011). Furthermore, *Karenia mikimotoi* challenged by the feeding of the cladoceran *Moina mongolica* responds by increasing the production of hemolytic toxins and the synthesis of EPA. Both the EPA and the hemolytic toxins decrease the survival rate of the grazer (Dang et al., 2015). Selander et al. (2015) showed that it is the lipids (copepodamides) of the predator that induce paralytic shellfish toxins in bloom-forming *A. minutum*.

In their study on DA production in *Pseudo-nitzschia seriata*, Tammilehto et al. (2015) revealed an intriguing grazer-induced response. When the toxic diatom was exposed either directly to copepods (*Calanus hyperboreus* and *C. finmarchicus*) or indirectly across a separating membrane, DA production increased significantly. Because this effect occurred without physical contact, it is likely mediated by waterborne chemical cues released by the copepods. Remarkably, the magnitude of this response exceeded that reported earlier for *Alexandrium* (Fig. 4.10). By contrast, Olson et al. (2006) found that copepod grazing has negligible effects on *Pseudo-nitzschia* populations. Such contradictory results highlight the limited understanding we still have of grazer–diatom interactions and underscore the need for further investigation.

4.4.4 Induction of Toxin Production by Predators

The defense hypothesis of toxin production in photoautotrophic organisms is strongly supported by research on *Microcystis aeruginosa*, which has been shown to possess inducible defenses that ramp up toxin production in the presence of zooplankton and fish. For example, Jang et al. (2004) demonstrated that potentially toxic *Microcystis* increases microcystin production in response to chemical signals or feeding activity from both a phytoplanktivorous fish (*Hypophthalmichthys molitrix*) and an omnivorous fish (*Carassius langsdorfi*), even though neither species fed

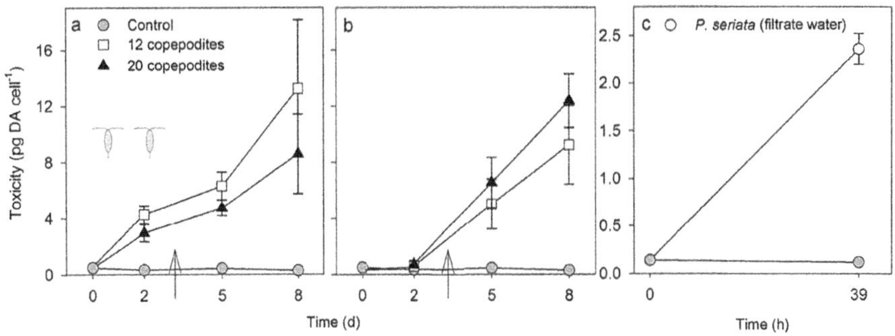

Fig. 4.10 Domoic acid (DA) cell quota (pg DA cell^{-1}, mean ± SD) of *Pseudo-nitzschia seriata* when grazed by two different densities of *Calanus* copepodites and in the control. (**a**) Flask A with copepodites and *P. seriata* ($N = 4$); (**b**) Flask B with only *P. seriata* cells separated from the copepodites by a 2 μm membrane ($N = 4$). The control contains *P. seriata* cells ($N = 3$). The arrows indicate the time when cells were added to ensure the copepodites had enough food; (**c**) toxicity of *P. seriata* (pg DA cell^{-1}, mean ± SD) when grown in the filtrate water where copepodites had been grazing on another diatom species (*Thalassiosira* sp.). Note the different scales in (**c**) (from Harðardóttir et al. (2015), credit to MDPI, Basel)

intensively on the cyanobacterium. Fish themselves can also respond actively to microcystin-producing cyanobacteria—most commonly by avoidance (Baganz et al., 2001, 2004)—thereby minimizing adverse effects such as intoxication (Ernst et al., 2007; Sotton et al., 2012; Godlewska et al., 2018), visual impairment (Gray et al. 2014), or food limitation (Reichwaldt et al., 2013).

Apart from being toxic, cyanotoxins can serve as infochemicals and contribute to specific responses of the species producing them. Cerbin et al. (2013) showed that *Aphanizomenon gracile* switches its growth mode in the presence of daphnids and invests more in trichome width to prevent filaments from being broken by grazers and consequently minimize predation pressure. In fact, thicker filaments are more harmful to *Daphnia* than thinner ones (Wejnerowski et al., 2015).

In support, Vilar et al. (2021) reported that infochemicals of *Daphnia gessneri* induce an upregulation of *sxt* genes and thereby an increase in saxitoxin production by the cyanobacterium *Raphidiopsis raciborskii*. Conversely, a growth loss may occur as a trade-off with enhanced toxin production. Van Gremberghe et al. (2009) noted substantial strain-specific variability in the toxin production response of the photoautotrophs.

Recent findings also suggest that interactions between toxic *Microcystis* and *Daphnia* are not limited to grazing but extend to reciprocal chemical communication. Exudates from grazers can trigger oxidative stress in *Microcystis*, leading to programmed cell death and the release of microcystins and secondary metabolites (Bojadzija Savic et al., 2019; Rzymski et al., 2020). This process may serve as both a protective strategy for the population and an anti-grazer weapon, creating a feedback loop where predator cues amplify the defensive arsenal of the prey.

Taken together, these studies reveal that cyanotoxins should not be seen as simple poisons, but rather as multifaceted ecological tools. They function simultaneously as defenses, infochemicals, and modulators of community dynamics—shaping the outcome of predator–prey interactions and influencing ecosystem-level processes. In this sense, toxin production represents a finely tuned strategy in the chemical arms race between phytoplankton and their consumers, balancing costs and benefits across a complex web of ecological relationships.

4.5 Higher Trophic Levels

Comprehensive overviews of disease and mortality events in marine mammals, seabirds, and fish associated with red tides have recently been provided by Würsig (2017), Ben-Gigirey et al. (2021), and Overstreet and Hawkins (2017). This naturally raises a critical question: What happens within the food web when top predators are weakened or even removed by mass mortality? Gray DiLeone and Ainsworth (2019) addressed this question using a fish community model, showing that *Karenia brevis* blooms can trigger strong top-down effects and trophic cascades (Fig. 4.11). When bloom-related mortality is applied, piscivores, planktivores, and detritivores experienced biomass increases, while invertebrate feeders and omnivores showed biomass decreases. These findings underscore that the consequences of red tides extend far beyond individual species, highlighting the need to consider entire ecosystems holistically when assessing the impacts of *K. brevis* and similar harmful algal blooms.

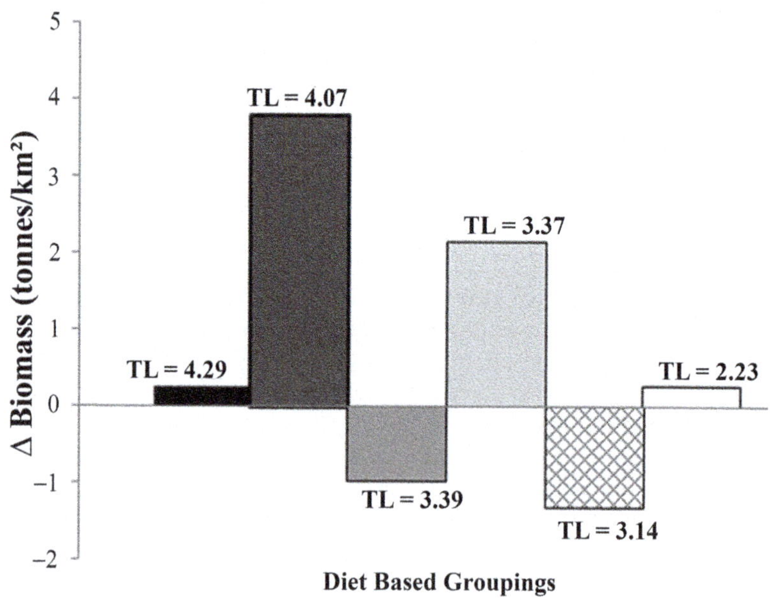

Fig. 4.11 Change in biomass across the food web in the presence and absence of *Karenia brevis* blooms. Trophic level (TL) values listed for each column represent the average trophic level of the functional groups included; from left to right: high trophic piscivores; low trophic piscivores; invertebrate eaters; planktivores; omnivores; and detritivores (from Gray DiLeone and Ainsworth (2019), with permission from Elsevier)

4.6 Concluding Remarks

As outlined above, the ecological significance of cyanotoxins is increasingly coming into focus. But what about marine phycotoxins? Are red tides—and by extension, other HABs—merely destructive forces in ecosystems? The structural complexity of marine HAB metabolites has not only revolutionized natural product chemistry but also made it challenging to assess their autecological roles. Progress in this area is therefore necessarily incremental.

A recent study by Song et al. (2024) offers thought-provoking evidence that HABs have often been viewed too narrowly. Red tide blooms off the southern coast of Korea, for example, are known to impact marine ecosystems and the aquaculture, but their effects on food web dynamics have been less explored. Zooplankton plays a crucial role in the food web, linking primary producers to higher predators, and interacts with red tide organisms in many ways (Fig. 4.12). By examining zooplankton communities over a 7-year period, spanning 3 years of red tide and four non-red tide years in Tongyeong, the authors revealed striking differences. In non-red tide years, zooplankton diversity followed predictable seasonal trends, increasing from June to early October. In red tide years, however, diversity remained consistently high but with a profound shift in species composition. Copepods declined while larvae of Branchiopoda, Echinodermata, Malacostraca, and Annelida increased. Diversity indices were significantly higher during red tides in all periods. Differences in zooplankton assemblages between periods were influenced by surface temperatures and the density of the red tide-causing alga *Margalefidinium* (formerly *Cochlodinium*) *polykrikoides*. Eight species emerged as indicator species and showed direct correlations with *M. polykrikoides*, seven of which were indicator species for years of red tide occurrence. These patterns suggest that zooplankton may be adapting to toxins and even exploiting blooms as cues for spawning. This long-term perspective provides the first comprehensive insight into zooplankton community shifts during red tide events, reframing them not only as ecological stressors but also as potential ecological drivers.

Still, such intriguing findings leave a fundamental question unresolved: Why do marine phytoplankton such as dinoflagellates, haptophytes, and diatoms invest in producing such energetically costly molecules in the first place? The answer must lie in the benefits to the algae themselves. Autecological studies, like those highlighted above, are beginning to shed light on the ecological functions of cyanobacterial toxins—but similar efforts are urgently needed for marine taxa. Yet the enormous diversity of toxin-producing species and the structural complexity of their metabolites make this one of the most formidable frontiers in HAB research.

Due to the wide variety of phycotoxin producers and the structural diversity of phycotoxins beyond microcystins, the following remarks may appear somewhat academic at present. Yet, with respect to phycotoxin production and release, one dimension remains critically underexplored: the interactome, i.e., the coevolutionary and functional interplay between toxigenic algae and their microbiomes (Du et al., 2022; Deng et al., 2023). As seen above, bacteria have been implicated in modulating DA dynamics, but the ecological consequences of such partnerships

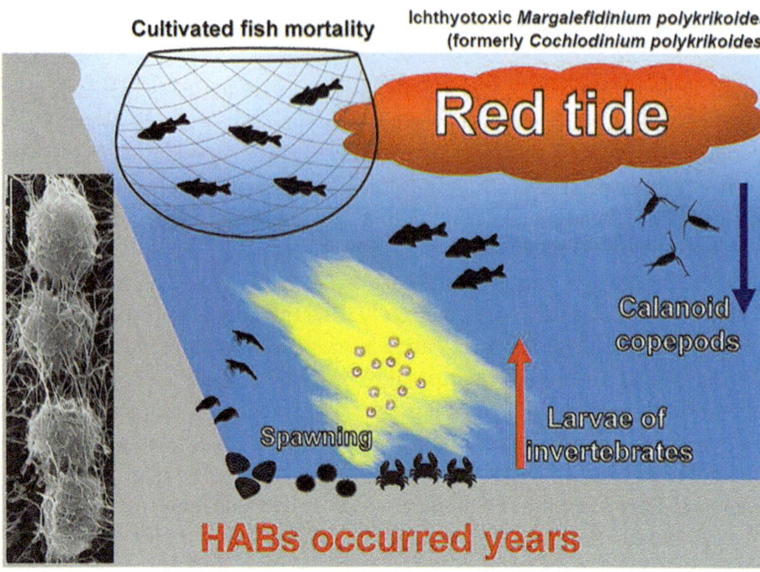

Fig. 4.12 HABs caused by *Margalefidinium polykrikoides* maintain or increase zooplankton diversity, increase primary production, and provide abundant food for zooplankton. Increased marine biodiversity suggests adaptation to spawning based on HABs cues (from Song et al. (2024), credit to Elsevier). A scanning electron micrograph of *M. polykrikoides*; the cell surface is masked by a filamentous structure that emerges from the cells (from Gómez et al. (2017), scale bar = 20 μm, with permission from Elsevier)

remain poorly understood. Insights from freshwater systems are already illuminating: Studies of *Microcystis* and its interactome suggest that successful long-term coevolution with associated bacteria may underpin its global dominance in lakes and reservoirs (Cook et al., 2020). Likewise, marine work points toward parallel collaborations, as evidenced by the emerging functional symbioses between diatoms and bacteria (Amin et al., 2012).

These are mainly epigenomics and, to a lesser extent, proteomics and metabolomics. Additionally, the interaction between gut microbiota and the life history traits of hosts consuming toxigenic algae warrants further scientific attention. If we take this one step further and conceptualize toxigenic algae, their associated bacteria, and their grazers or predators as a virtual holobiont, the picture becomes even more compelling. This systems-level perspective—already applied to fishes by Limborg et al. (2018)—has only begun to be touched upon in phytoplankton ecology. Crucially, several "omics" layers remain missing from this puzzle. While transcriptomics has provided a valuable starting point, epigenomics is virtually absent, and proteomics and metabolomics are still underrepresented. These gaps are critical because they may hold the key to explaining how toxin production is regulated in natural contexts, how interactions with microbial partners are stabilized, and how adaptive strategies are fine-tuned over ecological and evolutionary timescales (Roussel et al., 2023).

Finally, the interaction between gut microbiota and the life-history traits of consumers exposed to toxigenic algae represents another frontier with profound ecological implications (Claus et al., 2016). Predators may not simply ingest toxins; their microbiota could modulate toxicity, influence feeding behavior, or even shape long-term coevolutionary dynamics within food webs. In this sense, toxins should be viewed not only as chemical weapons or stress signals, but also as integral mediators of multispecies networks. Exploring these interdependencies may ultimately transform our understanding of HABs from a series of isolated harmful events into a broader framework of ecological interactions, evolutionary strategies, and ecosystem-level feedbacks.

References

Adolf JE, Bachvaroff T, Place AR (2008) Can cryptophyte abundance trigger toxic *Karlodinium veneficum* blooms in eutrophic estuaries? Harmful Algae 8(1):119–128. https://doi.org/10.1016/j.hal.2008.08.003

Adolf JE, Bachvaroff TR, Krupatkina DN, Nonogaki H, Brown PJP, Lewitus AJ, Harvey HR, Place AR (2006) Species specificity and potential roles of *Karlodinium micrum* toxin. Afr J Mar Sci 28(2):415–419. https://doi.org/10.2989/18142320609504189

Adolf JE, Krupatkina D, Bachvaroff T, Place AR (2007) Karlotoxin mediates grazing by *Oxyrrhis marina* on strains of *Karlodinium veneficum*. Harmful Algae 6(3):400–412. https://doi.org/10.1016/j.hal.2006.12.003

Alexova R, Fujii M, Birch D, Cheng J, Waite TD, Ferrari BC, Neilan BA (2011) Iron uptake and toxin synthesis in the bloom-forming *Microcystis aeruginosa* under iron limitation. Environ Microbiol 13(4):1064–1077. https://doi.org/10.1111/j.1462-2920.2010.02412.x

Allen JL, Ten-Hage L, Leflaive J (2018) Regulation of fatty acid production and release in benthic algae: could parallel allelopathy be explained with plant defence theories? Microb Ecol 75(3):609–621. https://doi.org/10.1007/s00248-017-1082-z

Amin SA, Parker MS, Armbrust EV (2012) Interactions between diatoms and bacteria. Microbiol Mol Biol Rev 76(3):667–684. https://doi.org/10.1128/MMBR.00007-12

Anderson DM, Cembella AD, Hallegraeff GM (2012) Progress in understanding harmful algal blooms: paradigm shifts and new technologies for research, monitoring, and management. Annu Rev Mar Sci 4:143–176. https://doi.org/10.1146/annurev-marine-120308-081121

Andras TD, Alexander TS, Gahlena A, Parry RM, Fernandez FM, Kubanek J, Wang MD, Hay ME (2012) Seaweed allelopathy against coral: surface distribution of a seaweed secondary metabolite by imaging mass spectrometry. J Chem Ecol 38(10):1203–1214. https://doi.org/10.1007/s10886-012-0204-9

Arakawa T, Ejima D, Kita Y, Tsumoto K (2006) Small molecule pharmacological chaperones: from thermodynamic stabilization to pharmaceutical drugs. Biochim Biophys Acta 1764(11):1677–1687. https://doi.org/10.1016/j.bbapap.2006.08.012

Auro ME, Cochlan WP (2013) Nitrogen utilization and toxin production by two diatoms of the *Pseudo-nitzschia pseudodelicatissima* complex: *P. cuspidata* and *P. fryxelliana*. J Phycol 49 (1):156–169. https://doi.org/10.1111/jpy.12033

Babica P, Hilscherová K, Bártová K, Bláha L, Maršálek B (2007) Effects of dissolved microcystins on growth of planktonic photoautotrophs. Phycologia 46(2):137–142. https://doi.org/10.2216/06-24.1

Baganz D, Staaks G, Pflugmacher S, Steinberg CEW (2004) Comparative study of microcystin-LR-induced behavioral changes of two fish species, *Danio rerio* and *Leucaspius delineatus*. Environ Toxicol 19(6):564–570. https://doi.org/10.1002/tox.20063

Baganz D, Staaks G, Steinberg C (2001) Changes in fish behaviour affected by microcystin-LR. In: Chorus I (ed) Cyanotoxins—occurrence, effects, controlling factors. Springer, Heidelberg, pp 261–280

Bar-Yosef Y, Sukenik A, Hadas O, Viner-Mozzini Y, Kaplan A (2010) Enslavement in the water body by toxic *Aphanizomenon ovalisporum*, inducing alkaline phosphatase in phytoplanktons. Curr Biol 20(17):1557–1561. https://doi.org/10.1016/j.cub.2010.07.032

Bates SS, Hubbard KA, Lundholm N, Montresor M, Leaw CP (2018) Pseudo-nitzschia, Nitzschia, and domoic acid: new research since 2011. Harmful Algae 79:3–43. https://doi.org/10.1016/j.hal.2018.06.001

Ben-Gigirey B, Soliño L, Bravo I, Rodríguez F, Casero MVM (2021) Paralytic and amnesic shellfish toxins impacts on seabirds, analyses and management. Toxins 13(7):454. https://doi.org/10.3390/toxins13070454

Berge T, Poulsen LK, Moldrup M, Daugbjerg N, Juel Hansen P (2012) Marine microalgae attack and feed on metazoans. ISME J 6(10):1926–1936. https://doi.org/10.1038/ismej.2012.29

Berry MA, Davis TW, Cory RM, Duhaime MB, Johengen TH, Kling GW, Marino JA, Den Uyl PA, Gossiaux D, Dick GJ, Denef VJ (2017) Cyanobacterial harmful algal blooms are a biological disturbance to Western Lake Erie bacterial communities. Environ Microbiol 19(3):1149–1162. https://doi.org/10.1111/1462-2920.13640

Beversdorf LJ, Miller TR, McMahon KD (2015) Long-term monitoring reveals carbon–nitrogen metabolism key to microcystin production in eutrophic lakes. Front Microbiol 6. https://doi.org/10.3389/fmicb.2015.00456

Bojadzija Savic G, Edwards C, Briand E, Lawton L, Wiegand C, Bormans M (2019) *Daphnia magna* exudates impact physiological and metabolic changes in *Microcystis aeruginosa*. Toxins 11(7):421. https://doi.org/10.3390/toxins11070421

Brand LE, Campbell L, Bresnan E (2012) *Karenia*: the biology and ecology of a toxic genus. Harmful Algae 14:156–178. https://doi.org/10.1016/j.hal.2011.10.020

Brandenburg K, Siebers L, Keuskamp J, Jephcott TG, Van de Waal DB (2020) Effects of nutrient limitation on the synthesis of N-rich phytoplankton toxins: a meta-analysis. Toxins 12(4):221. https://doi.org/10.3390/toxins12040221

Briand E, Yéprémian C, Humbert J-F, Quiblier C (2008) Competition between microcystin- and non-microcystin-producing *Planktothrix agardhii* (cyanobacteria) strains under different environmental conditions. Environ Microbiol 10(12):3337–3348. https://doi.org/10.1111/j.1462-2920.2008.01730.x

Camunas-Alberca SM, Moran-Garrido M, Sáiz J, Villaseñor A, Taha AY, Barbas C (2023) The role of oxylipins and their validation as biomarkers in the clinical context. TrAC Trends Anal Chem 164:117065. https://doi.org/10.1016/j.trac.2023.117065

Carrick HJ, Marble C, Tian YQ (2023) Differential responses for stream algal assemblages exposed to factorial N and P enrichment along an *in situ* DOC gradient. Front. Environ Sci 11. https://doi.org/10.3389/fenvs.2023.1066586

Casotti R, Mazza S, Brunet C, Vantrepotte V, Ianora A, Miralto A (2005) Growth inhibition and toxicity of the diatom aldehyde 2-trans, 4-trans-decadienal on *Thalassiosira weissflogii* (Bacillariophyceae). J Phycol 41(1):7–20. https://doi.org/10.1111/j.1529-8817.2005.04052.x

Cerbin S, Wejnerowski Ł, Dziuba M (2013) *Aphanizomenon gracile* increases in width in the presence of *Daphnia*. A defence mechanism against grazing? J Limnol 72 (3):505–511.e41. https://doi.org/10.4081/jlimnol.2013.e41

Chang FH (2011) Toxic effects of three closely-related dinoflagellates, *Karenia concordia*, *K. brevisulcata* and *K. mikimotoi* (Gymnodiniales, Dinophyceae) on other microalgal species. Harmful Algae 10(2):181–187. https://doi.org/10.1016/j.hal.2010.09.004

Chang FH, Uddstrom MJ, Pinkerton MH, Richardson KM (2008) Characterising the 2002 toxic *Karenia concordia* (Dinophyceae) outbreak and its development using satellite imagery on the north-eastern coast of New Zealand. Harmful Algae 7(4):532–544. https://doi.org/10.1016/j.hal.2007.11.004

Chapman IJ, Franklin DJ, Turner AD, McCarthy EJA, Esteban GF (2019) Predator–prey interactions between the ciliate *Blepharisma americanum* and toxic (*Microcystis* spp.) and non-toxic (*Chlorella vulgaris*, *Microcystis* sp.) photosynthetic microbes. Aquat Microb Ecol 83 (3):211–224. https://doi.org/10.3354/ame01913

Chaves-Fonnegra A, Castellanos L, Zea S, Duque C, Rodríguez J, Jiménez C (2008) Clionapyrrolidine A—A metabolite from the encrusting and excavating sponge *Cliona tenuis* that kills coral tissue upon contact. J Chem Ecol 34(12):1565–1574. https://doi.org/10.1007/s10886-008-9565-5

Claus SP, Guillou H, Ellero-Simatos S (2016) The gut microbiota: a major player in the toxicity of environmental pollutants? NPJ Biofilms Microb 2 (1):16003. https://doi.org/10.1038/npjbiofilms.2016.3

Cochlan WP, Bill BD, Cailipan AB, Trainer VL (2023) Domoic acid production by *Pseudo-nitzschia australis*: re-evaluating the role of macronutrient limitation on toxigenicity. Harmful Algae 125:102431. https://doi.org/10.1016/j.hal.2023.102431

Coll JC (1992) The chemistry and chemical ecology of octocorals (Coelenterata, Anthozoa, Octocorallia). Chem Rev 92(4):613–631. https://doi.org/10.1021/cr00012a006

Cook KV, Li C, Cai H, Krumholz LR, Hambright KD, Paerl HW, Steffen MM, Wilson AE, Burford MA, Grossart H-P, Hamilton DP, Jiang H, Sukenik A, Latour D, Meyer EI, Padisák J, Qin B, Zamor RM, Zhu G (2020) The global *Microcystis* interactome. Limnol Oceanogr 65(S1):S194–S207. https://doi.org/10.1002/lno.11361

Dang LX, Li Y, Liu F, Zhang Y, Yang WD, Li HY, Liu JS (2015) Chemical response of the toxic dinoflagellate *Karenia mikimotoi* against grazing by three species of zooplankton. J Eukaryot Microbiol 62(4):470–480. https://doi.org/10.1111/jeu.12201

Deng Y, Wang K, Hu Z, Hu Q, Tang YZ (2023) Toxic and non-toxic dinoflagellates host distinct bacterial communities in their phycospheres. Commun Earth Environ 4(1):263. https://doi.org/10.1038/s43247-023-00925-z

Dias F, Antunes JT, Ribeiro T, Azevedo J, Vasconcelos V, Leão PN (2017) Cyanobacterial allelochemicals but not cyanobacterial cells markedly reduce microbial community diversity. Front Microbiol 8:1495. https://doi.org/10.3389/fmicb.2017.01495

Dick GJ, Duhaime MB, Evans JT, Errera RM, Godwin CM, Kharbush JJ, Nitschky HS, Powers MA, Vanderploeg HA, Schmidt KC, Smith DJ, Yancey CE, Zwiers CC, Denef VJ (2021) The genetic and ecophysiological diversity of *Microcystis*. Environ Microbiol 23(12):7278–7313. https://doi.org/10.1111/1462-2920.15615

Ding Q, Liu K, Song Z, Sun R, Zhang J, Yin L, Pu Y (2020) Effects of microcystin-LR on metabolic functions and structure succession of sediment bacterial community under anaerobic conditions. Toxins 12(3):183. https://doi.org/10.3390/toxins12030183

Doucette G, King K, Thessen A, Dortch Q (2008) The effect of salinity on domoic acid production by the diatom *Pseudo-nitzschia multiseries*. Nova Hedwigia Beih 133:31–46

Drakare S (2002) Competition between picoplanktonic cyanobacteria and heterotrophic bacteria along crossed gradients of glucose and phosphate. Microb Ecol 44(4):327–335. https://doi.org/10.1007/s00248-002-1013-4

Driscoll WW, Espinosa NJ, Eldakar OT, Hackett JD (2013) Allelopathy as an emergent, exploitable public good in the bloom-forming microalga *Prymnesium parvum*. Evolution 67(6):1582–1590. https://doi.org/10.1111/evo.12030

Driscoll WW, Hackett JD, Ferrière R (2016) Eco-evolutionary feedbacks between private and public goods: evidence from toxic algal blooms. Ecol Lett 19(1):81–97. https://doi.org/10.1111/ele.12533

Du M, Wang J, Jin Y, Fan J, Zan S, Li Z (2022) Response mechanism of microbial community during anaerobic biotransformation of marine toxin domoic acid. Environ Res 215:114410. https://doi.org/10.1016/j.envres.2022.114410

Dziallas C, Grossart H-P (2011) Increasing oxygen radicals and water temperature select for toxic *Microcystis* sp. PLoS One 6(9):e25569. https://doi.org/10.1371/journal.pone.0025569

Ernst B, Hoeger SJ, O'Brien E, Dietrich DR (2007) Physiological stress and pathology in European whitefish (*Coregonus lavaretus*) induced by subchronic exposure to environmentally relevant densities of *Planktothrix rubescens*. Aquat Toxicol 82(1):15–26. https://doi.org/10.1016/j.aquatox.2007.01.007

Felpeto AB, Roy S, Vasconcelos VM (2018) Allelopathy prevents competitive exclusion and promotes phytoplankton biodiversity. Oikos 127(1):85–98. https://doi.org/10.1111/oik.04046

Feng B, Wang C, Wu X, Tian C, Tian Y, Xiao B (2019) Involvement of microcystins, colony size and photosynthetic activity in the benthic recruitment of *Microcystis*. J Appl Phycol 31(1):223–233. https://doi.org/10.1007/s10811-018-1508-0

Fernández-Herrera LJ, Band-Schmidt CJ, Zenteno-Savín T, Leyva-Valencia I, Hernández-Guerrero CJ, Hernández-Sandoval FE, Bustillos-Guzmán JJ (2022) Changes in toxin production, morphology and viability of *Gymnodinium catenatum* associated with allelopathy of *Chattonella marina* var. *marina* and Gymnodinium impudicum. Toxins 14 (9).616. https://doi.org/10.3390/toxins14090616

Figueredo CC, Giani A, Bird DF (2007) Does allelopathy contribute to *Cylindrospermopsis raciborskii* (cyanobacteria) bloom occurrence and geographic expansion? J Phycol 43(2):256–265. https://doi.org/10.1111/j.1529-8817.2007.00333.x

Fleury B, Coll J, Sammarco P (2006) Complementary (secondary) metabolites in a soft coral: sex-specific variability, inter-clonal variability, and competition. Mar Ecol 27(3):204–218. https://doi.org/10.1111/j.1439-0485.2006.00106.x

Fossat B, Porthé-Nibelle J, Sola F, Masoni A, Gentien P, Bodennec G (1999) Toxicity of fatty acid 18:5n3 from *Gymnodinium* cf. *mikimotoi*: II. Intracellular pH and K^+ uptake in isolated trout hepatocytes. J Appl Toxicol 19 (4):275–278. https://doi.org/10.1002/(SICI)1099-1263(199907/08)19:4<275::AID-JAT578>3.0.CO;2-B

Franzè G, Pierson JJ, Stoecker DK, Lavrentyev PJ (2018) Diatom-produced allelochemicals trigger trophic cascades in the planktonic food web. Limnol Oceanogr 63(3):1093–1108. https://doi.org/10.1002/lno.10756

Gallina AA, Brunet C, Palumbo A, Casotti R (2014) The effect of polyunsaturated aldehydes on *Skeletonema marinoi* (Bacillariophyceae): the involvement of reactive oxygen species and nitric oxide. Mar Drugs 12(7):4165–4187. https://doi.org/10.3390/md12074165

Gentien P, Lunven M, Lazure P, Youenou A, Crassous MP (2007) Motility and autotoxicity in *Karenia mikimotoi* (Dinophyceae). Philos Trans R Soc B 362(1487):1937–1946. https://doi.org/10.1098/rstb.2007.2079

Gerphagnon M, Macarthur DJ, Latour D, Gachon CM, Van Ogtrop F, Gleason FH, Sime-Ngando T (2015) Microbial players involved in the decline of filamentous and colonial cyanobacterial blooms with a focus on fungal parasitism. Environ Microbiol 17(8):2573–2587. https://doi.org/10.1111/1462-2920.12860

Ginn HP, Pearson LA, Neilan BA (2010) NtcA from *Microcystis aeruginosa* PCC 7806 is autoregulatory and binds to the microcystin promoter. Appl Environ Microbiol 76(13):4362–4368. https://doi.org/10.1128/AEM.01862-09

Glibert PM, Burkholder JM (2018) Causes of harmful algal blooms. In: Shumway SE, Burkholder JM, Morton SL (eds) Harmful algal blooms: a compendium desk reference. Wiley, pp 1–38. https://doi.org/10.1002/9781118994672.ch1

Glibert PM, Burkholder JM, Kana TM (2012) Recent insights about relationships between nutrient availability, forms, and stoichiometry, and the distribution, ecophysiology, and food web effects of pelagic and benthic *Prorocentrum* species. Harmful Algae 14:231–259. https://doi.org/10.1016/j.hal.2011.10.023

Godlewska M, Balk H, Kaczkowski Z, Jurczak T, Izydorczyk K, Długoszewski B, Jaskulska A, Gągała-Borowska I, Mankiewicz-Boczek J (2018) Night fish avoidance of *Microcystis* bloom revealed by simultaneous hydroacoustic measurements of both organisms. Fish Res 207:74–84. https://doi.org/10.1016/j.fishres.2018.05.025

Gómez F, Richlen ML, Anderson DM (2017) Molecular characterization and morphology of *Cochlodinium strangulatum*, the type species of *Cochlodinium*, and *Margalefidinium* gen. nov. for *C. polykrikoides* and allied species (Gymnodiniales, Dinophyceae). Harmful Algae 63:32–44. https://doi.org/10.1016/j.hal.2017.01.008

Graff van Creveld S (2018) The role of redox in cell fate regulation in marine diatom's response to environmental stresses. PhD thesis, Weizmann Institute of Science, Rehovot, Israel

Graff van Creveld S, Mizrachi A, Vardi A (2022) An ocean of signals: intracellular and extracellular signaling in diatoms. In: Falciatore A, Mock T (eds) The molecular life of diatoms. Springer Nature Switzerland AG, Cham, pp 641–678. https://doi.org/10.1007/978-3-030-92499-7_22

Graff van Creveld S, Rosenwasser S, Schatz D, Koren I, Vardi A (2015) Early perturbation in mitochondria redox homeostasis in response to environmental stress predicts cell fate in diatoms. ISME J 9(2):385–395. https://doi.org/10.1038/ismej.2014.136

Granéli E, Flynn K (2006) Chemical and physical factors influencing toxin content. In: Granéli E, Turner JT (eds) Ecology of harmful algae. Springer, Berlin Heidelberg, Berlin, Heidelberg, pp 229–241. https://doi.org/10.1007/978-3-540-32210-8_18

Gray DiLeone AM, Ainsworth CH (2019) Effects of *Karenia brevis* harmful algal blooms on fish community structure on the West Florida shelf. Ecol Model 392:250–267. https://doi.org/10.1016/j.ecolmodel.2018.11.022

Gray SM, Bieber FME, Mcdonnell LH, Chapman LJ, Mandrak NE (2014) Experimental evidence for species-specific response to turbidity in imperilled fishes. Aquat Conserv 24(4):546–560. https://doi.org/10.1002/aqc.2436

Griffith AW, Gobler CJ (2020) Harmful algal blooms: a climate change co-stressor in marine and freshwater ecosystems. Harmful Algae 91:101590. https://doi.org/10.1016/j.hal.2019.03.008

Grover JP, Hsu SB, Wang FB (2012) Competition between microorganisms for a single limiting resource with cell quota structure and spatial variation. J Math Biol 64(5):713–743. https://doi.org/10.1007/s00285-011-0426-4

Gu P, Li Q, Zhang H, Luo X, Zhang W, Zheng Z, Luo X (2020) Effects of cyanobacteria on phosphorus cycling and other aquatic organisms in simulated eutrophic ecosystems. Water (Switzerland) 12(8):2265. https://doi.org/10.3390/w12082265

Guerrini F, Ciminiello P, Dell'Aversano C, Tartaglione L, Fattorusso E, Boni L, Pistocchi R (2007) Influence of temperature, salinity and nutrient limitation on yessotoxin production and release by the dinoflagellate *Protoceratium reticulatum* in batch-cultures. Harmful Algae 6(5):707–717. https://doi.org/10.1016/j.hal.2007.02.006

Guisande C, Frangópulos M, Maneiro I, Vergara AR, Riveiro I (2002) Ecological advantages of toxin production by the dinoflagellate *Alexandrium minutum* under phosphorus limitation. Mar Ecol Prog Ser 225:169–176. https://doi.org/10.3354/meps225169

Guo X, Han T, Tan L, Zhao T, Zhu X, Huang W, Lin K, Zhang N, Wang J (2022) The allelopathy and underlying mechanism of *Skeletonema costatum* on *Karenia mikimotoi* integrating transcriptomics profiling. Aquat Toxicol 242:106042. https://doi.org/10.1016/j.aquatox.2021.106042

Hadas O, Pinkas R, Delphine E, Vardi A, Kaplan A, Sukenik A (1999) Limnological and ecophysiological aspects of *Aphanizomenon ovalisporum* bloom in Lake Kinneret, Israel. J Plankton Res 21(8):1439–1453. https://doi.org/10.1093/plankt/21.8.1439

Hagström JA, Granéli E, Moreira MOP, Odebrecht C (2011) Domoic acid production and elemental composition of two *Pseudo-nitzschia multiseries* strains, from the NW and SW Atlantic Ocean, growing in phosphorus or nitrogen-limited chemostat cultures. J Plankton Res 33(2):297–308. https://doi.org/10.1093/plankt/fbq102

Han L, Qiu J, Li A, Li D, Yang Y, Wang G, Li P (2023) Effects of marine phycotoxin dinophysistoxin-1 on the growth and cell cycle of *Isochrysis galbana*. Comp Biochem Phys C 273:109732. https://doi.org/10.1016/j.cbpc.2023.109732

Han M, Dong C, Ma S, Feng C, Lei C, Gu Z, Liu X (2021) Food web responses to a cyanobacterial bloom in a freshwater eutrophic lake. Water 13(9). https://doi.org/10.3390/w13091296

Harðardóttir S, Krock B, Wohlrab S, John U, Nielsen TG, Lundholm N (2018) Can domoic acid affect escape response in copepods? Harmful Algae 79:50–52. https://doi.org/10.1016/j.hal.2018.08.009

Harðardóttir S, Pančić M, Tammilehto A, Krock B, Møller EF, Nielsen TG, Lundholm N (2015) Dangerous relations in the Arctic marine food web: interactions between toxin producing *Pseudo-nitzschia* diatoms and *Calanus* copepodites. Mar Drugs 13(6):3809–3835. https://doi.org/10.3390/md13063809

Hardison DR, Sunda WG, Shea D, Litaker RW (2013) Increased toxicity of *Karenia brevis* during phosphate limited growth: ecological and evolutionary implications. PLoS One 8(3):e58545. https://doi.org/10.1371/journal.pone.0058545

Harris TD, Reinl KL, Azarderakhsh M, Berger SA, Berman MC, Bizic M, Bhattacharya R, Burnet SH, Cianci-Gaskill JA, Domis LNS, Elfferich I, Ger KA, Grossart H-PF, Ibelings BW, Ionescu D, Kouhanestani ZM, Mauch J, McElarney YR, Nava V, North RL, Ogashawara I, Paule-Mercado MCA, Soria-Píriz S, Sun X, Trout-Haney JV, Weyhenmeyer GA, Yokota K, Zhan Q (2024) What makes a cyanobacterial bloom disappear? A review of the abiotic and biotic cyanobacterial bloom loss factors. Harmful Algae 133:102599. https://doi.org/10.1016/j.hal.2024.102599

He H, Chen F, Li H, Xiang W, Li Y, Jiang Y (2010) Effect of iron on growth, biochemical composition and paralytic shellfish poisoning toxins production of *Alexandrium tamarense*. Harmful Algae 9(1):98–104. https://doi.org/10.1016/j.hal.2009.08.006

Helber SB, Hoeijmakers DJJ, Muhando CA, Rohde S, Schupp PJ (2018) Sponge chemical defenses are a possible mechanism for increasing sponge abundance on reefs in Zanzibar. PLoS One 13(6):e0197617. https://doi.org/10.1371/journal.pone.0197617

Holland A, Kinnear S (2013) Interpreting the possible ecological role(s) of cyanotoxins: compounds for competitive advantage and/or physiological aide? Mar Drugs 11(7):2239–2258. https://doi.org/10.3390/md11072239

Hu L, Shan K, Lin L, Shen W, Huang L, Gan N, Song L (2016) Multi-year assessment of toxic genotypes and microcystin concentration in northern Lake Taihu, China. Toxins 8(1). https://doi.org/10.3390/toxins8010023

Huang L, Liu SL, Zheng JW, Li HY, Liu JS, Yang WD (2015) *P*-glycoprotein and its inducible expression in three bivalve species after exposure to *Prorocentrum lima*. Aquat Toxicol 169:123–132. https://doi.org/10.1016/j.aquatox.2015.10.012

Husson B, Hernández-Fariñas T, Le Gendre R, Schapira M, Chapelle A (2016) Two decades of *pseudo-nitzschia* spp. blooms and king scallop (*Pecten maximus*) contamination by domoic acid along the French Atlantic and English Channel coasts: seasonal dynamics, spatial heterogeneity and interannual variability. Harmful Algae 51:26–39. https://doi.org/10.1016/j.hal.2015.10.017

Ibelings BW, Bruning K, de Jonge J, Wolfstein K, Pires LMD, Postma J, Burger T (2005) Distribution of microcystins in a lake foodweb: no evidence for biomagnification. Microb Ecol 49(4):487–500. https://doi.org/10.1007/s00248-004-0014-x

Islam S, Sabharwal T, Wu S, Bullock TJ, Mehdy MC (2020) Early dynamics of photosynthetic *Lhcf2* and *Lhcf15* transcription and mRNA stabilities in response to herbivory-related decadienal in *Phaeodactylum tricornutum*. Sci Rep 10(1):2029. https://doi.org/10.1038/s41598-020-58885-9

Jang MH, Ha K, Lucas MC, Joo GJ, Takamura N (2004) Changes in microcystin production by *Microcystis aeruginosa* exposed to phytoplanktivorous and omnivorous fish. Aquat Toxicol 68(1):51–59. https://doi.org/10.1016/j.aquatox.2004.02.002

Jankowiak JG, Gobler CJ (2020) The composition and function of microbiomes within *Microcystis* colonies are significantly different than native bacterial assemblages in two North American lakes. Front Microbiol 11. https://doi.org/10.3389/fmicb.2020.01016

Johnson MD, Edwards BR, Beaudoin DJ, Van Mooy BAS, Vardi A (2020) Nitric oxide mediates oxylipin production and grazing defense in diatoms. Environ Microbiol 22(2):629–645. https://doi.org/10.1111/1462-2920.14879

Kerimoglu O, Straile D, Peeters F (2012) Role of phytoplankton cell size on the competition for nutrients and light in incompletely mixed systems. J Theor Biol 300:330–343. https://doi.org/10.1016/j.jtbi.2012.01.044

Kim D, Hong S, Choi H, Choi B, Kim J, Khim JS, Park H, Shin KH (2019) Multimedia distributions, bioaccumulation, and trophic transfer of microcystins in the Geum River estuary, Korea: application of compound-specific isotope analysis of amino acids. Environ Int 133(Pt B):105194. https://doi.org/10.1016/j.envint.2019.105194

Kim JH, Jeong HJ, Lim AS, Rho JR, Lee SB (2016) Killing potential protist predators as a survival strategy of the newly described dinoflagellate *Alexandrium pohangense*. Harmful Algae 55:41–55. https://doi.org/10.1016/j.hal.2016.01.009

Kitchen SA, Bourdelais AJ, Taylor AR (2018) Interaction of a dinoflagellate neurotoxin with voltage-activated ion channels in a marine diatom. PeerJ 2018(4). https://doi.org/10.7717/peerj.4533

Komárek J, Komárková-Legnerová J, Sant'Anna CL, De Paiva Azevedo MT, Senna PAC (2002) Two common *Microcystis* species (Chroococcales, cyanobacteria) from tropical America, including *M. Panniformis* sp. nov. Cryptogam Algol 23(2):159–177

Kotak BG, Zurawell RW, Prepas EE, Holmes CFB (1996) Microcystin-LR concentration in aquatic food web compartments from lakes of varying trophic status. Can J Fish Aquat Sci 53(9):1974–1985. https://doi.org/10.1139/cjfas-53-9-1974

Kotaki Y, Koike K, Yoshida M, Van Thuoc C, Huyen NTM, Hoi NC, Fukuyo Y, Kodama M (2000) Domoic acid production in *Nitzschia* sp (Bacillariophyceae) isolated from a shrimp-culture pond in do son. Vietnam J Phycol 36(6):1057–1060. https://doi.org/10.1046/j.1529-8817.2000.99209.x

Kröger N, Poulsen N (2006) Biosilica nanofabrication in diatoms: the structures and properties of regulatory silaffins. MRS Online Proc Library 873(1):12. https://doi.org/10.1557/PROC-873-K1.2

Kubanek J, Hicks MK, Naar J, Villareal TA (2005) Does the red tide dinoflagellate *Karenia brevis* use allelopathy to outcompete other phytoplankton? Limnol Oceanogr 50(3):883–895. https://doi.org/10.4319/lo.2005.50.3.0883

Kuniyoshi TM, Gonzalez A, Lopez-Gomollon S, Valladares A, Bes MT, Fillat MF, Peleato ML (2011) 2-oxoglutarate enhances NtcA binding activity to promoter regions of the microcystin synthesis gene cluster. FEBS Lett 585(24):3921–3926. https://doi.org/10.1016/j.febslet.2011.10.034

Kurmayer R, Blom JF, Deng L, Pernthaler J (2014) Integrating phylogeny, geographic niche partitioning and secondary metabolite synthesis in bloom-forming *Planktothrix*. ISME J 9(4):909–921. https://doi.org/10.1038/ismej.2014.189

Le VV, Ko SR, Kang M, Park CY, Lee SA, Oh HM, Ahn CY (2022) The cyanobactericidal bacterium *Paucibacter aquatile* DH15 caused the decline of *Microcystis* and aquatic microbial community succession: a mesocosm study. Environ Pollut 311:119849. https://doi.org/10.1016/j.envpol.2022.119849

Leflaive J, Ten-Hage L (2009) Chemical interactions in diatoms: role of polyunsaturated aldehydes and precursors. New Phytol 184(4):794–805. https://doi.org/10.1111/j.1469-8137.2009.03033.x

Lelong A, Hégaret H, Soudant P, Bates SS (2012) *Pseudo-nitzschia* (Bacillariophyceae) species, domoic acid and amnesic shellfish poisoning: revisiting previous paradigms. Phycologia 51(2):168–216. https://doi.org/10.2216/11-37.1

Leunert F, Eckert W, Paul A, Gerhardt V, Grossart HP (2014) Phytoplankton response to UV-generated hydrogen peroxide from natural organic matter. J Plankton Res 36(1):185–197. https://doi.org/10.1093/plankt/fbt096

Leunert F, Grossart H-P, Gerhardt V, Eckert W (2013) Toxicant induced changes on delayed fluorescence decay kinetics of cyanobacteria and green algae: a rapid and sensitive biotest. PLoS One 8(4):e63127. https://doi.org/10.1371/journal.pone.0063127

Li B, Zhang X, Wu G, Qin B, Tefsen B, Wells M (2023) Toxins from harmful algal blooms: how copper and iron render chalkophore a predictor of microcystin production. Water Res 244:120490. https://doi.org/10.1016/j.watres.2023.120490

Li H, Barber M, Lu J, Goel R (2020) Microbial community successions and their dynamic functions during harmful cyanobacterial blooms in a freshwater lake. Water Res 185. https://doi.org/10.1016/j.watres.2020.116292

Li X, Yan T, Yu R, Zhou M (2019) A review of *Karenia mikimotoi*: bloom events, physiology, toxicity and toxic mechanism. Harmful Algae 90:101702. https://doi.org/10.1016/j.hal.2019.101702

Limborg MT, Alberdi A, Kodama M, Roggenbuck M, Kristiansen K, Gilbert MTP (2018) Applied hologenomics: feasibility and potential in aquaculture. Trends Biotechnol 36(3):252–264. https://doi.org/10.1016/j.tibtech.2017.12.006

Liu X, Bian Z, Hu S, Dickinson CF, Benjamin MM, Jia J, Tian Y, Place A, Hanna GS, Luesch H, Croot P, Reddy MM, Thomas OP, Hardiman G, Puglisi MP, Yang M, Zhong Z, Lemasters JJ, Korte JE, Waters AL, Heltzel CE, Williamson RT, Strangman WK, Valeriote F, Tius MA, DiTullio GR, Ferreira D, Alekseyenko A, Wang S, Hamann MT, Wang X (2024) The chemistry of phytoplankton. Chem Rev 124(23):13099–13177. https://doi.org/10.1021/acs.chemrev.4c00177

Long M, Peltekis A, González-Fernández C, Hégaret H, Bailleul B (2021) Allelochemicals of *Alexandrium minutum*: kinetics of membrane disruption and photosynthesis inhibition in a co-occurring diatom. Harmful Algae 103:101997. https://doi.org/10.1016/j.hal.2021.101997

Longo GO, Hay ME (2017) Seaweed allelopathy to corals: are active compounds on, or in, seaweeds? Coral Reefs 36(1):247–253. https://doi.org/10.1007/s00338-016-1526-9

Lundholm N, Hansen PJ, Kotaki Y (2005) Lack of allelopathic effects of the domoic acid-producing marine diatom *Pseudo-nitzschia multiseries*. Mar Ecol Prog Ser 288:21–33. https://doi.org/10.3354/meps288021

Lundholm N, Krock B, John U, Skov J, Cheng J, Pančić M, Wohlrab S, Rigby K, Nielsen TG, Selander E, Harðardóttir S (2018) Induction of domoic acid production in diatoms—types of grazers and diatoms are important. Harmful Algae 79:64–73. https://doi.org/10.1016/j.hal.2018.06.005

Lyck S, Gjølme N, Utkilen H (1996) Iron starvation increases toxicity of *Microcystis aeruginosa* CYA 228/1 (Chroococcales, Cyanophyceae). Phycologia 35 (Suppl):120–124. https://doi.org/10.2216/i0031-8884-35-6s-120.1

Maheswari U, Jabbari K, Petit JL, Porcel BM, Allen AE, Cadoret JP, De Martino A, Heijde M, Kaas R, La Roche J, Lopez PJ, Martin-Jézéquel V, Meichenin A, Mock T, Schnitzler Parker M, Vardi A, Armbrust EV, Weissenbach J, Katinka M, Bowler C (2010) Digital expression profiling of novel diatom transcripts provides insight into their biological functions. Genome Biol 11(8):R85. https://doi.org/10.1186/gb-2010-11-8-r85

Maibam C, Fink P, Romano G, Buia MC, Butera E, Zupo V (2015) *Centropages typicus* (Crustacea, Copepoda) reacts to volatile compounds produced by planktonic algae. Mar Ecol 36(3):819–834. https://doi.org/10.1111/maec.12254

Maibam C, Fink P, Romano G, Buia MC, Gambi MC, Scipione MB, Patti FP, Lorenti M, Butera E, Zupo V (2014) Relevance of wound-activated compounds produced by diatoms as toxins and infochemicals for benthic invertebrates. Mar Biol 161(7):1639–1652. https://doi.org/10.1007/s00227-014-2448-0

Mayali X, Azam F (2004) Algicidal bacteria in the sea and their impact on algal blooms. J Eukaryot Microbiol 51(2):139–144. https://doi.org/10.1111/j.1550-7408.2004.tb00538.x

McKindles KM, Manes MA, DeMarco JR, McClure A, McKay RM, Davis TW, Bullerjahn GS (2020) Dissolved microcystin release coincident with lysis of a bloom dominated by *Microcystis* spp. in western Lake Erie attributed to a novel cyanophage. Appl Environ Microbiol 86(22). https://doi.org/10.1128/aem.01397-20

Meyer KA, O'Neil JM, Hitchcock GL, Heil CA (2014) Microbial production along the West Florida shelf: responses of bacteria and viruses to the presence and phase of *Karenia brevis* blooms. Harmful Algae 38 (C):110-118. https://doi.org/10.1016/j.hal.2014.04.015

Meyer N, Pohnert G (2019) Isolate-specific resistance to the algicidal bacterium *Kordia algicida* in the diatom *Chaetoceros* genus. Bot Mar 62(6):527–535. https://doi.org/10.1515/bot-2019-0007

Meyer N, Rettner J, Werner M, Werz O, Pohnert G (2018) Algal oxylipins mediate the resistance of diatoms against algicidal bacteria. Mar Drugs 16(12):486. https://doi.org/10.3390/md16120486

Miesner AK, Lundholm N, Krock B, Nielsen TG (2016) The effect of *Pseudo-nitzschia seriata* on grazing and fecundity of *Calanus finmarchicus* and *Calanus glacialis*. J Plankton Res 38(3):564–574. https://doi.org/10.1093/plankt/fbw015

Miles CO, Wilkins AL, Stirling DJ, MacKenzie AL (2003) Gymnodimine C, an isomer of Gymnodimine B, from *Karenia selliformis*. J Agric Food Chem 51(16):4838–4840. https://doi.org/10.1021/jf030101r

Mitra A, Flynn KJ, Burkholder JM, Berge T, Calbet A, Raven JA, Granéli E, Glibert PM, Hansen PJ, Stoecker DK, Thingstad F, Tillmann U, Väge S, Wilken S, Zubkov MV (2014) The role of mixotrophic protists in the biological carbon pump. Biogeosciences 11(4):995–1005. https://doi.org/10.5194/bg-11-995-2014

Mitrovic SM, Fernández Amandi M, McKenzie L, Furey A, James KJ (2004) Effects of selenium, iron and cobalt addition to growth and yessotoxin production of the toxic marine dinoflagellate *Protoceratium reticulatum* in culture. J Exp Mar Biol Ecol 313(2):337–351. https://doi.org/10.1016/j.jembe.2004.08.014

Molgó J, Girard E, Benoit E (2007) Cyclic imines: an insight into this emerging Group of Bioactive Marine Toxins. In: Botana LM (ed) Phycotoxins: Chemistry and Biochemistry. Blackwell Publishing, pp 319–335. https://doi.org/10.1002/9780470277874.ch18

Mooney B, Dorantes-Aranda J, Place A, Hallegraeff G (2011) Ichthyotoxicity of gymnodinioid dinoflagellates: PUFA and superoxide effects in sheepshead minnow larvae and rainbow trout gill cells. Mar Ecol Prog Ser 426. https://doi.org/10.3354/meps09036

Mooney BD, Nichols PD, De Salas MF, Hallegraeff GM (2007) Lipid, fatty acid, and sterol composition of eight species of Kareniaceae (Dinophyta): chemotaxonomy and putative lipid phycotoxins. J Phycol 43(1):101–111. https://doi.org/10.1111/j.1529-8817.2006.00312.x

Morón-López J (2021) A holistic water monitoring approach for an effective ecosystem management. Ecohydrol Hydrobiol 21(3):549–554. https://doi.org/10.1016/j.ecohyd.2021.08.003

Motuhi S-E, Feizbakhsh O, Foll-Josselin B, Baratte B, Delehouzé C, Cousseau A, Fant X, Bulinski JC, Payri CE, Ruchaud S, Mehiri M, Bach S (2019) Neurymenolide A, a novel mitotic spindle poison from the new Caledonian Rhodophyta *Phacelocarpus neurymenioides*. Mar Drugs 17(2):93. https://doi.org/10.3390/md17020093

Mountfort D, Beuzenberg V, MacKenzie L, Rhodes L (2006) Enhancement of growth and gymnodimine production by the marine dinoflagellate, *Karenia selliformis*. Harmful Algae 5(6):658–664. https://doi.org/10.1016/j.hal.2006.02.001

Munday R, Towers NR, Mackenzie L, Beuzenberg V, Holland PT, Miles CO (2004) Acute toxicity of gymnodimine to mice. Toxicon 44(2):173–178. https://doi.org/10.1016/j.toxicon.2004.05.017

Nixdorf B, Fyson A, Krumbeck H (2001) Review: plant life in extremely acidic waters. Environ Exp Bot 46(3):203–211. https://doi.org/10.1016/S0098-8472(01)00104-6

Oh H-M, Lee SJ, Jang M-H, Yoon B-D (2000) Microcystin production by *Microcystis aeruginosa* in a phosphorus-limited chemostat. Appl Environ Microbiol 66(1):176–179. https://doi.org/10.1128/AEM.66.1.176-179.2000

Olson MB, Lessard EJ, Wong CHJ, Bernhardt MJ (2006) Copepod feeding selectivity on microplankton, including the toxigenic diatoms *pseudo-nitzschia* spp., in the coastal Pacific northwest. Mar Ecol Prog Ser 326:207–220. https://doi.org/10.3354/meps326207

Overstreet RM, Hawkins WE (2017) Diseases and mortalities of fishes and other animals in the Gulf of Mexico. In: Ward CH (ed) Habitats and biota of the Gulf of Mexico: before the Deepwater horizon oil spill - fish resources, fisheries, sea turtles, avian resources, marine mammals, diseases and mortalities, vol 2. SpringerOpen, pp 1589–1738. https://doi.org/10.1007/978-1-4939-3456-0_6

Paerl HW, Millie DF (1996) Physiological ecology of toxic aquatic cyanobacteria. Phycologia 35 (Suppl):160–167. https://doi.org/10.2216/i0031-8884-35-6s-160.1

Pal M, Yadav S, Kapley A, Qureshi A (2021) Impact of cyanobacterial bloom on microbiomes of freshwater lakes. J Biosci 46(4):96. https://doi.org/10.1007/s12038-021-00220-z

Pan Y, Subba Rao DV, Mann KH (1996) Changes in domoic acid production and cellular chemical composition of the toxigenic diatom *Pseudo-nitzschia multiseries* under phosphate limitation. J Phycol 32(3):371–381. https://doi.org/10.1111/j.0022-3646.1996.00371.x

Pančić M, Kiørboe T (2018) Phytoplankton defence mechanisms: traits and trade-offs. Biol Rev 93(2):1269–1303. https://doi.org/10.1111/brv.12395

Patin NV, Brown E, Chebli G, Garfield C, Kubanek J, Stewart FJ (2020) Microbial and chemical dynamics of a toxic dinoflagellate bloom. PeerJ 8. https://doi.org/10.7717/peerj.9493

Paz B, Vázquez JA, Riobó P, Franco JM (2006) Study of the effect of temperature, irradiance and salinity on growth and yessotoxin production by the dinoflagellate *Protoceratium reticulatum* in culture by using a kinetic and factorial approach. Mar Environ Res 62(4):286–300. https://doi.org/10.1016/j.marenvres.2006.04.066

Pednekar SM, Bates SS, Kerkar V, Matondkar SGP (2018) Environmental factors affecting the distribution of *Pseudo-nitzschia* in two monsoonal estuaries of Western India and effects of salinity on growth and domoic acid production by *P. pungens*. Estuar Coasts 41(5):1448–1462. https://doi.org/10.1007/s12237-018-0366-y

Perreault F, Matias MS, Oukarroum A, Matias WG, Popovic R (2012) Okadaic acid inhibits cell growth and photosynthetic electron transport in the alga *Dunaliella tertiolecta*. Sci Total Environ 414:198–204. https://doi.org/10.1016/j.scitotenv.2011.10.045

Pezzolesi L, Guerrini F, Ciminiello P, Dell'Aversano C, Iacovo ED, Fattorusso E, Forino M, Tartaglione L, Pistocchi R (2012) Influence of temperature and salinity on *Ostreopsis* cf. *ovata* growth and evaluation of toxin content through HR LC-MS and biological assays. Water Res 46 (1):82-92. https://doi.org/10.1016/j.watres.2011.10.029

Pichierri S, Accoroni S, Pezzolesi L, Guerrini F, Romagnoli T, Pistocchi R, Totti C (2017) Allelopathic effects of diatom filtrates on the toxic benthic dinoflagellate *Ostreopsis* cf. *ovata*. Mar Environ Res 131:116–122. https://doi.org/10.1016/j.marenvres.2017.09.016

Pichierri S, Pezzolesi L, Vanucci S, Totti C, Pistocchi R (2016) Inhibitory effect of polyunsaturated aldehydes (PUAs) on the growth of the toxic benthic dinoflagellate *Ostreopsis* cf. *ovata*. Aquat Toxicol 179:125–133. https://doi.org/10.1016/j.aquatox.2016.08.018

Pineda-Mendoza RM, Briones-Roblero CI, Gonzalez-Escobedo R, Rivera-Orduña FN, Martínez-Jerónimo F, Zúñiga G (2020) Seasonal changes in the bacterial community structure of three eutrophicated urban lakes in Mexico city, with emphasis on *Microcystis* spp. Toxicon 179:8–20. https://doi.org/10.1016/j.toxicon.2020.02.019

Pistocchi R (2016) Factors affecting algal toxicity. In: Rossini GP (ed) Toxins and biologically active compounds from microalgae, vol 1. CRC Press, Boca Raton, pp 75–96. https://doi.org/10.1201/b16569-9

Pohnert G, Boland W (2002) The oxylipin chemistry of attraction and defense in brown algae and diatoms. Nat Prod Rep 19(1):108–122. https://doi.org/10.1039/a806888g

Poulin RX, Hogan S, Poulson-Ellestad KL, Brown E, Fernández FM, Kubanek J (2018) *Karenia brevis* allelopathy compromises the lipidome, membrane integrity, and photosynthesis of competitors. Sci Rep 8(1). https://doi.org/10.1038/s41598-018-27845-9

Poulson-Ellestad K, McMillan E, Montoya JP, Kubanek J (2014) Are offshore phytoplankton susceptible to *Karenia brevis* allelopathy? J Plankton Res 36(5):1344–1356. https://doi.org/10.1093/plankt/fbu064

Poulson KL, Sieg RD, Prince EK, Kubanek J (2010) Allelopathic compounds of a red tide dinoflagellate have species-specific and context-dependent impacts on phytoplankton. Mar Ecol Prog Ser 416:69–78. https://doi.org/10.3354/meps08788

Prego-Faraldo MV, Valdiglesias V, Méndez J, Eirín-López JM (2013) Okadaic acid meet and greet: an insight into detection methods, response strategies and genotoxic effects in marine invertebrates. Mar Drugs 11(8):2829–2845. https://doi.org/10.3390/md11082829

Prince EK, Irmer F, Pohnert G (2013) Domoic acid improves the competitive ability of *Pseudo-nitzschia delicatissima* against the diatom *Skeletonema marinoi*. Mar Drugs 11(7):2398–2412. https://doi.org/10.3390/md11072398

Prince EK, Myers TL, Kubanek J (2008a) Effects of harmful algal blooms on competitors: Allelopathic mechanisms of the red tide dinoflagellate *Karenia brevis*. Limnol Oceanogr 53(2):531–541. https://doi.org/10.4319/lo.2008.53.2.0531

Prince EK, Myers TL, Naar J, Kubanek J (2008b) Competing phytoplankton undermines allelopathy of a bloom-forming dinoflagellate. Proc Royal Soc B Biol Sci 275(1652):2733–2741. https://doi.org/10.1098/rspb.2008.0760

Prince EK, Poulson KL, Myers TL, Sieg RD, Kubanek J (2010) Characterization of allelopathic compounds from the red tide dinoflagellate *Karenia brevis*. Harmful Algae 10(1):39–48. https://doi.org/10.1016/j.hal.2010.06.003

Rasher DB, Stout EP, Engel S, Kubanek J, Hay ME (2011) Macroalgal terpenes function as allelopathic agents against reef corals. Proc Natl Acad Sci USA 108(43):17726–17731. https://doi.org/10.1073/pnas.1108628108

Rastogi RP, Sinha RP, Moh SH, Lee TK, Kottuparambil S, Kim Y-J, Rhee J-S, Choi E-M, Brown MT, Häder D-P, Han T (2014) Ultraviolet radiation and cyanobacteria. J Photochem Photobio B 141:154–169. https://doi.org/10.1016/j.jphotobiol.2014.09.020

Raven JA (2010) Cyanotoxins: a poison that frees phosphate. Curr Biol 20(19):R850–R852. https://doi.org/10.1016/j.cub.2010.08.012

Reichwaldt ES, Song H, Ghadouani A (2013) Effects of the distribution of a toxic *Microcystis* bloom on the small scale patchiness of zooplankton. PLoS One 8(6):e66674. https://doi.org/10.1371/journal.pone.0066674

Remmel EJ, Hambright KD (2012) Toxin-assisted micropredation: experimental evidence shows that contact micropredation rather than exotoxicity is the role of *Prymnesium toxins*. Ecol Lett 15(2):126–132. https://doi.org/10.1111/j.1461-0248.2011.01718.x

Röder K, Hantzsche FM, Gebühr C, Miene C, Helbig T, Krock B, Hoppenrath M, Luckas B, Gerdts G (2012) Effects of salinity, temperature and nutrients on growth, cellular characteristics and yessotoxin production of *Protoceratium reticulatum*. Harmful Algae 15:59–70. https://doi.org/10.1016/j.hal.2011.11.006

Roussel A, Mériot V, Jauffrais T, Berteaux-Lecellier V, Lebouvier N (2023) OMICS approaches to assess dinoflagellate responses to chemical stressors. Biology 12(9):1234. https://doi.org/10.3390/biology12091234

Rue E, Bruland K (2001) Domoic acid binds iron and copper: a possible role for the toxin produced by the marine diatom *Pseudo-nitzschia*. Mar Chem 76(1):127–134. https://doi.org/10.1016/S0304-4203(01)00053-6

Rzymski P, Klimaszyk P, Jurczak T, Poniedziałek B (2020) Oxidative stress, programmed cell death and microcystin release in *Microcystis aeruginosa* in response to *Daphnia* grazers. Front Microbiol 11:1201. https://doi.org/10.3389/fmicb.2020.01201

Sabharwal T, Sathasivan K, Mehdy MC (2017) Defense related decadienal elicits membrane lipid remodeling in the diatom *Phaeodactylum tricornutum*. PLoS One 12(6):e0178761. https://doi.org/10.1371/journal.pone.0178761

Santos AA, Keim CN, Magalhães VF, Pacheco ABF (2022) Microcystin drives the composition of small-sized bacterioplankton communities from a coastal lagoon. Environ Sci Pollut Res. https://doi.org/10.1007/s11356-022-18613-4

Satake M, Honma D, Watanabe R, Oshima Y (2019) Alexandrolide, a diatom growth inhibitor isolated from the dinoflagellate *Alexandrium catenella*. Tetrahedron Lett 60(19):1341–1344. https://doi.org/10.1016/j.tetlet.2019.04.019

Satake M, Shoji M, Oshima Y, Naoki H, Fujita T, Yasumoto T (2002) Gymnocin-A, a cytotoxic polyether from the notorious red tide dinoflagellate, *Gymnodinium mikimotoi*. Tetrahedron Lett 43(33):5829–5832. https://doi.org/10.1016/S0040-4039(02)01171-1

Satake M, Tanaka Y, Ishikura Y, Oshima Y, Naoki H, Yasumoto T (2005) Gymnocin-B with the largest contiguous polyether rings from the red tide dinoflagellate, *Karenia* (formerly *Gymnodinium*) *mikimotoi*. Tetrahedron Lett 46(20):3537–3540. https://doi.org/10.1016/j.tetlet.2005.03.115

Schatz D, Keren Y, Hadas O, Carmeli S, Sukenik A, Kaplan A (2005) Ecological implications of the emergence of non-toxic subcultures from toxic *Microcystis* strains. Environ Microbiol 7(6):798–805. https://doi.org/10.1111/j.1462-2920.2005.00752.x

Schatz D, Keren Y, Vardi A, Sukenik A, Carmeli S, Börner T, Dittmann E, Kaplan A (2007) Towards clarification of the biological role of microcystins, a family of cyanobacterial toxins. Environ Microbiol 9(4):965–970. https://doi.org/10.1111/j.1462-2920.2006.01218.x

Scherer PI, Millard AD, Miller A, Schoen R, Raeder U, Geist J, Zwirglmaier K (2017) Temporal dynamics of the microbial community composition with a focus on toxic cyanobacteria and toxin presence during harmful algal blooms in two South German lakes. Front Microbiol 8 (Dec). https://doi.org/10.3389/fmicb.2017.02387

Schmidt KC, Jackrel SL, Smith DJ, Dick GJ, Denef VJ (2020) Genotype and host microbiome alter competitive interactions between *Microcystis aeruginosa* and *Chlorella sorokiniana*. Harmful Algae 99:101939. https://doi.org/10.1016/j.hal.2020.101939

Seki T, Satake M, Mackenzie L, Kaspar HF, Yasumoto T (1995) Gymnodimine, a new marine toxin of unprecedented structure isolated from New Zealand oysters and the dinoflagellate, *Gymnodinium* sp. Tetrahedron Lett 36(39):7093–7096. https://doi.org/10.1016/0040-4039(95)01434-J

Selander E, Kubanek J, Hamberg M, Andersson MX, Cervin G, Pavia H (2015) Predator lipids induce paralytic shellfish toxins in bloom-forming algae. Proc Natl Acad Sci USA 112(20):6395–6400. https://doi.org/10.1073/pnas.1420154112

Sevilla E, Martin-Luna B, Vela L, Bes MT, Fillat MF, Peleato ML (2008) Iron availability affects *mcyD* expression and microcystin-LR synthesis in *Microcystis aeruginosa* PCC7806. Environ Microbiol 10(10):2476–2483. https://doi.org/10.1111/j.1462-2920.2008.01663.x

Sha J, Song L, Liu C, Bai F, Zhang H, Gong Z, Li T, Liu J, Gao G, Jia Y (2025) The autotoxicity of dissolved organic matter from *Microcystis* may be a contributor to the decline of its bloom. Water Res 268:122710. https://doi.org/10.1016/j.watres.2024.122710

Shan K, Song L, Chen W, Li L, Liu L, Wu Y, Jia Y, Zhou Q, Peng L (2019) Analysis of environmental drivers influencing interspecific variations and associations among bloom-forming cyanobacteria in large, shallow eutrophic lakes. Harmful Algae 84:84–94. https://doi.org/10.1016/j.hal.2019.02.002

Shang L, Xu Y, Leaw CP, Lim PT, Wang J, Chen J, Deng Y, Hu Z, Tang YZ (2021) Potent allelopathy and non-PSTs, non-spirolides toxicity of the dinoflagellate *Alexandrium leei* to phytoplankton, finfish and zooplankton observed from laboratory bioassays. Sci Total Environ 780:146484. https://doi.org/10.1016/j.scitotenv.2021.146484

Sheng J, Malkiel E, Katz J, Adolf JE, Place AR (2010) A dinoflagellate exploits toxins to immobilize prey prior to ingestion. Proc Natl Acad Sci USA 107(5):2082–2087. https://doi.org/10.1073/pnas.0912254107

Silke J, O'Beirn F, Cronin M (2005) *Karenia*: an exceptional dinoflagellate bloom in western Irish waters, summer 2005. 21:44

Singh A, Thakur NL (2016) Significance of investigating allelopathic interactions of marine organisms in the discovery and development of cytotoxic compounds. Chem Biol Interact 243:135–147. https://doi.org/10.1016/j.cbi.2015.09.009

Singh A, Thakur NL (2021) Allelopathic interaction among rocky intertidal invertebrates: sponge *Cinachyrella* cf. *cavernosa* and Zooxanthellate zoanthids *Zoanthus sansibaricus*. Hydrobiologia 848 (19):4647–4659. https://doi.org/10.1007/s10750-021-04667-x

Sipler RE, McGuinness LR, Kirkpatrick GJ, Kerkhof LJ, Schofield OM (2014) Bacteriocidal effects of brevetoxin on natural microbial communities. Harmful Algae 38 (C):101–109. https://doi.org/10.1016/j.hal.2014.04.009

Sison-Mangus MP, Jiang S, Tran KN, Kudela RM (2014) Host-specific adaptation governs the interaction of the marine diatom, *pseudo-nitzschia* and their microbiota. ISME J 8(1):63–76. https://doi.org/10.1038/ismej.2013.138

Skovgaard A, Hansen PJ (2003) Food uptake in the harmful alga *Prymnesium parvum* mediated by excreted toxins. Limnol Oceanogr 48(3):1161–1166. https://doi.org/10.4319/lo.2003.48.3.1161

Śliwińska-Wilczewska S, Latała A (2018) Allelopathic activity of the bloom-forming picocyanobacterium *Synechococcus* sp. on the coexisting microalgae: the role of eutrophication. Intern Rev Hydrobiol 103 (3-4):37-47. https://doi.org/10.1002/iroh.201801940

Śliwińska-Wilczewska S, Maculewicz J, Barreiro Felpeto A, Latała A (2018) Allelopathic and bloom-forming picocyanobacteria in a changing world. Toxins 10(1):48. https://doi.org/10.3390/toxins10010048

Smayda TJ (1997) Harmful algal blooms: their ecophysiology and general relevance to phytoplankton blooms in the sea. Limnol Oceanogr 42 (5part2):1137-1153. https://doi.org/10.4319/lo.1997.42.5_part_2.1137

Sola F, Masoni A, Fossat B, Porthé-Nibelle J, Gentien P, Bodennec G (1999) Toxicity of fatty acid 18:5n3 from *Gymnodinium* cf. *mikimotoi*: I. morphological and biochemical aspects on *Dicentrarchus labrax* gills and intestine. J Appl Toxicol 19(4):279–284. https://doi.org/10.1002/(SICI)1099-1263(199907/08)19:4<279::AID-JAT579>3.0.CO;2-X

Song CU, Purnaningtyas DW, Choi H, Jeon D, Kim S, Hwang H, Kim CG, Lee YH, Eyun SI (2024) Do red tide events promote an increase in zooplankton biodiversity? Environ Pollut 361:124880. https://doi.org/10.1016/j.envpol.2024.124880

Sotton B, Guillard J, Bony S, Devaux A, Domaizon I, Givaudan N, Crespeau F, Huet H, Anneville O (2012) Impact of toxic cyanobacterial blooms on Eurasian perch (*Perca fluviatilis*): experimental study and in situ observations in a Peri-alpine lake. PLoS One 7(12):e52243. https://doi.org/10.1371/journal.pone.0052243

Stanier RY, Cohen-Bazire G (1977) Phototrophic prokaryotes: the cyanobacteria. Ann Rev Microbiol 31:225–274. https://doi.org/10.1146/annurev.mi.31.100177.001301

Steinberg CEW (2022a) Essential fatty acids—'*fueling versus controlling*'. In: Aquatic animal nutrition: organic macro- and micro-nutrients. Springer, Cham, pp 673–721. https://doi.org/10.1007/978-3-030-87227-4_27

Steinberg CEW (2022b) Lipids—'*the greasy, unhealthy stuff*'. In: Aquatic animal nutrition: organic macro- and micro-nutrients. Springer International Publishing, Cham, pp 531–582. https://doi.org/10.1007/978-3-030-87227-4_23

Steiner K, Wood SA, Puddick J, Hawes I, Dietrich DR, Hamilton DP (2017) A comparison of bacterial community structure, activity and microcystins associated with formation and breakdown of a cyanobacterial scum. Aquat Microb Ecol 80(3):243–256. https://doi.org/10.3354/ame01852

Stolte W, Riegman R (1995) Effect of phytoplankton cell size on transient-state nitrate and ammonium uptake kinetics. Microbiology 141(5):1221–1229. https://doi.org/10.1099/13500872-141-5-1221

Su X, Steinman AD, Tang X, Xue Q, Zhao Y, Xie L (2017) Response of bacterial communities to cyanobacterial harmful algal blooms in Lake Taihu, China. Harmful Algae 68:168–177. https://doi.org/10.1016/j.hal.2017.08.007

Sugg LM, Van Dolah FM (1999) No evidence for an allelopathic role of okadaic acid among ciguatera-associated dinoflagellates. J Phycol 35(1):93–103. https://doi.org/10.1046/j.1529-8817.1999.3510093.x

Sunesen I, Sar EA (2007) Marine diatoms from Buenos Aires coastal waters (Argentina). Ill potentially harmful genus *Asterionellopsis*, *Cerataulina*, *Ceratoneis* y *Leptocylindrus* (in Spanish). Rev Chil Hist Nat 80(4):493–507

Tammilehto A, Nielsen TG, Krock B, Møller EF, Lundholm N (2015) Induction of domoic acid production in the toxic diatom *Pseudo-nitzschia seriata* by calanoid copepods. Aquat Toxicol 159:52–61. https://doi.org/10.1016/j.aquatox.2014.11.026

Tan LT (2023) Impact of marine chemical ecology research on the discovery and development of new pharmaceuticals. Mar Drugs 21(3):174. https://doi.org/10.3390/md21030174

Teneva I, Velikova V, Belkinova D, Moten D, Dzhambazov B (2023) Allelopathic potential of the cyanotoxins microcystin-LR and cylindrospermopsin on green algae. Plants 12(6):1403. https://doi.org/10.3390/plants12061403

Ternon E, Pavaux AS, Marro S, Thomas OP, Lemée R (2018) Allelopathic interactions between the benthic toxic dinoflagellate *Ostreopsis* cf. *ovata* and a co-occurring diatom. Harmful Algae 75:35–44. https://doi.org/10.1016/j.hal.2018.04.003

Ternon E, Pavaux AS, Peltekis A, Gemin MP, Jauzein C, Bailleul B, Lemée R, Thomas OP (2022) Assessment of the allelochemical activity of *Ostreopsis* cf. *ovata* and the ovatoxins towards competitive benthic microalgae. Aquat Ecol 56 (2):475–491. https://doi.org/10.1007/s10452-022-09953-x

Terseleer N, Gypens N, Lancelot C (2013) Factors controlling the production of domoic acid by *pseudo-nitzschia* (Bacillariophyceae): a model study. Harmful Algae 24:45–53. https://doi.org/10.1016/j.hal.2013.01.004

Thorel M, Claquin P, Schapira M, Le Gendre R, Riou P, Goux D, Le Roy B, Raimbault V, Deton-Cabanillas A-F, Bazin P, Kientz-Bouchart V, Fauchot J (2017) Nutrient ratios influence variability in *pseudo-nitzschia* species diversity and particulate domoic acid production in the bay of seine (France). Harmful Algae 68:192–205. https://doi.org/10.1016/j.hal.2017.07.005

Trainer VL, Hickey BM, Lessard EJ, Cochlan WP, Trick CG, Wells ML, MacFadyen A, Moore SK (2009a) Variability of *pseudo-nitzschia* and domoic acid in the Juan de Fuca eddy region and its adjacent shelves. Limnol Oceanogr 54(1):289–308. https://doi.org/10.4319/lo.2009.54.1.0289

Trainer VL, Wells ML, Cochlan WP, Trick CG, Bill BD, Baugh KA, Beall BF, Herndon J, Lundholmf N (2009b) An ecological study of a massive bloom of toxigenic *Pseudo-nitzschia cuspidata* off the Washington State coast. Limnol Oceanogr 54(5):1461–1474. https://doi.org/10.4319/lo.2009.54.5.1461

Trick CG, Bill BD, Cochlan WP, Wells ML, Trainer VL, Pickell LD (2010) Iron enrichment stimulates toxic diatom production in high-nitrate, low-chlorophyll areas. Proc Natl Acad Sci USA 107(13):5887–5892. https://doi.org/10.1073/pnas.0910579107

Utkilen H, Gjølme N (1995) Iron-stimulated toxin production in *Microcystis aeruginosa*. Appl Environ Microbiol 61(2):797–800. https://doi.org/10.1128/aem.61.2.797-800.1995

Van Gremberghe I, Vanormelingen P, Van der Gucht K, Mancheva A, D'Hondt S, De Meester L, Vyverman W (2009) Influence of *daphnia* infochemicals on functional traits of *Microcystis* strains (*cyanobacteria*). Hydrobiologia 635(1):147–155. https://doi.org/10.1007/s10750-009-9907-5

Vanderploeg HA, Johengen TH, Liebig JR (2009) Feedback between zebra mussel selective feeding and algal composition affects mussel condition: did the regime changer pay a price for its success? Freshw Biol 54(1):47–63. https://doi.org/10.1111/j.1365-2427.2008.02091.x

Vanucci S, Guerrini F, Milandri A, Pistocchi R (2010) Effects of different levels of N- and P-deficiency on cell yield, okadaic acid, DTX-1, protein and carbohydrate dynamics in the benthic dinoflagellate *Prorocentrum lima*. Harmful Algae 9(6):590–599. https://doi.org/10.1016/j.hal.2010.04.009

Vardi A, Formiggini F, Casotti R, De Martino A, Ribalet F, Miralto A, Bowler C (2006) A stress surveillance system based on calcium and nitric oxide in marine diatoms. PLoS Biol 4(3):0411–0419. https://doi.org/10.1371/journal.pbio.0040060

Vieira C, Thomas OP, Culioli G, Genta-Jouve G, Houlbreque F, Gaubert J, De Clerck O, Payri CE (2016) Allelopathic interactions between the brown algal genus *Lobophora* (Dictyotales, Phaeophyceae) and scleractinian corals. Sci Rep 6(1):18637. https://doi.org/10.1038/srep18637

Vilar MCP, Rodrigues TFCP, Silva LO, Pacheco ABF, Ferrão-Filho AS, Azevedo SMFO (2021) Ecophysiological aspects and *sxt* genes expression underlying induced chemical defense in STX-producing *Raphidiopsis raciborskii* (cyanobacteria) against the zooplankter *Daphnia gessneri*. Toxins 13(6):406. https://doi.org/10.3390/toxins13060406

Visser PM, Ibelings BW, Mur LR (1995) Autunmal sedimentation of *Microcystis* spp. as result of an increase in carbohydrate ballast at reduced temperature. J Plankton Res 17 (5):919-933. https://doi.org/10.1093/plankt/17.5.919

Wang K, Razzano M, Mou X (2020) Cyanobacterial blooms alter the relative importance of neutral and selective processes in assembling freshwater bacterioplankton community. Sci Total Environ 706. https://doi.org/10.1016/j.scitotenv.2019.135724

Wang L, Zi J, Xu R, Hilt S, Hou X, Chang X (2017) Allelopathic effects of *Microcystis aeruginosa* on green algae and a diatom: evidence from exudates addition and co-culturing. Harmful Algae 61:56–62. https://doi.org/10.1016/j.hal.2016.11.010

Wasternack C (2007) Jasmonates: an update on biosynthesis, signal transduction and action in plant stress response, growth and development. Ann Bot 100(4):681–697. https://doi.org/10.1093/aob/mcm079

Watrous JD, Niiranen TJ, Lagerborg KA, Henglin M, Xu Y-J, Rong J, Sharma S, Vasan RS, Larson MG, Armando A, Mora S, Quehenberger O, Dennis EA, Cheng S, Jain M (2019) Directed non-targeted mass spectrometry and chemical networking for discovery of eicosanoids and related oxylipins. Cell Chem Biol 26(3):433–442.e434. https://doi.org/10.1016/j.chembiol.2018.11.015

Wei N, Hu C, Dittmann E, Song L, Gan N (2024) The biological functions of microcystins. Water Res 262:122119. https://doi.org/10.1016/j.watres.2024.122119

Wejnerowski L, Cerbin S, Dziuba MK (2015) Thicker filaments of *Aphanizomenon gracile* are more harmful to *Daphnia* than thinner *Cylindrospermopsis raciborskii*. Zool Stud 54 (Jan). https://doi.org/10.1186/s40555-014-0084-5

Welker M, Šejnohová L, Némethová D, von Döhren H, Jarkovský J, Maršálek B (2007) Seasonal shifts in chemotype composition of *Microcystis* sp. communities in the pelagial and the sediment of a shallow reservoir. Limnol Oceanogr 52(2):609–619

Wells ML, Trainer VL, Smayda TJ, Karlson BS, Trick CG, Kudela RM, Ishikawa A, Bernard S, Wulff A, Anderson DM, Cochlan WP (2015) Harmful algal blooms and climate change: learning from the past and present to forecast the future. Harmful Algae 49:68–93. https://doi.org/10.1016/j.hal.2015.07.009

Wilken S, Wiezer S, Huisman J, van Donk E (2010) Microcystins do not provide anti-herbivore defence against mixotrophic flagellates. Aquat Microb Ecol 59(3):207–216. https://doi.org/10.3354/ame01395

Windust AJ, Quilliam MA, Wright JLC, McLachlan JL (1997) Comparative toxicity of the diarrhetic shellfish poisons, okadaic acid, okadaic acid diol-ester and dinophysistoxin-4, to the diatom *Thalassiosira weissflogii*. Toxicon 35(11):1591–1603. https://doi.org/10.1016/S0041-0101(97)00047-0

Wolfram S, Wielsch N, Hupfer Y, Mönch B, Lu-Walther H-W, Heintzmann R, Werz O, Svatoš A, Pohnert G (2015) A metabolic probe-enabled strategy reveals uptake and protein targets of polyunsaturated aldehydes in the diatom *Phaeodactylum tricornutum*. PLoS One 10(10):e0140927. https://doi.org/10.1371/journal.pone.0140927

Würsig B (2017) Marine mammals of the Gulf of Mexico. In: Ward CH (ed) habitats and biota of the Gulf of Mexico: before the Deepwater horizon oil spill – fish resources, fisheries, sea turtles, avian resources, marine mammals, diseases and mortalities, vol 2. SpringerOpen, pp 1487–1588. https://doi.org/10.1007/978-1-4939-3456-0_5

Xie L, Rediske RR, Gillett ND, O'Keefe JP, Scull B, Xue Q (2016) The impact of environmental parameters on microcystin production in dialysis bag experiments. Sci Rep 6(1):38722. https://doi.org/10.1038/srep38722

Xu H, Zhao D, Zeng J, Jiao C, Yu Z, Wu QL (2020a) Distinct successional patterns and processes of free-living and particle-attached bacterial communities throughout a phytoplankton bloom. Freshw Biol 65(8):1363–1375. https://doi.org/10.1111/fwb.13505

Xu W, Tan L, Guo X, Wang J (2020b) Isolation of anti-algal substances from *Cylindrotheca closterium* and their inhibition activity on bloom-forming *Prorocentrum donghaiense*. Ecotox Environ Saf 190:110180. https://doi.org/10.1016/j.ecoenv.2020.110180

Yang I, Selander E, Pavia H, John U (2011) Grazer-induced toxin formation in dinoflagellates: a transcriptomic model study. Eur J Phycol 46 (1):66-73.Pii:934548314. https://doi.org/10.1080/09670262.2011.552194

Yang Y, Chen J, Chen X, Jiang Q, Liu Y, Xie S (2021) Cyanobacterial bloom induces structural and functional succession of microbial communities in eutrophic lake sediments. Environ Pollut 284. https://doi.org/10.1016/j.envpol.2021.117157

Yeung ACY, D'Agostino PM, Poljak A, McDonald J, Bligh MW, Waite TD, Neilan BA (2016) Physiological and proteomic responses of continuous cultures of *Microcystis aeruginosa* PCC 7806 to changes in iron bioavailability and growth rate. Appl Environ Microbiol 82(19):5918–5929. https://doi.org/10.1128/AEM.01207-16

Yoshida T, Takashima Y, Tomaru Y, Shirai Y, Takao Y, Hiroishi S, Nagasaki K (2006) Isolation and characterization of a cyanophage infecting the toxic cyanobacterium *Microcystis aeruginosa*. Appl Environ Microbiol 72(2):1239–1247. https://doi.org/10.1128/AEM.72.2.1239-1247.2006

Zabaglo K, Chrapusta E, Bober B, Kaminski A, Adamski M, Bialczyk J (2016) Environmental roles and biological activity of domoic acid: a review. Algal Res 13:94–101. https://doi.org/10.1016/j.algal.2015.11.020

Żak A, Kosakowska A (2015) The influence of extracellular compounds produced by selected Baltic cyanobacteria, diatoms and dinoflagellates on growth of green algae *Chlorella vulgaris*. Estuar Coast Shelf Sci 167:113–118. https://doi.org/10.1016/j.ecss.2015.07.038

Zhang H, Jia J, Chen S, Huang T, Wang Y, Zhao Z, Feng J, Hao H, Li S, Ma X (2018) Dynamics of bacterial and fungal communities during the outbreak and decline of an algal bloom in a drinking water reservoir. Int J Environ Res Publ Hlth 15(2). https://doi.org/10.3390/ijerph15020361

Zhang M, Lu T, Paerl HW, Chen Y, Zhang Z, Zhou Z, Qian H (2019) Feedback regulation between aquatic microorganisms and the bloom-forming cyanobacterium *Microcystis aeruginosa*. Appl Environ Microbiol 85:(21).e01362-19. https://doi.org/10.1128/AEM.01362-19

Zhang N, Guo X, Han T, Tan L, Xue Q, Wang J (2022) Isolation and identification of allelochemicals produced by *Phaeodactylum tricornutum* for *Prorocentrum donghaiense*. Algal Res 66:102808. https://doi.org/10.1016/j.algal.2022.102808

Zhang Q, Zhang Z, Lu T, Peijnenburg WJGM, Gillings M, Yang X, Chen J, Penuelas J, Zhu YG, Zhou NY, Su J, Qian H (2020) Cyanobacterial blooms contribute to the diversity of antibiotic-resistance genes in aquatic ecosystems. Commun Biol 3(1):737. https://doi.org/10.1038/s42003-020-01468-1

Zhang S, Zheng T, Lundholm N, Huang X, Jiang X, Li A, Li Y (2021) Chemical and morphological defenses of *Pseudo-nitzschia* multiseries in response to zooplankton grazing. Harmful Algae 104:102033. https://doi.org/10.1016/j.hal.2021.102033

Zhang W, Jeppesen E, Wang M, Xu X, Wang L (2017) Allelopathic effect boosts *Chrysosporum ovalisporum* dominance in summer at the expense of *Microcystis panniformis* in a shallow coastal water body. Environ Sci Pollut Res 24(5):4666–4675. https://doi.org/10.1007/s11356-016-8149-0

Zhang Y, Jiang H-B, Qiu B-S (2013) Effects of UVB radiation on competition between the bloom-forming cyanobacterium Microcystis aeruginosa and the Chlorophyceae Chlamydomonas microsphaera1. J Phycol 49(2):318–328. https://doi.org/10.1111/jpy.12038

Zhao H, Zhao X, Liu Y, Sun Y-F, Chen J, Liu Y, Miao X-Y, Wang H-J, Xu C, Liu Q, Yang W-Y, Steinberg CEW, MacIsaac HJ, Jeppesen E, Xie P, Paerl HW (2025) Synergistic effects of irradiation, nutrients, and cyanobacterial composition on microcystin production in Chinese Plateau Lakes. J Hazard Mater. under review

Zheng J-W, He G-H, Xu R-X, Wang X, Li H-Y, Yang W-D (2023) Systematic exploration of transcriptional responses of interspecies interaction between *Karenia mikimotoi* and *Prorocentrum shikokuense*. Harmful Algae 126:102441. https://doi.org/10.1016/j.hal.2023.102441

Zhu C, Zhang J, Nawaz MZ, Mahboob S, Al-Ghanim KA, Khan IA, Lu Z, Chen T (2019) Seasonal succession and spatial distribution of bacterial community structure in a eutrophic freshwater Lake, Lake Taihu. Sci Total Environ 669:29–40. https://doi.org/10.1016/j.scitotenv.2019.03.087

Zou Y, Wei XM, Weng HW, Li HY, Liu JS, Yang WD (2015) Expression profile of eight glutathione *S*-transferase genes in *Crassostrea ariakensis* after exposure to DSP toxins producing dinoflagellate *Prorocentrum lima*. Toxicon 105:45–55. https://doi.org/10.1016/j.toxicon.2015.08.018

Zou Y, Yamasaki Y, Matsuyama Y, Yamaguchi K, Honjo T, Oda T (2010) Possible involvement of hemolytic activity in the contact-dependent lethal effects of the dinoflagellate *Karenia mikimotoi* on the rotifer *Brachionus plicatilis*. Harmful Algae 9(4):367–373. https://doi.org/10.1016/j.hal.2010.01.005

Zuo J, Hu L, Shen W, Zeng J, Li L, Song L, Gan N (2021) The involvement of α-proteobacteria *Phenylobacterium* in maintaining the dominance of toxic *Microcystis* blooms in Lake Taihu. China Environ Microbiol 23(2):1066–1078. https://doi.org/10.1111/1462-2920.15301

The Tolerance Modes to Phycotoxins

5

It is a fascinating evolutionary arms race in which organisms producing potent toxins to deter predators and outcompete rivals are met by equally ingenious adaptations in their opponents to neutralize these biochemical weapons. This dynamic interplay is beautifully encapsulated by the seminal concept of the "Red Queen Hypothesis"[1] in evolutionary biology. According to this influential idea, species must perpetually evolve, adapt, and reproduce merely to maintain their position in an ever-shifting landscape of biotic conflict (Van Valen 1973; Solé 2022). This relentless cycle of adaptation underscores the remarkable resilience and creativity of life.

Across the tree of life—from the simplest invertebrates to complex vertebrates—individuals, populations, and entire species have evolved a diverse and sophisticated repertoire of strategies to tolerate or even resist the harmful effects of natural toxins. Some organisms merely tolerate sublethal doses, while others have developed biochemical or behavioral resistance that allows them to thrive in environments where toxins are pervasive. These adaptive mechanisms reveal an astonishing evolutionary creativity. Yet, they also highlight a striking gap in our current perspective: most research has concentrated narrowly on resistance to cyanotoxins, while overlooking the broader spectrum of natural toxin interactions across ecosystems.

Expanding our view to include the full diversity of tolerance and resistance strategies not only enriches our understanding of ecological and evolutionary dynamics but also carries practical implications. By learning from nature's own solutions, we may discover innovative ways to reduce the ecological risks of harmful blooms or even repurpose toxin-producing organisms as valuable resources. After all,

[1] Species must "run" or evolve in order to stay in the same place, or else go extinct, as the Red Queen said to Alice in Lewis Carroll's *Through the Looking-Glass* when explaining the nature of the Looking-Glass Land: "*Now, here, you see, it takes all the running you can do to stay in the same place.*" In biology, this hypothesis specifically states that the adaptive importance of genetic recombination (and sex) lies in its ability to generate genetic variation among offspring, which is important in the face of rapidly evolving environmental conditions, including coevolving parasites (Decaestecker and King 2018).

T. Lieke, C. E. W. Steinberg, *The Ecology of Phycotoxins*,
https://doi.org/10.1007/978-3-032-18380-4_5

cyanobacteria and other toxigenic algae are not just producers of dangerous metabolites—they are also rich reservoirs of carbon and nutrients. Understanding how their toxins can be tolerated, neutralized, or even harnessed opens the possibility of transforming what was once viewed solely as an ecological threat into a potential asset.

Some examples will shed light on this issue and show that current studies on resistance to cyanotoxins may be too narrow. The ultimate goal of this compilation is to find practical ways to utilize the carbon and nutrient content of cyanobacteria despite their toxins.

5.1 General Mechanisms

Before we can explore how animals across different trophic levels tolerate—or even thrive amid—phycotoxins, we first need to step back and look at the bigger picture: how these potent natural compounds move through aquatic ecosystems, accumulate in tissues, and shape the lives of the organisms that encounter them. Cyanotoxins, the most extensively studied group, provide a particularly illuminating example.

These toxins, produced by cyanobacteria, can enter food webs through multiple pathways—ingested directly by grazers or passed along through contaminated prey. As they move upward through trophic levels, they may accumulate or transform, often revealing unexpected patterns of resistance and vulnerability. Why can one species of zooplankton graze on toxic *Microcystis* with little consequence, while another collapses under its effects? Why do some fish tolerate high cyanotoxin levels, while others show signs of oxidative stress and liver damage? These questions remind us that toxin exposure is not simply a matter of dose—it's a story of adaptation, energy allocation, and survival.

Tolerating toxins is never free. Every detoxification enzyme synthesized, every transporter protein activated, and every repair pathway engaged diverts precious energy from other life functions such as growth, reproduction, or immunity. In this sense, tolerance is as much an evolutionary gamble as it is a physiological feat—a balancing act between resilience and cost.

In the following sections, we will unravel these dynamics step by step. We begin with the general processes of toxin accumulation and biomagnification, setting the stage for understanding how tolerance evolves and manifests. From there, we'll move through the astonishing diversity of tolerance strategies—from the selective feeding behaviors of invertebrates to the molecular adaptations and epigenetic plasticity of fish—each revealing a different facet of life in toxin-rich waters.

5.1.1 Selective Feeding

Comprehensive studies on feeding behavior were already carried out decades ago with species such as *Karenia*. For example, experiments by Collumb and Buskey (2004) showed that the copepod *Acartia tonsa* actively avoids ingesting *Karenia brevis*. Interestingly, this selective feeding behavior was not driven by brevetoxin or direct toxicity, but rather by a preference to exclude *K. brevis* from its diet.

Breier and Buskey (2007) argued that *K. brevis* is nutritionally deficient for *A. tonsa*, leading to reduced grazing rates and reproduction. The same observations were made in two species of rotifers given the opportunity to feed on *K. brevis* (Kubanek et al. 2007). In particular, *Brachionus ibericus* did not ingest *K. brevis* as a sole diet but tolerated *K. brevis* in a diet mixed with *Rhodomonas lens*, whereas the allopatric rotifer, *Brachionus plicatilis*, rejected *K. brevis* in either diet. The deterrence of *K. brevis* was mediated by its cellular organic extracts, rather than by exudates, with compounds other than brevetoxins implicated in this process. The finding that closely related rotifers respond differently to *K. brevis* and to its chemical defense suggests that a shared evolutionary history may have led *B. ibericus* to tolerate *K. brevis* in a mixed diet.

Huntley et al. (1986) and Sykes and Huntley (1987) reported that two copepod species, *Calanus pacificus* and *Paracalanus parvus*, refused to ingest *K. brevis*. Yet, even without direct consumption, they were still affected: toxins dissolved in the surrounding water triggered an accelerated heart rate and a loss of muscular control. Turner and Tester (1989) discovered that three copepod species, *Acartia tonsa*, *Oncaea venusta venusta*, and *Labidocera aestiva*, which naturally co-occur with *K. brevis*, readily consumed the toxic dinoflagellate without showing any obvious adverse effects. Contrary to that, the copepod species *Centropages typicus* and *Paracalanus quasimodo*, would not ingest *K. brevis* (from Brand et al. (2012), appropriately referenced).

Cohen, Tester et al. (2007) further showed that copepod species accumulate brevetoxins from *K. brevis* cells and from dissolved toxin exposure in a species-specific mann.

More recent research supports these findings and extends them. Prince et al. (2006) experimentally confirmed that the reduction in *A. tonsa* grazing on *K. brevis* is primarily due to nutritional inadequacy, as copepods increased ingestion rates to compensate for the poor food quality but still showed depressed fitness. Hong et al. (2012) employed digital holographic cinematography to quantify *A. tonsa* feeding behavior on toxic and non-toxic strains of *K. brevis* and *Karlodinium veneficum*, revealing aversion behaviors such as reduced grazing on toxic diets, highlighting the neurological effects of brevetoxins on feeding activity. This work illuminated behavioral mechanisms underlying reduced grazing beyond simple toxin avoidance. Furthermore, Kühn et al. (2023) demonstrated that environmental stressors, such as harbor noise, exacerbate feeding inhibition in *A. tonsa* exposed to toxic algae, revealing additional ecological contexts influencing copepod-algal interactions.

These phenotypic variations in copepod feeding and physiological responses underline the complexity of interactions between toxic phytoplankton and grazers. Deciphering the underlying mechanisms requires integrated studies that encompass genetic, epigenetic, proteomic, and metabolomic analyses to understand species-specific adaptation, toxin processing, and ecological outcomes in harmful algal bloom dynamics.

5.1.2 Accumulation and Biomagnification

Herbivorous and omnivorous finfish, as well as a variety of aquatic invertebrates, inevitably ingest phytoestrogens, phytotoxins, mycotoxins, phycotoxins and cyanotoxins when feeding on toxic plant or fungal material. This has been clearly demonstrated for microcystins (MCs), the most widespread and extensively studied cyanotoxins. The edible clam (*Corbicula leana*) accumulated MCs (up to 12.7 ± 2.5 μg g^{-1} dry weight (DW) of free MC and 4.2 ± 0.6 μg g^{-1} DW of covalently bound MC). The estimated MC concentration in *C. leana* exceeded the WHO's provisional tolerable daily intake of 0.04 μg kg^{-1} day^{-1} by far. But does this burden automatically transfer to higher trophic levels? And if so, under what circumstances? As illustrated in Fig. 5.1, maintaining such high MC concentrations requires continuous external uptake, since metabolism—oxidation by cytochrome P450 enzymes and particularly conjugation to GST—rapidly modifies and detoxifies the compounds, resulting in fast depuration of MCs and their metabolites via exporters (Pflugmacher et al. 1998; Humpage et al. 2005). In fact, using meta--analyses, Kozlowsky-Suzuki et al. (2012) demonstrated that MC biodilution is the prevailing process in aquatic food webs, consistent across groups of aquatic consumers, except for zooplankton and zooplanktivorous fish. Overall, this example clearly indicates that, for natural xenobiotics to reach higher trophic levels, they must not be subject to metabolism or depuration.

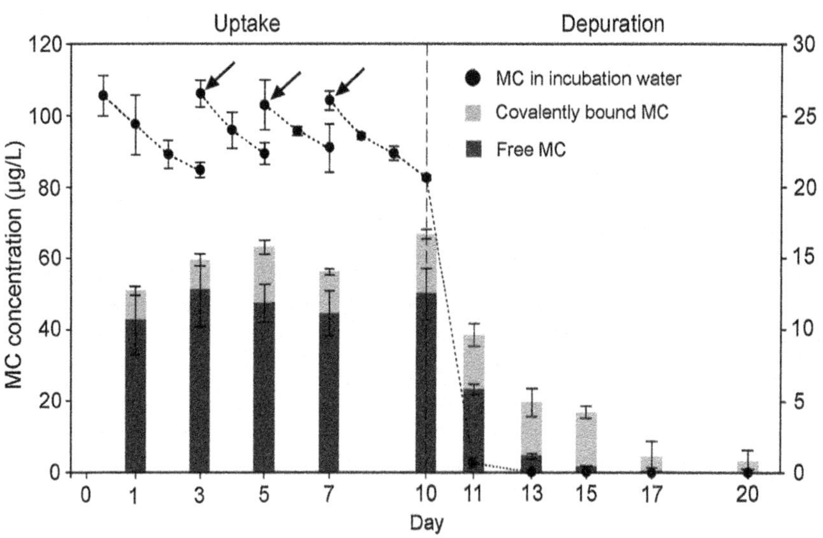

Fig. 5.1 Concentrations of microcystins (MCs) in incubation water, and of free MC, and covalently bound MC (Co-MC) accumulated in clams (*Corbicula leana*) during the uptake and depuration periods. Arrows indicate points at which the MC concentration and the water were renewed during the uptake period (from Pham et al. (2015), credit to Elsevier)

For organisms that do not feed directly on cyanobacteria or other toxic algae and fungi, the presence of toxins in their tissues can only be explained by transfer through the food web, where toxins are passed via prey–predator interactions, provided that the direct uptake of toxins from the surrounding environment is negligible. Of the toxins ingested with the food, only a small fraction may actually be absorbed into the body, while the rest is either excreted or metabolically transformed. Many studies have shown that cyanobacterial toxins bioaccumulate in aquatic biota, which may enhance the risk of exposure to biota higher up in the food web (biomagnification), as even small amounts produced by phytoplankton can accumulate to higher concentrations in aquatic organisms. Unequivocal reports on bioaccumulation and the related processes of bioconcentration and biomagnification of microcystins and other cyanotoxins in aquatic food webs are scarce (Ibelings and Havens 2008).

Using stable isotope analysis of carbon and nitrogen, Murby McQuaid (2019) assessed the roles of major plankton groups and zooplankton feeding behaviors in toxin accumulation. The results showed that cyanotoxin bioaccumulation in zooplankton depends on trophic level, species-specific feeding habits, seasonality, and the presence of toxic cyanobacteria, including small-sized picocyanobacteria. Flores et al. (2018) found a positive correlation between intracellular microcystin concentrations in water and microcystin levels in fish tissues, with omnivorous fish generally exhibiting higher toxin concentrations than carnivorous or planktivorous fish. However, results varied by tissue type and analytical method, underscoring the need for standardized and comprehensive toxin detection approaches to accurately assess cyanotoxin exposure risks through fish consumption.

Therefore, it remains uncertain whether cyanotoxins and other phycotoxins possess the physicochemical properties necessary to undergo significant biomagnification in food webs. Correspondingly, Walters et al. (2016) summarize these basic properties as (i) slow metabolization by animals and (ii) moderate hydrophobicity, that is, specifically, relatively high lipophilicity of log K_{OW} 6–8. However, low lipophilicity (high water solubility) applies to, e.g., microcystin-LR (2.18 at pH = 1 to −1.76 at pH = 10), anatoxin-a (1.12), domoic acid (−2.43), or saxitoxin (2.99) (Falk et al. 1991; De Maagd et al. 1999; Chan et al. 2005). Glidewell et al. (2025) reported that microcystins accumulate in aquatic organisms but do not biomagnify up the food chain, meaning toxin concentrations do not increase progressively at higher trophic levels. This suggests that while organisms at multiple trophic levels may be exposed to microcystins, the risks of amplified toxin concentrations at top predators may be limited. The findings underscore the significance of understanding species-specific accumulation and detoxification processes in assessing the ecological and health risks associated with cyanotoxins in freshwater ecosystems. Furthermore, several of these phycotoxins are subject to more or less intensive xenobiotic metabolism (Fig. 5.1). In sum, it is very unlikely that phycotoxins will sustainably biomagnify sensu stricto within food webs, as found for MCs in lake food webs (Ibelings et al. 2005). Nevertheless, very high toxin concentrations, far above the levels of daily allowance for humans, can be reached, and even lethal intoxication of wildlife can occur (Miller et al. 2010).

From a comprehensive literature review, Lassus et al. (2014) provide several general principles for most phycotoxins:

1. These toxins have an inhibitory effect on the feeding behavior of several bivalve species, particularly their clearance activity.
2. Contamination levels and bioaccumulation rate are influenced by toxin concentration per cell, algal strain geographical origin, bloom duration, and the relative proportion of non-toxic algae.
3. Bioaccumulation rates vary according to shellfish species, and therefore, the concept of sentinel species cannot be standardized. For example, although the blue mussel is a good indicator of PSP and DSP toxin contamination, this is not the case for domoic acid.
4. The digestive compartment appears to be a target organ, with some exceptions, especially for domoic acid.
5. The food supply improves detoxification performance, while temperature does not appear to play a significant role.
6. Zooplankton feeding on toxic phytoplankton should be considered an important source/entry point of toxicity/toxins into the marine aquatic food web, although biomagnification sensu stricto is only likely for highly lipophilic compounds.

In summary, while phycotoxins and cyanotoxins can bioaccumulate in aquatic organisms through trophic transfer, evidence for true biomagnification—where toxin concentrations increase substantially up the food web—is limited and toxin--specific. Many cyanotoxins, such as MCs, exhibit relatively low lipophilicity and are subject to metabolic transformation and excretion, which generally prevents sustained biomagnification in aquatic food webs. Nonetheless, toxin concentrations in organisms can reach levels that pose significant risks to wildlife and human consumers. Factors including toxin chemistry, species-specific physiology, environmental conditions, and the interplay of food web dynamics influence the complexity of bioaccumulation patterns. Ongoing research using advanced molecular and ecological approaches continues to refine our understanding of how phycotoxins move through ecosystems and impact higher trophic levels.

5.1.3 Biotransformation and Exporter Proteins

The resistance of consumers and the various underlying mechanisms against phycotoxins, especially cyanotoxins, have often been proposed, but proven only occasionally and unequivocally. For instance, zooplankton, including copepods, feed on a wide variety of phytoplankton but appear to avoid certain toxic species. In some habitats, copepods have evolved partial tolerance to toxins in their diets after many generations of exposure to toxic algae (Teegarden 1999; Avery and Dam 2007).

In addition, certain fish, particularly phytoplanktivorous species such as silver carp (*Hypophthalmichthys molitrix*), bighead carp (*H. nobilis*), and the omnivorous Nile tilapia (*Oreochromis niloticus niloticus*), show remarkable resistance to microcystin-LR (MC-LR) (Xie et al. 2004). These species can even exploit the nutrient and energy content of toxic cyanobacteria, turning a potential hazard into a resource. This raises an important question: Is their tolerance linked to the inducibility of specific glutathione *S*-transferase (GST) classes, which serve as key detoxification enzymes? To address this, Liang et al. (2007) compared MC-tolerant fish (silver carp and Nile tilapia) with the MC-sensitive grass carp (*Ctenopharyngodon idella*). Using semiquantitative mRNA expression analysis of alpha- and rho-class GSTs, they uncovered a striking contrast: The tolerant species exhibited inducible liver expression of either alpha- or rho-class GST, whereas the sensitive grass carp lacked inducible expression of both. These results strongly suggest that inducible, rather than constitutive, expression of liver GST genes represents a central mechanism of fish tolerance to microcystins.

Glisic et al. (2015) provided a comprehensive analysis of the GST superfamily in zebrafish, revealing that various GST isoforms—classified into alpha, mu, pi, zeta, and rho classes—play distinct and crucial physiological roles beyond xenobiotic metabolism. In particular, hepatic microsomal GST3 isoforms appear highly involved in phase II detoxification processes, highlighting the functional diversity and specialization of GSTs in zebrafish. These enzymes exhibit tissue-specific expression patterns and differ in substrate affinities, underlining their importance not only in detoxification but also in developmental and cellular functions (Glisic et al. 2016; Tierbach et al. 2018). Another picture emerges from the study of bighead carp, in which the pi-class GST isoform is highly expressed across all tissues examined, suggesting a central role in xenobiotic metabolism in this species (Li et al. 2010).

Collectively, research indicates that specific GST isoforms play a central role in fish resistance to microcystins. However, the dominant isoforms and underlying mechanisms vary by species, reflecting adaptive diversification in detoxification strategies. Beyond their role in conjugating glutathione to xenobiotics (e.g., GST3 with hepatic microsomal substrates), certain isoforms also inactivate endogenous compounds and modulate cellular signaling pathways, highlighting their multifunctionality and evolutionary divergence in enabling fish to cope with both environmental toxins and intrinsic physiological challenges.

Nevertheless, the question arises whether or not a comparable GST-based mechanism of resistance against phycotoxins also applies to invertebrates? Currently, most studies review the situation and confirm that GSTs are involved in the metabolism of cyanotoxins. However, there is no evidence yet that specific GST classes or the overexpression of GSTs contribute to or even facilitate resistance to cyanotoxins. More than two decades ago, the role of GSTs in microcystin metabolism was demonstrated through the identification of an enzymatically formed MC-LR–glutathione conjugate in a wide range of aquatic organisms. This conjugate was detected across diverse taxa—from plants such as *Ceratophyllum demersum* and invertebrates including *Dreissena polymorpha* and *Daphnia magna*, to fish eggs and adult fish like *Danio rerio* (Pflugmacher et al. 1998).

Besides the GST-based mechanism of toxin tolerance, Amé et al. (2009) showed that so-called multixenobiotic resistance transporters (MXRs) play a major role as an additional mechanism in the resistance of the one-sided livebearer (*Jenynsia (multidentata) lineata*) against microcystin. This is a membrane-bound group of P-glycoproteins (P-gp) that act as exporters of internalized xenobiotics and/or their metabolites. Upon exposure to MC-LR, transcriptional changes in P-gp were detected in the liver, gills, and even the brain of *J. lineata*. This response proved to be tissue-specific: Only in the gills was P-gp expression significantly elevated compared to controls, as confirmed by both real-time RT-PCR and Western blot analyses (Fig. 5.2). These findings clearly indicate that P-gp is an integral component of the defense system activated in response to MC-LR, at least in *J. lineata*.

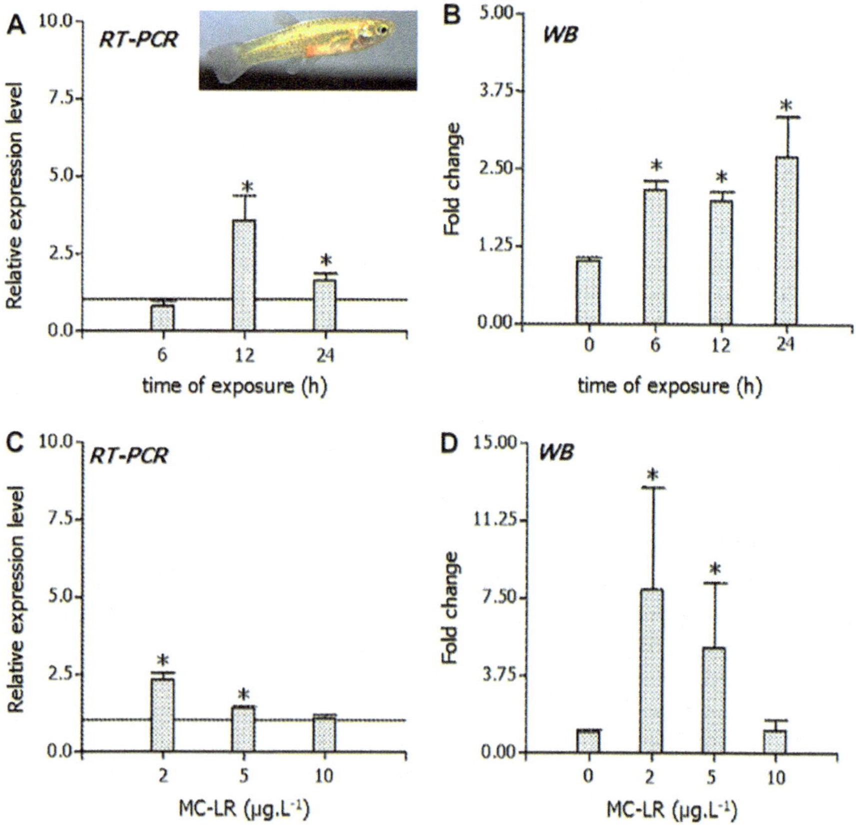

Fig. 5.2 *P-gp* expression in gills of *Jenynsia (multidentata) lineata* exposed to MC–LR at 2 μg L^{-1} for 6, 12, and 24 h by real-time RT-PCR (**a**) and by Western blot (WB) (**b**). *P-gp* expression in gills of *J. multidentata* exposed to MC–LR at 2, 5, and 10 μg L^{-1} for 24 h by RT-PCR (**c**) and by Western blot (**d**). The values are expressed as mean ± S.D. *The significance levels observed are $P < 0.05$ compared to the control group (from Amé et al. (2009), with permission from Elsevier; image credit to CHUCAO via Wikimedia, CC BY-SA 3.0)

The pioneering study by Amé et al. (2009) initiated several subsequent studies showing that exporters serve as a defense mechanism in vertebrates as well as invertebrates against cyanotoxins, in particular (Table 5.2), and phycotoxins and their metabolites, in general. P-glycoprotein multidrug transporters were active in the resistance of *Ruditapes philippinarum, Scapharca subcrenata, Tegillarca granosa, Perna viridis,* and *Mytilus edulis* to okadaic acid or its producer (*Prorocentrum lima*) (Whalen et al. 2010; Huang et al. 2014, 2015a).

5.1.3.1 Channel of Resistance

One impressive example of a tolerance mechanism is the "channel of resistance," which was also described as a resistance mechanism in aquatic animals. Coining this term, Howlett (2005) presents two striking examples of the Red Queen hypothesis. The soft-shell clam (*Mya arenaria*), a common bivalve along the Atlantic coast of North America, is known to accumulate saxitoxin when exposed to toxic dinoflagellate blooms. Bricelj et al. (2005) demonstrated that clams from areas affected by recurrent red tides are relatively resistant to the toxin and tend to accumulate it in their tissues. However, clams from unaffected areas have low resistance when exposed to the toxin in the laboratory. These differences are reflected in the sensitivity of isolated clam nerve trunks exposed to the toxin in vitro.

To uncover the molecular mechanism of resistance, Bricelj et al. (2005) sequenced the genomic region encoding a putative voltage-gated sodium channel, a key membrane protein that regulates ion flow. They discovered a single point mutation—a substitution of glutamic acid with aspartic acid—at a site already known to be critical for saxitoxin binding. Remarkably, when this mutation was introduced into a rat brain sodium channel, normal ion conductance was preserved. Still, the sensitivity of the channel to saxitoxin dropped dramatically due to a strong reduction in toxin-binding affinity at the pore. Saxitoxin, produced by red-tide algae, probably acts as a potent selective agent on clams, leading to genetic adaptation, with the target of selection being genetic variation at a single site in an ion channel.

However, this phenomenon is not unique to clams. Saxitoxin is related to another poison called tetrodotoxin (TTX). Tetrodotoxin is a potent neurotoxin that some newts (Salamandridae), such as *Taricha granulosa*, accumulate in their skin as a chemical defense against predators (Bucciarelli et al. 2017; Johnson et al. 2018). This has led to an evolutionary arms race between garter snakes (Thamnophis sirtalis) and newts, whereby the snakes have developed high resistance to the toxin, enabling them to survive encounters with their toxic prey. This coevolution exemplifies how predator–prey interactions can drive physiological adaptations and showcases the role of TTX as an effective antipredator defense in amphibians.

Geffeney et al. (2005) demonstrated that variation in the level of resistance of garter snakes coevolving with their newt prey can be attributed to molecular changes that affect the binding of TTX to the conserved voltage-gated sodium channels (Nav) of nerve and muscle, thereby causing paralysis.

Soong and Venkatesh (2006) specified that, given that the lethal effects of natural neurotoxins such as TTX and STX are mediated by their binding to the P-loop regions of Nav channels, crucial residues in the P-loop regions of Nav channels are obvious candidates for TTX and STX resistance in animals. Recent investigations into the molecular basis of TTX and STX resistance in pufferfish, garter snakes, and soft-shell clams have shown that diverse substitutions in different P-loop regions of skeletal muscle Na_v channels and neuronal Na_v channels are responsible for the convergent evolution of neurotoxin resistance in these animals.

5.1.3.2 Nuclear Hormone Receptor 96

Recent studies suggest that the three phases of response to toxicant exposure are not distinct, but rather coordinated by a master gene. Inspired by studies on the nuclear hormone receptor 96 (HR96) in *Daphnia pulex*, which acts as a toxicant receptor (Karimullina et al. 2012; Ji et al. 2023), Yuan et al. (2025) studied HR96 in *Magallana (Crassostrea) ariakensis* exposed to the toxic dinoflagellate *Prorocentrum lima*. After exposure to *P. lima*, the oyster nuclear receptor HR96 was activated, which induced the expression of genes related to three-phase metabolism and detoxification, including phase I (*cyp3a4* and *cyp3a24*), phase II (*gstκ1*), and phase III (*abcb1, abcb5, abcb6, abcg20, abcg,* and *abcc10*) (Fig. 5.3), which increased the

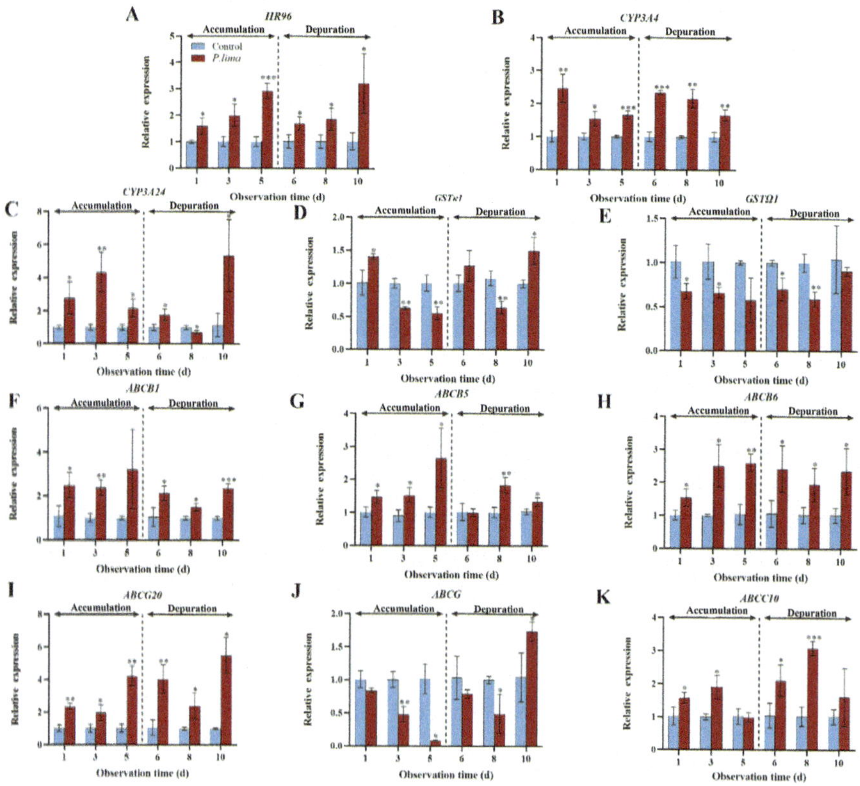

Fig. 5.3 Changes in mRNA levels of HR96 and metabolic detoxification-related genes in *Magallana ariakensis* after exposure to *Prorocentrum lima*. $N = 3$. * $P < 0.05$; ** $P < 0.01$; *** $P < 0.001$ (from Yuan et al. (2025), credit to Elsevier)

CYP450, GST, multidrug resistance-associated proteins (MRP) and P-gp protein content/activity, thereby alleviating the damage caused by diarrhetic shellfish toxins (DSTs) to shellfish.

5.1.4 Costs of Tolerance

Tolerance to phycotoxins likely carries significant costs, yet the evidence remains mixed. This complexity is well illustrated by contrasting examples from both marine and freshwater ecosystems.

In the copepod *Acartia hudsonica*, Avery and Dam (2007) showed that diet appears to influence development, since females have longer mean times to maturity when feeding on the toxic *Alexandrium catenella* dinoflagellate than on the standard diet. Note that females are more resistant than males (Avery et al. 2008). *Alexandrium* spp. produce a suite of neurotoxins (paralytic shellfish toxins, PSTs) that block sodium channels. A mutation in the sodium channel causes adaptation to PSTs in other organisms. A decade ago, Chen et al. (2015a) identified a mutation in the sodium channel of *A. hudsonica*, shedding light on a potential molecular mechanism of toxin tolerance. However, Finiguerra et al. (2015) showed that this channel is not involved in the observed tolerance. However, other ion channels, which are sensitive to PSTs, have not yet been studied. While some results suggest that the channel mutation influences the phenotypic response, there is no consistent evidence of an advantage, cost, or trade-off associated with the mutant sodium channel isoform in PSTs. There is no difference in ingestion and egg production rates on non-toxic food for PST-susceptible and tolerant *A. hudsonica* populations (Colin and Dam 2005; Colin and Dam 2007), suggesting that PST-adaptation incurs little or no cost.

In *Daphnia magna*, von Elert and Otte (2023) demonstrated that lower somatic growth rates are associated with a higher tolerance to MCs in the absence of MCs. If toxin tolerance is associated with biochemical reactions rather than an avoidance response, this trade-off should be expected, as these reactions are also associated with energy expenditure. The diverted energy is no longer available for regular traits, as has often been demonstrated with xenobiotics (Schmidt et al. 2005, 2006). This energetic trade-off means that resources devoted to detoxification reduce the energy available for growth and other vital functions, as consistently observed in *Daphnia* exposed to microcystins.

MCs induce oxidative stress in *D. magna*, a challenge that the organism mitigates through upregulation of catalase activity (Ortiz-Rodríguez et al. 2012). However, the authors demonstrated that the short-term increase in catalase activity, combined with the subsequent conjugation of MC via GST, occurs at the expense of somatic growth. Moreover, this acquired tolerance to microcystins is passed on to the offspring, resulting in similarly reduced somatic growth. However, mechanisms of transgenerational tolerance transfer are not discussed.

5.1.5 Adaptive Response

5.1.5.1 DNA Methylation

The environment induces epigenetic alteration and subsequent phenotypic variations that can be transmitted across generations. This phenomenon has often been referred to as "rapid evolution," as it increases the resilience of an organism's offspring and subsequent generations. It is recognized that epigenetic mechanisms play a vital role in the rapid evolution of organisms, e.g., the local fitness effects of the epigenome can result in the changes or rapid evolution of these patterns, thus impacting sustainability (Fig. 5.4). In this context, much interest has been focused on potential contributions and fitness consequences of transgenerational effects of epigenetic alterations to adaptive responses to environmental change and global warming, aimed at preventing farm populations from degradation and rescue natural populations from extinction (Wang and Shen 2023). Since specific studies on marine phycotoxins are lacking, we will draw attention to several general examples with aquatic animals.

Vogt (2017) outlines that, in general, the production of epigenetically diverse phenotypes from the same genome may explain why many clonal animals and plants can live in a wide range of habitats and geographical regions, despite their genetic uniformity. This phenomenon is often explained by the existence of so-called "general-purpose genotypes" or "superclones," which are believed to be

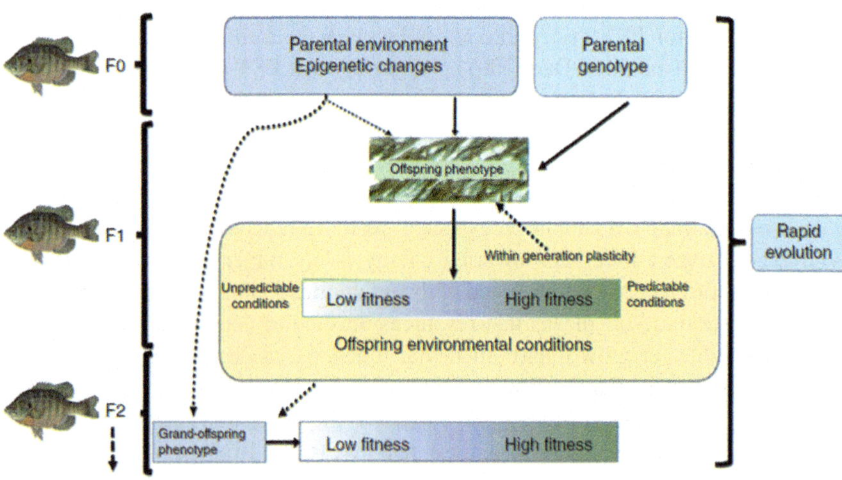

Fig. 5.4 Genetic and epigenetic effects of parents across generations. Offspring phenotype can be affected by the parental environment directly through body condition or indirectly via epigenetic changes (dashed arrows). The impact of environment-induced parental effects on offspring fitness is dependent on both genetic and nongenetic factors that interact within and across generations. This phenomenon can result in "rapid evolution," as it increases the resilience of the offspring of an organism and subsequent generations (from Wang and Shen (2023) modified from McGuigan et al. (2021), credit to the Royal Society with permission from Wiley & Sons)

capable of coping with a broad spectrum of environmental challenges (Parker et al. 2003; Lundmark 2010; Caron et al. 2014).

There is growing evidence that the importance of epigenetic diversity increases as genetic diversity decreases—in other words, epigenetic diversity compensates for low genetic diversity. This effect is most pronounced in invasive species that have successfully invaded a new area. Typically, only a few individuals are introduced or actively migrate into a new area, and it is necessary to understand how they successfully outcompete species already present in the area (Vogt 2023).

Didemnum vexillum is an invasive colonial ascidian that has become established in many coastal waters worldwide. Phylogenetic analyses reveal that *D. vexillum* populations consist of two distinct clades; clade B appears to be restricted to the native range (Japan), whereas clade A is found in many regions throughout the world, including New Zealand. The spread of *D. vexillum* clade A suggests that it may be intrinsically more invasive than clade B, despite low levels of genetic diversity compared to native populations. Indeed, introduced colonies show significantly reduced levels of DNA methylation, which is likely the mechanism for increased phenotypic plasticity and invasion success (Hawes et al. 2019).

An interesting animal example is the gynogenetic all-female fish *Chrosomus eos–neogaeus*, which occurs in North America in 14 genetically identical lineages originating from different hybridization events between the redbelly dace *C. eos* and the fine-scale dace *C. neogaeus* (inset in Fig. 5.5) (Angers and Schlosser 2007). Dating of hybridization events suggests a relatively recent origin (< 50,000 years ago). Each hybrid lineage apparently originated from a single hybrid zygote and is genetically uniform, except for random mutations that have accumulated over time. These lineages frequently display a large geographical distribution at the regional scale. The analysis of DNA methylation polymorphisms within such lineages revealed epigenetic similarity of individuals in a given lake but significant differences between lakes, suggesting that epigenetic phenotype diversification may be the mechanistic explanation of the "general-purpose genotype" (Massicotte and Angers 2012) (Vogt (2017), references adopted).

Recent analysis of DNA methylation in *C. eos–neogaeus* lineages from predictable and unpredictable environments (lakes vs. intermittent headwater streams) in Canada identified the relative contributions of SPV and EPV[2] to phenotype modification (Leung et al. 2016). A comparison of clones from the wild revealed that risk-spreading SPV prevails in unpredictable environments. In contrast, directional EPV is predominant in predictable environments (Fig. 5.5). Differences in environmental effects on epigenetic variation between sympatric lineages (Fig. 5.5, ET3 and ET4) and lineages reared in similar experimental conditions (Fig. 5.5, LR1 and ET2) further showed that the genotype strongly influences the epigenetic response to

[2] Phenotypic variation can be subdivided into genetic, stochastic developmental, and environmental components. The two epigenetic proportions are called "stochastic developmental phenotype variation" (SPV) and "environmentally induced phenotype variation" (EPV) in the following. Both epigenetic phenomena are closely intermingled but represent different mechanisms to generate phenotypic diversity (Vogt 2017).

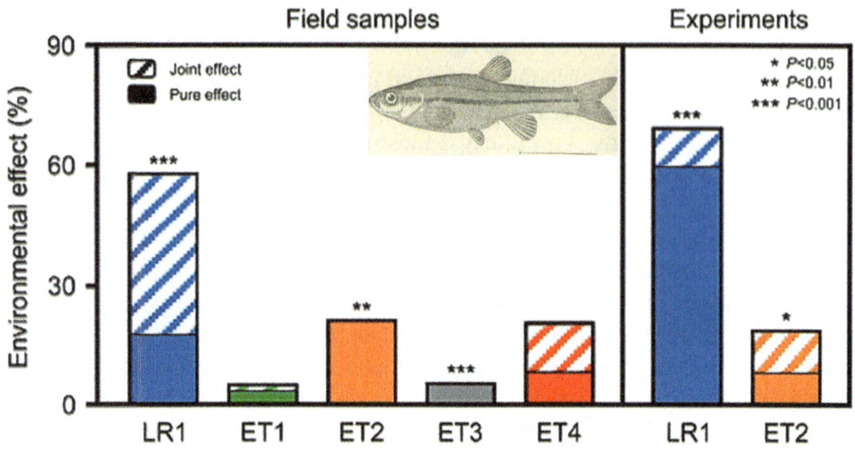

Fig. 5.5 Environmental effects on epigenetic variation in asexual *Chrosomus eos–neogaeus* under natural and experimental conditions. The environmental and genetic joint effect is separated from the pure environmental effect. Epigenetic differences were determined using MS-AFLP (methylation-sensitive amplified fragment length polymorphism), and genetic differences were assessed through microsatellite analysis. LR1 to ET4 refer to distinct hybrid lineages from the Laurentians region (LR) and Eastern Townships (ET) of southern Quebec (Canada). LR1 lineages are from environmentally stable lakes, whereas ET1–ET4 are from environmentally unstable streams. Experimental animals were sampled as larvae from the indicated sites and raised in common garden experiments for 5 months until they reached adulthood. P-values refer to the significance of the proportion of epigenetic variation explained uniquely by the environment. The graph illustrates the differences in environmental effects on epigenetic variation between predictable (LR1) and unpredictable environments (ET1–ET4) and among lineages, even when they occur in sympatry (ET3 and ET4). A comparison of LR1 and ET2 in the field and laboratory suggests rapid epigenetic response to environmental change (Vogt 2017), credit to the Oxford University Press; drawing of fine-scale dace (*Chrosomus* (*Phoxinus*) *neogaeus*) in public domain via Freshwater and Marine Image Bank at the University of Washington)

environmental signals. Common garden experiments revealed that the proportion of pure environmental effects can change considerably when clone members are transplanted into a new environment (Fig. 5.5, LR1 and ET2 under natural and experimental conditions). The example of *C. eos–neogaeus* demonstrates that SPV and EPV always act in conjunction but have different weights in different environments.

Lemmen et al. (2022) highlight the fundamentals of rapid heritable adaptation in the absence of initial standing genetic variation. Nongenetic inheritance allows environmentally induced phenotypes to be passed between generations via the transmission of factors (e.g., nutrients, anti-nutrients, such as phycotoxins, methylation patterns, small RNAs, or microbiomes) that modify phenotypes but not DNA sequences (Bonduriansky and Day 2018). Given the capacity of nongenetic inheritance to generate heritable phenotypic variation, it has been proposed to play an important role in adaptation (e.g., Jablonka et al. (1995), Danchin et al. (2011), O'Dea et al. (2016)). These mechanisms are expected to promote adaptation in clonal organisms (Verhoeven and Preite 2014) or populations with low genetic diversity (Vogt 2017).

To test if adaptation can rapidly occur in the absence of initial standing genetic variation and recombination in small populations, Lemmen et al. (2022) exposed replicate monoclonal populations of the microzooplankton *Brachionus calyciflorus* to a culturing regime that selected for phenotypic variants with elevated population growth with either high or low phosphorus food. Within six generations, the authors observed heritable local adaptation to nutrient limitation. More specifically, populations with a history of exposure to low P food exhibited higher population growth rates under low P food conditions than populations with a history of high P exposure. However, the capacity for such a response varies among clones.

Examples of true transgenerational epigenetic inheritance induced by parental genotype, physiology, or environment have become increasingly numerous in model animals. However, in most cases, the effects described have a limited duration, for example, typically spanning 3–4 generations in *Caenorhabditis elegans*. Characterized mechanisms commonly involve the inheritance of genome-associated epigenetic information via gametes, including histone modifications and small RNAs. Likely reasons for the limited duration of many transgenerational effects might be passive and active mechanisms that regulate small RNA populations and histone modifications across generations (Perez and Lehner 2019).

Recently, Wang and Shen (2023) summarized that, for natural populations, epigenetic mechanisms may play a role in the speciation process by increasing the range of phenotypes available for the action of natural selection and further facilitating the speed of speciation (Laland et al. 2014; Mendizabal et al. 2014; Pu and Zhan 2017; Jeremias et al. 2018). For example, epigenetic mechanisms potentially contributed to the establishment of the marbled crayfish as a new species (Jeremias et al. 2018) and increased the rapid phenotypic plasticity of the species (Weyrich et al. 2016). Considerable evidence shows that the short-term adaptation of species to environmental changes is partly modulated through epigenetic mechanisms. For instance, exposing zebrafish to crude oil led to adaptive responses in the F_1 generation, which was found to perform better in crude oil-contaminated water than in clean water (Bautista and Burggren 2019). Additionally, increased levels of heat shock protein 70 and enhanced heat stress tolerance were observed in a parthenogenetic population of *Artemia* following exposure to a nonlethal heat shock. The effects were transmitted to three successive nonexposed generations and correlated with changed DNA methylation and acetylated H3 and H4 (Norouzitallab et al. 2014; Jeremias et al. 2018) (appropriate references added).

Hutton and Brander (2023) draw attention to another interesting example: The Atlantic killifish (*Fundulus heteroclitus*) is one of the first species to show an evolutionary adaptive response to pollution that has enabled it to persist even in highly polluted waters (Whitehead et al. 2017). Killifish living in water contaminated with polychlorinated biphenyls and dioxins have decreased CYP450 activity, with a reduced risk of highly reactive epoxides that can damage DNA (Steinberg 2012). This reduced CYP450 is due to decreased global DNA methylation and expression of *dnmt1*, as well as increased expression of ten-eleven translocation (TET) genes, which are involved in demethylation (Wu and Zhang 2017), compared to populations living in unpolluted waters. The relationship between gene expression, epigenetic modification, and inheritance has been proposed as a route for rapid evolution. While literature is still sparse in terms of understanding the specific epigenetic mechanism that led to the type of adaptive inheritance required for rapid evolution, there is evidence to support its role (Brander et al. 2017; Luu et al. 2021) (appropriate references added).

In conclusion, Skinner (2015) suggests that epigenetic changes often precede and may even trigger or increase the likelihood of mutations that result in altered DNA sequence. Thus, epigenetic variation is the mechanism by which organisms can adapt to rapid environmental perturbations, such as pollution. However, these modifications on the surface of DNA or histones may, in turn, increase the likelihood that a genomic region will mutate, thereby facilitating adaptation in the long term. This suggests that the Lamarckian basis of inheritance facilitates rapid evolution while providing the stimulus for mutations that enable long-term Darwinian adaptation (Brander et al. 2017).

Although the underlying biomolecular mechanism is not documented, the phenomenon of adaptive response appears to be valid in the following examples. Zheng et al. (2011) reported differential responses of *Acartia hudsonica* populations to toxic and nutritionally deficient food algae. The authors examined the effects of *Alexandrium (fundyense) catenella* and *Phaeodactylum tricornutum* as sole foods, respectively, on egg production and egg hatching success of two geographically separated populations (Maine and Connecticut) of the copepod *A. hudsonica* in shared environment experiments after being reared under identical conditions for two generations. *P. tricornutum* is not toxic, but rather nutritionally insufficient to *A. hudsonica*. The Maine site regularly experiences toxic blooms of *Alexandrium* sp. while the Connecticut site does not. Over a 6-day period, *A. catenella* reduced the egg production rates of the Connecticut copepod population, but not the Maine population. In contrast, the diatom *P. tricornutum* reduced egg production in both populations. The results of this study are consistent with the hypothesis of local adaptation to toxic food, but not to nutritionally deficient food. This study confirms an earlier one (Colin and Dam 2002; Colin and Dam 2003) that examined populations from Massachusetts (USA) and New Brunswick (Canada).

Later, Kumari et al. (2022) exposed the tropical waterflea *Echinisca triserialis* to the oxylipin 2-trans,4-trans decadienal (DD), a common secondary metabolite in diatoms that co-occur with the waterfleas. Significantly higher mortality of maternal *E. triserialis* individuals and lower reproduction occurred in a DD-contaminated environment. Individuals of the F_1 generation produced in the DD medium exhibit significantly higher survival, reproduction, and population growth compared to the control. This work suggests that prolonged exposure to oxylipins favors the preexposed genotype within the community, indicating an intergenerational experiential benefit in diatom aldehyde-contaminated environments over time, a pattern consistent with the extension of maternal experience to the F_1 generation.

Schwarzenberger (2022) summarizes another transgenerational adaptation in cladocerans. On the genetic level, *Daphnia* shows target-specific responses to cyanobacterial protease inhibitors (PIs). In transcriptome studies, it has been observed that genes encoding digestive proteases are regulated depending on whether different *Daphnia* species have ingested PI-producing cyanobacteria or not. Protease genes are upregulated when *D. magna* grows on PI-producing cyanobacteria. This increase in gene expression proved to be adaptive: When *D. magna* mothers are fed with a trypsin-inhibitor-producing cyanobacterium, the increased protease gene expression is maternally transferred to their untreated offspring. In turn, the offspring show a higher somatic growth rate on dietary trypsin inhibitors in comparison to the offspring from naïve mothers (Schwarzenberger 2022).

Similarly, but using a different gene, Xu et al. (2022) showed that the offspring produced by *Daphnia sinensis* mothers pre-exposed to toxic *M. aeruginosa* have greater adaptability to *M. aeruginosa* than their mothers. Moreover, the *Hox* gene expressions of *D. sinensis* show obvious differences between clones under toxic *M. aeruginosa* stress. *Hox* genes function in the process of pattern formation along the anterior–posterior body axis (Bürglin 2013). Specifically, the downregulation of *abd-b* (*abdominal-b* gene) in the sensitive mothers may be responsible for a significant decrease in the survival rates at the end of the experiment under higher *M. aeruginosa* concentrations. However, how *abd-b* may modulate cyanotoxin sensitivity in *D. sinensis* remains unclear.

Nevertheless, it is very likely that these examples of adaptive response to cyanotoxins are epigenetically controlled (Asselman et al. 2017).

A final example of the transgenerational heritage of stress resistance from general ecology (Steinberg 2016): The waterflea *Moina macrocopa* lives in humic-rich Brazilian coastal lagoons, into which the Atlantic Ocean intrudes salt during storm events, placing the animals under osmotic stress. Mothers who have acquired multiple stress resistances through humic exposure (Steinberg 2012) pass this on to their daughters, allowing the daughters to grow better (Fig. 5.6a) and produce more offspring (Fig. 5.6b) under osmotic stress than the daughters of mothers without acquired stress resistance after humic exposure. This effect is also likely to be epigenetic, as the embryos were never in contact with humic substances—a hypothesis supported by the observation that exposure to humic substances leads to global DNA methylation in cladocerans (Fig. 5.6c) (Menzel et al. 2011).

5.1.5.2 Trained Immunity

Trained innate immunity is a form of nonspecific, long-term memory of the innate immune system. It results in a more robust response to subsequent infections after initial exposure. This "training" is triggered by a primary stimulus, such as a microbe, vaccine, or supplemented methyl donor. It involves epigenetic reprogramming and metabolic changes in immune cells, making them more alert and prepared for future threats. This process differs from adaptive immunity, which is antigen-specific and offers protection against different pathogens.

The role of epigenetic mechanisms, identified for synthetic xenobiotics in insects, is likely to extend to natural xenobiotics, such as phycotoxins, and their predators, leading to occasional, trained, and probably heritable tolerance in marine animals. Unlike mammals, lower vertebrates (such as fishes) and invertebrates do not develop

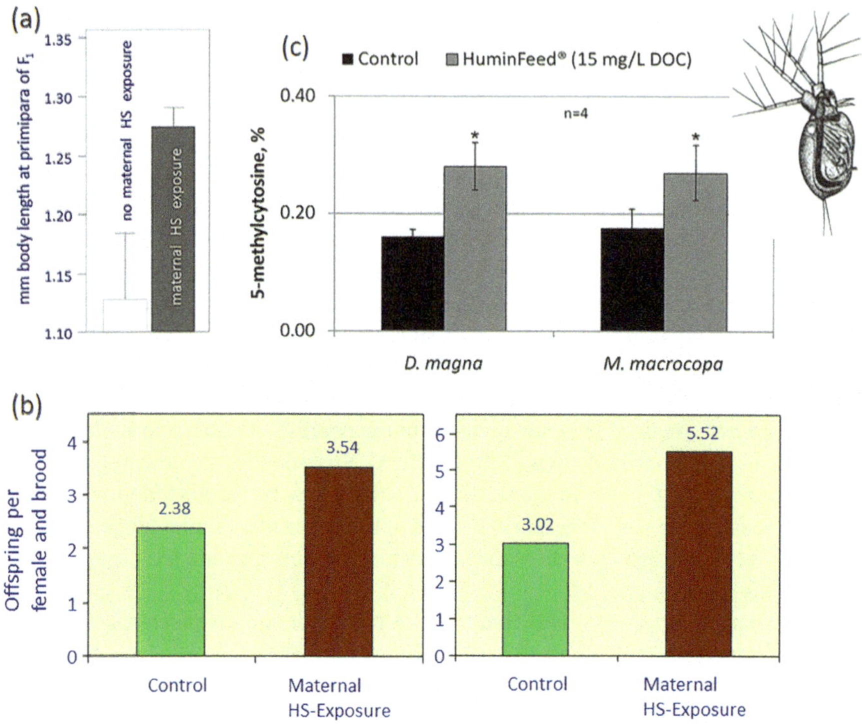

Fig. 5.6 (**a**) Salt-exposed *Moina macrocopa* daughters of humic-exposed mother grow better (modified from Suhett et al. (2011) with permission from Springer Nature); (**b**) increases in offspring number of salt-exposed *M. macrocopa* females if their mothers were exposed to humic substances (Dachsel and Steinberg unpublished); (**c**) DNA methylation in *Daphnia magna* and *M. macrocopa* after exposure to humic substances (from Menzel et al. (2011), with permission from Elsevier; drawing in public domain, credit to Keilhack (1909) via Biodiversity Heritage Library)

specific immunity when challenged with pathogens. However, these animals may benefit from trained immunity. Recent studies show that innate immunity can also be trained by allochthonous methyl donors. This has been proven with dietary supplements, such as β-glucans (Petit and Wiegertjes 2016; Roy et al. 2020), peptidoglycans (Lee and Kim 2025), mannan-oligosaccharides (MOS) (Wang et al. 2024), poly(I:C) (Lafont et al. 2017, 2020), or a combination of MOS with β-glucan or with inulin (Lu et al. 2019) and likely also with humic substances (Lieke et al. 2021; Zhao et al. 2025). It can be hypothesized here that low (hormetic) concentrations of several phycotoxins serve as methyl donors when metabolized. In light of the hormetic response of algae and invertebrates to synthetic xenobiotics (→Chap. 3), this hypothesis is worth testing experimentally.

Gonzalez-Romero et al. (2017) found that genes encoding histone variants (H2A.X, H2A.Z, and macroH2A) are potentially involved in maintaining genome integrity in eastern oysters (*Crassostrea virginica*) exposed to increasing concentrations of *Karenia brevis*. Although the authors studied the action of the methyl donor in the form of DNA methylation, they did not take the next step of looking for histone modifications via methylation. Nevertheless, one could hypothesize that methyl donors also modify histones and thereby train the oyster's immunity.

5.2 Marine Invertebrates

This section focuses on bivalves and, in particular, crustaceans such as copepods and cladocerans, which are keystone species in their ecosystems. Extensive research on these animals provides valuable model systems for understanding responses in other groups, including fish.

To start with an intriguing example, compared to other organisms, bivalve mollusks exhibit some tolerance to diarrhetic shellfish toxins (DSTs, mainly okadaic acid, OKA), and the DSTs toxicity to bivalves decrease with exposure time (Svensson et al. 2003; Prado-Alvarez et al. 2012, 2013; Prego-Faraldo et al. 2016). Using a mixture of DSTs, namely OA and dinophysistoxin (DTX) analogues, Yuan et al. (2023) identified a resistance mechanism in *Perna viridis* with long-chain polyunsaturated fatty acids as central components. The authors observed that docosahexaenoic acid (DHA, 22:6ω3) levels significantly decrease with the esterification of DSTs in the digestive gland after exposure to the DST-producing dinoflagellate *Prorocentrum lima*. The addition of DHA increases the esterification level of DSTs and increases the expression of Nrf2 pathway-related genes and enzyme activities, thereby mitigating the damage caused by DSTs to the digestive glands. Other PUFAs may also be directly or indirectly involved in the esterification process (Fig. 5.7). Levels of stearidonic acid (SDA, 18:4ω3) and eicosapentaenoic acid (EPA, 20:5ω3) increase significantly while levels of DHA decreases strongly. This may reflect a compensatory synthesis of EPA and SDA, key precursors for the endogenous production of long-chain PUFAs (Prasad et al. 2021; Tan and Zheng 2022). This hypothesis is particularly intriguing, as it points to an additional, potentially critical role for PUFAs beyond their well-established functions (refer to Steinberg (2022a, b, c, d, e, f, g, h, i)).

Also in *P. viridis*, Lv et al. (2023) demonstrated that ABCB and ABCC exporters respond to the toxic dinoflagellate *P. lima* by gene upregulation, first in hemocytes and later in gills and the digestive gland, thereby contributing significantly to the detoxification of *P. lima* toxins.

5.2.1 Saxitoxins

Saxitoxins (STXs) are a group of potent neurotoxins best known as the primary agents of paralytic shellfish poisoning (PSP). These toxins, produced by marine dinoflagellates and some freshwater cyanobacteria, act by blocking voltage-gated sodium channels in nerve and muscle cells, thereby inhibiting nerve signal

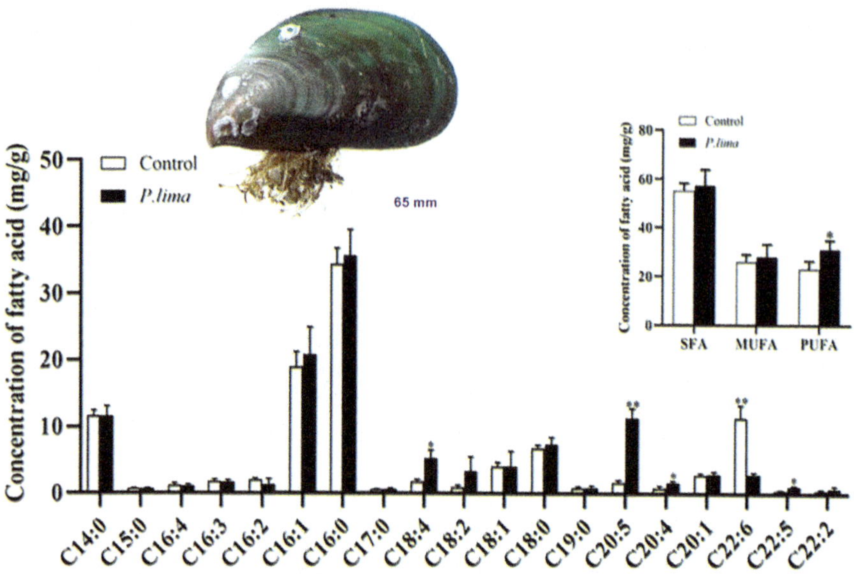

Fig. 5.7 Fatty acid concentration (mg g^{-1} dry weight) in the digestive gland of *Prorocentrum lima*-exposed mussels (*Perna viridis*) after 96 h of exposure. **SFA**, **MUFA**, and **PUFA** represent saturated fatty acids, monounsaturated fatty acids, and polyunsaturated fatty acids, respectively. Data are presented as mean ± SD ($N = 3$). Significant differences compared to the control are represented by asterisks (t-test, * $P < 0.05$. ** $P < 0.01$) (from Yuan et al. (2023), credit to MDPI, Basel). Image of *Perna viridis* showing the byssus, the downward-pointing beak, and the dark green color that turns brownish toward the umbo (credit to the USGS)

transmission and causing paralysis. Numerous studies have focused on saxitoxin accumulation and effects in crustaceans and mollusks, both of which are important vectors for toxin transfer in aquatic food webs and are significant as seafood species. This section explores the chemistry, toxicity, ecological roles, and bioaccumulation of saxitoxins, with an emphasis on the responses and mechanisms studied in crustaceans and mollusks.

5.2.1.1 Crustaceans

Several crustaceans and bivalves exhibit resistance to TTX and paralytic shellfish toxins (PSTs, which includes saxitoxins); however, their mechanisms of tolerance differ from those observed in the previously discussed species. For instance, the shore crab *Thalamita crenata* (Fig. 5.8a), which inhabits the Leizhou Peninsula, China, is tolerant to PST and TTX toxicity, with the hemolymph having neutralizing effects against the lethality of these toxins (Lin et al. 2012, 2015), as demonstrated for PST (Fig. 5.8a). Remarkably, the PST-binding ability of the protein differs among PST components, in descending order of neosaxitoxin, gonyautoxins 1 and 4, saxitoxin, and gonyautoxins 2 and 3, indicating a structure–activity relationship in PST binding.

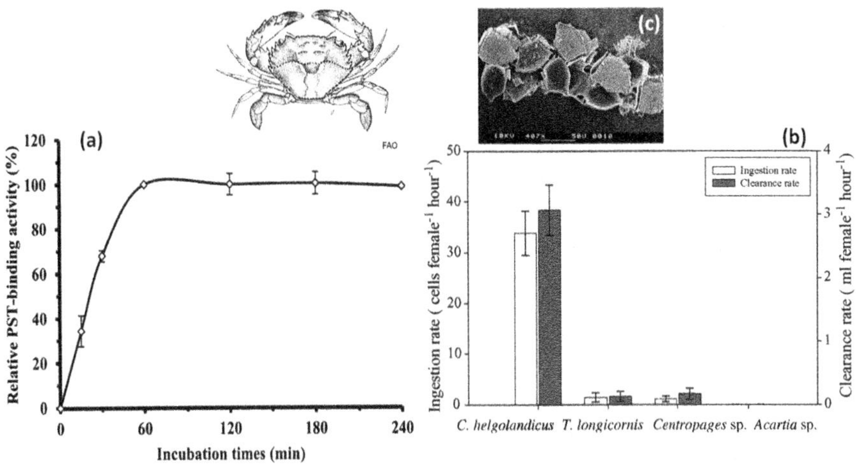

Fig. 5.8 (**a**) Effect of incubation time on the paralytic shellfish toxin (PST)-binding activity of the PST-binding protein in *Thalamita crenata*. The relative PST-binding activity for each sample is expressed as a percentage (%) of the bound-PST at 60 min. The results are shown as means ± SD of triplicates (from Lin et al. (2015a), with permission from Elsevier; image credit to FAO). (**b**) Mean (±SE) 24 h ingestion and clearance rates of *Dinophysis norvegica* by four copepod species prominent during the drift study. (**c**) Scanning electron photograph of fecal pellets from *Calanus helgolandicus* containing *D. norvegica* in grazing experiments on the natural plankton community during a drift study in the North Sea (from Jansen, Riser et al. (2006), with permission from Elsevier)

The sister species, *Calanus finmarchicus*, can also successfully feed on toxic phytoplankton. It co-occurs with *Alexandrium catenella* during seasonal blooms. At that time, *C. finmarchicus* population abundances are high, dominated by immature copepods preparing for diapause and by actively reproducing adults. High survival has been reported for copepods exposed to toxic *A. catenella*, but little is known about possible sublethal effects. In the study by Roncalli et al. (2016c), ingestion of the dinoflagellate did not affect copepod survival and grazing activity, as expected. However, there was a significant reduction in egg production and egg viability. After the 7-day experiment, total nauplius production by females on the low-dose and high-dose diets was reduced by 35–75% compared to the control females. Still, the question remains as to why animals can feed and grow on *Alexandrium* without immediate harm.

In a series of papers, Roncalli et al. approached the relevant answer in an exemplary manner, following the hypothesis that it is not just a single mechanism. The authors start with a transcriptomic analysis and found that the response after 5 days is characterized by downregulated genes involved in energy metabolism. Unexpectedly, detoxification is not a major component of the response. Instead, genes involved in digestion are consistently regulated, indicating that food assimilation may be affected. In sum, the energy budget of individuals may be affected by a reduced ability to build lipid reserves (Roncalli et al. 2016a).

In a companion paper, Roncalli et al. (2016b) confirmed that the tolerance to *A. catenella* by *C. finmarchicus* does not depend on the long-term upregulation of specific GSTs. In detail, the authors fed adult females one of three experimental diets for 5 days: control (100% *Rhodomonas* spp.), low dose of *A. catenella* (25% by volume, 75% *Rhodomonas* spp.), and high dose (100% *A. catenella*). Females showed an initial and transient response that includes the upregulation of two Sigma (σVI and VII) GSTs and downregulation of a single Delta (δI) GST. Changes in relative expression were already present at the first time point (0.5 days), underscoring an immediate response to the introduction of the dinoflagellate diet. Although detoxification likely contributes to the response of the copepod to a dinoflagellate diet, it does not appear to be the primary mechanism, as evidenced by only modest changes in gene expression.[3]

However, the most intriguing finding from the series of papers by Roncalli et al. is the reconsideration of "channels of resistance." Using transcriptomic approaches to search for the mechanism that confers resistance in *C. finmarchicus*, Roncalli et al. (2017) identified splice variants of Na_Vs that are predicted to be toxin-resistant. These are co-expressed with the putatively nonresistant form in all developmental stages. However, their expression does not respond to toxin challenge, nor is there any upregulation of genes involved in multi-xenobiotic resistance (MXR) or detoxification (phases I or II). Instead, adults consistently regulate genes encoding digestive enzymes, possibly to complement channel resistance by limiting the assimilation of toxins through the digestive process.

Additional support for a defense mechanism that limits absorption of the toxin comes from measurements of STX levels in *C. finmarchicus* (Hamasaki et al. 2003; Teegarden et al. 2003). In the clam *Mya arenaria,* which appears to possess a mutation in the Na_V channel as a mechanism of resistance to STX, retention efficiency in resistant individuals is high, ranging between 60% and 70% (Bricelj et al. 2005; Bricelj et al. 2010). This high retention efficiency—stemming from the organism's inability to eliminate the toxin actively—makes clams a significant risk factor for human PSP during red tide events (Bricelj et al. 2005). In contrast, retention efficiency is low in *C. finmarchicus*, ranging between 2% and 8% of the total ingested toxins (Hamasaki et al. 2003, Teegarden et al. 2003), which is consistent with a mechanism that limits the absorption of the toxin. The blue mussel *Mytilus edulis*, which also demonstrates relatively low STX accumulation (Bricelj et al. 1990), is believed to eliminate the toxin in part through bacterial action within its digestive tract (Bricelj et al. 1990).

[3] Lauritano et al. (2021). The authors point to a heterogeneous GST response among copepod species, even when exposed to the same stressor. This is partly due to the lack of a consistent methodology. Some papers analyze only bulk GST, while others differentiate different isoforms. This highlights the need to analyze and better understand the isoforms studied, as well as the environments in which they were sampled (e.g., coevolution with toxic algal species). To date, few studies have reported the specific subclass investigated; thus, future studies should make more effort for selecting the more relevant GST gene in quantitative expression analysis. On the other hand, it may well be that detoxification via the GST pathway is not the most important mechanism of action in general for the tolerance of some copepods to toxic phytoplankton.

Supporting this, several bacterial isolates from the digestive system of *M. edulis* have been shown to degrade paralytic shellfish toxins efficiently, with some capable of completely breaking down saxitoxin and related toxins within days (Donovan et al. 2008, 2009). Similar gut microbiome-mediated detoxification mechanisms have been well documented in insects (Engel and Moran 2013), suggesting that comparable processes could occur in copepods and other planktonic grazers, potentially aiding in their tolerance to toxin exposure. This highlights the crucial role of digestive tract bacteria in modulating toxin burdens in aquatic organisms and reducing toxic effects through biodegradation.

Subsequently, Roncalli et al. (2021) described an additional defense mechanism. The authors showed that ovothiol A has a strong antioxidant defense role in *C. finmarchicus* under oxidative stress, as demonstrated by its high expression in response to a toxic diet (Fig. 5.9). Ovothiol A is one of the most potent antioxidants, reacting with ROS and radicals significantly faster than other natural thiols (Shapiro and Hopkins 1991; Castellano and Seebeck 2018). For instance, it acts as a glutathione peroxidase, replacing the function of the enzyme in sea urchin eggs (Turner et al. 1988). The ovothiol system is more effective than egg catalase at destroying H_2O_2. In addition to their role as antioxidants, ovothiols have also been reported to have antiproliferative and anti-inflammatory activities (Russo et al. 2014; Castellano et al. 2016).

Based on laboratory and field work, Teegarden et al. (2003) conclude that copepods (*Acartia hudsonica, Eurytemora herdmani, Centropages hamatus*) feeding on *Alexandrium catenella* act as a sink; they regularly display very low retention of toxin, despite calculated high ingestion rates, across a continuum of conditions.

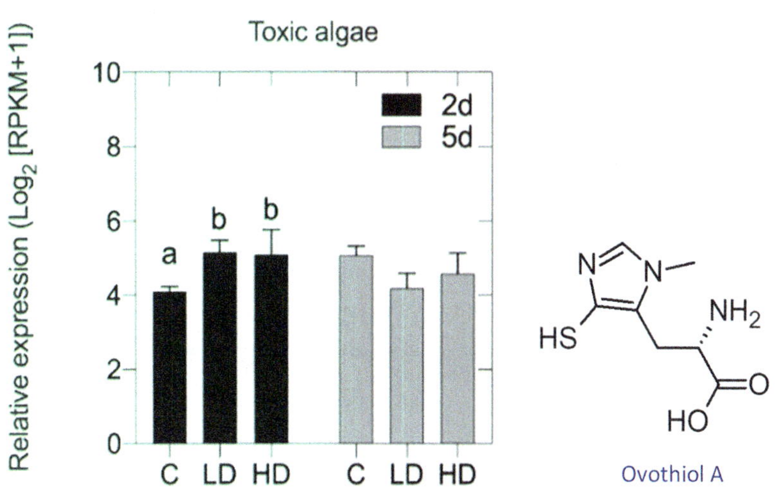

Fig. 5.9 *OvoA* expression in adult females of *Calanus finmarchicus* exposed for 2 days (2d) and 5 days (5d) to a control diet *Rhodomonas* sp. (C), low dose (LD), and high dose (HD) of the toxic dinoflagellate *Alexandrium catenella*. The relative expression was normalized by length reads per kilobase per million mapped reads (RPKM, Log_2), adding pseudocounts of 1 (from Roncalli et al. (2021), credit to MDPI, Basel)

Toxin analyses of field-collected zooplankton during blooms of *Alexandrium* spp. support the hypothesis that low levels of toxin are accumulated in zooplankton. This indicates that zooplankton effectively disperses saxitoxin into the surrounding environment. Even after fecal material is accounted for, ~90% of the toxin from cells removed by grazing activity remains unavailable to higher consumers. Overall, dispersal of toxins is another indirect mechanism of toxin resistance.

Turnbull et al. (2020) demonstrated that the southern rock lobster *Jasus edwardsii* employs a different strategy. *J. edwardsii* can rapidly accumulate PST in its hepatopancreas when feeding on highly toxic prey, without any noticeable impact on survival. This indicates that lobsters in the field will continue to feed during toxic blooms. PSTs are likely to be excreted through the feces, antennal glands, and gills.

Recently, an essential mechanism of resistance in copepods feeding on toxic *Alexandrium* sp. has come to the forefront, which deserves much more attention as it can be manipulated dietarily: namely, the gut microbiome. Yang et al. (2024) provided the first study of copepod gut microbial communities at the individual level, aiming to disentangle the roles and mechanisms of gut microbes on the detoxification of toxic dinoflagellates. The authors demonstrated that while differences between copepod species have a stronger influence than algal diet on the composition of their gut microbiota (Fig. 5.10), the functional capacity of the microbiota tells a different story—here, diet type plays the predominant role. Under a toxic diet, the number and taxonomy of enriched bacteria differ significantly between the two copepod species. Most of these taxa are reported to have the function of degrading algal toxins (e.g., species from the genera *Brevibacterium*, *Pseudomonas*, *Stenotrophomonas*, and *Alteromonadaceae*) and maintaining host homeostasis (e.g., species from *Rhodobacteraceae*, *Rhizobiaceae*, and *Parcubacteria*). This suggests a potential functional convergence of the gut microbiome in response to algal toxins among different copepod species. Further studies, such as transcriptome analysis, will be needed to identify the detailed gene and pathway regulation of the copepod gut microbiome.

5.2.1.2 Mollusks

In laboratory experiments, mollusks are typically exposed to toxin-producing microalgae until their tissue becomes contaminated, after which the food supply is halted to initiate the detoxification phase. This creates two distinct and separate periods: ingestion and detoxification. In natural ecosystems, however, these phases often overlap, as mollusks continue to feed while simultaneously processing toxins. As a result, detoxification in the wild may follow very different dynamics compared to extended detoxification periods observed under laboratory conditions (Leal and Cristiano 2024).

The snail *Chorus giganteus*, an important fishery resource, is capable of accumulating PST within its body, mainly in the digestive glands (Andrade-Villagrán et al. 2022). A diet consisting of prey intoxicated with PST has no apparent effect on the ingestion rate or total lipid content of this snail.

This consistent pattern across species highlights a convergent strategy among carnivorous gastropods, such as *Concholepas concholepas* (Compagnon et al. 1998)

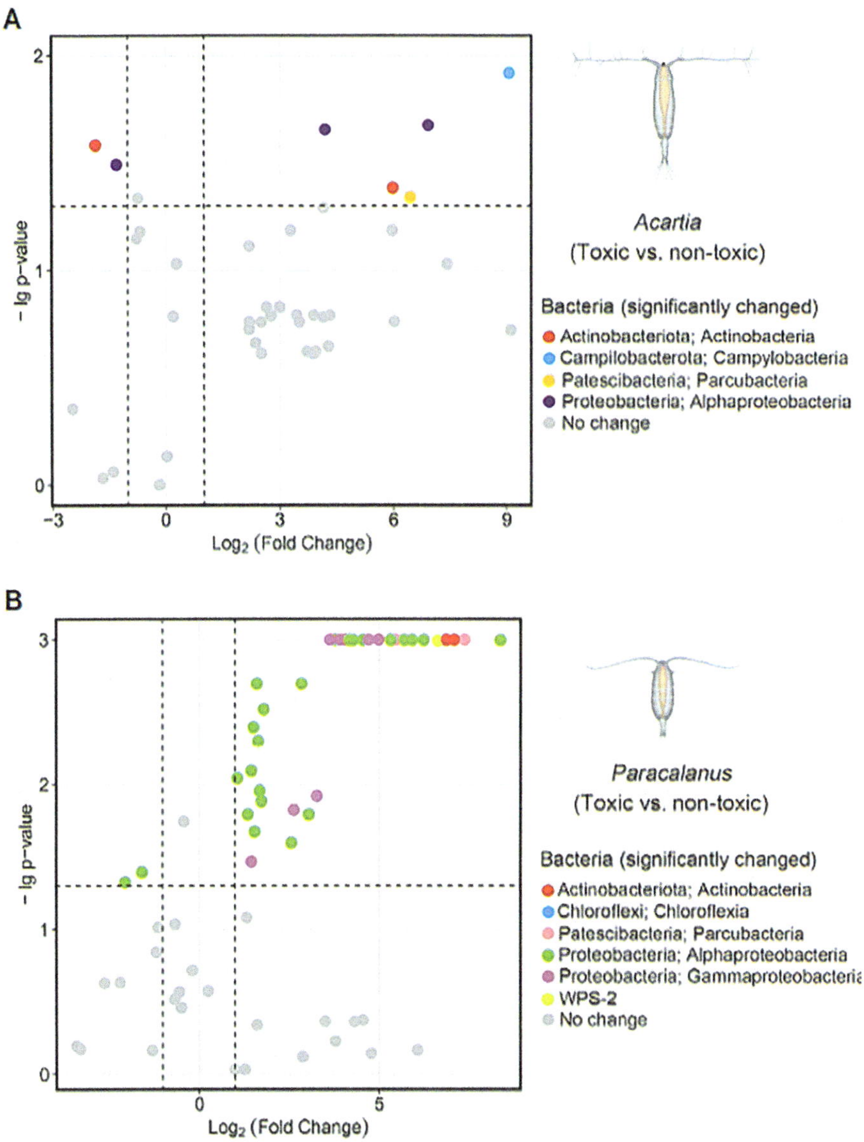

Fig. 5.10 Gut bacteria with significant changes under a toxic diet. A similarity analysis (SIMPER) was conducted to identify all the distinct ASVs between two groups (i.e., toxic and non-toxic) of GMC with significance (p-value) for *Acartia* sp. (**a**) and *Paracalanus* sp. (**b**). Fold changes of these distinct amplicon sequence variants (ASVs) were calculated based on their relative abundance in each group (i.e., toxic and non-toxic diet) and $\log_2$ transformed. The two dashed lines on the x-axis were $x = 1$ and $x = -1$, representing the increasing ($x > 1$) and decreasing ($x < -1$) trend of gut microbes after feeding toxic algae (vs. non-toxic), respectively. The dashed line on the y-axis was $y = 1.3$, and ASV with a p-value<0.05 (i.e., $-\lg(p\text{-value}) > 1.3$) was considered significantly changed and colored according to its taxonomy (Phylum, Class) (from Yang et al. (2024), credit to Frontiers Media)

and *Rapana venosa* (Ito et al. 2004)), where the digestive gland serves as the primary reservoir for PST, allowing them to withstand exposure while still posing a significant risk of toxin transfer through the food web.

However, the snails with accumulated PST exhibit an increase in total protein content (likely due to increased production of detoxification enzymes), together with a significant increase in the fatty acid profile, especially EPA and DHA. The increase in DHA content can be attributed to the indirect consumption of dinoflagellates (Leblond and Chapman 2000, Dalsgaard et al. 2003, Remize et al. 2020) and the autochthonous production of EPA and DHA by the snail itself (Kabeya et al. 2018; Monroig and Kabeya 2018). Unfortunately, the authors do not discuss the pros and cons of increased EPA and DHA levels or potential links to PST tolerance, nor do they discuss the possible long-term mechanisms and consequences of PST accumulation. However, it is clear that the increase in DHA content is in contrast to the response of *Perna viridis* discussed above; obviously, this valuable PUFA is not being used for toxin esterification.

A seminal work has just been published on the Yesso scallop (*Mizuhopecten* (*Patinopecten*) *yessoensis*) (Xun et al. 2020). The authors studied solute carriers (SLCs) in this bivalve. SLCs are membrane proteins that transport many endogenous and exogenous substances, including xenobiotic toxins. Exploring the SLCs related to PST accumulation provides insights into the mechanisms of PST bioavailability and adaptations of these species to PST. The authors identified SLC expansion in the genome, namely 673 SLCs in 48 subfamilies. After feeding the scallops with PST-producing dinoflagellates, 131 SLCs were regulated, and more than half of them are from expanded subfamilies. SLCs from the extended subfamily are found to be under positive selection and may be related to strain-specific adaptation to marine environments with algal-derived biotoxins. The expansion of stress-responsive genes is apparently a widespread mechanism for coping with environmental stress, as previously demonstrated in the Pacific oyster (*Magallana* (*Crassostrea*) *gigas*) (Zhang et al. 2012). In this context, it would be of great interest to determine whether other stress-responsive pathways, such as biotransformation phases I and II, are conspicuous by their gene expansion in the scallop. Exploring this possibility could reveal whether tolerance to saxitoxins depends not only on modifications to target molecules, such as sodium channels, but also on a broader genomic toolkit of detoxification, repair, and protective mechanisms. This concept was discussed above with respect to the nuclear hormone receptor 96. Such an integrated perspective would help explain why some bivalves act as toxin reservoirs with little physiological cost, while others remain highly vulnerable. More broadly, comparative genomic and transcriptomic studies across mollusks, crustaceans, and other invertebrates could uncover common strategies or lineage-specific solutions, shedding light on the evolutionary pressures that shape toxin resistance. Ultimately, advancing our knowledge of these molecular adaptations is not only critical for understanding the ecology and evolution of toxin–host interactions, but also for managing seafood safety and predicting how marine food webs may respond to intensifying harmful algal blooms in a changing ocean.

5.2.2 Domoic Acid

Among the various marine phycotoxins, domoic acid (DA) stands out due to its broad ecological and economic consequences. Produced by *Pseudo-nitzschia* species, DA accumulates through the food web, causing neurological impairment and mass mortalities across trophic levels from zooplankton to top predators, including humans. Because DA is distributed throughout the food web, a central question arises: How do organisms withstand repeated or chronic exposure to DA? The answer lies in diverse tolerance and resistance mechanisms that differ across taxa, ultimately shaping both individual survival and the stability of entire ecosystems.

Feeding, egg production, and egg hatching success of the copepods *Acartia tonsa* and *Temora longicornis* do not differ on diets of the toxic diatom *Pseudo-nitzschia multiseries* or the non-toxic diatom *Pseudo-nitzschia pungens* (Lincoln et al. 2001). The copepods accumulate relatively high levels of DA without immediate harm. In support, Maneiro et al. (2005) and Olson and Lessard (2010) show that at documented in situ concentrations, DA has little to no impact on the ability of micro- and mesozooplankton, including *Acartia clausii*, to grow at substantial rates compared to when feeding on optimal diets. This fact is also true for *Calanus* copepodites (Haroardóttir et al. 2015) and *Calanus finmarchicus* (Leandro et al. 2010). As one underlying mode of tolerance, Maneiro et al. (2005) and Leandro et al. (2010) mention a high depuration rate, which is consistent with the usual pattern observed in oysters (Lassus et al. 2000) and in copepods exposed to saxitoxin (Guisande et al. 2002). However, even high excretion rates do not adequately explain why immediate adverse effects do not occur when animals persistently ingest toxic prey.

Mafra et al. (2009a; b) presented experimental evidence for mechanisms contributing to low DA uptake by eastern oysters (*Crassostrea virginica*) feeding on *Pseudo-nitzschia* cells. Oysters filter increasing amounts of *P. multiseries* cells as cell density increases, but the production of pseudofeces limits ingestion. Oysters significantly reduce their clearance rate[4] when feeding on both toxic and non-toxic *P. multiseries* clones. It appears that DA intake by oysters from mono-specific *Pseudo-nitzschia* blooms is limited by a persistently reduced clearance rate and rejection of cells in pseudofeces. In the event of a multispecific bloom in which *Pseudo-nitzschia* spp. make a lesser contribution to the total food supply, no clearance rate inhibition is expected, and DA intake is therefore regulated by preferential rejection of *Pseudo-nitzschia* spp. cells in pseudofeces (Mafra et al. 2009a, b). Intriguingly, contamination experiments confirmed that *C. virginica* accumulates much less DA from *P. multiseries* than blue mussels (*Mytilus edulis*) under the same experimental conditions (Mafra et al. 2009a, b). Subsequently, Mafra et al. (2010) explained that feeding physiology (reduced clearance rate and selective rejection of *P. multiseries* cells from a mixed suspension) plays a major role in explaining the relatively low capacity of oysters, *C. virginica*, to accumulate DA during toxic blooms relative to other bivalves, most notably the blue mussel *M. edulis*. These

[4] Clearance rate: The volume of water cleared of particles per unit time.

findings highlight that not all bivalves accumulate DA to the same extent, and that species-specific feeding strategies—such as selective rejection or differential assimilation—can strongly influence the risk of toxin transfer through the food web.

Recent de novo assembled transcriptome studies comprehensively demonstrate the response of exposed animals to DA exposure. Pazos et al. (2017) presented a transcriptome analysis of *Mytilus galloprovincialis* after exposure to a DA-producing *Pseudo-nitzschia*. A total of 1158 differentially expressed genes (DEGs) were identified in 37 families. The overexpression of membrane transporters from the SLC family suggests their active involvement in the absorption and elimination of DA. This is in good agreement with the Yesso scallop study mentioned above. Other enzyme/protein families (sulfotransferases, aldo-keto reductases, carboxylesterases, C1q domain, and fibrinogen C-terminal globular domain)[5] are involved in DA detoxification, antioxidant response, and immunological processes.

Using two different scallop species, Ventoso et al. (2019, 2021) confirmed that low sensitivity to DA is not due to a single response/defense pathway; rather, it is the concerted interaction of multiple pathways (Fig. 5.11). In queen scallop (*Aequipecten opercularis*), a total of 10,144 genes were differentially expressed in the toxin-exposed scallops in relation to the controls. The results obtained are consistent with the hypothesis that exposure to DA-producing *Pseudo-nitzschia* causes oxidative stress in *A. opercularis*. Oxidative stress leads to impaired mitochondrial function and protein oxidation. In response, the queen scallop mounts a transcriptional defense by upregulating genes that help mitigate these effects. These include genes coding for proteasome components (to degrade oxidized proteins), mitochondrial ribosomal proteins (to promote mitochondrial biogenesis), and translocases of the outer and inner mitochondrial membranes (TOMMs and TIMMs, ensuring proper import of cytosolically synthesized preproteins). In addition, spliceosome components are upregulated to ensure efficient mRNA processing, while thioredoxins, glutaredoxins, and copper/zinc superoxide dismutase contribute to enzymatic and antioxidant activity (Fig. 5.11).

A significant number of upregulated genes code for proteins involved in the metabolism of xenobiotics (cytochromes P450, aldo-keto reductases,[6] glutathione transferases, and sulfotransferases) and transmembrane transport (solute carriers). On the contrary, genes coding for proteins with domains involved in immunological processes (C1q domain, C-type lectins, immunoglobulin domain, fibrinogen beta- and gamma-chain C-terminal globular domain, and tumor necrosis factors) are

---

[5] Sulfotransferases catalyze the transfer of a sulfo group from a donor molecule to an acceptor alcohol or amine and belong to phase II biotransformation (Steinberg 2012). Reductases catalyze the transfer of electrons from one molecule (electron donor) to another (electron acceptor). Carboxylesterase or carboxylic ester hydrolase catalyzes the cleavage of carboxylic esters into alcohol and carboxylate (Nelson and Cox 2005). C1q is the first subcomponent of the classical pathway of the complement system and an important link between innate and acquired immunity (Kishore et al. 2004). C-terminal globular domain of fibrinogen is involved in blood clotting (Akakura et al. 2006).

[6] It is noteworthy that DA with three carboxyl groups activates redox-sensitive phase I biotransformation enzymes, and not only phase II conjugating enzymes (transferases).

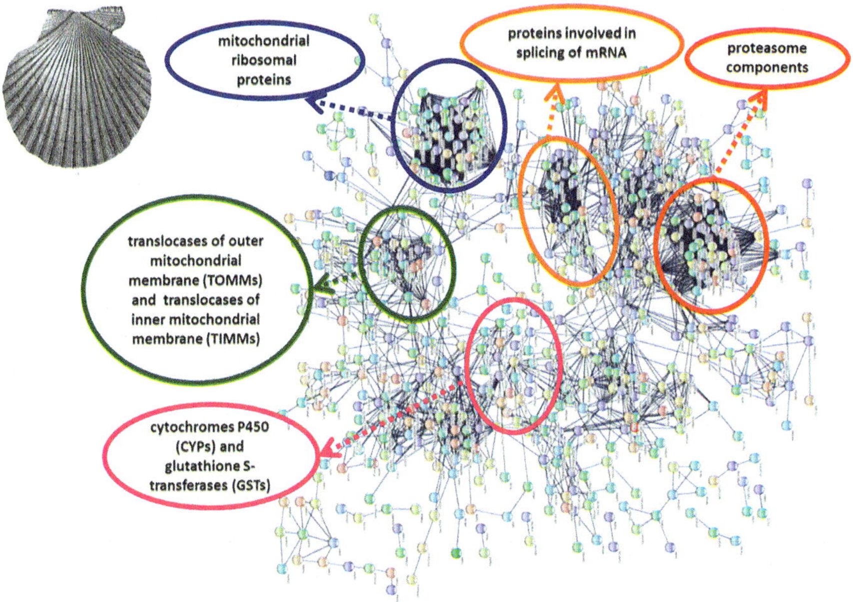

Fig. 5.11 Network showing interactions of proteins coded by genes upregulated in the queen scallop (*Aequipecten opercularis*) digestive gland after exposure to domoic acid-producing *Pseudo-nitzschia*. Some highly connected protein nodes are highlighted. Proteins are named according to the human protein names (from Ventoso et al. (2019), credit to MDPI, Basel; image in public domain, credit to Nyst (1878))

mainly downregulated—except for the C-type lectins. This study is among the few that provide a reasonably comprehensive overview of the pathways triggered by oxidative stress—yet notably, it does so without mentioning the central nuclear transcription factor NRF2.

A similar picture is obtained in the king scallop (*Pecten maximus*) after the injection of DA. Protein networks obtained from the upregulated genes are enriched in Gene Ontology (GO) terms, including vesicle-mediated transport, response to stress, signal transduction, immune system processes, RNA metabolic process, autophagy, lysosome, and oxidoreductase activity. On the other hand, networks obtained from downregulated genes are enriched in terms such as response to stress, immune system processes, ribosome biogenesis, signal transduction, mRNA processing, and oxidoreductase activity (Ventoso et al. 2021).

5.2.3 Okadaic Acid

A considerable number of studies have focused on the effects of okadaic acid (OkA), particularly in copepods and mollusks (bivalves), which represent two key groups in marine food webs. Bivalves, being primary filter feeders, are often the first to accumulate OkA and thus play a central role in its transfer through the trophic chain. Copepods, on the other hand, are crucial grazers linking phytoplankton to higher consumers, making them equally relevant for understanding the ecological impact of OkA. Together, these organisms provide valuable models for studying both the physiological consequences of toxin exposure and the potential tolerance or resistance mechanisms that may have evolved under natural bloom conditions.

5.2.3.1 Copepods

Despite the production of OkA by *Dinophysis norvegica*, several copepods use this alga as food. Jansen et al. (2006) compared the feeding behavior and egg production of three copepods during a dinoflagellate bloom in the North Sea. The general ability to ingest harmful algal species such as *Dinophysis norvegica* and *Ceratium furca* and subsequently produce fecal pellets (Fig. 5.8c) is present in *Calanus helgolandicus* and *Temora longicornis* (Fig. 5.8b), but not in *Acartia* sp. Among these, *C. helgolandicus* is markedly more efficient than *T. longicornis*. The high hatching success and the large proportion of incubated *C. helgolandicus* females producing eggs suggest that these copepods were in good physiological condition during the study. Yet, the key question remains unresolved: Which specific physiological traits set *C. helgolandicus* apart from other copepods and underpin its remarkable resistance to OkA?

Unlike the unnamed *Acartia* above, *Acartia bifilosa* is capable of feeding on *Dinophysis* spp. in the northern Baltic Sea, albeit at a lower rate (up to 64 cells (0.1 μg C) ind^{-1} day^{-1}) than *Eurytemora affinis* (up to 226 cells (0.50 μg C) ind^{-1} day^{-1}) (Setälä et al. 2009). This difference is unlikely to be explained by body size, as reproducing females of both species in the northern Baltic Sea are of comparable size (Viitasalo et al. 1995).

In addition to copepods, ciliates, including tintinnids and mixotrophic dinoflagellates, are also capable of taking nutritional advantage of *Dinophysis* (Reguera et al. 2012). For example, Maneiro et al. (2000) found that the population density of *Schmidingerella* (*Favella*) *serrata* increases significantly during *Dinophysis* blooms and decreases when the blooms collapse. *Schmidingerella* is apparently able to retain OkA without immediate harm.

5.2.3.2 Mollusks

So far, the interaction of OkA with zooplankton has shown that not all zooplankton species are equally susceptible to this biotoxin. The knowledge remains phenotypic—in contrast to bivalve mollusks. Prego-Faraldo et al. (2013) reviewed the state of the art of a decade ago. OkA is mainly absorbed in these animals and accumulated in the digestive gland either in a free form or (for the most part) associated with high-density soluble lipoproteins (Rossignoli and Blanco 2008). Upon exposure to OkA producers, lipopolysaccharide/β-1,3-glucan binding protein is upregulated (Romero-Geraldo et al. 2014). This association results in the sequestration of OkA, preventing its transportation to other tissues and hindering its elimination from the organism. On the contrary, free OkA is easily transported and quickly removed by means of different passive detoxification mechanisms, such as direct

OkA excretion through the gill or the digestive system (Svensson 2003; Blanco et al. 2007). Regulation of OkA depuration is insensitive to immediate environmental changes (Svensson and Förlin 2004) (from Prego-Faraldo et al. (2013), references added).

Prego-Faraldo et al. (2013) continue that, since most OkA is sequestered by lipoproteins in the digestive gland, a possible depuration mechanism may involve the association of high-density lipoproteins with ATP-binding cassette (ABC) transporters, similar to those involved in the removal of excess cholesterol from cells (Rossignoli and Blanco 2008; Rossignoli and Blanco 2010). This hypothesis is also supported by Xu et al. (2014), who assume that P-glycoproteins play a central role in the excretion of OkA by *Crassostrea ariakensis*.

However, it now seems clear that transformation processes, such as hydrolysis (during digestion) and, most importantly, acylation, are decisive for the metabolism of OkA, producing both free toxin and esters. Indeed, the rapid bioconversion of OkA to 7-*O*-acyl derivatives can be considered a defense mechanism against this biotoxin, although the hydrophobic nature of the esters may slow their elimination from tissues via body fluids (Suzuki et al. 2005). OkA is primarily transformed in the endoplasmic reticulum of bivalves (Rossignoli et al. 2011) (from Prego-Faraldo et al. (2013), references added). In addition to these pathways, Huang et al. (2015b), Dou et al. (2020), Martins et al. (2020), Martínez-Escauriaza et al. (2021), and Wei et al. (2019) reported that P-glycoprotein transporters, sodium-dependent solute carriers, GSTs, and several CYP450s are involved in OkA detoxification in *Perna viridis* and *M. galloprovincialis*. The CYP450s are tissue-specific. Finally, Dou et al. (2020) hypothesized that *P. viridis* relies primarily on GST and ABC transporters to counteract OkA during short-term exposure. However, prolonged exposure to *P. lima* activates the NRF2 pathway (→Box 2.1 in Steinberg (2024)) and inhibitors of apoptosis proteins, which, in turn, reduce the damage caused by OkA. Furthermore, OkA causes destabilization of the cytoskeleton and has adverse effects on the immune system.

In the meantime, some more comprehensive omics studies have been conducted. Suarez-Ulloa et al. (2015) assumed that the response to OkA found in *M. galloprovincialis* is consistent with the model of intracellular response to stress previously reported for bivalve mollusks (Anderson et al. 2015). Accordingly, the activation of energy production mechanisms observed in the present work could potentially produce harmful ROS, which, unless controlled by chaperones or eliminated in the proteasomes, would induce apoptosis. Indeed, the results show an upregulation of important chaperones (*hsp70*) and proteases (*cathepsins b* and *d*). Particularly, the strong upregulation of *cathepsins*, known to be activated in lysosomes (Kågedal et al. 2001), in conjunction with the activation of transport mechanisms, supports the lysosomal uptake hypothesis proposed by Svensson et al. (2003). In addition, the observed upregulation of antimicrobial peptides suggests the activation of immune defense mechanisms, operating alongside the broader environmental stress response. However, it remains unclear whether abiotic factors automatically trigger this immune response or whether it represents an opportunistic attack by pathogens present in the microbiota of the mussels (Suarez-Ulloa et al. 2015).

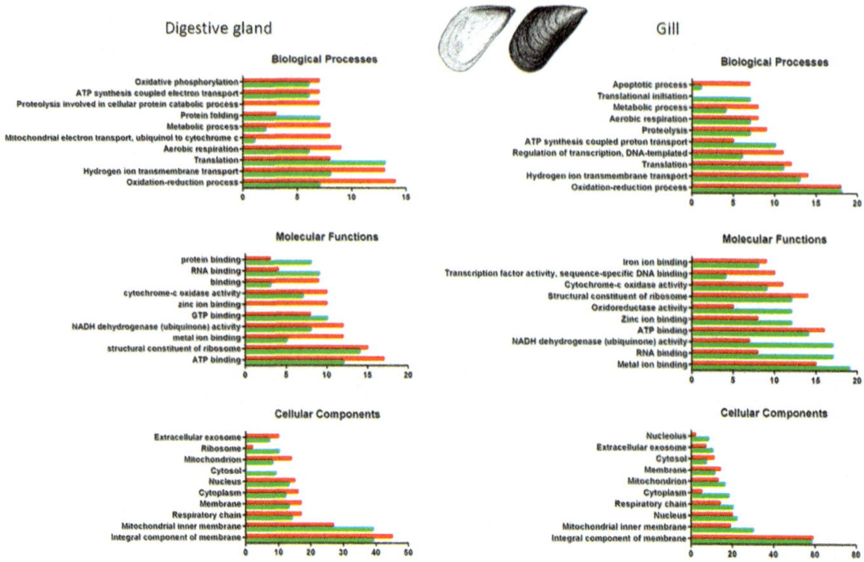

Fig. 5.12 Gene Ontology (GO) classification of differentially expressed genes from the digestive gland and gills of the Mediterranean mussel *Mytilus galloprovincialis* exposed to the DSP toxin--producing dinoflagellate *Prorocentrum lima*. Overrepresented and underrepresented biological processes, molecular functions, and cellular components are shown. Red and green bars represent the numbers of down- and upregulated genes, respectively. The length of the bars is determined by the number of genes identified within each subcategory. From Prego-Faraldo et al. (2018), credit to MDPI, Basel; image credit to FAO)

Later, the same laboratory elucidates the early transcriptional response of the digestive gland and the gill of *M. galloprovincialis* after exposure to *Prorocentrum lima* by de novo assembly based on the sequencing of eight cDNA libraries (Prego-Faraldo et al. 2018). A total of 2286 (in the digestive gland) and 4523 (in the gill) differentially expressed transcripts were obtained, indicating tissue-specific transcriptome responses (Fig. 5.12). The early response includes genes involved in defense, immunity, and metabolism, shedding some light on the resistance mechanisms that these organisms have against the harmful effects of diarrhetic shellfish toxins. In the digestive gland, *bd2*, *kazal dc*, and *nadh5* are upregulated, while *giy-yig dc* is downregulated. *Dyna* shows no changes. On the other hand, *alox15b*, *h_lectin dc*, *cpla2*, and *fuca* are upregulated, and *fibrinogen c_dc*[7] is downregulated in

[7] *bd2 big defensin 2*: defense response to bacteria; *nadh5 NADH dehydrogenase subunit 5*; *kazal dc Kazal domain containing protein*: The Kazal domain is an evolutionary conserved protein domain usually indicative of serine protease inhibitors (Rimphanitchayakit and Tassanakajon 2010). *giy-yig dc GIY-YIG domain containing protein*: an endonuclease domain (Fukui et al. 2018). *dyna dynactin-subunit-6-like*: dynactin is a multisubunit protein complex that is required for most types of cytoplasmic dynein activity in eukaryotes—dynein is a family of cytoskeletal motor proteins that move along microtubules in cells (Schroer 2004). *alox15b arachidonate 15-lipoxygenase B-like*: lipoxygenase family of structurally related nonheme iron dioxygenases involved in the

the gill. Remarkably, many of the genes that respond to these toxins also respond to other stimuli, indicating that the mussel defense response is to some extent nonspecific, which is advantageous for other potentially harmful compounds.

At the same time, Chi et al. (2018) focused on the transcriptomic analysis of detoxification- and immune-related genes in the gills of bay scallops (*Argopecten irradians*) in response to OkA. *Cyclic AMP-responsive element-binding proteins, acid phosphatase, toll-like receptors, nuclear erythroid 2-related factor*, and the *NADPH2 quinone reductase-related* genes are upregulated. In contrast, the expression of certain genes related to *glutathione S-transferase 1, C-type lectin, complement C1q tumor necrosis factor-related protein, superoxide dismutase 2*, and *fibrinogen C domain-containing protein* decreases. Of note, the response pattern of bay scallop gills exposed to OkA is quite similar, but not completely identical, to that of the queen scallop digestive gland following exposure to DA-producing *Pseudo-nitzschia* (Fig. 5.11).

Ye et al. (2022) compared the responses of three bivalve species after exposure to the DSP producer *Prorocentrum lima*. The transcriptional induction of Nrf2-targeted genes, such as *abcb1* and *gpx*, indicates the functionality of the Nrf2 pathway against DSP toxins in bivalves. However, the response pattern of the three species studied differs. The Pacific oyster (*Magallana gigas*) can protect itself against DSP toxins mainly by ABC transporters and esterification, while the mussel (*Mytilus unguiculatus*) and the clam (*Tegillarca granosa*) reduce the damage induced by DSP toxins mainly by regulating the expression of antioxidant genes.

Applying proteomics to Farrer's scallop (*Azumapecten* (*Chlamys*) *farreri*) exposed to OkA, Wang et al. (2021) demonstrated that OkA leads to significant changes in physiology and proteomics in addition to the toxicity mechanisms mentioned above, even though the concentration was far below the levels tolerated by the scallops. OkA also significantly alters biological processes such as amino acid biosynthesis, energy metabolism, cytoskeletal, and phagosome functions.

Huang et al. (2021) present one of the first epigenome studies in OkA-exposed *P. viridis*. 90 mature miRNAs and 33 potential new miRNAs are identified, and their expressions in digestive gland tissue are detected. After exposure to *P. lima*, the expression of some miRNAs in the digestive gland tissues changes, suggesting that miRNAs play an essential role in the response and detoxification of DSP toxins in *P. viridis*. Further analysis shows that *miR-71_5, miR-750_1*, and *novel_mir4* are significantly upregulated after 6 h of *P. lima* exposure. At the same time, their target genes, *nhl rc2*[8] and *C1q-like*, are downregulated (Fig. 5.13), indicating the involve-

production of fatty acid hydroperoxides (Brash et al. 1997). *h_lectin dc H_Lectin domain containing protein*: lectins are ubiquitous carbohydrate-binding proteins that interact with sugar moieties in a highly specific manner. H-type lectins represent a new group of lectins that were identified in invertebrates (Pietrzyk-Brzezinska and Bujacz 2020). *cpla2 cytosolic phospholipase A-2 like*: calcium-dependent enzyme that cleaves fatty acids, such as arachidonic acid, from phospholipids (Leslie 1997). *fuca alpha-L-fucosidase-like*: enzyme participates in n-glycan (fucoside) degradation; *fibrinogen c_dc fibrinogen_C domain containing protein* (→FN5).

[8] NHL Repeat Containing Protein 2 with anti-apoptotic properties (Polkoff et al. 2017).

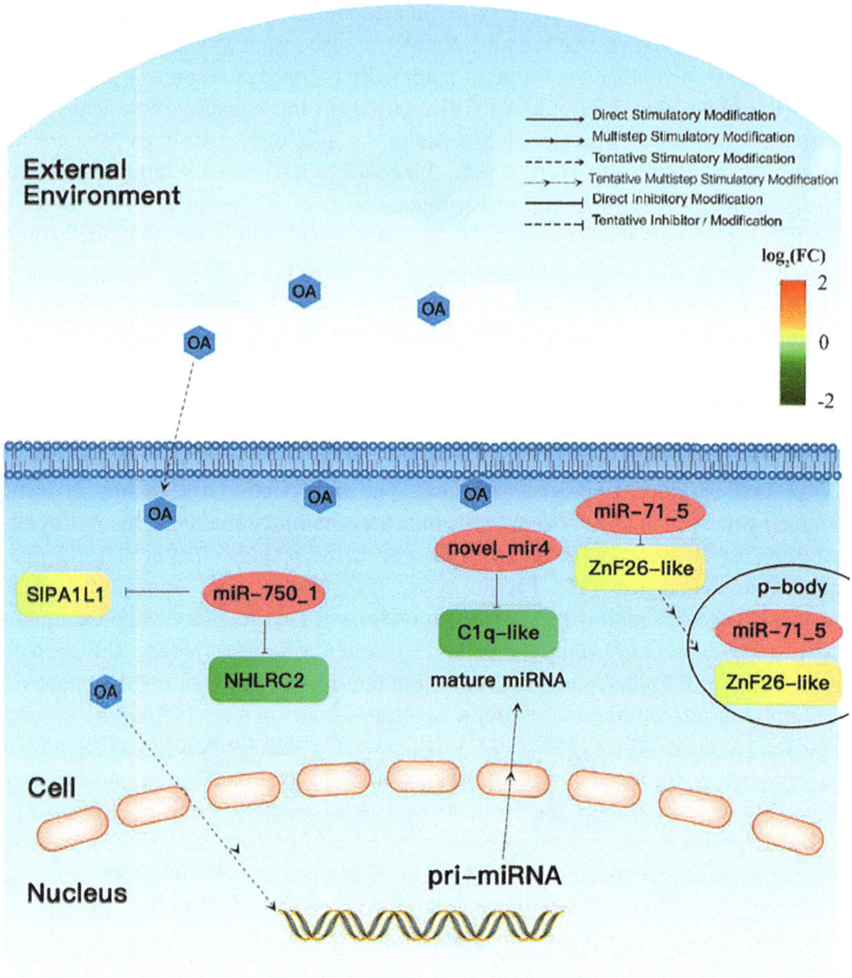

Fig. 5.13 Underlying response mechanism in the digestive gland of *Perna viridis* after exposure to *Prorocentrum lima*. Details in the text (from Huang et al. (2021) with permission from Elsevier)

ment of these miRNAs in apoptosis and immune stress during short-term exposure to DSP toxins. The downregulation of *ZnF26-like*[9] after 96 h of exposure indicates that *ZnF26-like* may contribute to the response of the immunity pathway, endocytosis, and MAPK signaling pathway to *P. lima* exposure in *P. viridis*.

[9]Zinc finger protein 26. Zinc finger proteins are a class of nucleic acid-binding proteins that have a specific protein domain: the zinc finger domain, in which Zn^{2+} is coordinately bound (Nelson and Cox 2005).

5.2.4 Oxylipins

Diatoms are the dominant component of the marine phytoplankton. Several diatoms produce secondary metabolites, specifically oxylipins, which have teratogenic effects on their primary predators, crustacean copepods. Asai et al. (2020) reported the de novo assembled transcriptome of *C. helgolandicus* feeding on the oxylipin-producing diatom *Skeletonema marinoi*. This study demonstrates that when *C. helgolandicus* females ingest *S. marinoi*, maternal folate metabolism is disrupted, likely impairing DNA synthesis and altering gene methylation. This, in turn, perturbs oogenesis and embryogenesis signaling pathways responsible for cellular proliferation and patterning, ultimately reducing embryo viability and causing teratogenic effects.

Also, based on a de novo assembled transcriptome analysis, Russo et al. (2020) studied females of the copepod *Temora stylifera* collected in the Gulf of Naples, Mediterranean Sea, over consecutive weeks, during which opposite naupliar survival was observed. The authors aimed at detecting key genes that may have influenced copepod reproductive potential in natural populations. Most of the significantly downregulated unigenes and isoforms were *A5 putative odorant-binding proteins*. This downregulation in *T. stylifera* may impair the detection of harmful prey, leading to unselective feeding, increased ingestion of oxylipin-producing diatoms, and negative effects on larval survival.

Although these studies are more concerned with toxicity than with the mechanism of tolerance, they clearly demonstrate that (de novo) whole-genome analyses are essential to identify both the mode of action of toxicity and tolerance.

5.2.5 Cyanotoxins

Pollution of estuarine and coastal ecosystems by toxic cyanobacteria and their associated cyanotoxins is a global issue that affects marine bivalves and other aquatic organisms. The effects of toxic cyanobacteria on marine and estuarine aquatic ecosystems are less well studied than those on freshwater ecosystems. Therefore, Oliveira et al. (2020) studied the physiological and metabolic responses of marine bivalves exposed to toxic cyanobacteria and found that species like *M. galloprovincialis* are capable of filter-feeding on toxic cyanobacteria, including the unicellular and colony-forming *M. aeruginosa* as well as the filamentous *Chrysosporum ovalisporum*. After 15 days of feeding on toxic cyanobacteria, marine mussels showed the ability to resist the potential toxicity of this diet. Surprisingly, there are no changes in animal weight, ability to produce byssus, or biochemical markers studied (total protein and glycogen), suggesting that the marine bivalve has its own defense mechanisms against cyanotoxins. Mussels can eliminate cyanotoxins such as cylindrospermopsin (CYN) during the clearance phase. This detoxification mechanism protects the animals and helps prevent the accumulation of harmful

CYN concentrations in their tissues. Subsequent proteomic analysis reveals changes, indicating that animals adjust their metabolism in response to diet. Some molecular responses are associated with defense mechanisms, such as overexpression of ENO1 (enolase 1—involved in energy metabolism), HSP90 (heat shock protein 90—a chaperone protein that assists other proteins in proper folding, stabilizes them against heat stress, and facilitates the degradation of damaged or misfolded proteins), and HNRNPA1 (heterogeneous nuclear ribonucleoprotein A1—an RNA-binding protein), which are linked to processes of protein stabilization, regulation of protein folding, and activity in cells (Oliveira et al. 2020).

By studying the diet of eastern oysters (*Crassostrea virginica*) growing in a eutrophic tributary of Chesapeake Bay, Maryland, USA, Weissberger and Glibert (2021) suggested that these oysters selectively reject cyanobacteria, especially pico-cyanobacteria, from their diet, which likely helps them avoid contamination with cyanotoxins. This selective feeding is consistent with previous findings showing that eastern oysters can discriminate between nutritious organic particles and less desirable components in the seston, actively excluding less nutritious or potentially harmful particles from ingestion, thereby maintaining a positive energy balance even in eutrophic environments rich in cyanobacteria (Newell and Jordan 1983; Galimany et al. 2017). Other studies reported that several bivalve species in an Indian lagoon, including eastern oysters, hooked mussels (*Ischadium recurvum*), charru mussels (*Mytella strigata*), green mussels (*Perna viridis*), Atlantic rangia (*Rangia cuneata*), and hard clams (*Mercenaria mercenaria*), were able to assimilate *Synechococcus*, as shown by enriched ^{15}N values in bivalve tissues (Galimany et al. 2020). In Florida Bay, USA, sponge mortality has been linked to *Synechococcus* blooms; however, experimental tests using three sponge species (*Biemna* spp., *Halichondria melanadocia*, and *Spheciospongia vesparium*) could not confirm any causal relationship (Lynch and Phlips 2000). This highlights a persistent gap in understanding: neither the causes of sponge mortality in situ nor the mechanisms by which some sponges tolerate cyanobacterial toxins in controlled conditions are well understood.

Overall, these studies underscore that marine organisms employ a range of strategies—from selective feeding to biochemical detoxification and proteomic adjustments—to cope with cyanotoxin exposure. At the same time, the ecological outcomes of cyanobacterial blooms in coastal and estuarine systems remain complex and context-dependent. The variable responses among taxa suggest that tolerance and resistance mechanisms are species-specific, often multifactorial, and shaped by both evolutionary history and local environmental pressures. Consequently, more comprehensive, ecosystem-level research is urgently needed to understand the interactions between toxic cyanobacteria and marine food webs, including how these interactions influence energy flow, species resilience, and overall ecosystem stability. This knowledge will be critical not only for assessing ecological risk but also for informing sustainable management of estuarine and coastal resources.

5.3 Freshwater Invertebrates

In freshwater ecosystems, cyanobacteria play roles far more important and complex than merely producing toxins and harming other organisms (Ferrão-Filho and Kozlowsky-Suzuki 2011). They are central players in nutrient cycling and food web dynamics. Besides alterations in the physical conditions of the habitat and the generation of new ecological niches and competition for space, Dudley et al. (1986) proposed the food source as a major ecological interrelationship among macroalgae, including cyanobacteria colonies, and macroinvertebrates in benthic stream ecosystems. Cyanobacteria are important primary producers in aquatic systems and can sometimes, under special environmental or seasonal circumstances, represent the main autotroph group and be the only food resource for freshwater invertebrates (Gérard and Lance 2019; Briland et al. 2020; Aboal et al. 2022). The interactions between invertebrates and cyanobacteria are therefore shaped by a delicate balance between risk and benefit: While toxins pose physiological challenges, cyanobacteria can also serve as an important food source, especially during blooms when other resources are scarce. Understanding how freshwater invertebrates tolerate, resist, or even exploit cyanobacteria provides key insights into ecosystem resilience and the evolutionary arms race between primary producers and consumers.

Calcareous oligotrophic rivers are colonized by a diverse cyanobacterial community for most of the year, especially in Mediterranean areas (Sabater et al. 2000; Aboal and Puig 2005). Cyanobacteria have been traditionally considered a poor food resource, not only because they may produce toxins or present morphologies that are unappealing or difficult to ingest, such as long or thick filaments or mucilage, but also because they are of low nutritional quality (Hart 1985; Ferrão-Filho 2009; Scott and Marcarelli 2012; Berezina et al. 2021; 2022a; b; c; d; e; f; g; h; i).

However, cyanobacteria provide macroinvertebrates with a variety of nutrients and bioactive macromolecules—including pigments, carbohydrates, lipids (including essential fatty acids), proteins, vitamins, and minerals—important for growth and maturation (Krivosheina 2008; Scott and Marcarelli 2012; Briland et al. 2020; Chen et al. 2021; Aboal et al. 2022). Yet, they lack two critical components: long-chain PUFAs and sterols. Intriguingly, Gerphagnon et al. (2019) demonstrated that chytrid parasites can enhance the biochemical quality of cyanobacterial organic matter by providing these missing compounds, thereby improving its nutritional value for consumers. And remarkably, *Ochromonas* (*gloeopara*) even prefers *Microcystis* over chlorophytes (Zhang et al. 2018a; Wei et al. 2022). Similarly, population growth of amoebae (*Cochliopodium* sp., *Vannella* sp., and *Vermamoeba vermiformis*) is supported by the common cyanobacterial bloom-formers *Dolichospermum* sp. and *Aphanizomenon flos-aquae*, while green algae generally do not support the growth of the amoebae tested (Weger et al. 2024). The testate amoeba *Penardochlamys* sp. grazes on two species of *Microcystis* in a hypereutrophic pond (Nishibe et al. 2004), and naked amoebae such as *Vannella* and *Cochliopodium* are capable of ingesting *Microcystis* cells as well (Van Wichelen et al. 2016). It appears that the potential for grazing of cyanobacteria by amoebae is relatively specific, with certain amoebae being capable of grazing certain cyanobacteria (Weger et al. 2024). However, none of these papers discuss potential underlying mechanisms ranging from microcystin degradation to toxin tolerance by appropriate gut microbiota.

Gut content analyses have confirmed that certain invertebrate grazers actively consume cyanobacteria (Foote 1993, Hollows et al. 2002, Frouz et al. 2004a, b, Krivosheina 2008, Clearwater et al. 2014, Toporowska et al. 2014, Hädicke, Rédei et al. 2017, Gérard and Lance 2019, Tierno de Figueroa and López-Rodríguez 2019, Aydın et al. 2021), and they even prefer filamentous species (Brock 1960; Foote 1993; Toporowska et al. 2014; Hädicke et al. 2017).

In many grazers, cyanobacteria may be ingested only incidentally, as they often occur as epiphytes on macrophytes or within complex biofilms, periphyton, and detritus (Krivosheina 2008; Catherine et al. 2013; Dionne et al. 2016; Wood et al. 2020; Chorus et al. 2021). This dichotomy raises an intriguing question: Can the simple detection of cyanobacteria in the gut contents of grazers be taken as evidence of nutritional use, or does it merely reflect incidental ingestion? To resolve this, enzyme activity analyses have been proposed as a more reliable approach to confirm whether invertebrates are able not only to ingest but also to digest cyanobacteria and assimilate their nutrients (Hädicke et al. 2017). Cases of undigested cyanobacteria being eliminated in feces are described, and the culture of such debris may inform about the digestibility of ingested cyanobacteria (Ward et al. 1985; Xue et al. 2016). Nevertheless, as certain cyanobacteria can fix atmospheric nitrogen, the nutritional value of cyanobacteria in some invertebrates, and also for the whole food web, has been revealed by studies of stable carbon and nitrogen isotopes (Salas and Dudgeon 2001; Chen and Xie 2005; Goedkoop et al. 2006; Dionne et al. 2016; Berezina et al. 2021). Studies of protein and lipids, especially fatty acid biomarkers, have also confirmed that cyanobacteria are essential for the survival of macroinvertebrates, particularly in winter when other food sources are unavailable (Aboal et al. 2005; Dionne et al. 2016). Benthic macroinvertebrates can eat cyanobacteria from benthos and plankton (Liu et al. 2016b; Stanković et al. 2020) (credit to Ubero-Pascal and Aboal (2024), appropriately referenced). While providing a comprehensive overview, this review touches on the underlying mechanisms only superficially. The same is true for the cyanobacterivorous protists discussed by Steinberg (2022a, b, c, d, e, f, g, h, i), which are described as mediating nutrient remineralization through the detoxification of secondary metabolites and providing a subsidy to the classical food web. However, mechanisms have only been mentioned recently. In Table 5.1, we therefore summarize selected existing studies on protists and invertebrates to highlight some potential tolerance or resistance mechanisms. The consumption of toxic cyanobacteria by zooplankton has often been observed and reported, especially in early phenotypic inventory studies. However, because these studies primarily focused on descriptive observations rather than mechanistic insights, they are not included in Table 5.1. Since the primordial response to biotic and abiotic changes is the respiratory burst (Steinberg 2012), activation of the antioxidant system should be found in any organism exposed to cyanotoxins. However, this response is rarely mentioned below. Regarding transgenerational increases in tolerance or resistance, only a single epigenetic mechanism has been documented to date (Lyu et al. 2024).

Table 5.1 Reported mechanisms of cyanotoxin tolerance in aquatic heterotrophic and mixotrophic protists and invertebrates

Species	Cyanotoxin/ Cyanobacterium	Effects/reported mechanism	References
Heterotrophic Protists			
Diphylleia rotans	*Microcystis aeruginosa*	Biodegradation assumed	Mohamed and Al-Shehri (2013)
Naegleria sp. strain W2	*Anabaena cylindrica, Dolichospermum flos-aquae, Cylindrospermum majus, Gloeotrichia echinulata, Leptolyngbya foveolarum*	Excretion without digestion	Xinyao et al. (2006)
Nassula sp.	*Planktothrix agardhii*	Biodegradation assumed	Combes et al. (2013)
Paramecium jenningsi	*Raphidiopsis raciborskii*		Mohamed et al. (2023)
P. multimicronucleatum	*M. aeruginosa*	Acidified food vacuoles?	Xu et al. (2021)
Mixotrophic Protists			
Ochromonas gloeopara	*M. aeruginosa*	*O-GlcNAc transferase*↑ (→Fig. 5.14)	Zhang et al. (2018b)
Ochromonas sp.	*M. aeruginosa* PCC7806	Fucoxanthin↑	Gao et al. (2022)
Poterioochromonas malhamensis	*Microcystis viridis*	Rapid excretion	Watanabe et al. (1996)
Poterioochromonas sp.	MC-LR, MC-RR	AOC↑ (mainly SOD↑)	Ou et al. (2005)
Poteriospumella lacustris	*Microcystis aeruginosa*	*Glutathione reductase*↑, *glutaredoxin*↑	Chen et al. (2024)
Zooplankton			
Bosmina longirostris	*M. aeruginosa; Dolichospermum flos-aquae*	Biodegradation assumed	Fulton (1988)
	M. aeruginosa	"*Genetically based variation in tolerance*"	Jiang et al. (2013a)
Brachionus calyciflorus	MC-LR	HSPs↑	Liang et al. (2020)
Chydorus sphaericus	*M. aeruginosa*	"*Genetically based variation in tolerance*"	Jiang et al. (2015)
Daphnia carinata offspring	*M. aeruginosa*	"*Adaptive phenotypic plasticity across generations*"	Jiang et al. (2013b)
D. galeata	MC	*T610*↑, *ubq*↑	Drugă et al. (2016)
D. longispina	*M. aeruginosa*	Phase II biotransformation↑	Wojtal--Frankiewicz et al. (2014)

(continued)

Table 5.1 (continued)

Species	Cyanotoxin/ Cyanobacterium	Effects/reported mechanism	References
D. magna	MC-LR	Glutathione-*S*--transferase↑	Pflugmacher et al. (1998)
	M. aeruginosa NIVA strain	Tolerance to protease inhibitors↑	Kuster and von Elert (2013)
	M. aeruginosa UTEX LB 2063 + PCC 7806	ABC transporters↓ (reduced MC uptake)	Schwarzenberger et al. (2014)
	Microcystis sp. BM25, (MC-free but protease inhibitor producer)	Positive selection on the loci of serine proteases	Schwarzenberger et al. (2021) and Lange et al. (2023)
	M. aeruginosa PCC7806	Arginine kinase↑	Lyu et al. (2015)
		Short-term: *δgst*↑ Long-term: *δgst*↓	Lyu et al. (2016a)
D. magna offspring	*M. aeruginosa* NIVA-CYA 228/1	Biodegradation assumed	Gustafsson et al. (2005)
		"*Phenotypic accommodation to stressful conditions spans generations*"	Radersma et al. (2018)
D. pulex	*M. aeruginosa*	"*Genetically based variation in tolerance*"	Jiang et al. (2015)
D. similoides	*M. aeruginosa*	Glutathione metabolism↑, protein processing in endoplasmic reticulum↑, ARA metabolism↑, *chitinase 10*↑, *chitin syn-thase 2*↑	Lyu et al. (2019)
D. sinensis	*M. aeruginosa*	Contrasting regulation of nuclear receptors in sensitive and tolerant clones	Xu et al. (2018b)
D. sinensis offspring		*Hox* genes differently expressed (→Box 5.1)	Xu et al. (2022)
Notodiaptomus iheringi	*M. aeruginosa* NPLJ-4	Detoxification assumed	Ger et al. (2016)
Simocephalus vetulus	*M. aeruginosa*	'*Local adaptation to toxic cyanobacteria on a microgeographic scale*'	Chen et al. (2015c)
Zoobenthos			
Corbicula fluminea	*M. aeruginosa*	Transporters↑, GST↑, AOC↑, lysosome & phagosome responses↑	Zhang et al. (2023a, b)
C. leana	MC-LR	P-glycoprotein exporter↑ assumed	Pham et al. (2015)

(continued)

Table 5.1 (continued)

Species	Cyanotoxin/ Cyanobacterium	Effects/reported mechanism	References
Dreissena polymorpha	MC-LR	Glutathione-*S*--transferase↑	Pflugmacher et al. (1998)
	MC-LR	P-glycoprotein exporter↑	Contardo-Jara et al. (2008)
Palaemon argentinus	MC-LR	P-glycoprotein exporter↑	Galanti et al. (2013)
Sinotaia (*Bellamya*) *aeruginosa*	*M. aeruginosa*	Gut microbiota: *Lactobacillus*↑, *Sphingobium*↑; MC-degraders *Pseudomonas*↑, *Ralstonia*↑	Liu et al. (2023), and Ren et al. (2023)

↑ upregulation; ↔ no obvious effect; *AOC* Antioxidant capacity, *ARA* arachidonic acid, *Glutaredoxin(s)* small redox enzymes that use glutathione as a cofactor, *GST* glutathione (*S*) transferase, *HSP* heat shock protein, *MC* microcystin, O-*GlcNAc transferase* (*O*-linked N-acetylglucosaminyltransferase) catalyzes the addition of the *O*-GlcNAc posttranslational modification to proteins, *SOD* superoxide dismutase; ***T610*** *ubiquitin-conjugating enzyme T610*; **ubq** *ubiquitin-conjugating enzyme*

Beyond this, the literature occasionally offers elegant formulations, yet these remain largely descriptive and do not uncover the underlying mechanisms.

Table 5.1 illustrates how each study contributes only a small tile to a large mosaic. In fact, different studies reveal very different biochemical mechanisms that may contribute to the tolerance of animals to cyanotoxins. Comprehensive studies of the defensome, such as those conducted by Zhang et al. (2012), are rare. It is unlikely that these mechanisms act in isolation; instead, several are likely to operate in concert, collectively shaping cyanotoxin tolerance. Furthermore, one essential tolerance mechanism is completely missing in the studies in this table: the buffering or mediating role of the gut microbiota. At this point, we can hypothesize that feeding appropriate prebiotics or probiotics alters the gut microbiome to enable it to tolerate higher cyanotoxin loads.

Nevertheless, as an intriguing perspective on the possible nutrition of farmed aquatic animals, Zhang et al. 2018a outline the following possible mechanism of MC tolerance in mixotrophic *Ochromonas gloeopara* (Fig. 5.14): Firstly, the expression levels of two essential chaperones—HSP70 and HSP90—increase. The HSP family plays a critical role in regulating protein modification and eliminating misfolded proteins. Secondly, increased serine/threonine protein phosphatase 2A catalytic subunit (PP2A), a type of serine/threonine phosphorylase involved in signal transduction and cell apoptosis, may be involved in repairing DNA and protein damage. MC normally causes histone phosphorylation by binding to the catalytic subunit of PP2A, leading to cell apoptosis (Klumpp and Krieglstein 2002). Thus, the increased transcription of PP2A in this study offset the adverse effects of MC-PP2A on cells and promoted DNA repair. The relative gene expression also indicates that PP2A is increased at an earlier stage (full activity in the degradation process) and

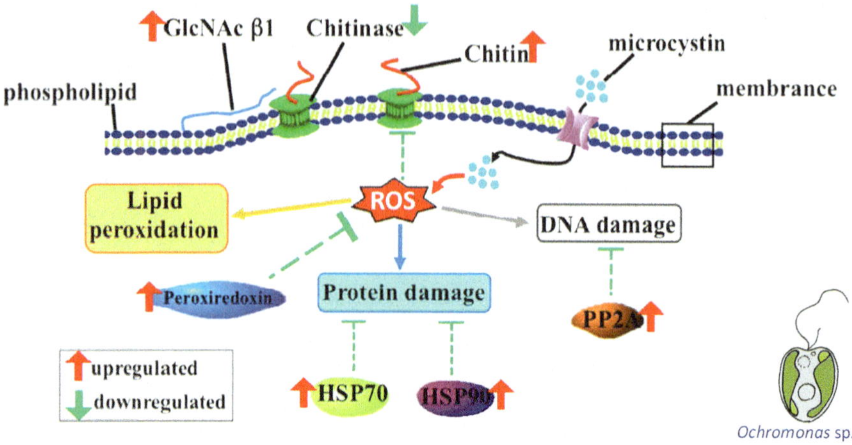

Fig. 5.14 The proposed transcriptome-based microcystin resistance mechanism in mixotrophic *Ochromonas*. *ROS* reactive oxygen species, *HSP* heat shock protein, *PP2A* serine/threonine--protein phosphatase 2A catalytic subunit, *GlcNAc β1* *N*-acetylglucosamine (from Zhang et al. (2018a), with permission from the American Chemical Society; drawing of *Ochromonas* sp. credit to Maulucioni via Wikimedia Commons)

then decreased with the decline of MC, which also indicates enhanced DNA repair when *Ochromonas* is fully active in MC degradation. Thirdly, the upregulated peroxiredoxin can directly reduce ROS as the first line of defense against oxidative stress (Chang et al. 2004). Overall, *O. gloeopara* enhances several repair capacities and ROS scavenging, thereby reducing MC damage to cells. This proposed mechanism is generally consistent with previous reports that *Daphnia* exposed to MC improve their antioxidant capacity and compensate for damaged proteins (Asselman et al. 2012, Lyu et al. 2016c, 2017). This topic will be revisited below.

O-GlcNAc transferase (OGT), which catalyzes the transfer of *N*-acetylglucosamine (GlcNAc β1) to serine/threonine, is upregulated by extracellular MC exposure (Fig. 5.14). Proteoglycan is then released and forms an extracellular matrix outside the cells (Kim et al. 2011). Chitinase, a hydrolase that breaks down chitin, is also downregulated by MC exposure. Flagellates do not have a cell wall, but chitin covers the cell membranes (Herth et al. 1977). Chitin is also a major component of the epidermis of crustaceans, insects, and many fungi, and is considered an antioxidant involved in ROS reduction and lifespan extension (Kurita 2006). Consequently, Zhang et al. (2018a) hypothesized that proteoglycan and chitin may act as extracellular barriers to protect *O. gloeopara* from externally dissolved MC. In light of the above considerations, *O. gloeopara* may enhance internal and external protective and repair mechanisms when grazing on toxic *Microcystis* and when exposed to dissolved MC (Fig. 5.14) (Zhang et al. (2018a), appropriately referenced). In a subsequent study, Zhang et al. (2020a) showed that cyanobacterial grazing also occurs on a larger scale: They showed that the mixotrophic addition of *Ochromonas* improves the harmful *Microcystis*-dominated phytoplankton community in in situ microcosms. Therefore, it seems worthwhile to test whether cyanobacteria-enriched diets supplemented with proteoglycans, chitin, or other prebiotics can improve the digestibility and nutritional value of cyanobacteria for aquatic animal nutrition.

We know from other studies that stress resistance is often associated with or mediated by duplication or multiplication of stress defense and biotransformation genes, such as in the Pacific oyster (Zhang et al. 2012). It would be desirable to look for similar phenomena in MC-resistant animals.

Zooplankton tolerance to cyanobacterial diets shows striking geographic variation: A crustacean species that thrives on such diets in one region may prove intolerant in another (Nandini and Rao 1997). For instance, the rotifer *Brachionus calyciflorus* is tolerant to *Cylindrospermopsis raciborskii* strains isolated from Brazil and readily consumes them (Soares et al. 2010). Yet, the very same rotifer species exhibits higher sensitivity when exposed to strains of *C. raciborskii* from other regions, highlighting that tolerance is not universal but strain- and location-dependent. This has been attributed to clone-related differences of the cyanobacterial species in question (Bolius et al. 2017) and the adaptation of the zooplankton taxa. However, it is also possible that this may be due to differences in the gut microflora (Macke et al. 2017) (from Nandini and Sarma (2023), credit to MDPI, Basel, appropriate references added) or adaptive response via epigenetics (→Sect. 5.1.5).

Numerous studies have described the dietary uptake and excretion kinetics of cyanotoxins in freshwater invertebrates, but only a few have addressed the underlying mechanisms of differential sensitivity. These toxicokinetic aspects are not the focus of this section; rather, the goal is to determine if there are additional mechanisms beyond those mentioned in the previous section. Mollusks and crustaceans, especially *Daphnia*, will be the main witnesses.

A cautionary note at the outset: Assuming that conjugation of glutathione to the phycotoxin is one key detoxification reaction, it is crucial not only to measure overall GST activity using generic substrates like CDNB,[10] but also to investigate specific isoforms. These isoforms can exhibit different sensitivities and roles in toxin metabolism (Gronwald and Plaisance 1998). For example, studies on the freshwater clam *Ruditapes philippinarum* exposed to microcystin-LR showed differential transcriptional responses of GST isoforms (mu, sigma, pi), indicating distinct temporal and tissue-specific functions in detoxification processes rather than a uniform GST response (Reis et al. 2015). This highlights the limitation of relying solely on total GST activity assays, such as those using CDNB, which may overlook isoform-specific involvement in tolerance mechanisms. Consequently, the absence of increased GST activity in CDNB assays, as reported by Pham et al. (2015) for *Corbicula leana* exposed to the toxic alga *Microcystis aeruginosa*, does not rule out the involvement of GST isoforms in microcystin tolerance that cannot be detected using CDNB. For example, in Pacific white shrimp farmed worldwide, *Ωgst* is the most responsive class after injection of a sublethal dose of a toxic *M. aeruginosa*

[10] 1-chloro-2,4,dinitrobenzene (CDNB) is a nonspecific substrate for α-, μ-, and π-class GSTs (Takamatsu and Inaba 1994).

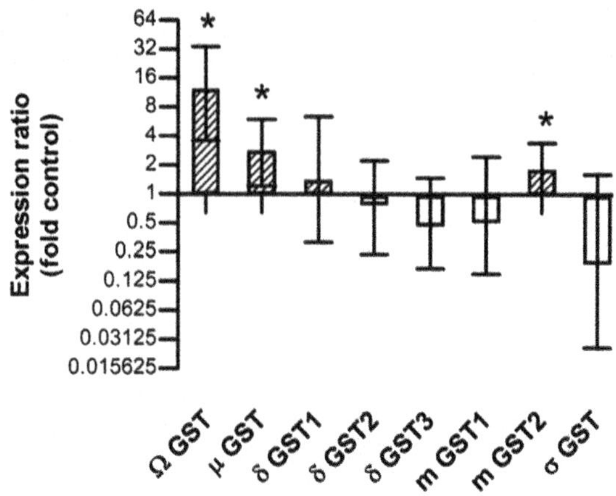

Fig. 5.15 Relative expression of gene transcripts from hepatopancreas of *Penaeus vannamei* 48 h postinoculation with 100 μg kg^{-1} of a *Microcystis aeruginosa* lysate. The data are expressed as expression ratio for each gene among MC-treated and control shrimp ($N = 8$ individual shrimps for each group). *Refers to significant differences (REST 2009 software Qiagen®; $P < 0.05$) among up- or downregulated genes. The target gene expression was normalized by the expression levels of the nonregulated reference genes 18S ribosomal and ATP synthase (M34362 and EU194608, respectively). Legend: Glutathione *S*-transferase isoforms (**GST** classes: *ΩGST* (omega), *μGST* (mu), *δGST* (delta, three isoforms) *microsomal GSTs* (mGST or MAPEG, two isoforms), and *σGST* (sigma)) (from Gonçalves-Soares et al. (2012), with permission from Elsevier)

extract (Fig. 5.15); however, this class does not respond to CDNB. Targeted isoform-level studies are thus essential to fully elucidate the molecular basis of phycotoxin resistance.

Another fact deserves mention: Since MCs possess multiple functional groups, studies on their biotransformation/detoxification deal almost exclusively with conjugating enzymes, primarily from the GST family (Fig. 5.15). Therefore, the involvement of redox-active enzymes, such as mixed-function oxidases (CYP450 enzymes), which place functional groups on the xenobiotic molecules, is not expected or studied in the biotransformation of cyanotoxins—a fundamental omission. This omission is also clearly visible in Tables 5.2 and 5.3.

As the above description of how *Aequipecten opercularis* responds to exposure to the phycotoxin DA, which has multiple functional groups, shows, it is evident that mixed-function oxidases (CYPs) play a key role in the detoxification process (Fig. 5.11). More recently, Cai et al. (2022) chronically exposed larvae of *Chironomus pallidivittatus* to MC-LR. The corresponding KEGG analysis identified that the differentially expressed genes belonged to fatty acid biosynthesis, biosynthesis of unsaturated fatty acids, fatty acid elongation, peroxisome proliferator-activated receptor (PPAR) signaling, oxidative phosphorylation, antigen processing and presentation, complement and coagulation cascade, phagosome, apoptosis, metabolism of xenobiotics by cytochrome P450, and drug metabolism-cytochrome P450 pathways (Fig. 5.16). Of note, both cytochrome P450 enzymes and members of the GST family are involved in detoxifying MC-LR (Cai et al. 2022).

Table 5.2 Summary of biochemical and physiological indicators reported or hypothesized to be involved in resistance to cyanotoxins in various aquatic invertebrates

Species/cyanotoxin		AO enzymes	CYPs	GST classes/ substrate	Exporter	Remarks	References
Anodonta cygnea, BMAA		CAT, GR, GPx	nc	CDNB	nc	–	Contardo-Jara et al. (2014)
Artemia salina MC-LR, NOD		nc	nc	CDNB, DCNB, EPNP	nc	–	Beattie et al. (2003)
Corbicula javanicus, BMAA		CAT, GR, GPx	nc	CDNB	nc	–	Contardo-Jara et al. (2014)
C. leana	Toxic *M. aeruginosa*	SOD, CAT	nc	CDNB	MXR assumed	GST not important	Pham et al. (2015)
	M. aeruginosa extract		nc	MCB[a]		–	Pham et al. (2016)
Cristaria plicata, MCs		*prx*, *Se-GPx*	nc	*σgst*	nc	*NRF2*↑	Wu et al. (2020)
Diplodon chilensis patagonicus, toxic *M. aeruginosa*		SOD, CAT	nc	CDNB	nc	–	Sabatini et al. (2011)
Dreissena polymorpha	Toxic *M. aeruginosa*	nc	nc	nc	nc	Clearance↑	Dionisio Pires et al. (2004)
	MC-LR	*ccat*↓	nnc	*πgst*↓	*p-gp*↔ P-gp↑	---	Contardo-Jara et al. (2008)
		CAT, GR, GPx	nc	CDNB	nc	–	Contardo-Jara et al. (2014)
		SOD, CAT	nc	CDNB	nc	–	Burmester et al. (2012)
Elliptio complanata	*Dolichospermum flosaquae*	Hemoprotein oxidase↔	nc	CDNB↔	nc	–	Gélinas et al. (2013)
	Toxic bloom	nc	nc	GST↓, method not mentioned	nc	LYZ↑	Gagné et al. (2018)
Hyriopsis cumingii, toxic *M. aeruginosa*		SOD↑, GPx↑, CAT↓	nc	CDNB↑	nc	LYZ↑	Hu et al. (2015)

(continued)

Table 5.2 (continued)

Species/cyanotoxin		AO enzymes	CYPs	GST classes/ substrate	Exporter	Remarks	References
Lymnaea stagnalis, toxic *Planktothrix agardhii*		CAT	nc	CDNB (sGST, mGST)↓	nc	–	Lance et al. (2016)
Magallana (*Crassostrea*) *gigas*, MC-LR *Mytilus edulis*, MC-LR		CAT↑, SOD↑, GPx↑, GR↑	nc	CDNB↑	nc	LPO↑, GSH↑	Kim et al. (2017a, b, 2018)
M. galloprovincialis	Toxic *M. aeruginosa*	nc	Nc	CDNB↑	nc	–	Fernandes et al. (2009)
	Toxic *M. aeruginosa* extract	nc	nc	*σ1gst*↑, *σ2gst*↔, *πgst*↔, *μgst*↔	nc	–	Antas et al. (2018)
Neohelice (*Chasmagnathus*) *granulata*	Toxic *M. aeruginosa* extract	TOSC↑	nc	CDNB↑	nc	LPO↔	Vinagre et al. (2003)
	Toxic *M. aeruginosa*	CAT_{tran}↑, SOD_{tran}↑	nc	$CDNB_{tran}$↑	nc	LPO↑	Pinho et al. (2005)
	Toxic *M. aeruginosa* extract	Nc	nc	CDNB(↑)	nc	LPO↑, glycogen↓	Dewes et al. (2006)
	Toxic *M. aeruginosa*	Sod↑	nc	CDNB↑	nc	LPO↑, GSH↓	Sabatini et al. (2015)
Neomysis awatschensis, MC-LR		CAT_{juv}↑, SOD_{juv}↑, GPx_{juv}↑, GR_{juv}↑; CAT_{ad}↑, ncSOD_{ad}↑	nc	$CDNB_{juv}$↑, $CDNB_{ad}$↔	nc	–	Min et al. (2018)
Palaemon argentinus		GR↑, CAT↓,	nc	mGST↑, sGST↑	nc	–	Galanti et al. (2013)
Penaeus vannamei, toxic *M. aeruginosa*		*cat*↑, CAT↑	nc	Various *gsts* (Fig. 5.15) GST↑	nc	–	Gonçalves-Soares et al. (2012)
Unio pictorum, MC-LR		Sod↓	No spot	GST↓	nc	Catchin↓, actin↓	Malécot et al. (2013)
U. tumidus, MC-LR		CAT, GR, GPx	nc	CDNB	nc	–	Contardo-Jara et al. (2014)
		SOD, CAT	nc	CDNB	nc	–	Burmester et al. (2012)

Venerupis (Ruditapes) philippinarum	MC-LR	SOD_{gills}↑	nc	CDNB↑, $\sigma 2gst_{Hepato}$↑	nc	–	Carneiro et al. (2015)
		nc	nc	$12h_{gills}$ $\sigma 1gst$↑, μgst↑ $12h_{Hepatop}$ $\sigma 1gst$↓, $\sigma 2gst$↓, πgst↓, μgst↓ $48h_{Hepatop}$ μgst↑	nc	–	Reis et al. (2015)
	Toxic *M. aeruginosa* extract	nc	nc	$\sigma 1gst$↑, $\sigma 2gst$↔, πgst↔, μgst↔	nc	–	Antas et al. (2018)

[a]This substrate is also isoform selective. Ublacker et al. (1991). MCB is a relatively poor substrate for the π isozyme. In addition, it does not seem to be equally sensitive to GSTs from different animals. Cook et al. (1991)

↑upregulation; ↓down-regulation; ↔ no obvious effect, *ad* adult, *AO* antioxidant enzymes, *CAT* catalase, *GPx* glutathione peroxidase, *GR* glutathione reductase; *GSH* glutathione, *GST substrates CDNB* 1-chloro-2,4-dinitrobenzene, *DCNB* 1-nitro-2,4-dichlorobenzene, *EPNP* 1,2-epoxy-3-(4-nitrophenoxy)propane (Fjellstedt et al. 1973; Habig et al. 1974), *juv* juvenile, *LPO* lipid peroxide, *LYZ* lysozyme, *MCB* monochlorobimane, *mGST* microsomal GST, *SOD* Mn superoxide dismutase, *nc* not considered, *NRF2* nuclear factor erythroid 2-related factor 2, *P-gp* P-glycoprotein, *rx* peroxiredoxin, *Se-GPx* Se glutathione peroxidases, *sGST* cytosolic GST, *SOD* superoxide dismutase, *TOSC* total oxyradical scavenging capacity, *tran* transient, *UCP2* uncoupling protein 2

Table 5.3 Summary of biochemical indicators reported or hypothesized to be involved in metabolism resistance to microcystins in various fish species

Species, cyanotoxin	AO enzyme	CYPs	GST classes/ substrate	Exporter	References
Corydoras paleatus, MC-RR	GPx, GR, POD, CAT	nc	CDNB	nc	Cazenave et al. (2006)
Cyprinus carpio, MC-LR	GPx, SOD, CAT	nc	CDNB (comm. kit)	nc	Li et al. (2003)
	nc	nc	α, ρ, mGST3 (Fig. 5.29)	nc	Fu and Xie (2006)
C. carpio, *M. aeruginosa* (RST9501)	GR	nc	π (CDNB)	nc	Amado et al. (2011)
Danio rerio, MC-LR	nc	nc	α, μ3, π1, π2, ρ1, τ1, ζ1	nc	Zhang et al. (2011)
	nc	nc	nc	Abcb4, *abcb4*↑ (Fig. 5.30)	Lu et al. (2015)
	nc	nc	CDNB	nc	Pflugmacher
D. rerio eggs, MC-LR	nc	nc	CDNB	nc	et al. (1998)
Hypophthalmichthys idella, MC-LR	nc	nc	α↔, ρ↔	nc	Liang et al. (2007)
H. molitrix, MC-LR	nc	nc	α↑, ρ↑	nc	
Jenynsia lineata, MC-LR	nc	nc	nc	P-glycoprotein↑	Amé et al. (2009)
Odontesthes hatcheri, MC-LR	nc	nc	CDNB	ABCc-like transporter	Bieczynski et al. (2013, 2016)
Oncorhynchus mykiss, MC-LR	nc	nc	CDNB	ABCc-like transporter	Bieczynski et al. (2014)
	GR(↑)	nc	GST↓, *gst-π*↔	*abcc2*↔,	Painefilú et al. (2022)
Oreochromis niloticus, MC-LR	GPx↔, UCP2↔	CYP1A↔	sGST↑	nc	Wang et al. (2006)
	nc	nc	α↑, ρ↑	nc	Liang et al. (2007)

Enzyme names always refer to their activity, and names of compounds to their contents. ↓ downregulation, ↑ upregulation, ↔ no obvious regulation, *ABCc*-like transporters ATP-binding cassette type C or multidrug resistance-associated proteins (MRP), *AO* antioxidant, *CAT* catalase, *CDNB* 1-chloro-2,4-dinitrobenzene (→FN10), *CYP1A* cytochrome P450 1A, *CYP26B1* retinoic acid hydroxylase, *GPx* glutathione peroxidase, *GR* glutathione reductase, *nc* not considered, *POD* peroxidase, *sGST* soluble GST, *SOD* superoxide dismutase, *UCP2* uncoupling protein 2

In a concomitant study, Davidović et al. (2022) further confirmed the pivotal role of *cyp450* genes in mediating the response of *Daphnia magna* to extracts of the aquatic strain *Nostoc* Z1. However, these extracts do not contain microcystins in pure form alone, but rather a mixture of other cyanotoxins—many of which are hydrophilic and already equipped with functional groups for conjugation

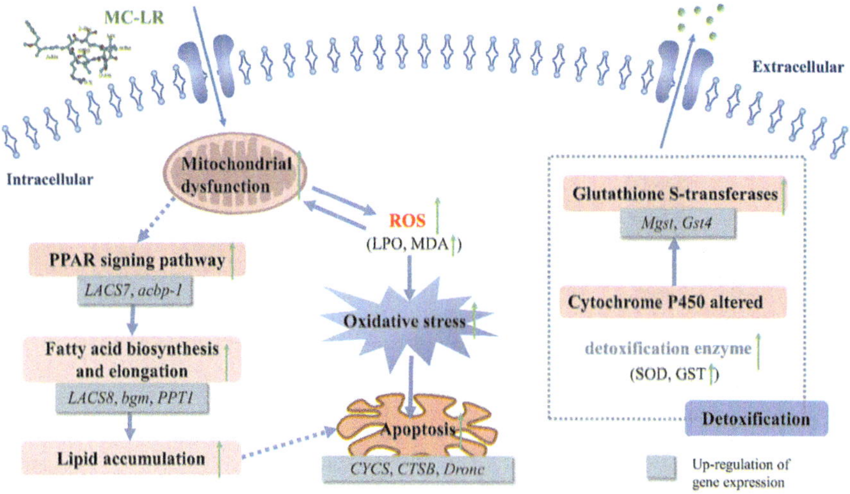

Fig. 5.16 The potential mechanism of lipid accumulation, cellular apoptosis, and detoxification in *Chironomus pallidivittatus* larvae under MC-LR stress. *GST* glutathione *S*-transferase, *LPO* lipid peroxidation, *MDA* malondialdehyde, *PPAR* peroxisome proliferator-activated receptor, *ROS* reactive oxygen species, *SOD* superoxide dismutase (from Cai et al. (2022), with permission from Elsevier)

reactions—making extensive phase I transformations theoretically unnecessary. Endpoints examined by Davidović et al. (2022) included gene expression of *cyp314*, *cyp360A8*, *gst*, *p-gp*, and *vtg*. Significant P-glycoprotein (*p-gp*) induction (Fig. 5.17) indicates the involvement of *p-gp*-mediated multixenobiotic resistance mechanism (MXR) in facilitating the excretion of toxic cyanobacterial compounds. Moreover, these strains induce an increase in the expression levels of *cyp360A8,* pointing out genes related to the detoxification of cyanotoxins. A significant reduction of *cyp314* and a slight induction of *gst* and *vtg* is also observed after exposure to *Nostoc* Z1 (Fig. 5.17). *Cyp314* is involved in molting, namely the final conversion of ecdysone to its effective form 20-hydroxyecdysone (Rewitz and Gilbert 2008). Changes in the expression level of *cyp314* are often used to assess contaminant potentials affecting *D. magna* reproduction and development (Le et al. 2010; Wang et al. 2016). In the current study of Davidović et al. (2022), the expression of *cyp314* increased when organisms were exposed to the extract (Fig. 5.17), indicating the potential of the extract to impair biosynthesis of the molting hormone and thus negatively affect *Daphnia* development and growth. *Cyp360A8* is another member of the cytochrome P450 superfamily, which has a major role in the xenobiotic transformation system in *D. magna* (Religia et al. 2021). The *cyp360A8* gene of *Daphnia* is also widely used as a sensitive biomarker of exposure to environmental xenobiotics (Aksakal and Arslan 2020; Bao et al. 2020; Ding et al. 2020). The expression of *cyp360A8* was significantly increased across all tested concentrations, showing a clear dose-dependent response to the extract. A similar upregulation was also reported by Xu et al. (2019), who analyzed the response of *D. magna* to a toxic strain of *M.*

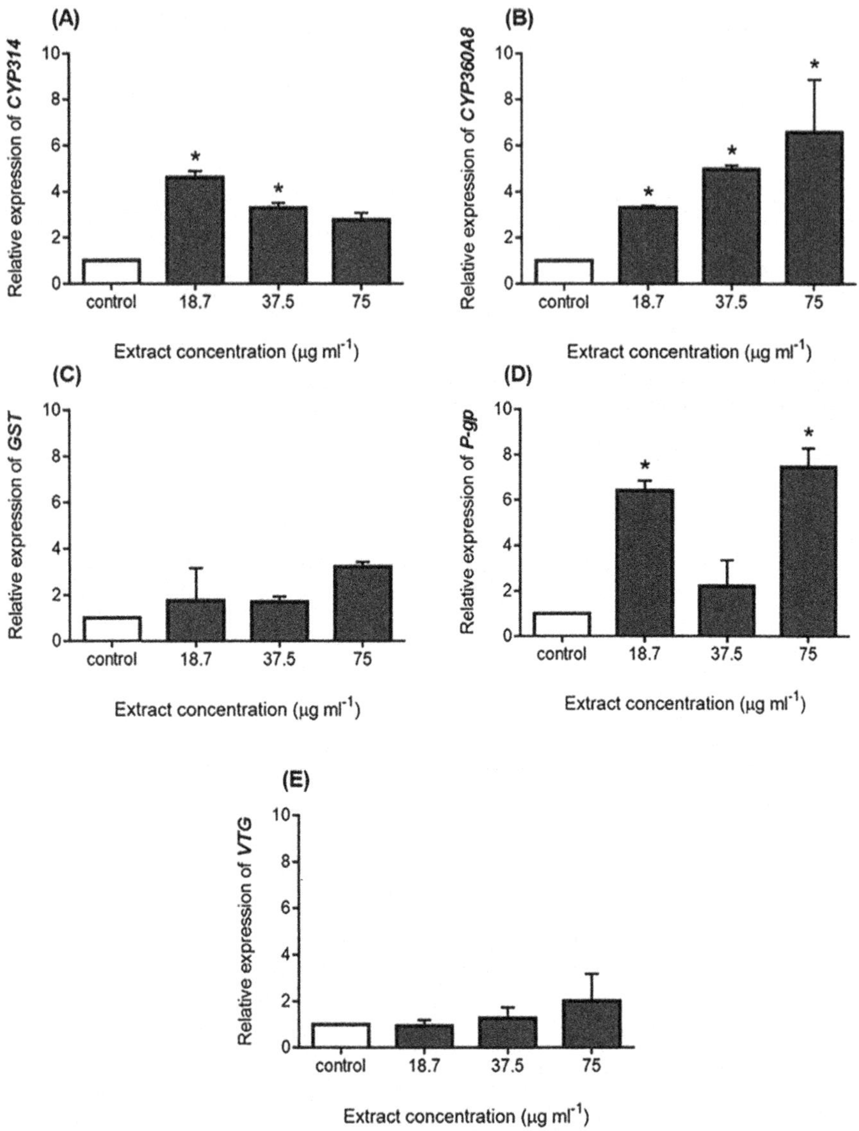

Fig. 5.17 Effects of three different concentrations of the cyanobacterial extract *Nostoc* Z1 on relative expression levels of five selected genes in *Daphnia magna*: (**a**) *cyp314*, (**b**) *cyp360A8*, (**c**) *gst*, (**d**) *p-gp*, and (**e**) *vtg*. Changes in the gene expression levels relative to the untreated DMSO control groups are presented as columns with mean ± SD values of two independent experiments. * indicates significant differences between the control and treated groups ($P < 0.05$) (from Davidović et al. (2022), with permission from Elsevier)

aeruginosa. Taken together, these findings highlight *cyp360A8* as a key component of the detoxification response of *Daphnia* to cyanobacterial toxins.[11]

5.3.1 Mollusks

Several mussel species exhibit remarkable tolerance to toxic cyanobacteria. Even more striking: Liu et al. (2014, 2015) demonstrated that *Sinanodonta woodiana* not only withstands exposure but also shows a feeding preference for the toxic *Microcystis aeruginosa* over the green algae *Tetradesmus* (*Scenedesmus*) *obliquus*. The toxic cyanobacteria actually provided more energy to the mussel for growth performance than the green alga. Similar results are reported from *Nodularia* (*Unio*) *douglasiae* and *Corbicula fluminea* (Liu et al. 2009). One possible explanation for this surprising preference is that *T. obliquus* and other coccal green algae have thick, resistant cell walls that make them difficult to digest within the muscles' gut.

In contrast, *M. aeruginosa* cells are more easily broken down, allowing bivalves not only to process them efficiently but also to capitalize on their nutrient content despite the presence of toxins. However, this does not explain the tolerance for MCs. Later, Kim et al. (2017a) generally confirmed the results of Liu et al. (2014, 2015) and added that the genus *Sinanodonta* has a relatively high detoxification capacity likely through the conjugation of MCs to glutathione or cysteine (Kondo et al. 1992; Pflugmacher et al. 1998).

Comparable to *S. woodiana*, Zhang et al. (2023a) showed that *Corbicula fluminea* grazes equally efficiently on green alga and *M. aeruginosa*, irrespective of whether the *M. aeruginosa* strains are toxic or non-toxic. MCs can be accumulated and depurated by *C. fluminea* efficiently. This study indicates that the hepatopancreas of *C. fluminea* is the primary target organ for MC-LR detoxification processes, as evidenced by a significant increase in GST activity.

In a companion study, Zhang et al. (2023b) investigated the mechanisms of MC tolerance in *C. fluminea* through a combination of transcriptional analysis, histopathological, and biochemical index analysis. Exposure to toxic *M. aeruginosa* causes dose-dependent histological damage in the hepatopancreas. The conjugation reaction catalyzed by GST is vulnerable to being activated by high concentrations of *M. aeruginosa* (10×10^5 cells mL^{-1}). Furthermore, the ROS scavenging process mediated by SOD and CAT is active in the initial stage of toxic *M. aeruginosa* exposure. The biotransformation and antioxidant defense system in *C. fluminea* is continuously activated after acute exposure to the high concentration of toxic *M. aeruginosa*. The GO analysis of DEGs supports this statement, since DEGs are significantly enriched in transporter activity, oxidant detoxification, and response to oxidative stress categories (Fig. 5.18). Additional support comes from the KEGG analysis: DEGs are also significantly annotated in several KEGG pathways involved in immunoreaction (lysosome and phagosome responses) and biotransformation (oxidation, co-oxidation, and conjugation). In a recent study, Hong et al. (2024) also

[11] Excerpted with permission from Elsevier and referenced accordingly.

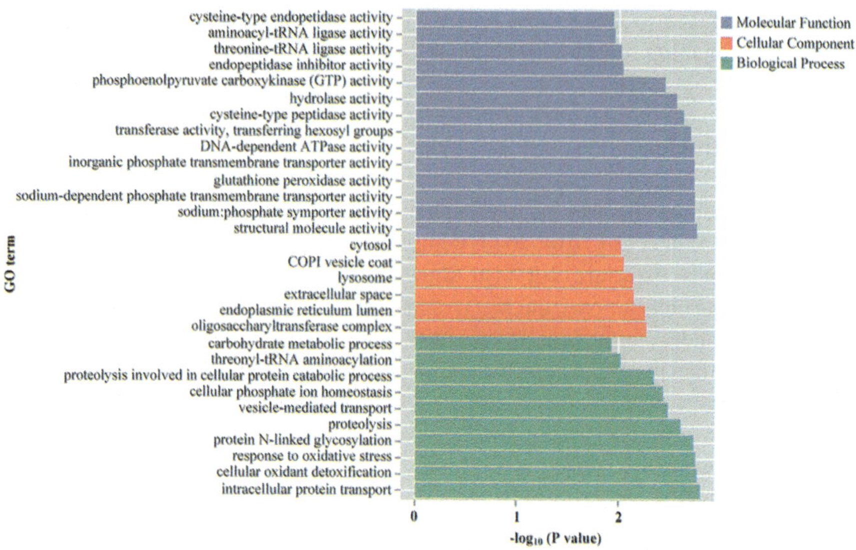

Fig. 5.18 Gene Ontology (GO) categories and the pattern of the differentially expressed genes (DEGs). The distribution of the GO categories was assigned into three categories: molecular functions (MF), cellular components (CC), and biological processes (BP). The lower P value represents the more significant enrichment (from Zhang et al. (2023a), credit to Elsevier)

reported active detoxification in exposed *C. fluminea*. Although the authors found histopathologic damage in all organ tissues that increased with increasing exposure concentrations, no mortality was observed.

To highlight the role of cytochrome P450 mixed-function oxygenases in MC detoxification, it is helpful to recall that maintaining redox balance is a central challenge for any organism. In mammals, cells have evolved a suite of adaptive mechanisms to counter oxidative stress, including the versatile cytochrome P450 system. These enzymes not only participate in the detoxification of xenobiotics but also modulate endogenous metabolic pathways, making them key players in the cellular defense against MC exposure (Motohashi and Yamamoto 2004). The transcription factor erythroid 2-related factor-2 (NRF2) is one of the most widely studied adaptive mechanisms, serving as the central regulator of cellular antioxidant response and regulating the expression of antioxidant and detoxification genes (Jaiswal 2004; Linker et al. 2011). It is also the most critical nuclear transcription factor that inhibits oxidative damage in cells and antagonizes the toxicity of xenobiotic compounds. Under normal conditions, Nrf2 is bound in the cytoplasm by Keap1 (Kelch-like ECH-associated protein 1). Keap1 acts as a cytoplasmic repressor and an adaptor for ubiquitin ligase complexes, targeting Nrf2 for continuous degradation via the proteasome. When cells are exposed to oxidative or electrophilic stress, reactive cysteine residues on Keap1 are modified. This disrupts the Keap1–Nrf2 interaction, allowing Nrf2 to accumulate and translocate into the nucleus. The promoters of target genes, including those encoding antioxidant proteins and phase II detoxification enzymes, are then activated by NRF2. This transcription factor binds to a specific DNA regulatory sequence known as the antioxidant/electrophilic response element (ARE/EpRE). The activation of this pathway initiates the transcription of a wide range of cytoprotective genes, enabling cells and tissues to counteract oxidative stress and limit damage from reactive oxygen species and electrophilic compounds (Rushmore et al. 1991; Maher and Yamamoto 2010). However, the recap is still incomplete. This becomes evident when examining the effects of xenobiotics on the constitutive expression of enzymes and transporters that promote drug metabolism in NRF2 knockout animals. Anwar-Mohamed et al. (2011) studied this phenomenon in the livers of C57Bl/6 mice. The knockdown of NRF2 disrupted the capacity for drug metabolism/transport. Given that NRF2 is a transcription factor of crucial chemopreventive capacity, the entire detoxification system will be malfunctioning due to its loss. Furthermore, NRF2 knockdown will lower the expression of phase I and II drug-metabolizing enzymes, as well as phase III drug transporters.

Consequently, overexpression of NRF2 protects against MC-induced hepatotoxicity in mice (Lu et al. 2014). NRF2 activation decreases MC-induced oxidative stress and liver injury. The protective effects of NRF2 depend on higher expression of genes involved in antioxidant defense, particularly the enhanced GSH synthesis, conjugation, and peroxide reduction enzymes. And vice versa: silencing of NRF2 in *Penaeus vannamei* decreased antioxidant capacity, and increased oxidative stress, apoptosis, and autophagy (Huang et al. 2022).

Studies have shown that estuarine and marine bivalves such as *Mytilus galloprovincialis* and *Ruditapes philippinarum* modulate their detoxification and oxidative stress defense systems after exposure to MCs (Carneiro et al. 2017; Kim et al. 2017b). In addition, marine and some freshwater bivalves can excrete and eliminate cyanobacteria after filtration through pseudofeces, rather than actually ingesting them. This defense mechanism often, but not always, attenuates or reduces the hazardous effects of toxic cyanobacteria. Paradoxically, the removal of cyanobacteria from bivalves into their pseudofeces can trigger the formation of cyanobacterial blooms. For example, a zebra mussel was reported to be responsible for the initiation of toxic blooms of *Microcystis aeruginosa* in Saginaw Bay (Lake Huron) and Lake Erie that were formed by the expelled live cyanobacterial cells in the mussel's pseudofeces (Vanderploeg et al. 2001).

So, where do we stand on cyanotoxin resistance in aquatic invertebrates? Table 5.2 compiles the current evidence—what we know, what remains uncertain, and which avenues deserve urgent study. Even a cursory glance at this table shows that the understanding of possible cyanotoxin resistance or tolerance in invertebrates is rather fragmentary and incomplete. Consequently, many papers conclude with vague sentences like this: "The responses of antioxidant and detoxification enzymes might thus contribute to the MC and cyanotoxin tolerance of the species studied."

Table 5.2 indicates a strong focus on conjugating enzymes as a possible mechanism of tolerance, often associated with the activity/expression of antioxidant enzymes. A complete approach, corresponding to the NRF2 pathway, is nowhere evident. It seems that different laboratories have their favorite methods that they seek to apply, but do not consider tolerance from the perspective of the exposed animal. Only one paper reports the expression of *nrf2*. Thus, explanations of MC tolerance in aquatic invertebrates remain incomplete. In addition, generalization of the results is complicated by the use of isolated MCs as well as cyanobacterial extracts and whole cyanobacteria, both of which contain more toxic secondary metabolites than MCs alone.

5.3.1.1 The *Sinotaia (Bellamya) aeruginosa* Story

Sinotaia (*Bellamya*) *aeruginosa* is a freshwater snail with a gill and an operculum (Fig. 5.19), belonging to the family Viviparidae. This aquatic gastropod is widely distributed across China and northern Vietnam, where it inhabits a variety of shallow freshwater environments and can reach remarkably high densities. As a keystone species, *S. aeruginosa* plays a crucial role in shaping water quality and structuring phytoplankton communities. Additionally, it holds cultural and economic significance, being a commonly consumed ingredient in Chinese cuisine. This species is characterized by a high tolerance to cyanobacteria as well as a strong depuration capacity (Wang et al. 2023) and a high recovery potential of cell organelles (Zhu et al. 2011); therefore, it can be consumed with minimal risk, even from freshwater bodies contaminated with cyanobacteria. Which remarkable physiological and molecular mechanisms enable this snail to thrive on toxic cyanobacteria while keeping its cells—and its consumers—safe?

As an avoidance strategy, Qiu et al. (2017) observed that blooms of toxic cyanobacteria cause *Sinotaia aeruginosa* to shift its diet from planktonic to benthic

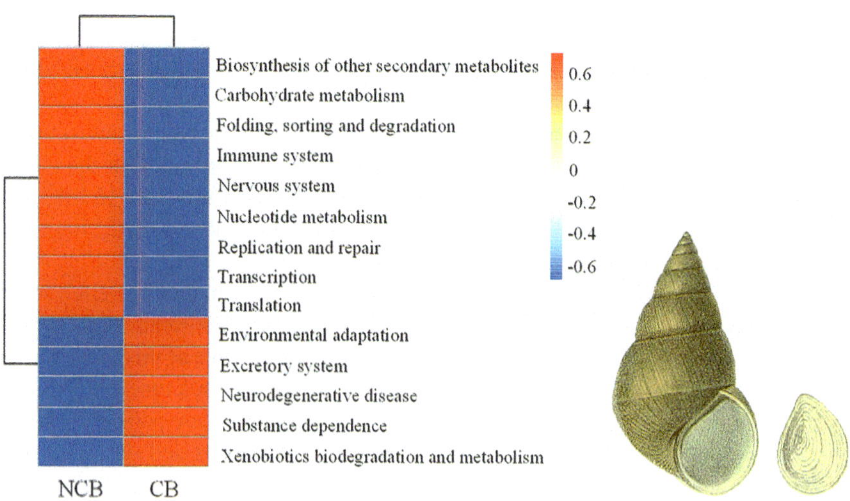

Fig. 5.19 Differences in KEGG functions in the *Sinotaia aeruginosa* gut microbial community during non-cyanobacterial bloom (NCB) and cyanobacterial bloom (CB) periods (level 2). ($P < 0.05$) (from Lyu et al. (2022), credit to Frontiers Media; drawing of an apertural view with operculum in public domain, credit to Kobelt et al. (1909) via Biodiversity Heritage Library)

material. More interestingly, internalized toxic cyanobacteria are likely to be wrapped in mucus and expelled as pseudofeces rather than directly ingested by *S. aeruginosa*. This is illustrated by the depleted $\delta^{13}C$ values in snails fed cyanobacteria. Interestingly, the cyanobacteria themselves are ^{13}C-enriched, so if the snails were actively assimilating the cyanobacteria, we would expect their $\delta^{13}C$ values to be enriched as well—yet the opposite pattern is observed.

The gut microbiota plays a central role in the tolerance of *S. aeruginosa* to toxic cyanobacteria. Lyu et al. (2022) reported that *Proteobacteria* and *Firmicutes* are the predominant phyla of gut microbes in *S. aeruginosa*. The composition and relative abundance of gut microbes are influenced by environmental conditions, particularly those associated with toxic cyanobacteria. The potential functions of the gut microbiome help the snail to meet its immunity and energy needs during cyanobacterial blooms. Specifically, functional metabolic pathways of the gut microbial community indicate a high abundance of genes related to metabolic pathways, environmental information processing, and genetic information processing. Nine pathways (e.g., carbohydrate metabolism and immune system) are more abundant during the non-cyanobacterial bloom (NCB) period, whereas five pathways (e.g., environmental adaptation and xenobiotics biodegradation and metabolism) are more abundant during the cyanobacterial bloom (CB) period (Fig. 5.19).

Yang et al. (2023a) continue to complete this picture: The microbiome results reveal that toxic cyanobacterial exposure alters the microbial composition, especially the enrichment of microcystin-degrading bacteria (*Pseudomonas*, *Lactobacillus*, and *Ralstonia*), and increases the diversity of gut microbiota. The relation analysis shows that the changes in gut microbiota are correlated with the expressions of antioxidant-, immune-, and metabolic-related DEGs.

5.3.2 Zooplankton with Particular Reference to Daphnia

Zooplankton have evolved a remarkable repertoire of strategies to cope with toxic cyanobacteria. These mechanisms range from behavioral avoidance to physiological and molecular adaptations. Bowník (2016)[12] reviewed the phenotypic studies. For example, some copepod species can selectively avoid ingestion of MC-producing strains of *Microcystis*. *Pseudodiaptomus forbesi* is capable of switching to alternative food sources, in contrast to the nonselective feeder *Eurytemora affinis*, which ingests toxic strains indiscriminately (Ger et al. 2010). Similarly, the tropical copepod *Notodiaptomus iheringi* shows the ability to discriminate between strains, consuming only MC-negative cyanobacteria when both toxic and non-toxic strains are present in the same diet (Ger and Panosso 2014). These observations highlight that feeding selectivity represents an essential first line of defense against cyanotoxin exposure.

Bownik (2016) continued his review by pointing out that large crustaceans appear to be more sensitive to toxic *M. aeruginosa* due to their higher grazing

[12] Excerpted with permission from Elsevier; appropriate references added.

activity and ingestion of higher amounts of cyanobacterial cells. As a result, in reservoirs affected by cyanobacterial blooms, larger crustaceans are replaced by smaller species. For instance, the number of large-bodied *Daphnia ambigua* decreases while the number of smaller copepods *Skistodiaptomus reighardi* increases (Fulton III and Paerl 1988). Studies performed in the Western basin of Lake Erie show that grazing rates of large-sized mesozooplankton (>200 μm-daphnids) decrease during blooms of toxic *Dolichospermum*, *Microcystis*, and *Raphidiopsis* genera. On the other hand, higher consumption rates of both toxic and non-toxic strains of cyanobacteria were observed in microzooplankton species (20–200 μm in size), indicating a more important role of the latter fraction of invertebrates in the natural control of cyanobacterial blooms (Davis et al. 2012).

Zooplankton are capable of consuming not only single-celled cyanobacteria but also filamentous forms that produce MCs. Oberhaus et al. (2007) found that *Daphnia pulicaria* feeds on toxic strains of *Planktothrix rubescens* and *P. agardhii* without inhibition of the filtration rate. In some species, reduction of grazing activity may prevent further ingestion of cyanotoxins (Łotocka 2001).

It is noteworthy that some zooplankton species preexposed to some other metabolites of cyanobacteria may change their sensitivity to cyanotoxins. Pretreatment of *D. magna* and *D. galeata* with purified cyanobacterial lipopolysaccharides (LPS) from *Microcystis* CYA 43 results in the increased resistance of these invertebrates to MC-LR (Lindsay et al. 2006). A study by Dionisio Pires et al. (2011) on *Daphnia galeata* exposed to MC-producing and bacterial LPS revealed that key life-history traits—including growth rate, age, and size at first reproduction, and mortality before first reproduction—remain unaffected; the only observed effect was a reduction in standard carbon content (Bownik 2016).

In *Simocephalus vetulus*, Chen et al. (2015a) tested the hypothesis that zooplankton evolves tolerance to cyanobacteria on a fine scale by comparing the intrinsic rates of population increase (r) from seven small isolated lakes with distances ranging from 2.2 to 31.7 km. The toxic cyanobacterium *Microcystis aeruginosa* (poor food) significantly depresses the r values, compared with a nutritious green alga (good food). The tolerance index to cyanobacteria (r_{poor}/r_{good}) of 24 clones from six populations differs significantly among clones and populations. The divergence in tolerance between populations is more significantly correlated with the difference in MC concentrations than the geographic distance between two sites (Fig. 5.20). These results indicate that local adaptation to toxic cyanobacteria in *S. vetulus* occurs on a micro-geographic scale.

So much for the phenotypic observations. Yet, the underlying biochemical—and especially biomolecular—mechanisms remain largely unexplored. When it comes to cyanotoxin tolerance or resistance, the best-studied invertebrates are planktonic Cladocera, particularly *Daphnia* spp., and current research is increasingly revisiting these studies to identify gaps and fill in the missing pieces of information.

Chislock et al. (2013) reported that genetically based tolerance to toxic *Microcystis* has been observed within species of *Daphnia* across habitats varying in nutrient enrichment (Sarnelle and Wilson 2005) and over time in a eutrophicated lake (Hairston et al. 1999, 2001). In a more recent study, Chislock et al. (2019)

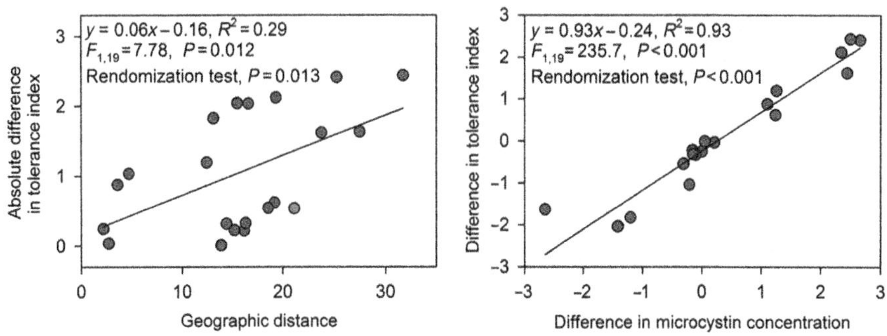

Fig. 5.20 Difference in the tolerance index of each pair of *Simocephalus vetulus* populations vs. geographic distance and difference in microcystin concentration in sediments as a proxy for long--term microcystin exposure of the animals (from Chen et al. (2015a), with permission from the Oxford University Press)

identified strong clonal selection mediated by eutrophication, resulting in a *Daphnia pulicaria* genotype tolerant of cyanobacteria over just 10 weeks. This selection is based on the existing intraspecific clonal variability of a given species (Steinberg Steinberg 2022a, b, c, d, e, f, g, h, i).

Recently, Xu et al. (2018b) presented a similarly intriguing example of intraspecific variability in a *Daphnia* species fed toxic *Microcystis*. The authors studied the effects of *M. aeruginosa* on the expression of nuclear receptor (NR) genes in *D. similoides sinensis*. This conserved superfamily is involved in the metabolism, reproduction, development, homeostasis, and resource allocation of organisms. Toxic *M. aeruginosa* can inhibit the growth and reproduction of *Daphnia* by involving NR genes. In total, 18 NRs are identified. A *M. aeruginosa* diet leads to downregulation of NR1 and NR2 subfamilies with marked differences between clone 1 and 2 on the one hand and clone 3 on the other hand, indicating differential adaptability and molecular mechanisms in individuals with different *Microcystis* tolerance genotypes. In clone 3, this results in reduced survival (Fig. 5.21). For recently identified epigenetics in cyanotoxin sensitivity in cladocerans, see below the example of *D. magna* and the action of *miR-210* (Fig. 5.26).

Recently, Lyu et al. (2025) reported that p38 MAPK determines the sensitivity of a clone of *Moina macrocopa* to toxic *Microcystis*: the knockdown of the MmMAPK gene leads to reduced lifespan and body size in *M. macrocopa*, with these effects being particularly pronounced in the presence of toxic *Microcystis*. Confirming this, knockdown of the p38 MAPK gene in *Penaeus vannamei* increases mortality when exposed to bacterial infection (Yan et al. 2013), suggesting a conserved role for MAPK pathways in mediating vulnerability to stress in different crustacean species.

However, the evolution of genetic tolerance or the clonal selection of tolerant genotypes cannot explain how the offspring of exposed mothers acquire increased tolerance to toxic cyanobacteria. In other words: What mechanisms underlie intergenerational improvements in toxin resistance? At least three potential modes of action may contribute to the rapid acquisition of tolerance: an adaptive

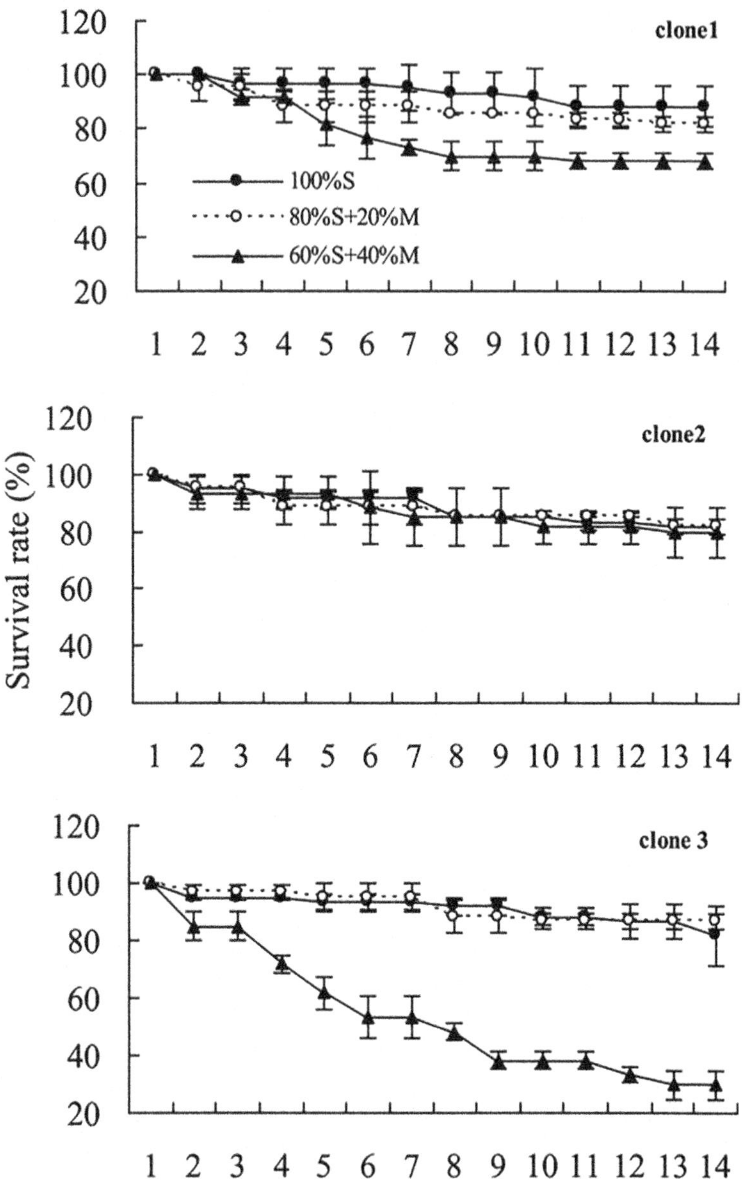

Fig. 5.21 Different survival rates (average ± SD) of three clones of *Daphnia similoides sinensis* under three food treatments (**100% S**, 100% *Scenedesmus* (currently accepted name is *Tetraedusmus*) *obliquus*; **20%M + 80%S**, 20% *Microcystis aeruginosa* and 80% *S. obliquus*; **40%M + 60%S**, 40% *M. aeruginosa* and 60% *S. obliquus*) over 14 weeks (from Xu et al. (2018b); with permission from Elsevier)

physiological response, modulation by the gut microbiome, and epigenetic mechanisms. These strategies are not mutually exclusive; rather, they may act synergistically, supporting or even reinforcing one another.

5.3.2.1 Gut Microbiome

Recent work highlights the crucial role of the gut microbiome in shaping the host's tolerance to cyanotoxins. In the following section, we discuss some intriguing findings on this topic.

One particularly striking example comes from the study by Gorokhova et al. (2021), who showed that copepods can consume cyanotoxin-producing prey without suffering adverse effects because their gut bacteria assist in detoxification. The authors examined the abundance of *mlrA* genes—unique markers of heterotrophic degraders of MCs and nodularins (NODs)—in the guts of Baltic copepods (*Acartia bifilosa* and *Eurytemora affinis*) during bloom season (August) and outside it (February). Remarkably, *mlrA* genes were expressed year-round, but their abundance was >tenfold higher in summer than in winter populations, and significantly greater in *Eurytemora* than in *Acartia*.

Further experiments by Gorokhova et al. provide empirical evidence that the presence of *mlrA* genes in the microbiome of copepods facilitates nutrient uptake and growth when they feed on phytoplankton mixtures with NOD-producing cyanobacteria. The microbiome responds to ingested dietary toxins by increasing the *mlrA* gene abundance (Fig. 5.22), and some adaptation occurs even in less tolerant species, such as *Acartia bifilosa*. Moreover, the growth potential assessed as RNA:DNA ratio is positively related to the total uptake of food comprised of cyanobacteria and *Rhodomonas salina*. Thus, the presence of heterotrophic MC- and NOD-degrading bacteria in the intestine serves as an adaptive mechanism that equips copepods with the ability to withstand, and even benefit from, cyanobacterial blooms. Nevertheless, it remains to be seen whether the introduction or increase of MlrA-carrying (intestinal) bacteria through breeding or genetic engineering techniques in fish or shrimp farming can increase the tolerance of farmed animals to microcystin. Pioneering studies with genetically modified microorganisms (e.g., *Saccharomyces cerevisiae*) are emerging and appear promising (Silva et al. 2023).

A cladoceran counterpart, however, without focus on heterotrophic microcystin degraders, is presented by Houwenhuyse et al. (2021). The authors studied whether *Daphnia magna* hosts perform better when receiving a microbiome from their source region than from another region when facing a stressful condition, in particular, exposure to the toxic cyanobacteria *Microcystis aeruginosa*. The authors demonstrate that most *D. magna* genotypes exhibit higher fitness in terms of survival and fecundity when they receive a sympatric microbiome than an allopatric one. These fitness benefits provided by the local microbiome are, however, genotype-dependent, supporting the hypothesis that different *Daphnia* genotypes vary in their tolerance to *M. aeruginosa* (Macke et al. 2017) and distinct genotypes select for distinct microbiomes (Goodrich et al. 2014). Beyond microcystin-degrading gut microorganisms, additional metabolic contributions from diverse microbiota may

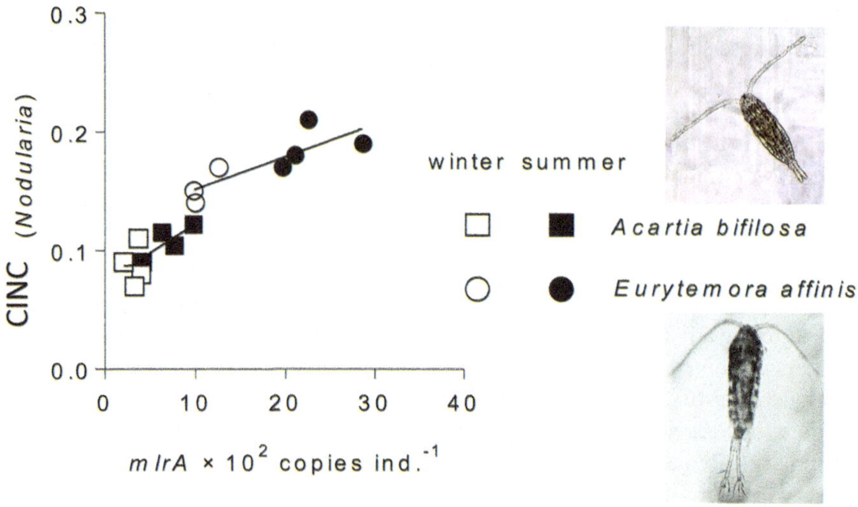

Fig. 5.22 Incorporation of *Nodularia spumigena* reflected in carbon incorporation index (CINC) by *Acartia bifilosa* and *Eurytemora affinis* in the feeding experiments conducted in summer (August) and winter (February) as a function of *mlrA* gene abundance in the microbiome of the copepods sampled from the same population as the experimental animals. The animals were exposed to a food mixture containing 35% of ^{13}C-labeled *N. spumigena* (by carbon mass). Both, species and seasonal effects, are significant (from Gorokhova et al. (2021), credit to Frontiers Media; image of *A. bifilosa* female credit to Vehmaa et al. (2013); image of *E. affinis* female credit to Natalia Sukhikh, Zoological Institute of RAS, Wikimedia)

underlie these differences in tolerance; however, the precise mechanisms remain elusive.

In addition to the MC-degrading gut bacteria mentioned in copepods, some intriguing details of how microbes beneficially influence the metabolism of hosts have been presented elsewhere; well understood are, for example, microbial-produced short-chain fatty acids (SCFAs) that boost host's immunity (Guilin et al. 2020, Steinberg 2022a, b, c, d, e, f, g, h, i).

As seen above, dysbiosis of the gut microbiota is a direct result of dietary toxins. A recent experiment in aquatic animals shows that the unbalanced gut microbiota can be at least partially reshaped by appropriate diets, as recently demonstrated in zebrafish larvae. Wu et al. (2024) investigated the protective effects of the probiotic *Lactobacillus rhamnosus* against MC-induced developmental toxicity. Probiotic supplementation significantly improved survival, hatching rates, and growth rates, and reduced malformation rates in MC-exposed larvae. *L. rhamnosus* alleviated MC-induced oxidative stress by reducing ROS levels and increasing glutathione levels and catalase activity. Probiotics also alleviated MC-induced lipid metabolism disorders by regulating key metabolites (triglycerides, cholesterol, bile acids, and free fatty acids) and gene expression (*ppara*, *pparβ*, *srebp1*, and nuclear receptor subfamily 1, group H, member 4 (*nr1h4*)). Moreover, 16S rRNA sequencing revealed that *L. rhamnosus* modulated the structure and diversity of the gut

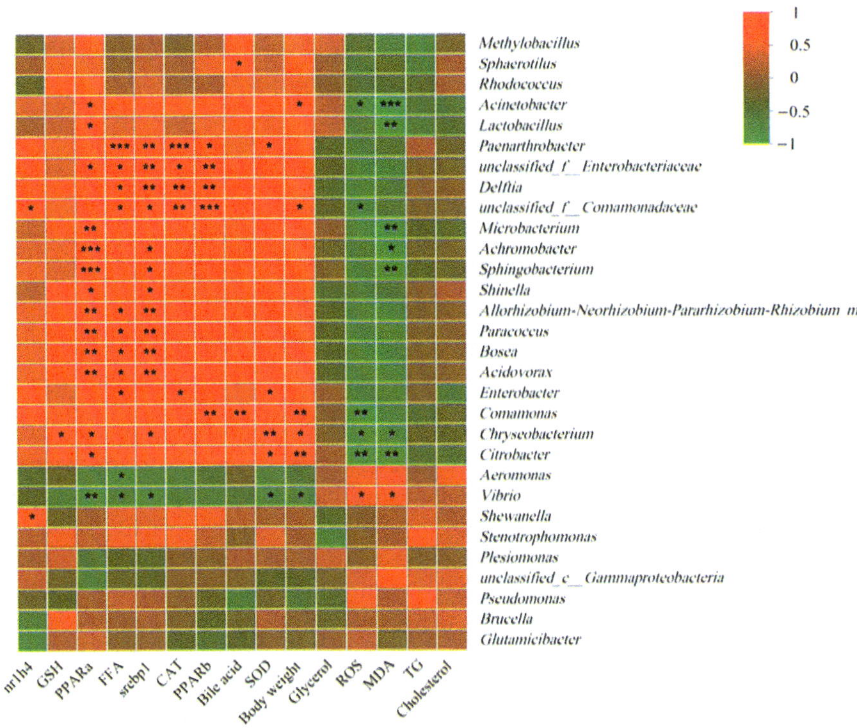

Fig. 5.23 Correlation analysis of the gut microbiome, host physiological parameters, and gene expression levels in zebrafish larvae exposed to MC-LR with or without probiotic supplementation: Heatmap of Spearman's correlation coefficients between dominant bacterial genera and biochemical markers, as well as gene expression levels. Red represents a positive correlation, while green represents a negative correlation. * $P < 0.05$, ** $P < 0.01$, *** $P < 0.001$ (from Wu et al. (2024), credit to MDPI, Basel)

microbiome in MC-exposed larvae, promoting beneficial genera such as *Shewanella* and *Enterobacter* and inhibiting potential pathogens such as *Vibrio*. Significant correlations were found between gut microbiota composition and host antioxidant and lipid metabolism parameters (Fig. 5.23).

5.3.2.2 Epigenetics

In addition to the above discussion about adaptive responses, we refer here to epigenetics within a broader spectrum. However, evidence regarding cyanotoxin tolerance in zooplankton is still very limited. Most research to date has focused on DNA methylation and microRNA-mediated regulation, with other key epigenetic processes, such as histone modifications and chromatin remodeling, largely unexplored in this context. For a thorough overview of epigenetic processes in aquaculture species, see Piferrer and Wang (2023), who highlight the growing importance of epigenetics in regulating traits such as growth, reproduction, disease resistance, and stress response. Recent studies suggest that environmental exposure to cyanotoxins may induce changes in DNA methylation patterns and microRNA expression in aquatic organisms, potentially modulating gene expression linked to toxin metabolism and defense; however, direct evidence in zooplankton remains scarce (Liu et al. 2022b). Furthermore, epigenetic mechanisms may contribute to population-level adaptation to cyanobacterial toxins, as zooplankton exposed to persistent toxic blooms show enhanced tolerance, possibly mediated by heritable epigenetic modifications (Jiang et al. 2015; Lin et al. 2025). In the following sections, we will discuss studies addressing DNA methylation and the effects of microRNAs in more detail; however, other potential mechanisms, such as histone modifications or chromatin remodeling, are still largely unexplored.

DNA Methylation

A comprehensive epigenetic study was conducted by Asselman et al. 2017 and focuses on genome-wide methylation patterns in *D. magna* fed toxic *Microcystis aeruginosa*. The authors attribute tolerance to *Microcystis* in different *Daphnia* genotypes to a differential sensitivity to protease inhibitors. Thus, methylation may potentially play a role in adaptation to environmental stress by inducing polymorphisms and genetic variation, similar to what has been reported in host–parasite interactions. Comparably, methylation may also play a role in micro-evolutionary effects, as manifested by changes in methylation upon exposure to environmental stress that can be passed on to subsequent generations. For example, cyanobacterial exposure has been reported to potentially influence the genetic population structure of *Daphnia*, particularly affecting protease genes. These micro-evolutionary effects may be modulated, at least in part, through DNA methylation, highlighting a possible epigenetic mechanism by which *Daphnia* adapts to toxic cyanobacterial diets (Asselman et al. (2017) and references therein).

Epigenetic changes can be passed to future generations. More recently, Feiner et al. (2022) showed that microcystin-induced DNA methylation is inherited across generations in *D. magna*. The authors identified between 70 and 225 differentially methylated CpG positions (DMPs) in F_1 individuals whose mothers were exposed to microcystin. About half of these environmentally induced DMPs persist into the F_4 generation. The top ten GO terms are shown in Fig. 5.24.

In contrast, treatment with MC shows that widespread hypomethylation upon exposure is almost completely reset after one generation. These results suggest that environmentally induced DNA methylation is non-randomly and stably inherited across generations in *Daphnia*, making epigenetic inheritance a putative factor in the eco-evolutionary dynamics of freshwater communities. Other environmental stresses, such as temperature, lasted up to the F_3 generation.

Interestingly, not in fish or *Daphnia*, but in small mammals, Chen et al. (2019) reported an epigenetic inactivation of microcystin-LR-induced hepatocarcinogenesis. This effect is mediated by LHX6 through the Wnt/β-catenin and P53 signaling pathways,[13] highlighting a sophisticated epigenetic defense mechanism in mammals.

[13] The **LHX6** gene belongs to the LIM homeobox gene family and encodes a LIM homologous domain transcription factor that binds DNA and proteins to regulate and participate in embryonic

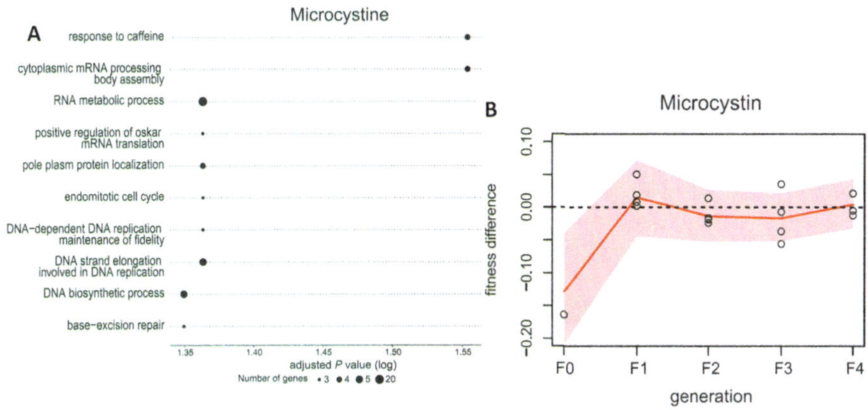

Fig. 5.24 (**a**) Putative functional roles of transgenerationally inherited differentially methylated CpG positions (DMPs) in MC-exposed *Daphnia magna*. Top ten significantly enriched GO terms of biological processes for genes containing transgenerationally inherited DMPs. The size of each GO term represents the number of genes it represents, and the position along the x-axis indicates its significance (log-transformed adjusted P-value). (**b**) Fitness effects of MC as an environmental stressor: Difference in lifetime reproductive output of exposed *Daphnia* per generation compared to control conditions (marked as dashed line). Red lines indicate mean values and shaded areas indicate 95% credible intervals. Observed values are shown as black circles. The fitness effects of MC disappeared after generation F_1 (from Feiner et al. (2022), credit to Elsevier)

In particular, the authors hypothesized that the LHX6 gene inhibits the migration of malignant transformed L02 cells through the Wnt/β-catenin pathway and promotes apoptosis through the P53 pathway. As shown in Fig. 5.25, exposure to MC-LR results in the hypermethylation of hepatocytes, thereby inhibiting the expression of the LHX6 gene. The LHX6 gene may regulate cell migration by regulating the expression of MMP7 through the Wnt/β-catenin signaling pathway. What's more, the LHX6 gene may regulate cell apoptosis by regulating the expression of Bax through the P53 signaling pathway.

Overall, this study demonstrates that LHX6 gene expression is regulated by DNA methylation and can inhibit proliferation, invasion, and migration through the Wnt/β-catenin and P53 signaling pathways during MC-LR-induced hepatocarcinogenesis.

---

development, especially the differentiation and development of the central nervous system (Grigoriou et al. 1998). Both the Wnt/β-catenin signaling pathway and the P53 signaling pathway play crucial roles in the initiation and maintenance of human hepatocellular carcinoma. The **Wnt/β-catenin** signaling pathway comprises a family of proteins that play critical roles in embryonic development and adult tissue homeostasis. Deregulation of Wnt/β-catenin signaling often leads to a variety of serious diseases, including cancer and non-cancer diseases (Liu et al. 2022a). The **p53 pathway** is a complex cellular stress response network with multiple diverse inputs and downstream outputs relevant to its role as a tumor suppressor pathway (Hernández Borrero and El-Deiry 2021).

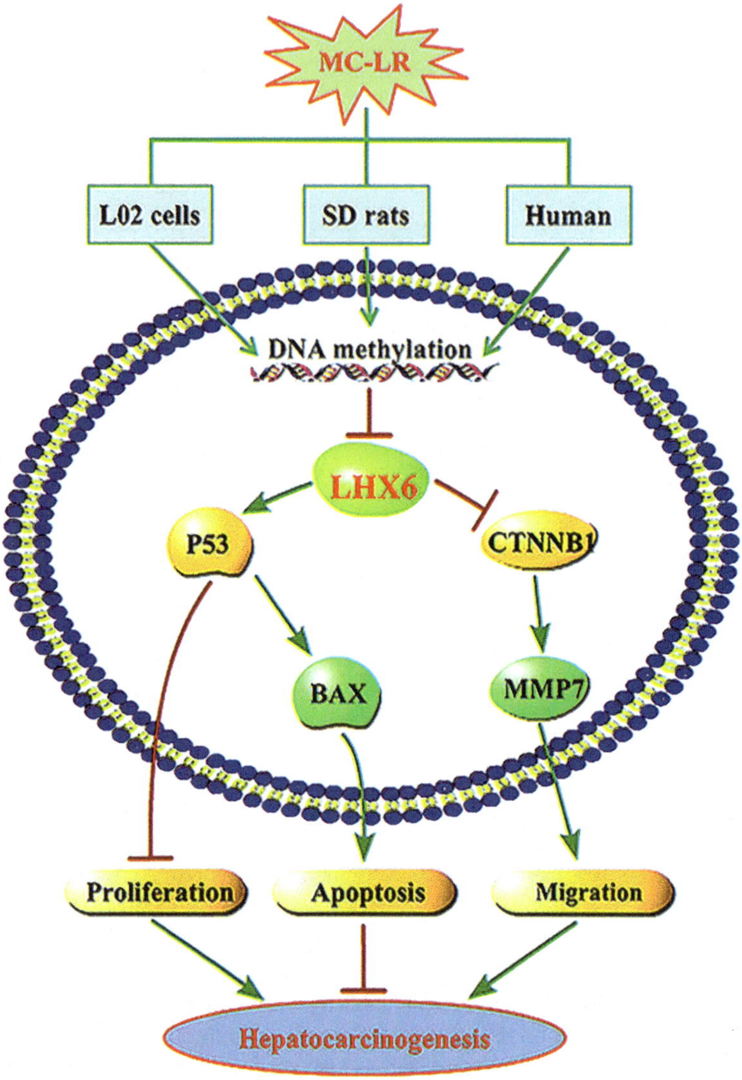

Fig. 5.25 Possible molecular mechanisms of the LHX6 gene during MC-LR-induced hepatocarcinogenesis (from Chen et al. (2019), with permission from Elsevier)

MicroRNAs

MiRNAs provide an efficient, flexible, and remarkably powerful way to fine-tune gene expression, allowing organisms to regulate a myriad of biological processes with precision. The vast diversity of miRNA species, their numerous complementary mRNA targets, and their temperature-sensitive binding dynamics together create a highly versatile regulatory RNA network. This network provides organisms with broad and flexible control over mRNA expression, particularly under stress conditions. Moreover, miRNAs offer an energetically cost-efficient strategy: Instead of investing resources into synthesizing nonessential proteins—or into producing additional negative regulators to suppress them—cells can simply terminate translation via miRNAs. This makes them an especially effective tool for stressed and resource-limited cells. These features give cells the ability to quickly modify their phenotype in response to changes in the environment, and thus it is no surprise that miRNAs have shown significant involvement in adaptations to various environmental stresses, including dietary cyanotoxins (Hadj-Moussa et al. (2022), and references therein).

Because there are only a few studies on microRNA in relation to cyanotoxins, we will summarize those on fish and invertebrates here. Ahkin Chin Tai and Freeman (2020) provided a comprehensive review of zebrafish as a vertebrate model for studying miRNA alterations induced by toxin exposure, highlighting its value in aquatic toxicology research. Their work revealed that various environmental toxins, including cyanotoxins, can lead to significant changes in the expression profiles of numerous miRNAs in zebrafish, which in turn mediate physiological and developmental effects such as impaired vascularization and neurobehavioral changes. This review emphasizes the importance of miRNAs as regulatory molecules in toxicity mechanisms and identifies the zebrafish system as a powerful tool to unravel miRNA-mediated toxicological pathways.

Since miRNAs play such a central role in cyanobacterial intoxication, the use of antisense microRNAs seems feasible in the long term, provided that the biomolecular and epigenetic basis of the toxic effect in aquatic animals is extended and robust. Stenvang et al. (2012) reviewed the development and application of antimiR oligonucleotides, including single-stranded RNAs (antagomirs), discussing their mechanism of action, delivery methods, therapeutic efficacy, and potential toxicities. This work provides critical insights into how these molecules can be used to inhibit specific miRNAs, such as *miR-122*, with implications for manipulating miRNA activity in disease and environmental toxicity contexts, including aquatic organisms. Krützfeldt et al. (2005) pioneered the design of a cholesterol-linked antagomir targeting *miR-122*. They demonstrated the effective knockdown of this liver-abundant microRNA in vivo, laying the foundation for its therapeutic use. Tsai et al. (2012) demonstrated that *miR-122* plays a critical role in liver homeostasis and hepatocarcinogenesis. In genetically improved farmed tilapia exposed to cadmium, Qiang et al. (2017) found that *miR-122* promotes hepatic antioxidant defense by directly targeting a metallothionein gene. Similarly, Cao et al. (2023) investigated the miRNA profiles of genetically improved farmed tilapia exposed to environmental stressors, revealing complex miRNA-mediated regulatory networks involved in stress responses. Their findings underscore the adaptive role of miRNAs in modulating gene expression related to detoxification, antioxidant defense, and cellular protection, highlighting the potential for miRNA-targeted strategies to enhance resilience in aquaculture species. Remarkably, *miR-122* also plays a central, not yet fully understood role in the development of MC-LR toxicity in mammals (Xu et al. 2018a; Wang et al. 2020; Wang et al. 2022) as well as in MC-sensitive and

MC-tolerant fish species (Li et al. 2013; Florczyk et al. 2016; Feng et al. 2017; Liu et al. 2020).

Together, these studies enhance our understanding by highlighting how miRNAs—most notably *miR-122*—act as key regulators of toxin response pathways in aquatic animals. Looking ahead, exploring the effects of dietary administration of *antagomir-122* in MC-containing aquafeeds would not only provide valuable scientific insights but could also open innovative avenues for aquaculture, particularly in the sustainable use of novel protein sources.

Specific studies on how miRNAs can promote tolerance to cyanotoxins in aquatic invertebrates are just beginning to emerge (Lyu et al. 2024) (Fig. 5.26). These authors revealed a novel mechanism of cyanobacterial toxicity on *Daphnia magna* fitness through miRNA–gene interactions. They observed that a poor *Microcystis aeruginosa* diet significantly increases the expression of *miR-210* compared to a nutritious diet, and this effect is stronger in the sensitive strain than in the tolerant strain. Subsequently, the downstream target genes *cdhr5*, *hyou1*, and *mlh3* are repressed. These genes are respectively involved in variations in intestinal structure, intestinal self-repair capacity, and the maintenance of genetic integrity.[14] Moreover, this paper illuminates the role of *miR-210* regulation in dictating clonal differences in *Daphnia*'s tolerance to cyanobacteria, thereby shedding light on microRNAs' contribution to the adaptive responses of *D. magna* to toxic cyanobacterial blooms.

MiR-210 is one of the most extensively studied hypoxia-inducible miRNAs (Cicchillitti et al. 2012), and there is evidence that *miR-210* plays a crucial role in cell apoptosis. The regulation and function of *miR-210* extends beyond the cellular response to hypoxia. Studying the marine medaka (*Oryzias melastigma*), Tse et al. (2015) found that *miR-210* plays an anti-apoptotic role in ovarian follicular cells under hypoxia. On the other hand, inhibition of *miR-210* promotes apoptotic cell death and the expression of apoptotic markers such as caspase 3 in follicular cells under hypoxic treatment, supporting the regulatory role of *miR-210* in ovarian cell apoptosis.

Notably, the regulation and function of *miR-210* goes still further: It appears to be an even more general stress-responsive biomarker, as shown in another *Daphnia*, in *Penaeus vannamei*, and in finfish: Chen et al. (2015b) reported that *miR-210* is responsive to hypoxia and cadmium in *D. pulex* and propose that hypoxia and Cd induce *miR-210* via the same HIF1α-modulated pathway. The same pathway is postulated for *P. vannamei* and European eel in response to hypoxia (Bertucci et al. 2019; Wang et al. 2019). Taken together, these results suggest that *miR-210* may be a promising marker of general environmental stress.

[14] CDHR5 is a homeodomain protein known for its tumor-suppressor activity in the gastrointestinal tissues (van Roy 2014). *Hyou1* encodes a protein that localizes to the endoplasmic reticulum (ER) via a carboxy-terminal peptide sequence and to mitochondria via an amino-terminal targeting signal. This protein is a member of the heat shock protein 70 family, which is known to accumulate in the ER under conditions such as hypoxia and other stressors, and therefore plays a crucial role in the processes of protein folding and secretion within the ER (Liu et al. 2016a). *Mlh3* is a DNA mismatch repair gene associated with mammalian microsatellite instability (Lipkin et al. (2000).

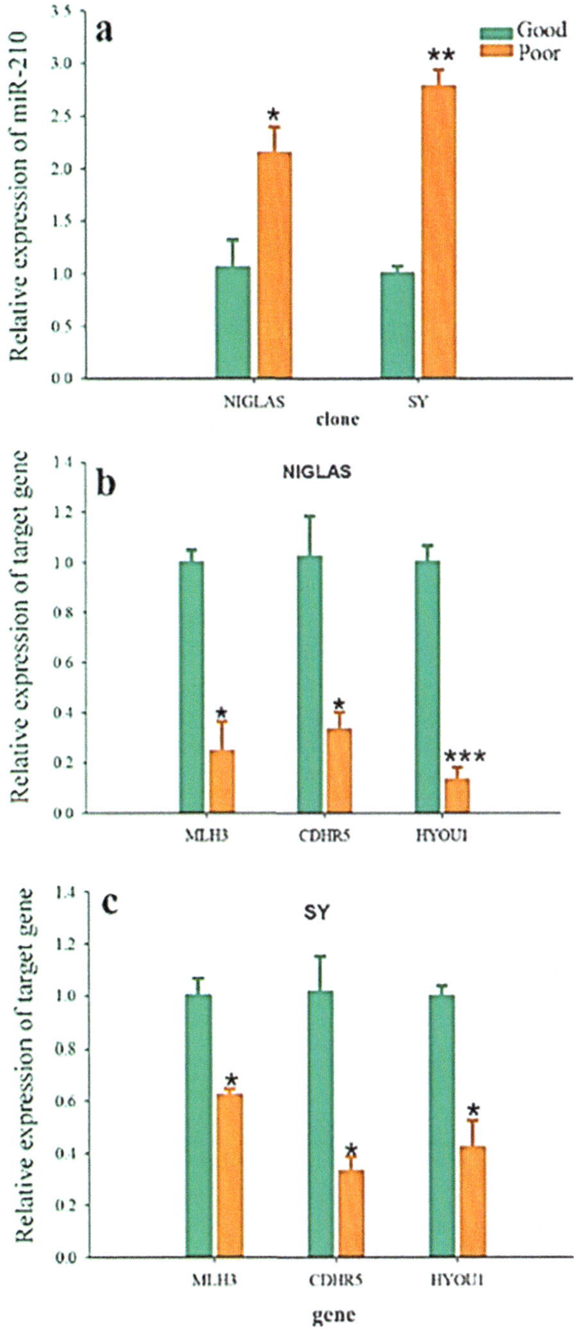

Fig. 5.26 *miR-210* expression (**a**) of *Daphnia magna* (**NIGLAS**, tolerant, and **SY**, sensitive, clones) following 48 h of exposure to toxic *Microcystis aeruginosa*. Expression of *miR-210* target genes *mlh3*, *cdhr5*, and *hyou1* in *Daphnia magna* NIGLAS clone (**b**) and SY clone (**c**) following 48 h of exposure to toxic *Microcystis aeruginosa*. Significant differences between food treatment within each clone are indicated as *$P < 0.05$, **$P < 0.01$, ***$P < 0.001$. 'GOOD' (green bar) refers to the diet control, while 'POOR' (orange bar) indicates the *M. aeruginosa* diet (from Lyu et al. (2024), with permission from the American Chemical Society).

As expected, *miR-210* is not the only microRNA involved in the development of MC toxicity in *D. magna*. In a companion paper, Fan et al. (2024) showed that when exposed to *Microcystis*, *dm-traf4* expression is markedly downregulated. The *tumor necrosis factor receptor-associated factor 4* (*traf4*) is central in innate immunity. Notably, *miR-4443* shows a negative correlation with *dm-traf4* expression under the threat of *M. aeruginosa* in both *D. magna* strains, indicating that posttranscriptional modification of the TRAF4 gene is involved in the *D. magna* response to *M. aeruginosa* stress. The suppression of the TRAF4 gene by *miR-4443* ultimately leads to reduced cellular immune resistance in unfavorable environments (Ruan et al. 2022).

5.3.2.3 Transgenerational Effects

In addition to the examples of adaptive responses mentioned in Box 5.1, several papers specifically report an increase in transgenerational tolerance following dietary consumption of cyanobacteria over more than one generation, as in the *Daphnia* example described above (Fig. 5.24), often without clearly demonstrating the biochemical or biomolecular mechanisms involved. Instead, a few hypotheses are put forward that leave room for future research. A recent empirical paper provides evidence for an underlying mechanism. Overall, fish studies are scarce.

For pragmatic reasons, most studies have been conducted with short-lived, mostly clonal invertebrates. Low maternal food availability and quality can cause daphnids to produce larger but fewer offspring, thereby increasing their fitness under stressful conditions. Li and Jiang (2014) showed that this pattern also increases the tolerance of daphnids to toxic cyanobacteria. This suggests that maternal exposure to cyanobacteria can shape trophic interactions between zooplankton and toxic cyanobacteria by improving offspring performance, and confirms previous studies (Gustafsson et al. 2005; Sarnelle et al. 2010; Ortiz-Rodríguez et al. 2012; Jiang et al. 2013b; Schwarzenberger and von Elert 2013).

Some cladoceran species/strains become more tolerant to MCs if they are preexposed to toxic strains or even non-toxic cyanobacteria (Guo and Xie 2006). For example, *Daphnia magna* exhibits increased survival rates after prior exposure to a toxic strain of *Microcystis* sp., suggesting that pretreatment can enhance tolerance to MCs (Gustafsson and Hansson 2004). Furthermore, neonates from parents exposed to MCs are less sensitive to these cyanotoxins than those from the untreated mothers, suggesting that adaptation to MC-producing strains is transferred to the offspring (Gustafsson et al. 2005). Higher survival rates of neonates correlate with higher activities of detoxifying enzymes, and it appears that this type of stimulation is transferred from mothers exposed to these cyanotoxins to their offspring (Ortiz-Rodríguez et al. 2012). Offspring from nonexposed and 1-day-exposed mothers display increased mortality when exposed to MC compared with offspring from 7-day-exposed mothers. Survival is positively correlated with the elevated activity of GST, malate dehydrogenase, and catalase, indicating maternal transfer of activation factors. However, increased survival occurs at the expense of individual growth. In this study, the transgenerational mechanism remains obscure. Later, Lyu et al. (2016b) confirmed findings of Guo and Xie (2006) showing that maternal

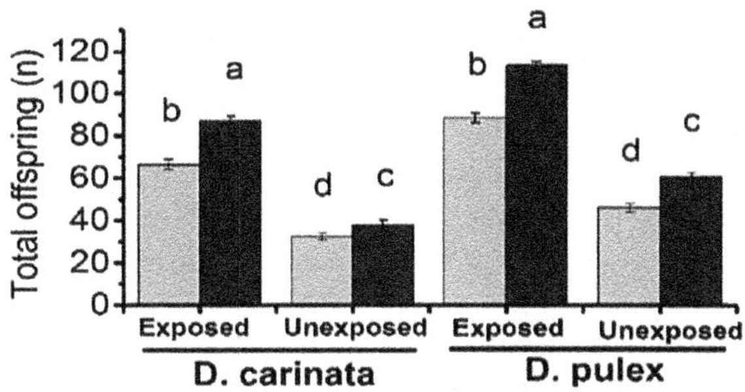

Fig. 5.27 Total offspring (F_1) of *Daphnia* mothers fed either a 4:1 mixture of coccal green algae (*Chlorella vulgaris*) and toxic *Microcystis aeruginosa* (exposed) or coccal green algae (unexposed) and raised in low (2.5 mg L^{-1} Ca, gray) or high (10.0 mg L^{-1} Ca, black) calcium medium. Different letters indicate a significant difference ($P < 0.05$) between neonates from exposed and unexposed treatments at both low and high Ca levels. Error bars indicate ±SD, ($N = 10$) (from Akbar et al. (2017); credit to Frontiers Media)

consumption of non-toxic *Microcystis* by *Daphnia magna* induces tolerance to toxic *Microcystis* in offspring.

Using clones of two *Daphnia* species, Akbar et al. (2017) highlighted the importance of maternal calcium (Ca) in improving offspring resistance to toxic *Microcystis*. Offspring growth and survival increase when mothers are raised in high Ca culture medium and fed increasing concentrations of *M. aeruginosa* (Fig. 5.27). As their exoskeleton is heavily reinforced with Ca, freshwater crustaceans have a high specific Ca requirement relative to other aquatic organisms.

Lyu et al. (2015) highlighted another transgenerational mechanism of microcystin tolerance. Individuals may develop increased tolerance by enhancing energy production, thereby allocating more energy to detoxification processes. Arginine kinase (AK) serves as a core componenet in temporal and spatial adenosine triphosphate (ATP) buffering in cells with high fluctuating energy requirements. To study the energetic response of a single *Daphnia magna* clone exposed to a toxic strain of *Microcystis aeruginosa* (PCC7806), Lyu et al. (2015) checked the transcript level of Dm-AK. Actually, Dm-AK transcript level and ATP content in the presence of *M. aeruginosa* are significantly lower than those in the control diet containing only a nutritious chlorophyte. Notably, the two parameters in neonates whose mothers have been previously exposed to *M. aeruginosa* are significantly higher than those of mothers fed on pure chlorophytes. Overall, Dm-AK plays a crucial role in the coupling of energy production and utilization, as well as in the tolerance of *D. magna* to toxic cyanobacteria.

The transgenerational tolerance of aquatic invertebrates to anatoxin-a has also been studied. Schwarzenberger and Martin-Creuzburg (2021) investigated the effects of the anatoxin-a-producing *Tychonema bourrellyi* on life-history parameters and gene expression of nicotine-acetylcholine receptors (nAChRs or NAR in the cited paper), the direct targets of anatoxin-a, using several *D. magna* clones. The authors used juvenile somatic growth rates as a measure of fitness and analyzed gene expression. Exposure to 100% *Tychonema* reduces the growth rates of the clones and causes an upregulation of NAR gene expression. When *Tychonema* makes up 50% of the diet, none of the clones showed reduced growth, and only one exhibited an increase in NAR gene expression. This increased NAR gene expression can be maternally transmitted, and offspring from exposed mothers show a higher growth rate when fed 50% *Tychonema* compared with control offspring. This transgenerational increase in tolerance to anatoxin-a is phenotypically similar to that found for microcystin.

Using a carbamate insecticide rather than cyanotoxins, Ishimota et al. (2024) performed multigenerational tests to elucidate chemical tolerance mechanisms in the non-target aquatic organism, *Daphnia magna*, and showed that epigenetics and ATP-Binding Cassette transporters (ABC transporters) act in concert. After 15 generations of exposure to sublethal and lethal levels of pirimicarb, *D. magna* acquires chemical tolerance. Model analyses indicate that *abc* expression levels are associated with pirimicarb tolerance, whereas mRNA levels of acetylcholinesterase and glutathione transferase are unlikely to be associated. In addition, DNA methylation levels are positively correlated with pirimicarb tolerance (based on EC_{50} values, Fig. 5.28): Methylated DNA levels in pirimicarb-tolerant clones decrease with exposure to a demethylating reagent. It is clear that cladocerans acquire chemical tolerance under the control of epigenetic changes that contribute to long-term adaptation to chemical exposure. It is worthwhile to investigate whether a similar mechanism applies to transgenerational tolerance to cyanotoxins.

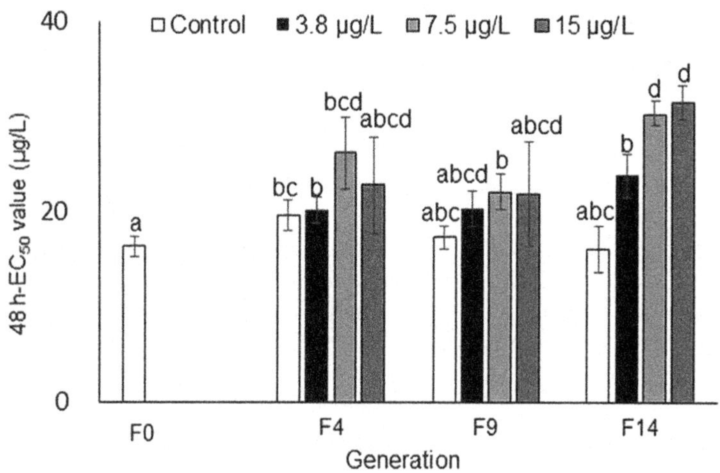

Fig. 5.28 Generational changes in 48 h EC_{50} values in *Daphnia magna* (< 24 h old) exposed to lethal or sublethal concentrations (0 [control], 3.8, 7.5, and 15 μg L^{-1}) of pirimicarb solutions for 15 generations (F0–F14). EC_{50} values were estimated by two independent experiments. Error bars show 95% confidence intervals for the EC_{50} value. Letters a–d represent significant differences. The 48 h EC_{50} values in the high concentration groups (7.5 and 15 μg L^{-1}) in F14 were higher than those in the control ($P < 0.05$), indicating decreased sensitivity (from Ishimota et al. (2024), with permission from Wiley & Sons).

5.4 Fishes

A few cyprinid and tilapia species demonstrate notable tolerance to toxic cyanobacteria, particularly those that are phytoplanktivorous, such as silver carp (*Hypophthalmichthys molitrix*), bighead carp (*Hypophthalmichthys nobilis*), an Indian major carp (*Labeo rohita*), and Nile tilapia (*Oreochromis niloticus niloticus*). These species not only tolerate the presence of toxic cyanobacteria but also utilize the nutrients and energy contained within these harmful algae (Xie et al. 2004; Zhao et al. 2006; Das et al. 2013). Notably, in the Indian major carp, diets supplemented with *M. aeruginosa* have been shown to enhance immunity and survival following infection with *Aeromonas hydrophila*; however, the study did not confirm the toxic activity of the *Microcystis* material used, raising questions about whether the carp ingested significant amounts of cyanotoxins (Das et al. 2013).

Mechanistic insights into this tolerance are beginning to emerge. Shen et al. (2005) suggested that in silver carp, most cyanotoxins are excreted in the feces rather than being biodegraded or accumulated. However, the role of efflux transporters like P-glycoproteins remains unexamined in this species, contrasting with research in *D. magna* in the above study with pirimicarb (Fig. 5.28).

Further elucidation comes from Liang et al. (2007), who compared MC-tolerant silver carp and Nile tilapia with the MC-sensitive grass carp (*Ctenopharyngodon idella*). They found that inducible liver expression of specific glutathione *S*-transferase (GST) isoforms, primarily alpha or rho class, correlates with MC tolerance; in contrast, sensitive grass carp lack such inducibility. This establishes inducible GST expression as a critical molecular mechanism underpinning fish tolerance to MCs. Zhang et al. (2011) added the classes zeta, mu, theta, and pi to the catalog of MC-responsive genes (Table 5.3). However, the resistance to MCs in zebrafish or Mangrove rivulus (*Kryptolebias hermaphroditus*) cannot be compared with that of silver carp or Nile tilapia.

Subsequent studies suggest that exporter proteins are also involved in MC tolerance: Amé et al. (2009) showed that P-glycoprotein expression responds to MC exposure in the fish *Jenynsia lineata*. However, the comparison of this resistant species with a related sensitive fish species and its resistance mechanisms is lacking.

In common carp, Fu and Xie (2006) studied nine glutathione transferase genes, including cytosolic GSTs (rho, mu, theta, alpha, and pi), mitochondrial GST (kappa), and microsomal GSTs (mGST1, mGST2, and mGST3). After MC-LR exposure, transcription of GST alpha, rho, and mGST3 isoforms increased at 6 h poststimulation; the transcription of mu, theta, and mGST2 isoforms remained relatively stable; and all the GST isoforms except GST kappa and rho returned to their original levels at 72 h (Fig. 5.29).

Table 5.3 provides a brief overview of the current state of information in the few available papers. A quick glance at this table indicates that knowledge about possible cyanotoxin resistance in fish species is even more patchy and incomplete than in invertebrates. To our knowledge, there is no work investigating the entire NRF2 pathway or even the *nrf2* gene or phase I biotransformation genes alone. Five papers consider exporters. To the extent that GST isoforms have been studied, it is clear

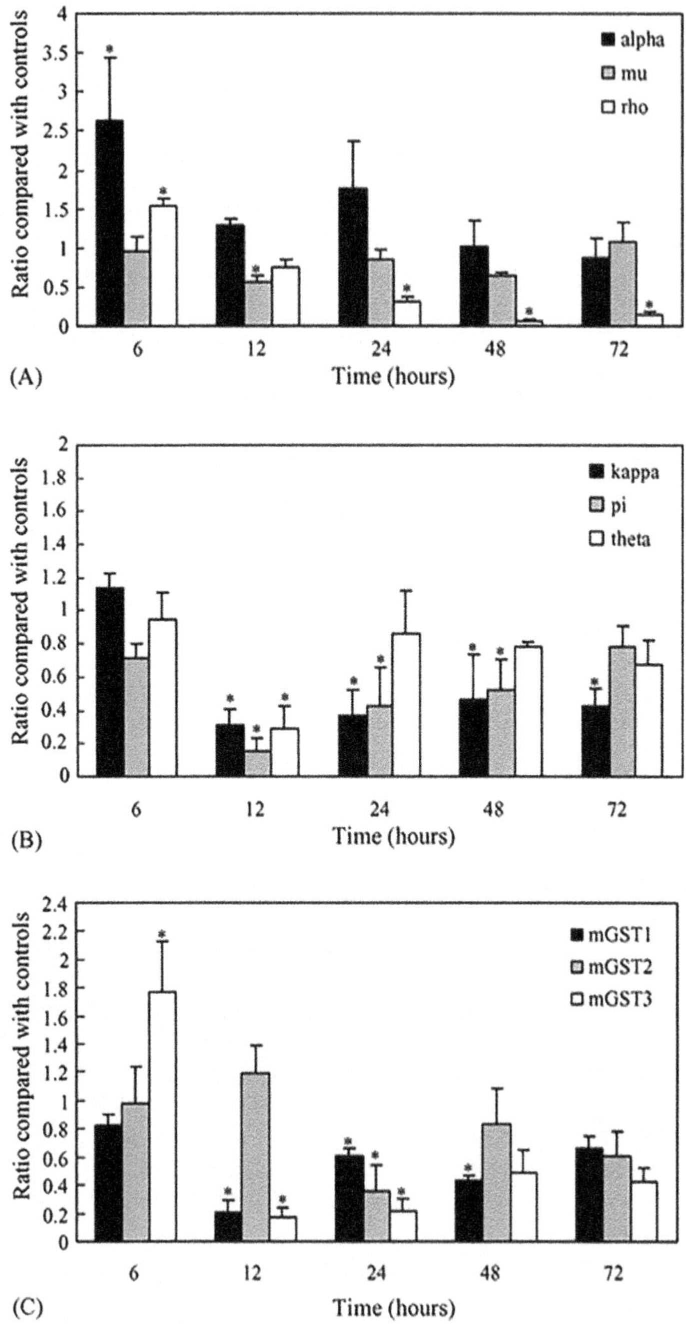

Fig. 5.29 The temporal changes in GSTs transcripts after microcystin stimulation in common carp, compared with controls. (**a**) **black** alpha GST; **gray** mu GST; and **white** rho GST; (**b**) **black** kappa GST; **gray** pi GST; and **white** theta GST; (**c**) **black** mGST1; **gray** mGST2; and **white** mGST3. *Significant change ($P < 0.05$) (from Fu and Xie (2006), with permission from Elsevier)

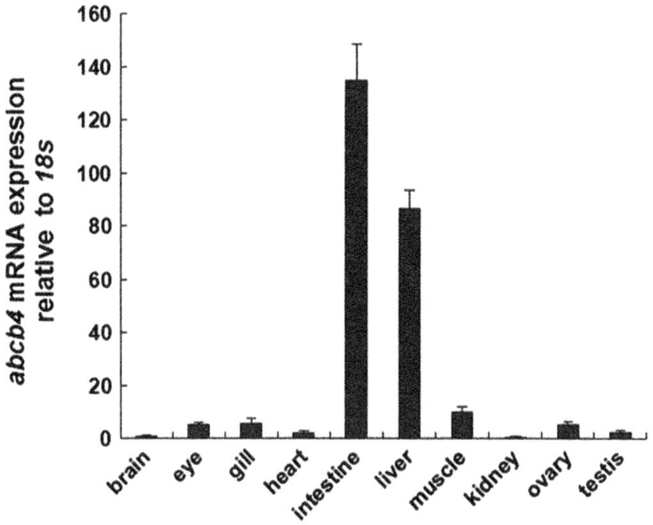

Fig. 5.30 Transcriptional expression of the zebrafish *abcb4* gene in adult tissues. Brain, eyes, gills, intestine, liver, muscle, ovaries, and testes from three individuals were collected for total RNA isolation and subjected to qRT-PCR analysis. Hearts or kidneys from ten individuals were pooled and used for total RNA isolation and subjected to qRT-PCR analysis. Values are given as mean ± SD ($N = 3$) (from Lu et al. (2015) with permission from Elsevier)

that it is not just one or a few isoforms whose job it is to conjugate MCs; alpha (α) and rho (ρ) isozymes, however, seem to be mostly involved in this conjugation. Often, no proper dose–response curves are recorded, but only a single concentration is tested; therefore, the conclusions obtained, for example, regarding the involvement of antioxidant and biotransformation enzymes, can only be considered preliminary.

In addition to oxidation/reduction and conjugation, the third part of the biotransformation machinery is excretion via relatively nonspecific exporter proteins (Table 5.3). Exporters are found in all tissues, but are particularly abundant in the most exposed tissues, as demonstrated in zebrafish (Fig. 5.30) (Lu et al. 2015).

5.4.1 Adaptive Response

Specific studies on the adaptive response of fish to cyanotoxins are not yet available. However, examples such as the northern redbelly dace and finescale dace species complex (*Chrosomus eos–neogaeus*) provide a first valuable model for understanding potential adaptive mechanisms (→Sect. 5.1.5). While direct evidence in fish remains limited, molecular and epigenetic regulatory pathways identified in invertebrate model systems suggest that similar adaptive responses to cyanotoxins may develop in fish species exposed to persistent cyanobacterial blooms. This underlines the need for further focused research on fish adaptive tolerance, leveraging species such as *Chrosomus eos–neogaeus* as natural case studies.

5.4.2 Gut Microbiome

In contrast to the growing body of research on aquatic invertebrates and even small mammals fed diets containing toxic cyanobacteria, comparable studies in fish remain relatively scarce. This section therefore highlights studies in small mammals that shed light on the potential impacts of cyanobacterial diets, offering clues to what might also occur in fishes through modulation of the gut microbiome: gastroenteritis (Lin et al. 2015b), non-alcoholic fatty liver disease (Sarkar et al. 2019), altered hepatic lipid metabolism linked to gut microbial community structure (Zhang et al. 2016), and even tumor development via microbial dysbiosis (Lee et al. 2020).

As one of the first, Lin et al. (2015b) reported that the composition of bacterial and fungal functional genes in the rat intestine changes significantly after 1 week of exposure to MC-LR, with genes involved in carbon degradation, including chitin, starch, and limonene metabolism, being relatively enriched. These genes are mainly derived from fungal and bacterial pathogens. Additionally, a large number of significantly enriched genes relevant to the degradation of aromatic carbon compounds were identified. Dysbiosis of bacterial and fungal flora indicates invasion by pathogens. The enriched gene functions may be associated with acute gastroenteritis induced by MC-LR.

Of note, Guilin et al. (2020) showed that MC-LR exposure induces a reduction in gut microbial diversity in mice. More than 30 genera decreased, especially those producing short-chain fatty acids (SCFAs), such as *Alistipes* and *Ruminococcus*. Consequently, the concentrations of SCFAs significantly decreased after MCLR exposure, with butyrate decreasing by approximately fivefold. This striking decline in SCFA-producing bacteria and the associated drop in butyrate levels underscore how profoundly MC-LR can disrupt host–microbiome interactions, providing a mechanistic link to the adverse effects described above.

Intriguingly, Li et al. (2019) found that zebrafish possess a core microbiota that is resistant to MC-LR, and exposure to MC-LR does not significantly modulate this core microbiota but causes only minor changes in its overall structure. In fact, treatment with 20 μg L^{-1} MC-LR alters the characteristics of the microbiota with high abundance. The identified phylotypes, including *Actinobacteria*, *Lactobacillus*, and some opportunistic pathogens, show both an increasing risk of pathogen invasion and a tendency to develop resistance via potential probiotics. In contrast, Zhang et al. (2020b) reported that MC-LR reduces bacterial richness and diversity and significantly alters microbial composition at the phylum and genus levels in the gut of *Procambarus clarkii*. Potentially probiotic lactic acid bacteria do not undergo significant changes in the intestine.

While evidence from small mammals clearly demonstrates that cyanobacterial toxins such as MC-LR can disrupt gut microbial diversity, reduce beneficial SCFA-producing bacteria, and trigger a cascade of adverse physiological effects, studies in fish remain sparse and urgently needed. The few available reports suggest that fish may harbor a relatively stable core microbiota, with only subtle changes in community structure under toxin exposure, yet findings in other aquatic animals (e.g., crayfish) point to pronounced shifts in diversity and composition. Together, these observations highlight both the resilience and vulnerability of aquatic gut microbiomes to cyanotoxins, underscoring the urgent need for more targeted research in fish to clarify potential risks for aquaculture and natural ecosystems.

5.4.3 Epigenetics

In general, species can reduce their risk of extinction in unpredictable environments by diversifying individual phenotypes, with epigenetic mechanisms playing a central role. Various sources of epigenetic variation can be shared across individuals, creating a reservoir of phenotypic diversity. Because both environmentally induced and stochastic epigenetic changes can generate variation in traits, they are expected to be key drivers of plasticity and bet-hedging strategies (Leung et al. 2016). Through such strategies, species effectively "hedge their bets," buffering against environmental fluctuations and enhancing their chances of survival.

Shahmohamadloo et al. (2022) demonstrated intraspecific variability in the proteomics of rainbow trout and lake trout exposed to MCs. This variability may be a result of differences in mechanistic responses to the toxicity, biotransformation, and excretion of MCs (see above and Adamovský et al. (2007)) and subsequent epimutations (Anastasiadi and Piferrer 2019). Whether the individual differences in this example are due to epigenetic mechanisms has not been elucidated.

Recently, a growing number of epigenetic studies have highlighted how environmental stressors, including those beyond phycotoxins, can shape gene expression and phenotypic traits, revealing the broader relevance of epigenetic mechanisms in adaptation and resilience. A link between the rainbow/sea trout example and these epigenetics-based examples is very likely. Therefore, we will discuss some intriguing examples of environmental stressors in the following.

In wild three-spined sticklebacks (*Gasterosteus aculeatus*), Heckwolf et al. (2020) found two different epigenetic information channels involved in salinity adaptation: (i) inducible epigenetic marks that enable transgenerational plasticity and (ii) non-inducible epigenetic marks resulting from random epimutations shaped by selection. Wild populations differing in salinity tolerance reveal differential methylation in genes enriched for osmoregulatory processes, with a majority being non-inducible by salinity. This indicates that they are the result of either direct selection or associated genomic divergence at regulatory sites. Two-thirds of the remaining inducible methylated sites increase in similarity to patterns detected in wild populations from corresponding salinities. The level of similarity accentuates over consecutive generations, indicating a mechanism of transgenerational plasticity. While the selection-based information channel assumes random variation from epimutation that is subsequently shaped by selection or drift, the detection-based information channel allows a directional response in the form of transgenerational plasticity.

Fellous et al. (2022) presented another intriguing example in sticklebacks, exploring how epigenetic reprogramming—through DNA methylation and hydroxymethylation—shapes gene expression across gametogenesis and embryogenesis when fish are exposed to three different ocean warming scenarios. The authors found that parental acclimation to ocean warming leads to dynamic and temperature-sensitive reprogramming throughout offspring development. Both global methylation/hydroxymethylation and expression of genes involved in epigenetic modifications are strongly and differentially affected by the increased warming scenarios. A comparison of transcriptomic profiles from gonads, mature gametes, and early embryonic stages shows sex-specific accumulation and temperature sensitivity of several epigenetic regulators. DNA methyltransferase induction is primarily maternally inherited (indicating maternal control of remethylation), whereas both parents shape the induction of several histone-modifying enzymes.

Although hydrogen sulfide is toxic to most organisms, the shortfin molly (*Poecilia mexicana*) has adapted to survive in environments with high levels that exceed toxicity thresholds by orders of magnitude. Kelley et al. (2021) examined epigenetic changes in response to this environmental stressor by assessing DNA methylation alterations in the nucleated red blood cells of fish. The authors compared epimutations between the sexes, as well as field and laboratory populations, and succeeding generations reared in non-sulfidic environments. For both the wild-caught (F_0) and the laboratory-reared (F_2) fish, comparisons of the sulfidic and non-sulfidic populations reveals evidence for significant differential DNA methylation regions (DMRs). Strikingly, over 80% of these DMRs persisted across generations, demonstrating stable epigenetic inheritance even after the environmental stressor was removed. This example illustrates how epigenetic marks can convey a generational memory of past environmental challenges. Taken together, the observed phenotypic plasticity and molecular alterations in the epigenome and transcriptome underscore the critical role of epigenetic mechanisms in enabling organisms to sense, respond to, and thrive under changing environmental conditions.

5.5 Concluding Remarks

Intraspecific variation, which can evolve rapidly in multiple species within an ecosystem as well as in aquaculture, is an important driver of community structure and ecosystem function (Rudman et al. 2015; Anastasiadi and Piferrer 2019; Wong et al. 2022; Yang et al. 2022). Individuals that are well acclimated or carry specific epimutations—exhibiting low microcystin accumulation and minimal adverse effects—could serve as foundational strains in aquaculture, providing stock that can tolerate cyanobacterial proteins to some degree.

In addition to this more or less classical breeding approach, major remarks on the tolerance of aquatic animals to phycotoxins have already been made elsewhere (Steinberg 2025a, b, c), where cyanobacteria in the food chain and their potential as a food source, as well as the mitigation of mycotoxin loads in aquafeeds, were discussed. The discussion touched on prophylactic nutrition, including algicidal bacteria, antisense non-coding RNAs, prebiotics, and probiotics, as well as physicochemical and biochemical measures such as the use of adsorbents or dietary antioxidants. The role of functional feeds will be discussed in detail elsewhere (Steinberg forthcoming).

To harness biomolecular tools, particularly antisense non-coding RNAs, for mitigating the effects of cyanotoxins, a deeper focus on epigenetic mechanisms is essential. This also raises the intriguing question of whether such tolerance can be transmitted to subsequent generations. Likewise, research on the role of the gut microbiota in mediating and mitigating the effects of cyanotoxins is still in its infancy. It warrants significant expansion, offering promising avenues to enhance resilience in both wild and farmed aquatic organisms.[15]

To briefly mention, adsorbents such as biochar have also been used against microcystins, for example, by Lin et al. (2022, 2023). In addition to the fact that several probiotic strains are capable of degradation, adsorption, or nonspecific removal (Halttunen et al. 2008; Nybom et al. 2012; Chaosuan et al. 2019; Singh et al. 2024), dietary probiotics are emerging as an effective tool against cyanobacteria and their toxins in the gut of aquatic animals (Zhao et al. 2017; Yang et al. 2023b). As is often the case, insights from studies in small mammals can offer valuable lessons for aquaculture scientists and practitioners.

This chapter concludes with a graph from human medicine that illustrates the various pathways by which drugs and dietary components can counteract cyanotoxin toxicity when the above-mentioned biomolecular, prebiotic, or probiotic approaches are not suitable. One or more of these approaches may be worth testing in aquatic animal nutrition. With the focus on hepatotoxicity, Mashayekhi-Sardoo et al. (2024) recently reviewed the existing literature and created an educational graph (Fig. 5.31) that highlights the high diversity of interference points and indicates where polyphenolic and other supplements in aquafeed may be effective. MC-LR enters and accumulates in hepatocytes via the organic anion-transporting polypeptide superfamily (OATPs). In zebrafish, rainbow trout, and skate, the OATPs associated with this function are *oatp1d1* (expressed in liver, brain, and intestine)

[15] To emphasize this point, we cite the concern of Feng et al. (2023) who studied marine copepods, the major pelagic invertebrate algae predators that are able to tolerate several dietary phytotoxins. Although probiotics have received attention in the aquaculture industry, there has been a marked lack of research on copepod-associated microorganisms with probiotic potential and their impact on aquaculture activities.

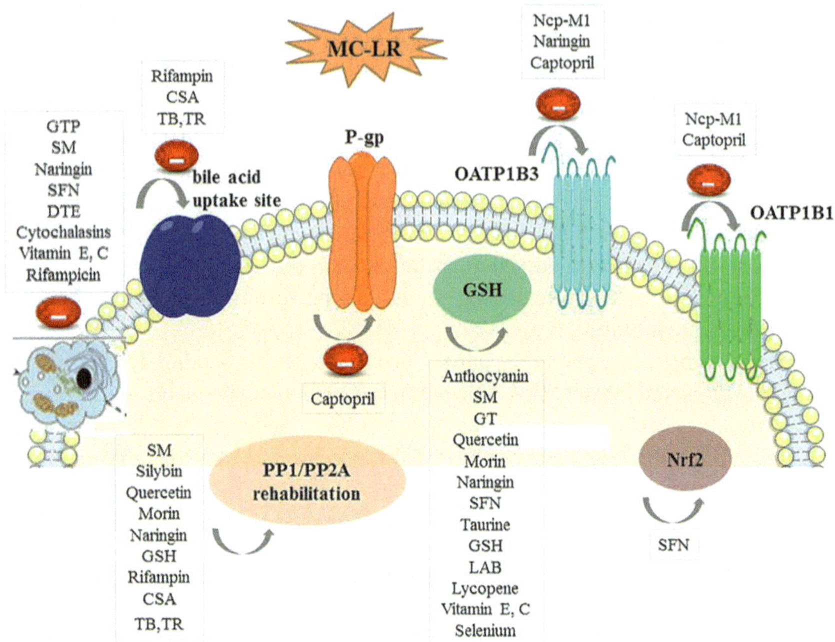

Fig. 5.31 The mechanisms of protective effects of various compounds against MC-LR-induced hepatotoxicity. *Ncp-M1* non-toxic nostocyclopeptide-M1, *OTAP* organicanion-transporting polypeptide, *P-gp* P-glycoprotein, *CSA* cyclosporin A, *TB* trypan blue, *TR* trypan red, *SM* silymarin, *GTP* green tea polyphenols, *SFN* sulforaphane, *DTE* dithioerythritol, *GT* green tea, *LAB* lactic acid bacteria, and *Nrf2* nuclear factor erythroid 2-related factor 2 (from Mashayekhi-Sardoo et al. (2024), with permission from Elsevier)

and *oatp1f* (in kidney) (Boaru et al. 2006; Steiner et al. 2016). In general, the transport of MCs by OATPs is highly dependent on the specific MC congener.

Once internalized in the liver, MC-LR inhibits protein phosphatases (PP1 and PP2A). Consequently, MC-LR disrupts many signaling pathways and induces oxidative stress, which damages cellular macromolecules. Some protective agents, especially plants rich in flavonoids and natural supplements, as well as chemoprotectants, have been shown to reduce MC-LR hepatotoxicity. Many of these protective agents have been reviewed as dietary supplements by Steinberg (2024).

Specifically, the reviewed agents exert their hepatoprotective effects against MC-LR by first blocking the OATP transporters (non-toxic nostocyclopeptide-M1, captopril, and naringin), then inhibiting MC-LR uptake (naringin, cyclosporin A, silymarin, and captopril), and finally restoring protein phosphatase activity (silybin, quercetin, morin, naringin, rifampin, captopril, azo dyes (trypan blue, trypan red)).[16]

[16] Pharmaceuticals: **Nostocyclopeptide M1** (also **Nep-M1**) is novel cyanobacterial cyclic peptide that blocks the hepatotoxic action of microcystin and nodularin (Herfindal et al. 2011). **Captopril** is an angiotensin-converting enzyme inhibitor used for the treatment of hypertension and some types of congestive heart failure (Vidt et al. 1982). **Naringin** is a flavanone-7-*O*-glycoside between

As we reach the end of this chapter, it is worthwhile to take a step back and reflect on the double-edged nature of microcystins in aquaculture. On the one hand, these toxins threaten the health of fish and invertebrates through hepatotoxicity, oxidative stress, and disruption of cellular signaling. On the other hand, cyanobacteria—the very organisms that produce these toxins—also carry valuable nutrients, offering a rich source of proteins, lipids, and other bioactive compounds. The challenge lies in navigating this delicate balance: how can we harness the nutritional potential of cyanobacteria while protecting cultured animals from harm?

Emerging strategies provide hope. Probiotics, prebiotics, adsorbents, and protective dietary supplements have demonstrated the ability to mitigate microcystin toxicity, and research into gut microbiomes, epigenetics, and detoxification pathways continues to expand our understanding. Yet, the complexity of microcystin dynamics—uptake, metabolism, and tissue-specific effects—requires tools that can handle this intricacy.

Here, artificial intelligence emerges as a fascinating avenue. By integrating multi-omics data, toxicological profiles, and nutritional composition, AI-driven models can predict safe inclusion rates of cyanobacterial biomass in aquafeed, optimize combinations of protective dietary agents, and even guide the development of new formulations that transform a traditionally hazardous component into a valuable and functional nutrient source. In this way, AI does not merely help to respond to risks—it enables proactive, intelligent feed design, transforming cyanobacteria from a toxin threat into a sustainable and valuable resource for aquaculture. Looking forward, the synergy of molecular biology, microbiome research, and AI-guided feed design promises a future where the risks of cyanobacteria can be managed intelligently. At the same time, their nutritional potential is fully exploited—ultimately supporting both the health of cultured species and the sustainability of aquaculture systems.

---

the flavanone naringenin and the disaccharide neohesperidose. It occurs naturally in citrus fruits and inhibits some drug-metabolizing cytochrome P450 enzymes (Fuhr and Kummert 1995). **Cyclosporin** is a calcineurin inhibitor, used as an immunosuppressant medication (Lei et al. 2011). **Rifampin** (also **rifampicin**) is an ansamycin antibiotic used to treat several types of bacterial infections (Lee et al. 2017). **Morin** is a yellow chemical compound that can be isolated from *Maclura pomifera* (Osage orange), *Maclura tinctoria* (old fustic), and from the leaves of *Psidium guajava* (common guava) (Rattanachaikunsopon and Phumkhachorn 2007). Morin was found to be a weak inhibitor of fatty acid synthase (Tian 2006). Prevention of MC-LR uptake into hepatocytes is the most likely chemoprotective mechanism of the azo dyes **trypan blue** and **trypan red** (Runnegar et al. 1995). **Cytochalasins** are fungal metabolites that have the ability to bind to actin filaments and block polymerization and the elongation of actin. As a result, they can change cellular morphology, inhibit cellular processes such as cell division, and even cause cells to undergo apoptosis (Haidle and Myers 2004).

References

Aboal M, Puig MÁ (2005) Intracellular and dissolved microcystin in reservoirs of the river Segura basin, Murcia, SE Spain. Toxicon 45(4):509–518. https://doi.org/10.1016/j.toxicon.2004.12.012

Aboal M, Puig MÁ, Asencio AD (2005) Production of microcystins in calcareous Mediterranean streams: the Alharabe River, Segura River basin in south-East Spain. J Appl Phycol 17(3):231–243. https://doi.org/10.1007/s10811-005-2999-z

Aboal M, Belando MD, Ubero N, González-Silvera D, López-Jiménez JA (2022) Photoautotrophs and macroinvertebrate trophic relations in calcareous semiarid streams: the role of cyanobacteria. Sci Total Environ 838:156206. https://doi.org/10.1016/j.scitotenv.2022.156206

Adamovský O, Kopp R, Hilscherová K, Babica P, Palíková M, Pašková V, Navrátil S, Maršálek B, Bláha L (2007) Microcystin kinetics (bioaccumulation and elimination) and biochemical responses in common carp (*Cyprinus carpio*) and silver carp (*Hypophthalmichthys molitrix*) exposed to toxic cyanobacterial blooms. Environ Toxicol Chem 26(12):2687–2693. https://doi.org/10.1897/07-213.1

Ahkin Chin Tai JK, Freeman JL (2020) Zebrafish as an integrative vertebrate model to identify miRNA mechanisms regulating toxicity. Toxicol Rep 7:559–570. https://doi.org/10.1016/j.toxrep.2020.03.010

Akakura N, Hoogland C, Takada YK, Saegusa J, Ye X, Liu FT, Cheung AT, Takada Y (2006) The COOH-terminal globular domain of fibrinogen gamma chain suppresses angiogenesis and tumor growth. Cancer Res 66(19):9691–9697. https://doi.org/10.1158/0008-5472.can-06-1686

Akbar S, Du J, Jia Y, Tian X (2017) The importance of calcium in improving resistance of *Daphnia* to *Microcystis*. PLoS One 12(4). https://doi.org/10.1371/journal.pone.0175881

Aksakal FI, Arslan H (2020) Detoxification and reproductive system-related gene expression following exposure to $Cu(OH)_2$ nanopesticide in water flea (*Daphnia magna* Straus 1820). Environ Sci Pollut Res 27(6):6103–6111. https://doi.org/10.1007/s11356-019-07414-x

Amado LL, Garcia ML, Ramos PB, Yunes JS, Monserrat JM (2011) Influence of a toxic *Microcystis aeruginosa* strain on glutathione synthesis and glutathione-*S*-transferase activity in common carp *Cyprinus carpio* (Teleostei: Cyprinidae). Arch Environ Contam Toxicol 60(2):319–326. https://doi.org/10.1007/s00244-010-9594-2

Amé MV, Baroni MV, Galanti LN, Bocco JL, Wunderlin DA (2009) Effects of microcystin-LR on the expression of P-glycoprotein in *Jenynsia multidentata*. Chemosphere 74(9):1179–1186. https://doi.org/10.1016/j.chemosphere.2008.11.068

Anastasiadi D, Piferrer F (2019) Epimutations in developmental genes underlie the onset of domestication in farmed European sea bass. Mol Biol Evol 36(10):2252–2264. https://doi.org/10.1093/molbev/msz153

Anderson K, Taylor DA, Thompson EL, Melwani AR, Nair SV, Raftos DA (2015) Meta-analysis of studies using suppression subtractive hybridization and microarrays to investigate the effects of environmental stress on gene transcription in oysters. PLoS One 10(3):e0118839. https://doi.org/10.1371/journal.pone.0118839

Andrade-Villagrán PV, Agüero MJ, Navarro JM, Urzúa Á (2022) The paralytic shellfish toxin effect on bioenergetic constituents of the fishery resource *Chorus giganteus* (Gastropoda: Muricidae). Mar Environ Res 180. https://doi.org/10.1016/j.marenvres.2022.105735

Angers B, Schlosser IJ (2007) The origin of *Phoxinus eos-neogaeus* unisexual hybrids. Mol Ecol 16(21):4562–4571. https://doi.org/10.1111/j.1365-294X.2007.03511.x

Antas P, Carneiro M, Reis B, Castelo-Branco R, Azevedo J, Urbatzka R, Campos A, Vasconcelos V, Martins JC (2018) GST transcriptional changes induced by a toxic *Microcystis aeruginosa* strain in two bivalve species during exposure and recovery phases. Ecotoxicology 27(9):1272–1280. https://doi.org/10.1007/s10646-018-1980-y

Anwar-Mohamed A, Degenhardt OS, El Gendy MAM, Seubert JM, Kleeberger SR, El-Kadi AOS (2011) The effect of Nrf2 knockout on the constitutive expression of drug metabolizing enzymes and transporters in C57Bl/6 mice livers. Toxicol In Vitro 25(4):785–795. https://doi.org/10.1016/j.tiv.2011.01.014

Asai S, Sanges R, Lauritano C, Lindeque PK, Esposito F, Ianora A, Carotenuto Y (2020) De novo transcriptome assembly and gene expression profiling of the copepod *Calanus helgolandicus* feeding on the PUA-producing diatom *Skeletonema marinoi*. Mar Drugs 18(8):392. https://doi.org/10.3390/md18080392

Asselman J, De Coninck DIM, Glaholt S, Colbourne JK, Janssen CR, Shaw JR, De Schamphelaere KAC (2012) Identification of pathways, gene networks, and paralogous gene families in *Daphnia pulex* responding to exposure to the toxic cyanobacterium *Microcystis aeruginosa*. Environ Sci Technol 46(15):8448–8457. https://doi.org/10.1021/es301100j

Asselman J, De Coninck DIM, Beert E, Janssen CR, Orsini L, Pfrender ME, Decaestecker E, De Schamphelaere KAC (2017) Bisulfite sequencing with *daphnia* highlights a role for epigenetics in regulating stress response to *Microcystis* through preferential differential methylation of serine and threonine amino acids. Environ Sci Technol 51(2):924–931. https://doi.org/10.1021/acs.est.6b03870

Avery DE, Dam HG (2007) Newly discovered reproductive phenotypes of a marine copepod reveal the costs and advantages of resistance to a toxic dinoflagellate. Limnol Oceanogr 52(5):2099–2108. https://doi.org/10.4319/lo.2007.52.5.2099

Avery DE, Altland KK, Dam HG (2008) Sex-related differential mortality of a marine copepod exposed to a toxic dinoflagellate. Limnol Oceanogr 53(6):2627–2635. https://doi.org/10.4319/lo.2008.53.6.2627

Aydın GB, Öterler B, Çamur Elipek B, Güher H (2021) The comparative gut content analysis of some Chironomidae larvae living in the freshwaters at northern Thrace region of Turkey. J Limnol Freshw Fisher Res 7(1):14–23. https://doi.org/10.17216/limnofish.741780

Bao S, Pan B, Wang L, Cheng Z, Liu X, Zhou Z, Nie X (2020) Adverse effects in *Daphnia magna* exposed to e-waste leachate: assessment based on life trait changes and responses of detoxification-related genes. Environ Res 188:109821. https://doi.org/10.1016/j.envres.2020.109821

Bautista NM, Burggren WW (2019) Parental stressor exposure simultaneously conveys both adaptive and maladaptive larval phenotypes through epigenetic inheritance in the zebrafish (*Danio rerio*). J Exp Biol 222(17). https://doi.org/10.1242/jeb.208918

Beattie KA, Ressler J, Wiegand C, Krause E, Codd GA, Steinberg CEW, Pflugmacher S (2003) Comparative effects and metabolism of two microcystins and nodularin in the brine shrimp *Artemia salina*. Aquat Toxicol 62(3):219–226. https://doi.org/10.1016/S0166-445X(02)00091-7

Berezina NA, Tiunov AV, Tsurikov SM, Kurbatova SA, Korneva LG, Makarova OS, Bykova SN (2021) Cyanobacteria as a food source for invertebrates: results of a model experiment. Russ J Ecol 52(3):247–252. https://doi.org/10.1134/S1067413621030036

Bertucci A, Pierron F, Ye T, Gonzalez P, Couture P, Baudrimont M (2019) Identification and expression of microRNAs in European eels *Anguilla Anguilla* from two natural sites with different pollution levels. Environ Pollut 250:274–283. https://doi.org/10.1016/j.envpol.2019.04.029

Bieczynski F, Bianchi VA, Luquet CM (2013) Accumulation and biochemical effects of microcystin-LR on the Patagonian pejerrey (*Odontesthes hatcheri*) fed with the toxic cyanobacteria *Microcystis aeruginosa*. Fish Physiol Biochem 39(5):1309–1321. https://doi.org/10.1007/s10695-013-9785-7

Bieczynski F, De Anna JS, Pirez M, Brena BM, Villanueva SSM, Luquet CM (2014) Cellular transport of microcystin-LR in rainbow trout (*Oncorhynchus mykiss*) across the intestinal wall: possible involvement of multidrug resistance-associated proteins. Aquat Toxicol 154:97–106. https://doi.org/10.1016/j.aquatox.2014.05.003

Bieczynski F, Torres WDC, Painefilu JC, Castro JM, Bianchi VA, Frontera JL, Paz DA, González C, Martín A, Villanueva SSM, Luquet CM (2016) Alterations in the intestine of Patagonian silverside (*Odontesthes hatcheri*) exposed to microcystin-LR: changes in the glycosylation pattern of the intestinal wall and inhibition of multidrug resistance proteins efflux activity. Aquat Toxicol 178:106–117. https://doi.org/10.1016/j.aquatox.2016.07.016

Blanco J, Mariño C, Martín H, Acosta CP (2007) Anatomical distribution of diarrhetic shellfish poisoning (DSP) toxins in the mussel *Mytilus galloprovincialis*. Toxicon 50(8):1011–1018. https://doi.org/10.1016/j.toxicon.2007.09.002

Boaru DA, Dragoş N, Schirmer K (2006) Microcystin-LR induced cellular effects in mammalian and fish primary hepatocyte cultures and cell lines: a comparative study. Toxicology 218(2–3):134–148. https://doi.org/10.1016/j.tox.2005.10.005

Bolius S, Wiedner C, Weithoff G (2017) High local trait variability in a globally invasive cyanobacterium. Freshw Biol 62(11):1879–1890. https://doi.org/10.1111/fwb.13028

Bonduriansky R, Day T (2018) Extended heredity: a new understanding of inheritance and evolution. Princeton University Press

Bownik A (2016) Harmful algae: effects of cyanobacterial cyclic peptides on aquatic invertebrates-a short review. Toxicon 124:26–35. https://doi.org/10.1016/j.toxicon.2016.10.017

Brand LE, Campbell L, Bresnan E (2012) *Karenia*: the biology and ecology of a toxic genus. Harmful Algae 14:156–178. https://doi.org/10.1016/j.hal.2011.10.020

Brander SM, Biales AD, Connon RE (2017) The role of epigenomics in aquatic toxicology. Environ Toxicol Chem 36(10):2565–2573. https://doi.org/10.1002/etc.3930

Brash AR, Boeglin WE, Chang MS (1997) Discovery of a second 15*S*-lipoxygenase in humans. Proc Natl Acad Sci 94(12):6148–6152. https://doi.org/10.1073/pnas.94.12.6148

Breier CF, Buskey EJ (2007) Effects of the red tide dinoflagellate, *Karenia brevis*, on grazing and fecundity in the copepod *Acartia tonsa*. J Plankton Res 29(2):115–126. https://doi.org/10.1093/plankt/fbl075

Bricelj VM, Lee JH, Cembella AD, Anderson DM (1990) Uptake kinetics of paralytic shellfish toxins from the dinoflagellate *Alexandrium fundyense* in the mussel *Mytilus edulis*. Mar Ecol Prog Ser 63:177–188

Bricelj VM, Connell L, Konoki K, MacQuarrie SP, Scheuer T, Catterall WA, Trainer VL (2005) Sodium channel mutation leading to saxitoxin resistance in clams increases risk of PSP. Nature 434(7034):763–767. https://doi.org/10.1038/nature03415

Bricelj VM, MacQuarrie SP, Doane JAE, Connell LB (2010) Evidence of selection for resistance to paralytic shellfish toxins during the early life history of soft-shell clam (*Mya arenaria*) populations. Limnol Oceanogr 55(6):2463–2475. https://doi.org/10.4319/lo.2010.55.6.2463

Briland RD, Stone JP, Manubolu M, Lee J, Ludsin SA (2020) Cyanobacterial blooms modify food web structure and interactions in western Lake Erie. Harmful Algae 92. https://doi.org/10.1016/j.hal.2019.03.004

Brock EM (1960) Mutualism between the midge *Cricotopus* and the alga *Nostoc*. Ecology 41(3):474–483. https://doi.org/10.2307/1933322

Bucciarelli GM, Shaffer HB, Green DB, Kats LB (2017) An amphibian chemical defense phenotype is inducible across life history stages. Sci Rep 7(1):8185. https://doi.org/10.1038/s41598-017-08154-z

Bürglin TR (2013) Homeobox genes. In: Maloy S, Hughes K (eds) Brenner's encyclopedia of genetics. Academic, San Diego, pp 503–508

Burmester V, Nimptsch J, Wiegand C (2012) Adaptation of freshwater mussels to cyanobacterial toxins: response of the biotransformation and antioxidant enzymes. Ecotoxicol Environ Saf 78:296–309. https://doi.org/10.1016/j.ecoenv.2011.11.037

Cai S, Shu Y, Tian C, Wang C, Fang T, Xiao B, Wu X (2022) Effects of chronic exposure to microcystin-LR on life-history traits, intestinal microbiota and transcriptomic responses in *Chironomus pallidivittatus*. Sci Total Environ 823:153624. https://doi.org/10.1016/j.scitotenv.2022.153624

Cao Q, Zhang H, Li T, He L, Zong J, Shan H, Huang L, Zhang Y, Liu H, Jiang J (2023) Profiling miRNAs of teleost fish in responses to environmental stress: a review. Biology 12(3):388. https://doi.org/10.3390/biology12030388

Carneiro M, Reis B, Azevedo J, Campos A, Osório H, Vasconcelos V, Martins JC (2015) Glutathione transferases responses induced by microcystin-LR in the gills and hepatopancreas of the clam *Venerupis philippinarum*. Toxins 7(6):2096–2120. https://doi.org/10.3390/toxins7062096

Carneiro M, Antas P, Reis B, Azevedo J, Osório H, Campos A, Vasconcelos V, Martins JC (2017) Modulation of hepatic glutathione transferases isoenzymes in three bivalve species exposed to purified microcystin-LR and *Microcystis* extracts. Toxicon 137:150–157. https://doi.org/10.1016/j.toxicon.2017.07.003

Caron V, Ede FJ, Sunnucks P (2014) Unravelling the paradox of loss of genetic variation during invasion: superclones may explain the success of a clonal invader. PLoS One 9(6):e97744. https://doi.org/10.1371/journal.pone.0097744

Castellano I, Seebeck FP (2018) On ovothiol biosynthesis and biological roles: from life in the ocean to therapeutic potential. Nat Prod Rep 35(12):1241–1250. https://doi.org/10.1039/C8NP00045J

Castellano I, Migliaccio O, D'Aniello S, Merlino A, Napolitano A, Palumbo A (2016) Shedding light on ovothiol biosynthesis in marine metazoans. Sci Rep 6(1):21506. https://doi.org/10.1038/srep21506

Catherine Q, Susanna W, Isidora E-S, Mark H, Aurélie V, Jean-François H (2013) A review of current knowledge on toxic benthic freshwater cyanobacteria—ecology, toxin production and risk management. Water Res 47(15):5464–5479. https://doi.org/10.1016/j.watres.2013.06.042

Cazenave J, Bistoni MDLA, Pesce SF, Wunderlin DA (2006) Differential detoxification and antioxidant response in diverse organs of *Corydoras paleatus* experimentally exposed to microcystin-RR. Aquat Toxicol 76(1):1–12. https://doi.org/10.1016/j.aquatox.2005.08.011

Chan IOM, Lam PKS, Cheung RHY, Lam MHW, Wu RSS (2005) Application of solid phase microextraction in the determination of paralytic shellfish poisoning toxins. Analyst 130(11):1524–1529. https://doi.org/10.1039/b506324h

Chang TS, Cho CS, Park S, Yu S, Sang WK, Sue GR (2004) Peroxiredoxin III, a mitochondrion-specific peroxidase, regulates apoptotic signaling by mitochondria. J Biol Chem 279(40):41975–41984. https://doi.org/10.1074/jbc.M407707200

Chaosuan N, Pongswat S, Chorum M (2019) Removal of cyanotoxin microcystin-LR by *lactobacillus plantarum*. Malays Appl Biol 48(4):75–82

Chen J, Xie P (2005) Tissue distributions and seasonal dynamics of the hepatotoxic microcystins-LR and -RR in two freshwater shrimps, *Palaemon modestus* and *Macrobrachium nipponensis*, from a large shallow, eutrophic lake of the subtropical China. Toxicon 45(5):615–625. https://doi.org/10.1016/j.toxicon.2005.01.003

Chen L, Zhang H, Finiguerra M, Bobkov Y, Bouchard C, Avery DE, Anderson PAV, Lin S, Dam HG (2015a) A novel mutation from gene splicing of a voltage-gated sodium channel in a marine copepod and its potential effect on channel function. J Exp Mar Biol Ecol 469:131–142. https://doi.org/10.1016/j.jembe.2015.04.003

Chen S, McKinney GJ, Nichols KM, Colbourne JK, Sepúlveda MS (2015b) Novel cadmium responsive microRNAs in *Daphnia pulex*. Environ Sci Technol 49(24):14605–14613. https://doi.org/10.1021/acs.est.5b03988

Chen Y, Zhang L, Jiang X (2015c) Fine-scale local adaptation of cyanobacterial tolerance in *Simocephalus vetulus* (Crustacea: Cladocera). J Plankton Res 37(4):764–772. https://doi.org/10.1093/plankt/fbv047

Chen H-q, Zhao J, Li Y, Huang Y-j, Chen D-j, He L-x, Wang L-q, Zheng C-f, Wang J, Cao J, Shu W-q, Liu J-y, Liu W-b (2019) Epigenetic inactivation of LHX6 mediated microcystin-LR induced hepatocarcinogenesis via the Wnt/β-catenin and P53 signaling pathways. Environ Pollut 252:216–226. https://doi.org/10.1016/j.envpol.2019.05.049

Chen L, Giesy JP, Adamovsky O, Svirčev Z, Meriluoto J, Codd GA, Mijovic B, Shi T, Tuo X, Li SC, Pan BZ, Chen J, Xie P (2021) Challenges of using blooms of *Microcystis* spp. in animal feeds: a comprehensive review of nutritional, toxicological and microbial health evaluation. Sci Total Environ 764. https://doi.org/10.1016/j.scitotenv.2020.142319

Chen S, Xie W, Lin X, Zhou H, Teng S, Jiang Z, Yao L, Xu H (2024) Controlling toxic *Microcystis* blooms: the power of a novel microalgal predator *Poteriospumella lacustris* in water safety improvement. J Clean Prod 441. https://doi.org/10.1016/j.jclepro.2024.141011

Chi C, Giri SS, Jun JW, Kim SW, Kim HJ, Kang JW, Park SC (2018) Detoxification- and immune-related transcriptomic analysis of gills from bay scallops (*Argopecten irradians*) in response to algal toxin okadaic acid. Toxins 10(8). https://doi.org/10.3390/toxins10080308

Chislock MF, Sarnelle O, Jernigan LM, Wilson AE (2013) Do high concentrations of microcystin prevent *daphnia* control of phytoplankton? Water Res 47(6):1961–1970. https://doi.org/10.1016/j.watres.2012.12.038

Chislock MF, Kaul RB, Durham KA, Sarnelle O, Wilson AE (2019) Eutrophication mediates rapid clonal evolution in *Daphnia pulicaria*. Freshw Biol 64(7):1275–1283. https://doi.org/10.1111/fwb.13303

Chorus I, Fastner J, Welker M (2021) Cyanobacteria and cyanotoxins in a changing environment: concepts, controversies, challenges. Water 13(18). https://doi.org/10.3390/w13182463

Cicchillitti L, Di Stefano V, Isaia E, Crimaldi L, Fasanaro P, Ambrosino V, Antonini A, Capogrossi MC, Gaetano C, Piaggio G, Martelli F (2012) Hypoxia-inducible factor 1-α induces miR-210 in normoxic differentiating myoblasts. J Biol Chem 287(53):44761–44771. https://doi.org/10.1074/jbc.M112.421255

Clearwater SJ, Wood SA, Phillips NR, Parkyn SM, Van Ginkel R, Thompson KJ (2014) Toxicity thresholds for juvenile freshwater mussels *Echyridella menziesii* and crayfish *Paranephrops planifrons*, after acute or chronic exposure to *Microcystis* sp. Environ Toxicol 29(5):487–502. https://doi.org/10.1002/tox.21774

Cohen JH, Tester PA, Forward RB Jr (2007) Sublethal effects of the toxic dinoflagellate *Karenia brevis* on marine copepod behavior. J Plankton Res 29(3):301–315. https://doi.org/10.1093/plankt/fbm016

Colin SP, Dam HG (2002) Latitudinal differentiation in the effects of the toxic dinoflagellate *Alexandrium* spp. on the feeding and reproduction of populations of the copepod *Acartia hudsonica*. Harmful Algae 1(1):113–125. https://doi.org/10.1016/S1568-9883(02)00007-0

Colin SP, Dam HG (2003) Effects of the toxic dinoflagellate *Alexandrium fundyense* on the copepod *Acartia hudsonica*: a test of the mechanisms that reduce ingestion rates. Mar Ecol Prog Ser 248:55–65. https://doi.org/10.3354/meps248055

Colin SP, Dam HG (2005) Testing for resistance of pelagic marine copepods to a toxic dinoflagellate. Evol Ecol 18(4):355–377. https://doi.org/10.1007/s10682-004-2369-3

Colin SP, Dam HG (2007) Comparison of the functional and numerical responses of resistant versus non-resistant populations of the copepod *Acartia hudsonica* fed the toxic dinoflagellate *Alexandrium tamarense*. Harmful Algae 6(6):875–882. https://doi.org/10.1016/j.hal.2007.05.003

Collumb CJ, Buskey EJ (2004) Effects of the toxic red tide dinoflagellate (*Karenia brevis*) on survival, fecal pellet production and fecundity of the copepod *Acartia tonsa*. In: Steidinger KA, Landsberg JH, Omas CR, Vargo GA (eds) Harmful algae 2002. Florida Fish and Wildlife Conservation Commission, Florida Institute of Oceanography, and Intergovernmental Oceanographic Commission of UNESCO, St. Pete Beach, pp 44–46

Combes A, Dellinger M, CadelSix S, Amand S, Comte K (2013) Ciliate *Nassula* sp. grazing on a microcystin-producing cyanobacterium (*Planktothrix agardhii*): impact on cell growth and in the microcystin fractions. Aquat Toxicol 126:435–441. https://doi.org/10.1016/j.aquatox.2012.08.018

Compagnon D, Lembeye G, Marcos N, Ruiz-Tagle N, Lagos N (1998) Accumulation of paralytic shellfish poisoning toxins in the bivalve *Aulacomya ater* and two carnivorous gastropods concholepas concholepas and *Argobuccinum ranelliformes* during an *Alexandrium catenella* bloom in Southern Chile. J Shellfish Res 17(1):67–73

Contardo-Jara V, Pflugmacher S, Wiegand C (2008) Multi-xenobiotic-resistance a possible explanation for the insensitivity of bivalves towards cyanobacterial toxins. Toxicon 52(8):936–943. https://doi.org/10.1016/j.toxicon.2008.09.005

Contardo-Jara V, Otterstein SKB, Downing S, Downing TG, Pflugmacher S (2014) Response of antioxidant and biotransformation systems of selected freshwater mussels (*Dreissena polymorpha*, *Anadonta cygnea*, *Unio tumidus*, and *Corbicula javanicus*) to the cyanobacterial neurotoxin β-*N*-methylamino-L-alanine. Toxicol Environ Chem 96(3):451–465. https://doi.org/10.1080/02772248.2014.945452

Cook JA, Iype SN, Mitchell JB (1991) Differential specificity of monochlorobimane for isozymes of human and rodent glutathione *S*-transferases. Cancer Res 51(6):1606–1612

Dalsgaard J, St. John M, Kattner G, Müller-Navarra D, Hagen W (2003) Fatty acid trophic markers in the pelagic marine environment. Adv Mar Biol 46:225–340

Danchin É, Charmantier A, Champagne FA, Mesoudi A, Pujol B, Blanchet S (2011) Beyond DNA: integrating inclusive inheritance into an extended theory of evolution. Nat Rev Genet 12(7):475–486. https://doi.org/10.1038/nrg3028

Das BK, Pradhan J, Sahu S, Marhual NP, Mishra BK, Eknath AE (2013) *Microcystis aeruginosa* (Kütz) incorporated diets increase immunity and survival of Indian major carp *Labeo rohita* (Ham.) against *Aeromonas hydrophila* infection. Aquac Res 44(6):918–927. https://doi.org/10.1111/j.1365-2109.2012.03098.x

Davidović PG, Blagojević DJ, Lazić GG, Simeunović JB (2022) Gene expression changes in *Daphnia magna* following waterborne exposure to cyanobacterial strains from the genus *Nostoc*. Harmful Algae 115:102232. https://doi.org/10.1016/j.hal.2022.102232

Davis TW, Koch F, Marcoval MA, Wilhelm SW, Gobler CJ (2012) Mesozooplankton and microzooplankton grazing during cyanobacterial blooms in the western basin of Lake Erie. Harmful Algae 15:26–35. https://doi.org/10.1016/j.hal.2011.11.002

De Maagd PGJ, Hendriks AJ, Seinen W, Sijm DTHM (1999) pH-dependent hydrophobicity of the cyanobacteria toxin microcystin-LR. Water Res 33(3):677–680. https://doi.org/10.1016/S0043-1354(98)00258-9

Decaestecker E, King K (2018) Red queen dynamics. Encyclop Ecol:185–192

Dewes LJ, Sandrini JZ, Monserrat JM, Yunes JS (2006) Biochemical and physiological responses after exposure to microcystins in the crab *Chasmagnathus granulatus* (Decapoda, Brachyura). Ecotoxicol Environ Saf 65(2):201–208. https://doi.org/10.1016/j.ecoenv.2005.07.013

Ding R, Liu S, He C, Nie X (2020) Paracetamol affects the expression of detoxification- and reproduction-related genes and alters the life traits of *Daphnia magna*. Ecotoxicology 29(4):398–406. https://doi.org/10.1007/s10646-020-02199-z

Dionisio Pires LM, Karlsson KM, Meriluoto JAO, Kardinaal E, Visser PM, Siewertsen K, Van Donk E, Ibelings BW (2004) Assimilation and depuration of microcystin-LR by the zebra mussel, *Dreissena polymorpha*. Aquat Toxicol 69(4):385–396. https://doi.org/10.1016/j.aquatox.2004.06.004

Dionisio Pires LM, Sarpe D, Brehm M, Ibelings BW (2011) Potential synergistic effects of microcystins and bacterial lipopolysaccharides on life history traits of *Daphnia galeata* raised on low and high food levels. Aquat Toxicol 104(3–4):230–242. https://doi.org/10.1016/j.aquatox.2011.05.001

Dionne K, Dufresne F, Nozais C (2016) Variation in $\delta^{13}C$ and $\delta^{15}N$ trophic enrichment factors among *Hyalella azteca* amphipods from different lakes. Hydrobiologia 781(1):217–230. https://doi.org/10.1007/s10750-016-2846-z

Donovan CJ, Ku JC, Quilliam MA, Gill TA (2008) Bacterial degradation of paralytic shellfish toxins. Toxicon 52(1):91–100. https://doi.org/10.1016/j.toxicon.2008.05.005

Donovan CJ, Garduño RA, Kalmokoff M, Ku JC, Quilliam MA, Gill TA (2009) *Pseudoalteromonas* bacteria are capable of degrading paralytic shellfish toxins. Appl Environ Microbiol 75(21):6919–6923. https://doi.org/10.1128/aem.01384-09

Dou M, Jiao YH, Zheng JW, Zhang G, Li HY, Liu JS, Yang WD (2020) *De novo* transcriptome analysis of the mussel *Perna viridis* after exposure to the toxic dinoflagellate *Prorocentrum lima*. Ecotoxicol Environ Saf 192. https://doi.org/10.1016/j.ecoenv.2020.110265

Drugă B, Turko P, Spaak P, Pomati F (2016) Cyanobacteria affect fitness and genetic structure of experimental *Daphnia* populations. Environ Sci Technol 50(7):3416–3424. https://doi.org/10.1021/acs.est.5b05973

Dudley TL, Cooper SD, Hemphill N (1986) Effects of macroalgae on a stream invertebrate community. J N Am Benthol Soc 5(2):93–106. https://doi.org/10.2307/1467864

Engel P, Moran NA (2013) The gut microbiota of insects - diversity in structure and function. FEMS Microbiol Rev 37(5):699–735. https://doi.org/10.1111/1574-6976.12025

Falk M, Seto PF, Walter JA (1991) Solubility of domoic acid in water and in non-aqueous solvents. Can J Chem Revue Canadienne De Chimie 69(11):1740–1744. https://doi.org/10.1139/v91-255

Fan Y, Huang J, Huang S, Lyu K (2024) Molecular identification of TRAF4 gene in aquatic keystone species *Daphnia magna* and its response to toxic cyanobacteria challenge. Chemosphere 369. https://doi.org/10.1016/j.chemosphere.2024.143856

Feiner N, Radersma R, Vasquez L, Ringnér M, Nystedt B, Raine A, Tobi EW, Heijmans BT, Uller T (2022) Environmentally induced DNA methylation is inherited across generations in an aquatic keystone species. iScience 25(5):104303. https://doi.org/10.1016/j.isci.2022.104303

Fellous A, Wegner KM, John U, Mark FC, Shama LNS (2022) Windows of opportunity: ocean warming shapes temperature-sensitive epigenetic reprogramming and gene expression across gametogenesis and embryogenesis in marine stickleback. Glob Chang Biol 28(1):54–71. https://doi.org/10.1111/gcb.15942

Feng Y, Ma J, Xiang R, Li X (2017) Alterations in microRNA expression in the tissues of silver carp (*Hypophthalmichthys molitrix*) following microcystin-LR exposure. Toxicon 128:15–22. https://doi.org/10.1016/j.toxicon.2017.01.016

Feng J, Mazzei M, Di Gregorio S, Niccolini L, Vitiello V, Ye Y, Guo B, Yan X, Buttino I (2023) Marine copepods as a microbiome hotspot: revealing their interactions and biotechnological applications. Water 15(24). https://doi.org/10.3390/w15244203

Fernandes S, Welker M, Vasconcelos VM (2009) Changes in the GST activity of the mussel *Mytilus galloprovincialis* during exposure and depuration of microcystins. J Exp Zool A Ecol Genet Physiol 311A(3):226–230. https://doi.org/10.1002/jez.524

Ferrão-Filho AS (2009) Bioaccumulation of cyanotoxins and their effects on aquatic organisms (in Portuguese). Oecol Bras 13:272–312

Ferrão-Filho ADS, Kozlowsky-Suzuki B (2011) Cyanotoxins: bioaccumulation and effects on aquatic animals. Mar Drugs 9(12):2729–2772. https://doi.org/10.3390/md9122729

Finiguerra M, Avery DE, Dam HG (2015) Determining the advantages, costs, and trade-offs of a novel sodium channel mutation in the copepod *Acartia hudsonica* to paralytic shellfish toxins (PST). PLoS One 10(6). https://doi.org/10.1371/journal.pone.0130097

Fjellstedt TA, Allen RH, Duncan BK, Jakoby WB (1973) Enzymatic conjugation of epoxides with glutathione. J Biol Chem 248(10):3702–3707. https://doi.org/10.1016/S0021-9258(19)43983-5

Florczyk M, Brzuzan P, Krom J, Woźny M, Łakomiak A (2016) miR-122-5p as a plasma biomarker of liver injury in fish exposed to microcystin-LR. J Fish Dis 39(6):741–751. https://doi.org/10.1111/jfd.12406

Flores NM, Miller TR, Stockwell JD (2018) A global analysis of the relationship between concentrations of microcystins in water and fish. Front Mar Sci 5. https://doi.org/10.3389/fmars.2018.00030

Foote BA (1993) biology of *Hyadina albovenosa* (Diptera, Ephydridae), a consumer of cyanobacteria. Proc Entomol Soc Wash 95:377–382

Frouz J, Ali A, Lobinske RJ (2004a) Algal food selection and digestion by larvae of the pestiferous chironomid *Chironomus* iunder laboratory conditions. J Am Mosq Control Assoc 20(4):458–461

Frouz J, Ali A, Lobinske RJ (2004b) Laboratory evaluation of six algal species for larval nutritional suitability of the pestiferous midge *Glyptotendipes paripes* (Diptera: Chironomidae). J Econ Entomol 97(6):1884–1890. https://doi.org/10.1093/jee/97.6.1884

Fu J, Xie P (2006) The acute effects of microcystin LR on the transcription of nine glutathione S-transferase genes in common carp *Cyprinus carpio* L. Aquat Toxicol 80(3):261–266. https://doi.org/10.1016/j.aquatox.2006.09.003

Fuhr U, Kummert AL (1995) The fate of naringin in humans: a key to grapefruit juice-drug interactions? Clin Pharmacol Therapeut 58(4):365–373. https://doi.org/10.1016/0009-9236(95)90048-9

Fukui K, Harada A, Wakamatsu T, Minobe A, Ohshita K, Ashiuchi M, Yano T (2018) The GIY-YIG endonuclease domain of *Arabidopsis* MutS homolog 1 specifically binds to branched DNA structures. FEBS Lett 592(24):4066–4077. https://doi.org/10.1002/1873-3468.13279

Fulton RS III (1988) Resistance to blue-green algal toxins by *Bosmina longirostris*. J Plankton Res 10(4):771–778. https://doi.org/10.1093/plankt/10.4.771

Fulton RS III, Paerl HW (1988) Effects of the blue-green alga *Microcystis aeruginosa* on zooplankton competitive relations. Oecologia 76(3):383–389. https://doi.org/10.1007/BF00377033

Gagné F, Gélinas M, Fortier M, Fournier M (2018) The effects of cyanobacterial blooms on the immune system of *Elliptio complanata* in urban and agricultural areas in the Yamaska River watershed. Invertebr Surviv J 15:39–51

Galanti LN, Amé MV, Wunderlin DA (2013) Accumulation and detoxification dynamic of cyanotoxins in the freshwater shrimp *Palaemonetes argentinus*. Harmful Algae 27:88–97. https://doi.org/10.1016/j.hal.2013.05.007

Galimany E, Lunt J, Freeman CJ, Reed S, Segura-García I, Paul VJ (2017) Feeding behavior of eastern oysters *Crassostrea virginica* and hard clams *Mercenaria mercenaria* in shallow estuaries. Mar Ecol Prog Ser 567:125–137. https://doi.org/10.3354/meps12050

Galimany E, Lunt J, Freeman CJ, Houk J, Sauvage T, Santos L, Lunt J, Kolmakova M, Mossop M, Domingos A, Phlips EJ, Paul VJ (2020) Bivalve feeding responses to microalgal bloom species in the Indian River Lagoon: the potential for top-down control. Estuar Coasts 43(6):1519–1532. https://doi.org/10.1007/s12237-020-00746-9

Gao L, Wang F, Wu T, Zhou Q, Zhang D, Hou X, Zhang X, Ye J, Ma Y, Wang Z, Song L, Chen W (2022) New insights into the production of fucoxanthin by mixotrophic cultivation of *Ochromonas* and *Microcystis aeruginosa*. Bioresour Technol 363. https://doi.org/10.1016/j.biortech.2022.127922

Geffeney SL, Fujimoto E, Brodie ED 3rd, Brodie ED Jr, Ruben PC (2005) Evolutionary diversification of TTX-resistant sodium channels in a predator-prey interaction. Nature 434(7034):759–763. https://doi.org/10.1038/nature03444

Gélinas M, Fortier M, Lajeunesse A, Fournier M, Gagnon C, Gagné F (2013) Energy status and immune system alterations in *Elliptio complanata* after ingestion of cyanobacteria *Anabaena flos-aquae*. Ecotoxicology 22(3):457–468. https://doi.org/10.1007/s10646-012-1039-4

Ger KA, Panosso R (2014) The effects of a microcystin-producing and lacking strain of *Microcystis* on the survival of a widespread tropical copepod (*Notodiaptomus iheringi*). Hydrobiologia 738(1):61–73. https://doi.org/10.1007/s10750-014-1915-4

Ger KA, Arneson P, Goldman CR, Teh SJ (2010) Species specific differences in the ingestion of *Microcystis* cells by the calanoid copepods *Eurytemora affinis* and *Pseudodiaptomus forbesi*. J Plankton Res 32(10):1479–1484. https://doi.org/10.1093/plankt/fbq071

Ger KA, Leitao E, Panosso R, Havens K (2016) Potential mechanisms for the tropical copepod *Notodiaptomus* to tolerate *Microcystis* toxicity. J Plankton Res 38(4):843–854. https://doi.org/10.1093/plankt/fbw036

Gérard C, Lance E (2019) Decline of freshwater gastropods exposed to recurrent interacting stressors implying cyanobacterial proliferations and droughts. Aquat Ecol 53(1):79–96. https://doi.org/10.1007/s10452-019-09674-8

Gerphagnon M, Agha R, Martin-Creuzburg D, Bec A, Perriere F, Rad-Menéndez C, Gachon CMM, Wolinska J (2019) Comparison of sterol and fatty acid profiles of chytrids and their hosts reveals trophic upgrading of nutritionally inadequate phytoplankton by fungal parasites. Environ Microbiol 21(3):949–958. https://doi.org/10.1111/1462-2920.14489

Glidewell D, Beyer JE, Hambright KD (2025) Microcystins bioaccumulate but do not biomagnify in an experimental aquatic food chain. Harmful Algae 141:102768. https://doi.org/10.1016/j.hal.2024.102768

Glisic B, Mihaljevic I, Popovic M, Zaja R, Loncar J, Fent K, Kovacevic R, Smital T (2015) Characterization of glutathione-S-transferases in zebrafish (*Danio rerio*). Aquat Toxicol 158:50–62. https://doi.org/10.1016/j.aquatox.2014.10.013

Glisic B, Hrubik J, Fa S, Dopudj N, Kovacevic R, Andric N (2016) Transcriptional profiles of glutathione-*S*-transferase isoforms, Cyp, and AOE genes in atrazine-exposed zebrafish embryos. Environ Toxicol 31(2):233–244. https://doi.org/10.1002/tox.22038

Goedkoop W, Åkerblom N, Demandt MH (2006) Trophic fractionation of carbon and nitrogen stable isotopes in *Chironomus riparius* reared on food of aquatic and terrestrial origin. Freshw Biol 51(5):878–886. https://doi.org/10.1111/j.1365-2427.2006.01539.x

Gonçalves-Soares D, Zanette J, Yunes JS, Yepiz-Plascencia GM, Bainy ACD (2012) Expression and activity of glutathione *S*-transferases and catalase in the shrimp *Litopenaeus vannamei* inoculated with a toxic *Microcystis aeruginosa* strain. Mar Environ Res 75:54–61. https://doi.org/10.1016/j.marenvres.2011.07.007

Gonzalez-Romero R, Suarez-Ulloa V, Rodriguez-Casariego J, Garcia-Souto D, Diaz G, Smith A, Pasantes JJ, Rand G, Eirin-Lopez JM (2017) Effects of Florida red tides on histone variant expression and DNA methylation in the eastern oyster *Crassostrea virginica*. Aquat Toxicol 186:196–204. https://doi.org/10.1016/j.aquatox.2017.03.006

Goodrich JK, Waters JL, Poole AC, Sutter JL, Koren O, Blekhman R, Beaumont M, Van Treuren W, Knight R, Bell JT, Spector TD, Clark AG, Ley RE (2014) Human genetics shape the gut microbiome. Cell 159(4):789–799. https://doi.org/10.1016/j.cell.2014.09.053

Gorokhova E, El-Shehawy R, Lehtiniemi M, Garbaras A (2021) How copepods can eat toxins without getting sick: gut bacteria help zooplankton to feed in cyanobacteria blooms. Front Microbiol 11. https://doi.org/10.3389/fmicb.2020.589816

Grigoriou M, Tucker AS, Sharpe PT, Pachnis V (1998) Expression and regulation of *Lhx6* and *Lhx7*, a novel subfamily of LIM homeodomain encoding genes, suggests a role in mammalian head development. Development 125(11):2063–2074. https://doi.org/10.1242/dev.125.11.2063

Gronwald JW, Plaisance KL (1998) Isolation and characterization of glutathione *S*-transferase isozymes from sorghum. Plant Physiol 117(3):877–892. https://doi.org/10.1104/pp.117.3.877

Guilin Z, Pengyu Z, Wei L, Fengqi H, Chen F, Yu Y, Yue H, Yuting T (2020) Reduction of gut microbial diversity and short chain fatty acids in BALB/c mice exposure to microcystin-LR. Ecotoxicology 29(9):1347–1357. https://doi.org/10.1007/s10646-020-02254-9

Guisande C, Frangópulos M, Carotenuto Y, Maneiro I, Riveiro I, Ruth Vergara A (2002) Fate of paralytic shellfish poisoning toxins ingested by the copepod *Acartia clausi*. Mar Ecol Prog Ser 240:105–115. https://doi.org/10.3354/meps240105

Guo N, Xie P (2006) Development of tolerance against toxic *Microcystis aeruginosa* in three cladocerans and the ecological implications. Environ Pollut 143(3):513–518. https://doi.org/10.1016/j.envpol.2005.11.044

Gustafsson S, Hansson L-A (2004) Development of tolerance against toxic cyanobacteria in *daphnia*. Aquat Ecol 38(1):37–44. https://doi.org/10.1023/B:AECO.0000020985.47348.5e

Gustafsson S, Rengefors K, Hansson L-A (2005) Increased consumer fitness following transfer of toxin tolerance to offspring via maternal effects. Ecology 86(10):2561–2567. https://doi.org/10.1890/04-1710

Habig WH, Pabst MJ, Jakoby WB (1974) Glutathione S transferases. The first enzymatic step in mercapturic acid formation. J Biol Chem 249(22):7130–7139

Hädicke CW, Rédei D, Kment P (2017) The diversity of feeding habits recorded for water boatmen (Heteroptera: Corixoidea) world-wide with implications for evaluating information on the diet of aquatic insects. Eur J Entomol 114:147–159. https://doi.org/10.14411/eje.2017.020

Hadj-Moussa H, Hawkins LJ, S. K.B. (2022) Role of microRNAs in extreme animal survival strategies. In: Allmer J, Yousef M (eds) miRNomics: microRNA biology and computational analysis, vol 2257. Humana, New York, pp 311–347

Haidle AM, Myers AG (2004) An enantioselective, modular, and general route to the cytochalasins: synthesis of L-696,474 and cytochalasin B. Proc Natl Acad Sci 101(33):12048–12053. https://doi.org/10.1073/pnas.0402111101

Hairston NG Jr, Lampert W, Cáceres CE, Holtmeier CL, Weider LJ, Gaedke U, Fischer JM, Fox JA, Post DM (1999) Rapid evolution revealed by dormant eggs. Nature 401(6752):446. https://doi.org/10.1038/46731

Hairston NG Jr, Holtmeier CL, Lampert W, Weider LJ, Post DM, Fischer JM, Cáceres CE, Fox JA, Gaedke U (2001) Natural selection for grazer resistance to toxic cyanobacteria: evolution of phenotypic plasticity? Evolution 55(11):2203–2214. https://doi.org/10.1111/j.0014-3820.2001.tb00736.x

Halttunen T, Collado MC, El-Nezami H, Meriluoto J, Salminen S (2008) Combining strains of lactic acid bacteria may reduce their toxin and heavy metal removal efficiency from aqueous solution. Lett Appl Microbiol 46(2):160–165. https://doi.org/10.1111/j.1472-765X.2007.02276.x

Hamasaki K, Takahashi T, Uye SI (2003) Accumulation of paralytic shellfish poisoning toxins in planktonic copepods during a bloom of the toxic dinoflagellate *Alexandrium tamarense* in Hiroshima Bay, western Japan. Mar Biol 143(5):981–988. https://doi.org/10.1007/s00227-003-1131-7

Haroardóttir S, Pančić M, Tammilehto A, Krock B, Møller EF, Nielsen TG, Lundholm N (2015) Dangerous relations in the arctic marine food web: interactions between toxin producing *pseudo-nitzschia* diatoms and *Calanus* copepodites. Mar Drugs 13(6):3809–3835. https://doi.org/10.3390/md13063809

Hart DD (1985) Grazing insects mediate algal interactions in a stream benthic community. Oikos 44(1):40–46. https://doi.org/10.2307/3544041

Hawes NA, Amadoru A, Tremblay LA, Pochon X, Dunphy B, Fidler AE, Smith KF (2019) Epigenetic patterns associated with an ascidian invasion: a comparison of closely related clades in their native and introduced ranges. Sci Rep 9(1). https://doi.org/10.1038/s41598-019-49813-7

Heckwolf MJ, Meyer BS, Häsler R, Höppner MP, Eizaguirre C, Reusch TBH (2020) two different epigenetic information channels in wild three-spined sticklebacks are involved in salinity adaptation. Sci Adv 6(12):eaaz1138. https://doi.org/10.1126/sciadv.aaz1138

Herfindal L, Myhren L, Kleppe R, Krakstad C, Selheim F, Jokela J, Sivonen K, Døskeland SO (2011) Nostocyclopeptide-M1: a potent, nontoxic inhibitor of the hepatocyte drug transporters OATP1B3 and OATP1B1. Mol Pharm 8(2):360–367. https://doi.org/10.1021/mp1002224

Hernández Borrero LJ, El-Deiry WS (2021) Tumor suppressor p53: biology, signaling pathways, and therapeutic targeting. Biochim Biophys Acta Rev Cancer 1876(1):188556. https://doi.org/10.1016/j.bbcan.2021.188556

Herth W, Kuppel A, Schnepf E (1977) Chitinous fibrils in the lorica of the flagellate chrysophyte *Poteriochromonas stipitata* (Syn. *Ochromonas malhamensis*). J Cell Biol 73(2):311–321. https://doi.org/10.1083/jcb.73.2.311

Hollows JW, Townsend CR, Collier KJ (2002) Diet of the crayfish *Paranephrops zealandicus* in bush and pasture streams: insights from stable isotopes and stomach analysis. N Z J Mar Freshw Res 36(1):129–142. https://doi.org/10.1080/00288330.2002.9517076

Hong J, Talapatra S, Katz J, Tester PA, Waggett RJ, Place AR (2012) Algal toxins alter copepod feeding behavior. PLoS One 7(5):e36845. https://doi.org/10.1371/journal.pone.0036845

Hong Z, Chen X, Hu J, Chang X, Qian Y (2024) Adverse effects of *Microcystis aeruginosa* exudates on the filtration, digestion, and reproduction organs of benthic bivalve *Corbicula fluminea*. Sci Rep 14(1):10934. https://doi.org/10.1038/s41598-024-61333-7

Houwenhuyse S, Stoks R, Mukherjee S, Decaestecker E (2021) Locally adapted gut microbiomes mediate host stress tolerance. ISME J 15(8):2401–2414. https://doi.org/10.1038/s41396-021-00940-y

Howlett R (2005) Evolutionary biology: channels of resistance. Nature 434(7034):716. https://doi.org/10.1038/434716a

Hu M, Wu F, Yuan M, Li Q, Gu Y, Wang Y, Liu Q (2015) Antioxidant responses of triangle sail mussel *Hyriopsis cumingii* exposed to harmful algae *Microcystis aeruginosa* and hypoxia. Chemosphere 139:541–549. https://doi.org/10.1016/j.chemosphere.2015.07.074

Huang L, Wang J, Chen WC, Li HY, Liu JS, Tao J, Yang WD (2014) P-glycoprotein expression in *Perna viridis* after exposure to *Prorocentrum lima*, a dinoflagellate producing DSP toxins. Fish Shellf Immunol 39(2):254–262. https://doi.org/10.1016/j.fsi.2014.04.020

Huang L, Liu SL, Zheng JW, Li HY, Liu JS, Yang WD (2015a) *P*-glycoprotein and its inducible expression in three bivalve species after exposure to *Prorocentrum lima*. Aquat Toxicol 169:123–132. https://doi.org/10.1016/j.aquatox.2015.10.012

Huang L, Zou Y, Weng HW, Li HY, Liu JS, Yang WD (2015b) Proteomic profile in *Perna viridis* after exposed to *Prorocentrum lima*, a dinoflagellate producing DSP toxins. Environ Pollut 196:350–357. https://doi.org/10.1016/j.envpol.2014.10.019

Huang JH, Jiao YH, Li L, Li DW, Li HY, Yang WD (2021) Small RNA analysis of *Perna viridis* after exposure to *Prorocentrum lima*, a DSP toxins-producing dinoflagellate. Aquat Toxicol 239. https://doi.org/10.1016/j.aquatox.2021.105950

Huang Y, Li Q, Yuan Y, Zhang Z, Jiang B, Yang S, Jian J (2022) Silencing of *Nrf2* in *Litopenaeus vannamei*, decreased the antioxidant capacity, and increased apoptosis and autophagy. Fish Shellf Immunol 122:257–267. https://doi.org/10.1016/j.fsi.2022.02.010

Humpage AR, Fontaine F, Froscio S, Burcham P, Falconer IR (2005) Cylindrospermopsin genotoxicity and cytotoxicity: role of cytochrome P-450 and oxidative stress. J Toxicol Environ Health Part A 68(9):739–753. https://doi.org/10.1080/15287390590925465

Huntley M, Pf S, Rohan S, Marin V (1986) Chemically-mediated rejection of dinoflagellate prey by the copepods *Calanus pacificus* and *Paracalanus parvus*: mechanism, occurrence and significance. Mar Ecol Prog Ser 28:105–120. https://doi.org/10.3354/meps028105

Hutton SJ, Brander SM (2023) Epigenetics in aquatic toxicology. In: Piferrer F, Wang H (eds) Epigenetics in aquaculture. Wiley & Sons, Chichester, pp 301–323

Ibelings BW, Havens KE (2008) Cyanobacterial toxins: a qualitative meta-analysis of concentrations, dosage and effects in freshwater, estuarine and marine biota. In: Hudnell HK (ed) Cyanobacterial harmful algal blooms: state of the science and research needs. Springer, New York, pp 675–732

Ibelings BW, Bruning K, de Jonge J, Wolfstein K, Pires LMD, Postma J, Burger T (2005) Distribution of microcystins in a lake foodweb: no evidence for biomagnification. Microb Ecol 49(4):487–500. https://doi.org/10.1007/s00248-004-0014-x

Ishimota, M., M. Kodama, N. Tomiyama and K. Ohyama (2024). "Chemical tolerance related to the ABC transporter gene and DNA methylation in cladocera (*Daphnia magna*)." Environ Toxicol n/a(n/a).DOI: https://doi.org/10.1002/tox.24077

Ito K, Asakawa M, Beppu R, Takayama H, Miyazawa K (2004) PSP-toxicification of the carnivorous gastropod *Rapana venosa* inhabiting the estuary of Nikoh River, Hiroshima Bay, Hiroshima prefecture, Japan. Mar Pollut Bull 48(11):1116–1121. https://doi.org/10.1016/j.marpolbul.2003.12.020

Jablonka E, Oborny B, Molnár I, Kisdi É, Hofbauer J, Czárán T (1995) The adaptive advantage of phenotypic memory in changing environments. Philos Trans R Soc Lond B Biol Sci 350(1332):133–141. https://doi.org/10.1098/rstb.1995.0147

Jaiswal AK (2004) Nrf2 signaling in coordinated activation of antioxidant gene expression. Free Radic Biol Med 36(10):1199–1207. https://doi.org/10.1016/j.freeradbiomed.2004.02.074

Jansen S, Riser CW, Wassmann P, Bathmann U (2006) Copepod feeding behaviour and egg production during a dinoflagellate bloom in the North Sea. Harmful Algae 5(1):102–112. https://doi.org/10.1016/j.hal.2005.06.006

Jeremias G, Barbosa J, Marques SM, Asselman J, Gonçalves FJM, Pereira JL (2018) Synthesizing the role of epigenetics in the response and adaptation of species to climate change in freshwater ecosystems. Mol Ecol 27(13):2790–2806. https://doi.org/10.1111/mec.14727

Ji M, Vandenhole M, De Beer B, De Rouck S, Villacis-Perez E, Feyereisen R, Clark RM, Van Leeuwen T (2023) A nuclear receptor HR96-related gene underlies large trans-driven differences in detoxification gene expression in a generalist herbivore. Nat Commun 14(1):4990. https://doi.org/10.1038/s41467-023-40778-w

Jiang X, Li Q, Liang H, Zhao S, Zhang L, Zhao Y, Chen L, Yang W, Xiang X (2013a) Clonal variation in growth plasticity within a *Bosmina longirostris* population: the potential for resistance to toxic cyanobacteria. PLoS One 8(9). https://doi.org/10.1371/journal.pone.0073540

Jiang X, Yang W, Zhao S, Liang H, Zhao Y, Chen L, Li R (2013b) Maternal effects of inducible tolerance against the toxic cyanobacterium *Microcystis aeruginosa* in the grazer *Daphnia carinata*. Environ Pollut 178:142–146. https://doi.org/10.1016/j.envpol.2013.03.017

Jiang X, Liang H, Chen Y, Xu X, Huang D (2015) Microgeographic adaptation to toxic cyanobacteria in two aquatic grazers. Limnol Oceanogr 60(3):947–956. https://doi.org/10.1002/lno.10067

Johnson PTJ, Calhoun DM, Stokes AN, Susbilla CB, McDevitt-Galles T, Briggs CJ, Hoverman JT, Tkach VV, de Roode JC (2018) Of poisons and parasites-the defensive role of tetrodotoxin against infections in newts. J Anim Ecol 87(4):1192–1204. https://doi.org/10.1111/1365-2656.12816

Kabeya N, Fonseca MM, Ferrier DEK, Navarro JC, Bay LK, Francis DS, Tocher DR, Castro LFC, Monroig Ó (2018) genes for de novo biosynthesis of omega-3 polyunsaturated fatty acids are widespread in animals. Sci Adv 4(5). https://doi.org/10.1126/sciadv.aar6849

Kågedal K, Johansson U, Ollinger K (2001) The lysosomal protease cathepsin D mediates apoptosis induced by oxidative stress. FASEB J 15(9):1592–1594. https://doi.org/10.1096/fj.00-0708fje

Karimullina E, Li Y, Ginjupalli GK, Baldwin WS (2012) *Daphnia* HR96 is a promiscuous xenobiotic and endobiotic nuclear receptor. Aquat Toxicol 116-117:69–78. https://doi.org/10.1016/j.aquatox.2012.03.005

Keilhack L (1909) Die Süsswasserfauna Deutschlands. Eine Exkursionsfauna. G. Fischer, Jena

Kelley JL, Tobler M, Beck D, Sadler-Riggleman I, Quackenbush CR, Arias Rodriguez L, Skinner MK (2021) Epigenetic inheritance of DNA methylation changes in fish living in hydrogen sulfide-rich springs. Proc Natl Acad Sci USA 118(26). https://doi.org/10.1073/pnas.2014929118

Kim SH, Turnbull J, Guimond S (2011) Extracellular matrix and cell signalling: the dynamic cooperation of integrin, proteoglycan and growth factor receptor. J Endocrinol 209(2):139–151. https://doi.org/10.1530/JOE-10-0377

Kim MS, Lee YJ, Ha SY, Kim BH, Hwang SJ, Kwon JT, Choi JW, Shin KH (2017a) Accumulation of microcystin (LR, RR and YR) in three freshwater bivalves in *Microcystis aeruginosa* bloom using dual isotope tracer. Mar Drugs 15(7). https://doi.org/10.3390/md15070226

Kim YD, Kim WJ, Shin YK, Lee DH, Kim YJ, Kim JK, Rhee JS (2017b) Microcystin-LR bioconcentration induces antioxidant responses in the digestive gland of two marine bivalves *Crassostrea gigas* and *Mytilus edulis*. Aquat Toxicol 188:119–129. https://doi.org/10.1016/j.aquatox.2017.05.003

Kim BM, Haque MN, Lee DH, Nam SE, Rhee JS (2018) Comparative toxicokinetics and antioxidant response in the microcystin-LR-exposed gill of two marine bivalves, *Crassostrea gigas* and *Mytilus edulis*. J Shellfish Res 37(3):497–506. https://doi.org/10.2983/035.037.0305

Kishore U, Ghai R, Greenhough TJ, Shrive AK, Bonifati DM, Gadjeva MG, Waters P, Kojouharova MS, Chakraborty T, Agrawal A (2004) Structural and functional anatomy of the globular domain of complement protein C1q. Immunol Lett 95(2):113–128. https://doi.org/10.1016/j.imlet.2004.06.015

Klumpp S, Krieglstein J (2002) Serine/threonine protein phosphatases in apoptosis. Curr Opin Pharmacol 2(4):458–462. https://doi.org/10.1016/S1471-4892(02)00176-5

Kobelt W, Martini FHW, Chemnitz JH (1909) Die Gattung *Paludina* Lam. (Vivipara Montfort) : neue Folge. Verlag von Bauer und Raspe (Emil Küster), Nürnberg

Kondo F, Ikai Y, Oka H, Okumura M, Ishikawa N, Harada K, Matsuura K, Murata H, Suzuki M (1992) Formation, characterization, and toxicity of the glutathione and cysteine conjugates of toxic heptapeptide microcystins. Chem Res Toxicol 5(5):591–596. https://doi.org/10.1021/tx00029a002

Kozlowsky-Suzuki B, Wilson AE, Ferrão-Filho A d S (2012) Biomagnification or biodilution of microcystins in aquatic foodwebs? Meta-analyses of laboratory and field studies. Harmful Algae 18:47–55. https://doi.org/10.1016/j.hal.2012.04.002

Krivosheina MG (2008) On insect feeding on cyanobacteria. Paleontol J 42(6):596–599. https://doi.org/10.1134/S003103010806004X

Krützfeldt J, Rajewsky N, Braich R, Rajeev KG, Tuschl T, Manoharan M, Stoffel M (2005) Silencing of microRNAs in vivo with 'antagomirs'. Nature 438(7068):685–689. https://doi.org/10.1038/nature04303

Kubanek J, Snell TW, Pirkle C (2007) Chemical defense of the red tide dinoflagellate *Karenia brevis* against rotifer grazing. Limnol Oceanogr 52(3):1026–1035. https://doi.org/10.4319/lo.2007.52.3.1026

Kühn S, King F, Heubel K (2023) Decreased feeding rates of the copepod *Acartia tonsa* when exposed to playback harbor traffic noise. Front Mar Sci 10. https://doi.org/10.3389/fmars.2023.1134792

Kumari S, Dhanker R, Kumar R (2022) Does maternal exposure to diatom polyunsaturated aldehyde enhance offspring fitness in limnetic zooplankton *Echinisca triserialis* (Crustacea: Claodcera)? Fundam Appl Limnol 195(4):325–341. https://doi.org/10.1127/fal/2022/1455

Kurita K (2006) Chitin and chitosan: functional biopolymers from marine crustaceans. Mar Biotechnol 8(3):203–226. https://doi.org/10.1007/s10126-005-0097-5

Kuster CJ, von Elert E (2013) Interspecific differences between *D. pulex* and *D. magna* in tolerance to cyanobacteria with protease inhibitors. PLoS One 8(5):e62658. https://doi.org/10.1371/journal.pone.0062658

Lafont M, Petton B, Vergnes A, Pauletto M, Segarra A, Gourbal B, Montagnani C (2017) Long-lasting antiviral innate immune priming in the Lophotrochozoan Pacific oyster, *Crassostrea gigas*. Sci Rep 7(1):13143. https://doi.org/10.1038/s41598-017-13564-0

Lafont M, Vergnes A, Vidal-Dupiol J, d. Lorgeril J, Gueguen Y, Haffner P, Petton B, Chaparro C, Barrachina C, Destoumieux-Garzon D, Mitta G, Gourbal B, Montagnani C (2020) A sustained immune response supports long-term antiviral immune priming in the Pacific oyster, *Crassostrea gigas*. MBio 11(2):e02777–e02719. https://doi.org/10.1128/mbio.02777-19

Laland K, Uller T, Feldman M, Sterelny K, Müller GB, Moczek A, Jablonka E, Odling-Smee J, Wray GA, Hoekstra HE, Futuyma DJ, Lenski RE, Mackay TFC, Schluter D, Strassmann JE (2014) Does evolutionary theory need a rethink? Nature 514(7521):161–164. https://doi.org/10.1038/514161a

Lance E, Desprat J, Holbech BF, Gérard C, Bormans M, Lawton LA, Edwards C, Wiegand C (2016) Accumulation and detoxication responses of the gastropod *Lymnaea stagnalis* to single and combined exposures to natural (cyanobacteria) and anthropogenic (the herbicide RoundUp® flash) stressors. Aquat Toxicol 177:116–124. https://doi.org/10.1016/j.aquatox.2016.05.024

Lange J, Berges AC, von Elert E (2023) Multiclonal study of *Daphnia magna* with respect to adaptation to toxic cyanobacteria. Limnol Oceanogr 68(2):514–524. https://doi.org/10.1002/lno.12291

Lassus P, Bardouil M, Massselin P, Naviner M, Truquet P (2000) Comparative efficiencies of different non-toxic microalgal diets in detoxification of PSP-contaminated oysters (*Crassostrea gigas* Thunberg). J Nat Toxins 9(1):1–12

Lassus P, Bordeau P, Marcaillou C, Soudant P (2014) Phycotoxins: seafood contamination, detoxification and processing. In: Rossini GP (ed) Toxins and biologically active compounds from microalgae: biological effects and risk management, vol 2. CRC Press

Lauritano C, Carotenuto Y, Roncalli V (2021) Glutathione S-transferases in marine copepods. J Mar Sci Eng 9(9). https://doi.org/10.3390/jmse9091025

Le T-H, Lim E-S, Lee SK, Choi Y-W, Kim Y-H, Min J (2010) Effects of glyphosate and methidathion on the expression of the Dhb, Vtg, Arnt, CYP4 and CYP314 in *Daphnia magna*. Chemosphere 79(1):67–71. https://doi.org/10.1016/j.chemosphere.2009.12.067

Leal JF, Cristiano MLS (2024) Why are bivalves not detoxified? Curr Opin Food Sci 57. https://doi.org/10.1016/j.cofs.2024.101162

Leandro LF, Teegarden GJ, Roth PB, Wang Z, Doucette GJ (2010) The copepod *Calanus finmarchicus*: a potential vector for trophic transfer of the marine algal biotoxin, domoic acid. J Exp Mar Biol Ecol 382(2):88–95. https://doi.org/10.1016/j.jembe.2009.11.002

Leblond JD, Chapman PJ (2000) Lipid class distribution of highly unsaturated long chain fatty acids in marine dinoflagellates. J Phycol 36(6):1103–1108. https://doi.org/10.1046/j.1529-8817.2000.00018.x

Lee KM, Kim KH (2025) Long-term and non-specific immune memory induced by muramyl dipeptide and its protective effect against *vibrio anguillarum* in rainbow trout (*Oncorhynchus mykiss*). Develop Comparat Immunol 167:105386. https://doi.org/10.1016/j.dci.2025.105386

Lee C-Y, Huang C-H, Lu P-L, Ko W-C, Chen Y-H, Hsueh P-R (2017) Role of rifampin for the treatment of bacterial infections other than mycobacteriosis. J Infect 75(5):395–408. https://doi.org/10.1016/jinf.2017.08.013

Lee J, Lee S, Mayta A, Mrdjen I, Weghorst C, Knobloch T (2020) Microcystis toxin-mediated tumor promotion and toxicity lead to shifts in mouse gut microbiome. Ecotoxicol Environ Saf 206:111204. https://doi.org/10.1016/j.ecoenv.2020.111204

Lei H, Luo J, Tong L, Peng L-q, Qi Y, Jia Z-g, Wei Q (2011) Quercetin binds to calcineurin at a similar region to cyclosporin a and tacrolimus. Food Chem 127(3):1169–1174. https://doi.org/10.1016/j.foodchem.2011.01.119

Lemmen KD, Verhoeven KJF, Declerck SAJ (2022) Experimental evidence of rapid heritable adaptation in the absence of initial standing genetic variation. Funct Ecol 36(1):226–238. https://doi.org/10.1111/1365-2435.13943

Leslie CC (1997) Properties and regulation of cytosolic phospholipase A_2. J Biol Chem 272(27):16709–16712. https://doi.org/10.1074/jbc.272.27.16709

Leung C, Breton S, Angers B (2016) Facing environmental predictability with different sources of epigenetic variation. Ecol Evol 6(15):5234–5245. https://doi.org/10.1002/ece3.2283

Li Q, Jiang X (2014) Offspring tolerance to toxic *Microcystis aeruginosa* in *Daphnia pulex* shaped by maternal food availability and age. Fundam Appl Limnol 185(3–4):315–319. https://doi.org/10.1127/fal/2014/0704

Li X, Liu Y, Song L, Liu J (2003) Responses of antioxidant systems in the hepatocytes of common carp (*Cyprinus carpio* L.) to the toxicity of microcystin-LR. Toxicon 42(1):85–89. https://doi.org/10.1016/S0041-0101(03)00104-1

Li G, Xie P, Li H, Chen J, Hao L, Xiong Q (2010) Quantitative profiling of mRNA expression of glutathione S-transferase superfamily genes in various tissues of bighead carp (*Aristichthys nobilis*). J Biochem Mol Toxicol 24(4):250–259. https://doi.org/10.1002/jbt.20333

Li X, Ma J, Fang Q, Li Y (2013) Transcription alterations of microRNAs, cytochrome P4501A1 and 3A65, and AhR and PXR in the liver of zebrafish exposed to crude microcystins. Toxicon 73:17–22. https://doi.org/10.1016/j.toxicon.2013.07.002

Li J, Chen C, Zhang T, Liu W, Wang L, Chen Y, Wu L, Hegazy AM, El-Sayed AF, Zhang X (2019) μEvaluation of microcystin-LR absorption using an *in vivo* intestine model and its effect on zebrafish intestine. Aquat Toxicol 206:186–194. https://doi.org/10.1016/j.aquatox.2018.11.014

Liang XF, Li GG, He S, Huang Y (2007) Transcriptional responses of alpha- and rho-class glutathione *S*-transferase genes in the liver of three freshwater fishes intraperitoneally injected with microcystin-LR: relationship of inducible expression and tolerance. J Biochem Mol Toxicol 21(5):289–298. https://doi.org/10.1002/jbt.20188

Liang Y, Gao T, Shao L, Min Y, Yang J (2020) Effects of microcystin-LR and nitrite on the lifespan, reproduction, and heat shock responses of rotifer *Brachionus calyciflorus* at different temperatures. Aquat Sci 82(4). https://doi.org/10.1007/s00027-020-00748-6

Lieke T, Steinberg CEW, Bittmann S, Behrens S, Hoseinifar SH, Meinelt T, Knopf K, Kloas W (2021) Fulvic acid accelerates hatching and stimulates antioxidative protection and the innate immune response in zebrafish larvae. Sci Total Environ 796. https://doi.org/10.1016/j.scitotenv.2021.148780

Lin H, Nagashima Y, Jiang P, Qin X, Lu Y, Zhang C (2012) Screening for toxicity and resistance to paralytic shellfish toxin of shore crabs inhabiting at Leizhou peninsula, China. Mar Environ Res 78:48–52. https://doi.org/10.1016/j.marenvres.2012.04.002

Lin H, Zhang C, Liao J, Yang F, Zhong S, Jiang P, Chen X, Nagashima Y (2015a) Neutralizing effect of hemolymph from the shore crab, *Thalamita crenata*, on paralytic shellfish toxins. Toxicon 99:51–57. https://doi.org/10.1016/j.toxicon.2015.03.010

Lin J, Chen J, He J, Chen J, Yan Q, Zhou J, Xie P (2015b) Effects of microcystin-LR on bacterial and fungal functional genes profile in rat gut. Toxicon 96:50–56. https://doi.org/10.1016/j.toxicon.2015.01.011

Lin W, Guo H, Yang L, Kuang Y, Li D, Yang P, Li L (2022) Alleviation of microcystin-LR-induced hepatic lipidosis and apoptosis in zebrafish by use of rice straw-derived biochar. Ecotoxicol Environ Saf 229. https://doi.org/10.1016/j.ecoenv.2021.113054

Lin W, Ouyang K, He Y, Yang H, Kuang Y, Li D, Li L (2023) Combined effects of microcystin-LR and rice straw-derived biochar on the hepatic antioxidant capacity of zebrafish: insights from LC-MS/MS-based metabolomics analysis. Sci Total Environ 904. https://doi.org/10.1016/j.scitotenv.2023.166830

Lin Q, Xiao X, Feng X, Lei L (2025) Does *Microcystis* exposure history facilitate adaptation of *Diaphanosoma dubium* to expanding filamentous cyanobacterium *Raphidiopsis raciborskii*? Harmful Algae 145. https://doi.org/10.1016/j.hal.2025.102847

Lincoln JA, Turner JT, Bates SS, Léger C, Gauthier DA (2001) Feeding, egg production, and egg hatching success of the copepods *Acartia tonsa* and *Temora longicornis* on diets of the toxic diatom *Pseudo-nitzschia multiseries* and the non-toxic diatom *Pseudo-nitzschia pungens*. Hydrobiologia 453-454:107–120. https://doi.org/10.1023/A:1013163816771

Lindsay J, Metcalf JS, Codd GA (2006) Protection against the toxicity of microcystin-LR and cylindrospermopsin in *Artemia salina* and *daphnia* spp. by pre-treatment with cyanobacterial lipopolysaccharide (LPS). Toxicon 48(8):995–1001. https://doi.org/10.1016/j.toxicon.2006.07.036

Linker RA, Lee D-H, Ryan S, van Dam AM, Conrad R, Bista P, Zeng W, Hronowsky X, Buko A, Chollate S, Ellrichmann G, Brück W, Dawson K, Goelz S, Wiese S, Scannevin RH, Lukashev M, Gold R (2011) Fumaric acid esters exert neuroprotective effects in neuroinflammation via activation of the Nrf2 antioxidant pathway. Brain 134(3):678–692. https://doi.org/10.1093/brain/awq386

Lipkin SM, Wang V, Jacoby R, Banerjee-Basu S, Baxevanis AD, Lynch HT, Elliott RM, Collins FS (2000) *MLH3*: a DNA mismatch repair gene associated with mammalian microsatellite instability. Nat Genet 24(1):27–35. https://doi.org/10.1038/71643

Liu Y, Xie P, Wu X-P (2009) Grazing on toxic and non-toxic *Microcystis aeruginosa* PCC7820 by *Unio douglasiae* and *Corbicula fluminea*. Limnology 10(1):1–5. https://doi.org/10.1007/s10201-008-0255-3

Liu Y, Hao A, Iseri Y, Kuba T, Zhang Z (2014) A comparison of the mussel *Anodonta woodiana*'s acute physiological responses to different algae diets. J Clean Energy Technol 2(2):126–131. https://doi.org/10.7763/JOCET.2014.V2.106

Liu Y, Hao A, Iseri Y, Li C, Zhang Z, Kuba T (2015) Possible survival mechanism of *Sinanodonta woodiana* exposed to microcystin. J JSCE 3(1):259–267. https://doi.org/10.2208/journalofjsce.3.1_259

Liu C-L, Zhong W, He Y-Y, Li X, Li S, He K-L (2016a) Genome-wide analysis of tunicamycin-induced endoplasmic reticulum stress response and the protective effect of endoplasmic reticulum inhibitors in neonatal rat cardiomyocytes. Mol Cell Biochem 413(1):57–67. https://doi.org/10.1007/s11010-015-2639-0

Liu LP, Su XM, Chen TY, Li K, Zhan J, Egna H, Diana J (2016b) Evidence of rapid transfer and bioaccumulation of microcystin-LR poses potential risk to freshwater prawn *Macrobrachium rosenbergii* (de man). Aquac Res 47(10):3088–3097. https://doi.org/10.1111/are.12759

Liu X, Hao Y, Peng L, Liu Y, Wei N, Liang Q (2020) MiR-122 is involved in immune response by regulating Interleukin-15 in the orange-spotted grouper (*Epinephelus coioides*). Fish Shellf Immunol 106:404–409. https://doi.org/10.1016/j.fsi.2020.08.021

Liu J, Xiao Q, Xiao J, Niu C, Li Y, Zhang X, Zhou Z, Shu G, Yin G (2022a) Wnt/β-catenin signalling: function, biological mechanisms, and therapeutic opportunities. Signal Transduct Target Ther 7(1):3. https://doi.org/10.1038/s41392-021-00762-6

Liu Z, Zhou T, Gao D (2022b) Genetic and epigenetic regulation of growth, reproduction, disease resistance and stress responses in aquaculture. Front Genet 13. https://doi.org/10.3389/fgene.2022.994471

Liu H, Yang X, Yang W, Zheng Z, Zhu J (2023) Gut microbiota of freshwater gastropod (*Bellamya aeruginosa*) assist the adaptation of host to toxic cyanobacterial stress. Toxins 15(4). https://doi.org/10.3390/toxins15040252

Łotocka M (2001) Toxic effect of cyanobacterial blooms on the grazing activity of *Daphnia magna* Straus. Oceanologia 43(4):441–453

Lu YF, Liu J, Wu KC, Qu Q, Fan F, Klaassen CD (2014) Overexpression of Nrf2 protects against microcystin-induced hepatotoxicity in mice. PLoS One 9(3):e93013. https://doi.org/10.1371/journal.pone.0093013

Lu, X., Y. Long, R. Sun, B. Zhou, L. Lin, S. Zhong and Z. Cui (2015). "Zebrafish Abcb4 is a potential efflux transporter of microcystin-LR." Comparat Biochem Physiol Part C Toxicol Pharmacol 167(0): 35–42.DOI: https://doi.org/10.1016/j.cbpc.2014.08.005

Lu J, Bu X, Xiao S, Lin Z, Wang X, Jia Y, Wang X, Qin JG, Chen L (2019) Effect of single and combined immunostimulants on growth, anti-oxidation activity, non-specific immunity and resistance to *Aeromonas hydrophila* in Chinese mitten crab (*Eriocheir sinensis*). Fish Shellf Immunol 93:732–742. https://doi.org/10.1016/j.fsi.2019.08.027

Lundmark M (2010) Otiorhynchus sulcatus, an autopolyploid general-purpose genotype species? Hereditas 147(6):278–282. https://doi.org/10.1111/j.1601-5223.2010.02198.x

Luu I, Ikert H, Craig PM (2021) Chronic exposure to anthropogenic and climate related stressors alters transcriptional responses in the liver of zebrafish (*Danio rerio*) across multiple generations. Comparat Biochem Physiol Part C Toxicol Pharmacol 240:108918. https://doi.org/10.1016/j.cbpc.2020.108918

Lv JJ, Yuan KK, Lu GX, Li HY, Kwok HF, Yang WD (2023) Responses of ABCB and ABCC transporters to the toxic dinoflagellate *Prorocentrum lima* in the mussel *Perna viridis*. Aquat Toxicol 254. https://doi.org/10.1016/j.aquatox.2022.106368

Lynch TC, Phlips E (2000) Filtration of the bloom-forming cyanobacteria *Synechococcus* by three sponge species from Florida Bay, U.S.A. Bull Mar Sci 67:923–936

Lyu K, Zhang L, Zhu X, Cui G, Wilson AE, Yang Z (2015) Arginine kinase in the cladoceran *Daphnia magna*: cDNA sequencing and expression is associated with resistance to toxic *Microcystis*. Aquat Toxicol 160:13–21. https://doi.org/10.1016/j.aquatox.2014.12.023

Lyu K, Gu L, Li B, Lu Y, Wu C, Guan H, Yang Z (2016a) Stress-responsive expression of a glutathione S-transferase (delta) gene in waterflea *Daphnia magna* challenged by microcystin-producing and microcystin-free *Microcystis aeruginosa*. Harmful Algae 56:1–8. https://doi.org/10.1016/j.hal.2016.04.009

Lyu K, Guan H, Wu C, Wang X, Wilson AE, Yang Z (2016b) Maternal consumption of non-toxic *Microcystis* by *Daphnia magna* induces tolerance to toxic *Microcystis* in offspring. Freshw Biol 61(2):219–228. https://doi.org/10.1111/fwb.12695

Lyu K, Meng Q, Zhu X, Dai D, Zhang L, Huang Y, Yang Z (2016c) Changes in iTRAQ-based proteomic profiling of the cladoceran *Daphnia magna* exposed to microcystin-producing and microcystin-free *Microcystis aeruginosa*. Environ Sci Technol 50(9):4798–4807. https://doi.org/10.1021/acs.est.6b00101

Lyu K, Gu L, Wang H, Zhu X, Zhang L, Sun Y, Huang Y, Yang Z (2019) Transcriptomic analysis dissects the mechanistic insight into the *Daphnia* clonal variation in tolerance to toxic *Microcystis*. Limnol Oceanogr 64(1):272–283. https://doi.org/10.1002/lno.11038

Lyu T, Zhu J, Yang X, Yang W, Zheng Z (2022) Responses of gut microbial community composition and function of the freshwater gastropod *Bellamya aeruginosa* to cyanobacterial bloom. Front Microbiol 13. https://doi.org/10.3389/fmicb.2022.906278

Lyu K, Wu Y, Li J, Yang Z (2024) MicroRNA *mir-210* modulates the water flea *Daphnia magna* response to cyanobacterial stress. Environ Sci Technol 58(42):18520–18530. https://doi.org/10.1021/acs.est.4c04190

Lyu K, Fan Y, Zhou X, Hou Y, Yang Z (2025) p38 MAPK determines the sensitivity of the aquatic keystone species *Moina macrocopa* to toxic *Microcystis*: insights into potential biomarker applications. Environ Pollut 366. https://doi.org/10.1016/j.envpol.2024.125458

Macke E, Callens M, De Meester L, Decaestecker E (2017) host-genotype dependent gut microbiota drives zooplankton tolerance to toxic cyanobacteria. Nat Commun 8(1). https://doi.org/10.1038/s41467-017-01714-x

Mafra LL Jr, Bricelj VM, Ouellette C, Léger C, Bates SS (2009a) Mechanisms contributing to low domoic acid uptake by oysters feeding on *Pseudo-nitzschia* cells. I. Filtration and pseudofeces production. Aquat Biol 6(1–3):201–212. https://doi.org/10.3354/ab00121

Mafra LL Jr, Bricelj VM, Ward JE (2009b) Mechanisms contributing to low domoic acid uptake by oysters feeding on *pseudo-nitzschia* cells. II. Selective rejection. Aquat Biol 6(1–3):213–226. https://doi.org/10.3354/ab00122

Mafra LL Jr, Bricelj VM, Ouellette C, Bates SS (2010) Feeding mechanics as the basis for differential uptake of the neurotoxin domoic acid by oysters, *Crassostrea virginica*, and mussels, *Mytilus edulis*. Aquat Toxicol 97(2):160–171. https://doi.org/10.1016/j.aquatox.2010.01.009

Maher J, Yamamoto M (2010) The rise of antioxidant signaling—the evolution and hormetic actions of Nrf2. Toxicol Appl Pharmacol 244(1):4–15. https://doi.org/10.1016/j.taap.2010.01.011

Malécot M, Guével B, Pineau C, Holbech BF, Bormans M, Wiegand C (2013) Specific proteomic response of *Unio pictorum* mussel to a mixture of glyphosate and microcystin-LR. J Proteome Res 12(11):5281–5292. https://doi.org/10.1021/pr4006316

Maneiro I, Frangópulos M, Guisande C, Fernández M, Reguera B, Riveiro I (2000) Zooplankton as a potential vector of diarrhetic shellfish poisoning toxins through the food web. Mar Ecol Prog Ser 201:155–163

Maneiro I, Iglesias P, Guisande C, Riveiro I, Barreiro A, Zervoudaki S, Granéli E (2005) Fate of domoic acid ingested by the copepod *Acartia clausi*. Mar Biol 148(1):123–130. https://doi.org/10.1007/s00227-005-0054-x

Martínez-Escauriaza R, Lozano V, Pérez-Parallé ML, Blanco J, Sánchez JL, Pazos AJ (2021) Expression analyses of genes related to multixenobiotic resistance in *Mytilus galloprovincialis* after exposure to okadaic acid-producing *Dinophysis acuminata*. Toxins 13(9). https://doi.org/10.3390/toxins13090614

Martins JC, Domínguez-Pérez D, Azevedo C, Braga AC, Costa PR, Osório H, Vasconcelos V, Campos A (2020) Molecular responses of mussel *Mytilus galloprovincialis* associated to accumulation and depuration of marine biotoxins okadaic acid and dinophysistoxin-1 revealed by shotgun proteomics. Front Mar Sci 7. https://doi.org/10.3389/fmars.2020.589822

Mashayekhi-Sardoo H, Rezaee R, Riahi-Zanjani B, Karimi G (2024) Alleviation of microcystin-leucine arginine -induced hepatotoxicity: an updated overview. Toxicon 243. https://doi.org/10.1016/j.toxicon.2024.107715

Massicotte R, Angers B (2012) General-purpose genotype or how epigenetics extend the flexibility of a genotype. Genet Res Int 2012:317175. https://doi.org/10.1155/2012/317175

Mendizabal I, Keller TE, Zeng J, Yi SV (2014) Epigenetics and evolution. Integr Comp Biol 54(1):31–42. https://doi.org/10.1093/icb/icu040

Menzel S, Bouchnak R, Menzel R, Steinberg CEW (2011) Dissolved humic substances initiate DNA-methylation in cladocerans. Aquat Toxicol 105(3–4):640–642. https://doi.org/10.1016/j.aquatox.2011.08.025

Miller MA, Kudela RM, Mekebri A, Crane D, Oates SC, Tinker MT, Staedler M, Miller WA, Toy-Choutka S, Dominik C, Hardin D, Langlois G, Murray M, Ward K, Jessup DA (2010) Evidence for a novel marine harmful algal bloom: cyanotoxin (microcystin) transfer from land to sea otters. PLoS One 5(9). https://doi.org/10.1371/journal.pone.0012576

Min BH, Ravikumar Y, Lee DH, Choi KS, Kim BM, Rhee JS (2018) Age-dependent antioxidant responses to the bioconcentration of microcystin-LR in the mysid crustacean, *Neomysis awatschensis*. Environ Pollut 232:284–292. https://doi.org/10.1016/j.envpol.2017.09.050

Mohamed ZA, Al-Shehri AM (2013) Grazing on *Microcystis aeruginosa* and degradation of microcystins by the heterotrophic flagellate *Diphylleia rotans*. Ecotoxicol Environ Saf 96:48–52. https://doi.org/10.1016/j.ecoenv.2013.06.015

Mohamed ZA, Elnour RO, Alamri S, Hashem M (2023) *Paramecium jenningsi* effectively grazes on toxic *Raphidiopsis raciborskii* and degrades cylindrospermopsin: implications for control harmful cyanobacterial blooms. Ecohydrol Hydrobiol 23(4):614–622. https://doi.org/10.1016/j.ecohyd.2023.04.004

Monroig Ó, Kabeya N (2018) Desaturases and elongases involved in polyunsaturated fatty acid biosynthesis in aquatic invertebrates: a comprehensive review. Fish Sci 84(6):911–928. https://doi.org/10.1007/s12562-018-1254-x

Motohashi H, Yamamoto M (2004) Nrf2–Keap1 defines a physiologically important stress response mechanism. Trends Mol Med 10(11):549–557. https://doi.org/10.1016/j.molmed.2004.09.003

Murby McQuaid AL (2019) The bioaccumulation of cyanotoxins in aquatic food webs. PhD doctoral dissertations, University of New Hampshire, Durham

Nandini S, Rao TR (1997) Somatic and population growth in selected cladoceran and rotifer species offered the cyanobacterium it *Microcystis aeruginosa* as food. Aquat Ecol 31(3):283–298. https://doi.org/10.1023/A:1009940127959

Nandini S, Sarma SSS (2023) Experimental studies on zooplankton-toxic cyanobacteria interactions: a review. Toxics 11(2):176. https://doi.org/10.3390/toxics11020176

Nelson DL, Cox MM (2005) Lehninger – principles of biochemistry. W. H. Freeman, New York

Newell RIE, Jordan SJ (1983) Preferential ingestion of organic material by the American oyster *Crassostrea virginica*. Mar Ecol Prog Ser 13(1):47–53

Nishibe Y, Manage PM, Kawabata Z i, Nakano S-i (2004) Trophic coupling of a testate amoeba and *Microcystis* species in a hypertrophic pond. Limnol 5(2):71–76. https://doi.org/10.1007/s10201-004-0114-9

Norouzitallab P, Baruah K, Vandegehuchte M, Van Stappen G, Catania F, Bussche JV, Vanhaecke L, Sorgeloos P, Bossier P (2014) Environmental heat stress induces epigenetic transgenerational inheritance of robustness in parthenogenetic *Artemia* model. FASEB J 28(8):3552–3563. https://doi.org/10.1096/fj.14-252049

Nybom SMK, Dziga D, Heikkilä JE, Kull TPJ, Salminen SJ, Meriluoto JAO (2012) Characterization of microcystin-LR removal process in the presence of probiotic bacteria. Toxicon 59(1):171–181. https://doi.org/10.1016/j.toxicon.2011.11.008

Nyst PH (1878) Conchyliologie des terrains tertiaires de la Belgique. 1. Terrain Pliocène Scaldisien. Atlas. F. Hayez, Bruxelles

O'Dea RE, Noble DWA, Johnson SL, Hesselson D, Nakagawa S (2016) The role of non-genetic inheritance in evolutionary rescue: epigenetic buffering, heritable bet hedging and epigenetic traps. Environ Epigenet 2(1). https://doi.org/10.1093/eep/dvv014

Oberhaus L, Gélinas M, Pinel-Alloul B, Ghadouani A, Humbert J-F (2007) Grazing of two toxic *Planktothrix* species by *Daphnia pulicaria*: potential for bloom control and transfer of microcystins. J Plankton Res 29(10):827–838. https://doi.org/10.1093/plankt/fbm062

Oliveira F, Diez-Quijada L, Turkina MV, Morais J, Felpeto AB, Azevedo J, Jos A, Camean AM, Vasconcelos V, Martins JC, Campos A (2020) Physiological and metabolic responses of marine mussels exposed to toxic cyanobacteria *Microcystis aeruginosa* and *Chrysosporum ovalisporum*. Toxins 12(3):196. https://doi.org/10.3390/toxins12030196

Olson MB, Lessard EJ (2010) The influence of the *pseudo-nitzschia* toxin, domoic acid, on microzooplankton grazing and growth: a field and laboratory assessment. Harmful Algae 9(6):540–547. https://doi.org/10.1016/j.hal.2010.04.002

Ortiz-Rodríguez R, Dao TS, Wiegand C (2012) Transgenerational effects of microcystin-LR on *Daphnia magna*. J Exp Biol 215(16):2795–2805. https://doi.org/10.1242/jeb.069211

Ou D, Song L, Gan N, Chen W (2005) Effects of microcystins on and toxin degradation by *Poterioochromonas* sp. Environ Toxicol 20:373–380. https://doi.org/10.1002/tox.20114

Painefilú JC, González C, Cárcamo JG, Bianchi VA, Luquet CM (2022) Microcystin-LR modulates multixenobiotic resistance proteins in the middle intestine of rainbow trout, *Oncorhynchus mykiss*. Aquat Toxicol 253:106327. https://doi.org/10.1016/j.aquatox.2022.106327

Parker IM, Rodriguez J, Loik ME (2003) An evolutionary approach to understanding the biology of invasions: local adaptation and general-purpose genotypes in the weed *Verbascum thapsus*. Conserv Biol 17(1):59–72. https://doi.org/10.1046/j.1523-1739.2003.02019.x

Pazos AJ, Ventoso P, Martínez-Escauriaza R, Pérez-Parallé ML, Blanco J, Triviño JC, Sánchez JL (2017) Transcriptional response after exposure to domoic acid-producing *Pseudo-nitzschia* in the digestive gland of the mussel *Mytilus galloprovincialis*. Toxicon 140:60–71. https://doi.org/10.1016/j.toxicon.2017.10.002

Perez MF, Lehner B (2019) Intergenerational and transgenerational epigenetic inheritance in animals. Nat Cell Biol 21(2):143–151. https://doi.org/10.1038/s41556-018-0242-9

Petit J, Wiegertjes GF (2016) Long-lived effects of administering β-glucans: indications for trained immunity in fish. Dev Comp Immunol 64:93–102. https://doi.org/10.1016/j.dci.2016.03.003

Pflugmacher S, Wiegand C, Oberemm A, Beattie KA, Krause E, Codd GA, Steinberg CEW (1998) Identification of an enzymatically formed glutathione conjugate of the cyanobacterial hepatotoxin microcystin-LR: the first step of detoxication. Biochim Biophys Acta Gen Subj 1425(3):527–533. https://doi.org/10.1016/S0304-4165(98)00107-X

Pham TL, Shimizu K, Dao TS, Hong-Do LC, Utsumi M (2015) Microcystin uptake and biochemical responses in the freshwater clam *Corbicula leana* P. exposed to toxic and non-toxic *Microcystis aeruginosa*: evidence of tolerance to cyanotoxins. Toxicol Rep 2:88–98. https://doi.org/10.1016/j.toxrep.2015.01.012

Pham TL, Shimizu K, Kanazawa A, Gao Y, Dao TS, Utsumi M (2016) Microcystin accumulation and biochemical responses in the edible clam *Corbicula leana* P. exposed to cyanobacterial crude extract. J Environ Sci (China) 44:120–130. https://doi.org/10.1016/j.jes.2015.09.018

Pietrzyk-Brzezinska AJ, Bujacz A (2020) H-type lectins—structural characteristics and their applications in diagnostics, analytics and drug delivery. Int J Biol Macromol 152:735–747. https://doi.org/10.1016/j.ijbiomac.2020.02.320

Piferrer F, Wang H (2023) Epigenetics in aquaculture. Wiley & Sons, Chichester

Pinho GLL, Moura da Rosa C, Maciel FE, Bianchini A, Yunes JS, Proença LAO, Monserrat JM (2005) Antioxidant responses and oxidative stress after microcystin exposure in the hepatopancreas of an estuarine crab species. Ecotoxicol Environ Saf 61(3):353–360. https://doi.org/10.1016/j.ecoenv.2004.11.014

Polkoff KM, Lotti SN, Beever JE, Wheeler MB (2017) 206 Crispr/Cas9-mediated repair of the NHLRC2 locus in beef cattle. Reprod Fertil Dev 29(1):212–212. https://doi.org/10.1071/RDv29n1Ab206

Prado-Alvarez M, Flórez-Barrós F, Sexto-Iglesias A, Méndez J, Fernandez-Tajes J (2012) Effects of okadaic acid on haemocytes from *Mytilus galloprovincialis*: a comparison between field and laboratory studies. Mar Environ Res 81:90–93. https://doi.org/10.1016/j.marenvres.2012.08.011

Prado-Alvarez M, Flórez-Barrós F, Méndez J, Fernandez-Tajes J (2013) Effect of okadaic acid on carpet shell clam (*Ruditapes decussatus*) haemocytes by in vitro exposure and harmful algal bloom simulation assays. Cell Biol Toxicol 29(3):189–197. https://doi.org/10.1007/s10565-013-9246-1

Prasad P, Anjali P, Sreedhar RV (2021) Plant-based stearidonic acid as sustainable source of omega-3 fatty acid with functional outcomes on human health. Crit Rev Food Sci Nutr 61(10):1725–1737. https://doi.org/10.1080/10408398.2020.1765137

Prego-Faraldo MV, Valdiglesias V, Méndez J, Eirín-López JM (2013) Okadaic acid meet and greet: an insight into detection methods, response strategies and genotoxic effects in marine invertebrates. Mar Drugs 11(8):2829–2845. https://doi.org/10.3390/md11082829

Prego-Faraldo MV, Valdiglesias V, Laffon B, Mendez J, Eirin-Lopez JM (2016) Early genotoxic and cytotoxic effects of the toxic dinoflagellate *Prorocentrum lima* in the mussel *Mytilus galloprovincialis*. Toxins 8(6):159. https://doi.org/10.3390/toxins8060159

Prego-Faraldo MV, Martínez L, Méndez J (2018) RNA-seq analysis for assessing the early response to DSP toxins in *Mytilus galloprovincialis* digestive gland and gill. Toxins 10(10). https://doi.org/10.3390/toxins10100417

Prince EK, Lettieri L, McCurdy KJ, Kubanek J (2006) Fitness consequences for copepods feeding on a red tide dinoflagellate: deciphering the effects of nutritional value, toxicity, and feeding behavior. Oecologia 147(3):479–488. https://doi.org/10.1007/s00442-005-0274-2

Pu C, Zhan A (2017) Epigenetic divergence of key genes associated with water temperature and salinity in a highly invasive model ascidian. Biol Invasions 19(7):2015–2028. https://doi.org/10.1007/s10530-017-1409-1

Qiang J, Tao YF, He J, Xu P, Bao JW, Sun YL (2017) miR-122 promotes hepatic antioxidant defense of genetically improved farmed tilapia (GIFT, *Oreochromis niloticus*) exposed to cadmium by directly targeting a metallothionein gene. Aquat Toxicol 182:39–48. https://doi.org/10.1016/j.aquatox.2016.11.009

Qiu H, Lu K, Zheng Z, Wang J, Zhu J (2017) Blooms of toxic cyanobacteria cause the gastropod *Bellamya aeruginosa* to shift its diet from planktic to benthic material. Int Rev Hydrobiol 102(3–4):90–99. https://doi.org/10.1002/iroh.201601859

Radersma R, Hegg A, Noble DWA, Uller T (2018) Timing of maternal exposure to toxic cyanobacteria and offspring fitness in *Daphnia magna*: implications for the evolution of anticipatory maternal effects. Ecol Evol 8(24):12727–12736. https://doi.org/10.1002/ece3.4700

Rattanachaikunsopon P, Phumkhachorn P (2007) Bacteriostatic effect of flavonoids isolated from leaves of *Psidium guajava* on fish pathogens. Fitoterapia 78(6):434–436. https://doi.org/10.1016/j.fitote.2007.03.015

Reguera B, Velo-Suárez L, Raine R, Park MG (2012) Harmful *Dinophysis* species: a review. Harmful Algae 14:87–106. https://doi.org/10.1016/j.hal.2011.10.016

Reis B, Carneiro M, Machado J, Azevedo J, Vasconcelos V, Martins JC (2015) Transcriptional responses of glutathione transferase genes in *Ruditapes philippinarum* exposed to microcystin-LR. Int J Mol Sci 16(4):8397–8414. https://doi.org/10.3390/ijms16048397

Religia P, Nguyen ND, Nong QD, Matsuura T, Kato Y, Watanabe H (2021) Mutation of the cytochrome P450 *CYP360A8* gene increases sensitivity to paraquat in *Daphnia magna*. Environ Toxicol Chem 40(5):1279–1288. https://doi.org/10.1002/etc.4970

Remize M, Planchon F, Loh AN, Le Grand F, Lambert C, Bideau A, Bidault A, Corvaisier R, Volety A, Soudant P (2020) Identification of polyunsaturated fatty acids synthesis pathways in the toxic dinophyte *Alexandrium minutum* using ^{13}C-labelling. Biomolecules 10(10). https://doi.org/10.3390/biom10101428

Ren X, Zhang J, Huang Y, Yang W, Lu K, Zhu J (2023) Toxic cyanobacteria induce coupled changes in gut microbiota and co-metabolite of freshwater gastropods. Environ Pollut 338. https://doi.org/10.1016/j.envpol.2023.122651

Rewitz KF, Gilbert LI (2008) *Daphnia* Halloween genes that encode cytochrome P450s mediating the synthesis of the arthropod molting hormone: evolutionary implications. BMC Evol Biol 8(1):60. https://doi.org/10.1186/1471-2148-8-60

Rimphanitchayakit V, Tassanakajon A (2010) Structure and function of invertebrate Kazal-type serine proteinase inhibitors. Dev Comp Immunol 34(4):377–386. https://doi.org/10.1016/j.dci.2009.12.004

Romero-Geraldo RDJ, Garciá-Lagunas N, Hernańdez-Saavedra NY (2014) Effects of *in vitro* exposure to diarrheic toxin producer *Prorocentrum lima* on gene expressions related to cell cycle regulation and immune response in *Crassostrea gigas*. PLoS One 9(5):e97181. https://doi.org/10.1371/journal.pone.0097181

Roncalli V, Cieslak MC, Lenz PH (2016a) Transcriptomic responses of the calanoid copepod *Calanus finmarchicus* to the saxitoxin producing dinoflagellate *Alexandrium fundyense*. Sci Rep 6. https://doi.org/10.1038/srep25708

Roncalli V, Jungbluth MJ, Lenz PH (2016b) Glutathione *S*-transferase regulation in *Calanus finmarchicus* feeding on the toxic dinoflagellate *Alexandrium fundyense*. PLoS One 11(7). https://doi.org/10.1371/journal.pone.0159563

Roncalli V, Turner JT, Kulis D, Anderson DM, Lenz PH (2016c) The effect of the toxic dinoflagellate *Alexandrium fundyense* on the fitness of the calanoid copepod *Calanus finmarchicus*. Harmful Algae 51:56–66. https://doi.org/10.1016/j.hal.2015.11.003

Roncalli V, Lenz PH, Cieslak MC, Hartline DK (2017) Complementary mechanisms for neurotoxin resistance in a copepod. Sci Rep 7(1). https://doi.org/10.1038/s41598-017-14545-z

Roncalli V, Lauritano C, Carotenuto Y (2021) First report of OvoA gene in marine arthropods: a new candidate stress biomarker in copepods. Mar Drugs 19(11). https://doi.org/10.3390/md19110647

Rossignoli AE, Blanco J (2008) Cellular distribution of okadaic acid in the digestive gland of *Mytilus galloprovincialis* (Lamarck, 1819). Toxicon 52(8):957–959. https://doi.org/10.1016/j.toxicon.2008.09.003

Rossignoli AE, Blanco J (2010) Subcellular distribution of okadaic acid in the digestive gland of *Mytilus galloprovincialis*: first evidences of lipoprotein binding to okadaic acid. Toxicon 55(2–3):221–226. https://doi.org/10.1016/j.toxicon.2009.07.022

Rossignoli AE, Fernández D, Regueiro J, Mariño C, Blanco J (2011) Esterification of okadaic acid in the mussel *Mytilus galloprovincialis*. Toxicon 57(5):712–720. https://doi.org/10.1016/j.toxicon.2011.02.003

Roy S, Bossier P, Norouzitallab P, Vanrompay D (2020) Trained immunity and perspectives for shrimp aquaculture. Rev Aquac 12(4):2351–2370. https://doi.org/10.1111/raq.12438

Ruan X, Zhang R, Li R, Zhu H, Wang Z, Wang C, Cheng Z, Peng H (2022) The research progress in physiological and pathological functions of TRAF4. Front Oncol 12. https://doi.org/10.3389/fonc.2022.842072

Rudman SM, Rodriguez-Cabal MA, Stier A, Sato T, Heavyside J, El-Sabaawi RW, Crutsinger GM (2015) Adaptive genetic variation mediates bottom-up and top-down control in an aquatic ecosystem. Proc R Soc B Biol Sci 282(1812):20151234. https://doi.org/10.1098/rspb.2015.1234

Runnegar M, Berndt N, Kaplowitz N (1995) Microcystin uptake and inhibition of protein phosphatases: effects of chemoprotectants and self-inhibition in relation to known hepatic transporters. Toxicol Appl Pharmacol 134(2):264–272. https://doi.org/10.1006/taap.1995.1192

Rushmore TH, Morton MR, Pickett CB (1991) The antioxidant responsive element. Activation by oxidative stress and identification of the DNA consensus sequence required for functional activity. J Biol Chem 266(18):11632–11639. https://doi.org/10.1016/S0021-9258(18)99004-6

Russo GL, Russo M, Castellano I, Napolitano A, Palumbo A (2014) Ovothiol isolated from sea urchin oocytes induces autophagy in the Hep-G2 cell line. Mar Drugs 12(7):4069–4085. https://doi.org/10.3390/md12074069

Russo E, Lauritano C, d'Ippolito G, Fontana A, Sarno D, von Elert E, Ianora A, Carotenuto Y (2020) RNA-Seq and differential gene expression analysis in *Temora stylifera* copepod females with contrasting non-feeding nauplii survival rates: an environmental transcriptomics study. BMC Genomics 21(1). https://doi.org/10.1186/s12864-020-07112-w

Sabater S, Guasch H, Romaní A, Muñoz I (2000) Stromatolitic communities in Mediterranean streams: adaptations to a changing environment. Biodivers Conserv 9(3):379–392. https://doi.org/10.1023/A:1008954801397

Sabatini SE, Brena BM, Luquet CM, San Julián M, Pirez M, Carmen Ríos de Molina MD (2011) Microcystin accumulation and antioxidant responses in the freshwater clam *Diplodon chilensis patagonicus* upon subchronic exposure to toxic *Microcystis aeruginosa*. Ecotoxicol Environ Saf 74(5):1188–1194. https://doi.org/10.1016/j.ecoenv.2011.03.012

Sabatini SE, Brena BM, Pirez M, de Molina MDCR, Luquet CM (2015) Oxidative effects and toxin bioaccumulation after dietary microcystin intoxication in the hepatopancreas of the crab *Neohelice (Chasmagnathus) granulata*. Ecotoxicol Environ Saf 120:136–141. https://doi.org/10.1016/j.ecoenv.2015.05.041

Salas M, Dudgeon D (2001) Stable-isotope determination of mayfly (Insecta: Ephemeroptera) food sources in three tropical Asian streams. Arch Hydrobiol 151:7–32. https://doi.org/10.1127/archiv-hydrobiol/151/2001/17

Sarkar S, Kimono D, Albadrani M, Seth RK, Busbee P, Alghetaa H, Porter DE, Scott GI, Brooks B, Nagarkatti M, Nagarkatti P, Chatterjee S (2019) Environmental microcystin targets the microbiome and increases the risk of intestinal inflammatory pathology via NOX2 in underlying murine model of nonalcoholic fatty liver disease. Sci Rep 9(1):8742. https://doi.org/10.1038/s41598-019-45009-1

Sarnelle O, Wilson AE (2005) Local adaptation of *Daphnia pulicaria* to toxic cyanobacteria. Limnol Oceanogr 50(5):1565–1570. https://doi.org/10.4319/lo.2005.50.5.1565

Sarnelle O, Gustafsson S, Hansson L-A (2010) Effects of cyanobacteria on fitness components of the herbivore *Daphnia*. J Plankton Res 32(4):471–477. https://doi.org/10.1093/plankt/fbp151

Schmidt K, Staaks GB, Pflugmacher S, Steinberg CE (2005) Impact of PCB mixture (Aroclor 1254) and TBT and a mixture of both on swimming behavior, body growth and enzymatic biotransformation activities (GST) of young carp (*Cyprinus carpio*). Aquat Toxicol 71(1):49–59. https://doi.org/10.1016/j.aquatox.2004.10.012

Schmidt K, Pflugmacher S, Staaks GBO, Steinberg CEW (2006) The influence of tributyltin chloride and polychlorinated biphenyls on swimming behavior, body growth, reproduction, and activity of biotransformation enzymes in *Daphnia magna*. J Freshw Ecol 21(1):109–120. https://doi.org/10.1080/02705060.2006.9664103

Schroer TA (2004) Dynactin. Annu Rev Cell Dev Biol 20:759–779. https://doi.org/10.1146/annurev.cellbio.20.012103.094623

Schwarzenberger A (2022) Negative effects of cyanotoxins and adaptative responses of *daphnia*. Toxins 14(11):770. https://doi.org/10.3390/toxins14110770

Schwarzenberger A, Martin-Creuzburg D (2021) *Daphnia*'s adaptive molecular responses to the cyanobacterial neurotoxin anatoxin-α are maternally transferred. Toxins 13(5):326. https://doi.org/10.3390/toxins13050326

Schwarzenberger A, von Elert E (2013) Cyanobacterial protease inhibitors lead to maternal transfer of increased protease gene expression in *Daphnia*. Oecologia 172(1):11–20. https://doi.org/10.1007/s00442-012-2479-5

Schwarzenberger A, Sadler T, Motameny S, Ben-Khalifa K, Frommolt P, Altmuller J, Konrad K, von Elert E (2014) Deciphering the genetic basis of microcystin tolerance. BMC Genomics 15:776. https://doi.org/10.1186/1471-2164-15-776

Schwarzenberger A, Ilić M, von Elert E (2021) *Daphnia* populations are similar but not identical in tolerance to different protease inhibitors. Harmful Algae 106:102062. https://doi.org/10.1016/j.hal.2021.102062

Scott JT, Marcarelli AM (2012) Cyanobacteria in freshwater benthic environments. In: Whitton BA (ed) Ecology of Cyanobacteria II: their diversity in space and time. Springer Netherlands, Dordrecht, pp 271–289

Setälä O, Sopanen S, Autio R, Erler K (2009) Grazing and food selection of the calanoid copepods *Eurytemora affinis* and *Acartia bifilosa* feeding on plankton assemblages containing *Dinophysis* spp. Boreal Environ Res 14(5):837–849

Shahmohamadloo RS, Ortiz Almirall X, Simmons DBD, Poirier DG, Bhavsar SP, Sibley PK (2022) Fish tissue accumulation and proteomic response to microcystins is species-dependent. Chemosphere 287. https://doi.org/10.1016/j.chemosphere.2021.132028

Shapiro BM, Hopkins PB (1991) Ovothiols: biological and chemical perspectives. Adv Enzymol Relat Areas Mol Biol 64:291–316. https://doi.org/10.1002/9780470123102.ch6

Shen Q, Hu J, Li D, Wang G, Liu Y (2005) Investigation on intake, accumulation and toxicity of microcystins to silver carp. Fresenius Environ Bull 15:1124–1128

Silva FDG, Lopes DD, Hector RE, Nascimento MTD, Miguel TDÁ, Kuroda EK, Nóbrega GMDAD, Harada KI, Hirooka EY (2023) Microcystin-detoxifying recombinant *Saccharomyces cerevisiae* expressing the *mlrA* gene from *Sphingosinicella microcystinivorans* B9. Microorganisms 11(3). https://doi.org/10.3390/microorganisms11030575

Singh A, Mazumder A, Das S, Tyagi P, Chaitanya MVNL (2024) Probiotics in action: enhancing immunity and combatting diseases for optimal health. J Nat Remed 24(6):1153–1167. https://doi.org/10.18311/jnr/2024/35894

Skinner MK (2015) Environmental epigenetics and a unified theory of the molecular aspects of evolution: a neo-Lamarckian concept that facilitates neo-Darwinian evolution. Genome Biol Evol 7(5):1296–1302. https://doi.org/10.1093/gbe/evv073

Soares MCS, Lürling M, Huszar VLM (2010) Responses of the rotifer *Brachionus calyciflorus* to two tropical toxic cyanobacteria (*Cylindrospermopsis raciborskii* and *Microcystis aeruginosa*) in pure and mixed diets with green algae. J Plankton Res 32(7):999–1008. https://doi.org/10.1093/plankt/fbq042

Solé, R. (2022). "Revisiting Leigh van Valen's "a new evolutionary law" (1973)." Biol TheoryDOI: https://doi.org/10.1007/s13752-021-00391-w

Soong TW, Venkatesh B (2006) Adaptive evolution of tetrodotoxin resistance in animals. Trends Genet 22(11):621–626. https://doi.org/10.1016/j.tig.2006.08.010

Stanković N, Kostić I, Jovanović B, Savić-Zdravković D, Matić S, Bašić J, Cvetković T, Simeunović J, Milošević D (2020) Can phytoplankton blooming be harmful to benthic organisms? The toxic influence of *anabaena* sp. and *chlorella* sp. on *Chironomus riparius* larvae. Sci Total Environ 729:138666. https://doi.org/10.1016/j.scitotenv.2020.138666

Steinberg CEW (2012) Stress ecology–environmental stress as ecological driving force and key player in evolution. Springer, Dordrecht

Steinberg CEW (2016) Environmental stress, the overlooked ecological driving force (in German). Sitzungsberichte der Gesellschaft der Naturforschenden Freunde zu Berlin (Neue Folge) 52:137–163

Steinberg CEW (2022a) Essential fatty acids—'*fueling versus controlling*'. In: Aquatic animal nutrition: organic macro- and micro-nutrients. Springer, Cham, pp 673–721

Steinberg CEW (2022b) Fatty acids—'*fueling versus steering*'. In: Aquatic animal nutrition: organic macro- and micro-nutrients. Springer, Cham, pp 633–672

Steinberg CEW (2022c) Intraspecific variability—'*the apple may be a PineApple*'. In: Aquatic animal nutrition: organic macro- and micro-nutrients. Springer, Cham, pp 1037–1062

Steinberg CEW (2022d) Oligosaccharides—'*sweet or healthy promises*'. In: Aquatic animal nutrition: organic macro- and micro-nutrients. Springer, Cham, pp 455–472

Steinberg CEW (2022e) Protein sparing by carbohydrates—'*life's useful luxury*'. In: Aquatic animal nutrition: organic macro- and micro-nutrients. Springer, Cham, pp 377–392

Steinberg CEW (2022f) Protein sparing by lipids—'*learning from wild conspecifics*'. In: Aquatic animal nutrition: organic macro- and micro-nutrients. Springer, Cham, pp 599–632

Steinberg CEW (2022g) Regulatory potential of carbohydrates—'*life's little luxury controls*'. In: Aquatic animal nutrition: organic macro- and micro-nutrients. Springer, Cham, pp 435–454

Steinberg CEW (2022h) Starch—'*gluey promise*'. In: Aquatic animal nutrition: organic macro- and micro-nutrients. Springer, Cham, pp 473–507

Steinberg CEW (2022i) Trophic transfer of PUFAs—'*vital ones reach top predators*'. In: Aquatic animal nutrition: organic macro- and micro-nutrients. Springer, Cham, pp 773–790

Steinberg CEW (2024) Aquatic animal nutrition – plant preparations. Springer Nature Switzerland AG

Steinberg CEW (2025a) Cyanobacteria and the food chain: "cyanobacteria are also not eaten with pleasure". In: Aquatic animal nutrition: plant compounds and dietary obstacles. Springer Nature Switzerland, Cham, pp 459–554

Steinberg CEW (2025b) Mycotoxins: "*another* Noli me Tangere". In: Aquatic animal nutrition: plant compounds and dietary obstacles. Springer Nature Switzerland, Cham, pp 555–632

Steinberg CEW (2025c) Phyto- and Mycosterols: "benefit or risk?". In: Aquatic animal nutrition: plant compounds and dietary obstacles. Springer Nature Switzerland, Cham, pp 417–457

Steinberg CEW (forthcoming) Aquatic animal nutrition - prebiotics. Springer Nature Switzerland AG

Steiner K, Zimmermann L, Hagenbuch B, Dietrich D (2016) Zebrafish Oatp-mediated transport of microcystin congeners. Arch Toxicol 90(5):1129–1139. https://doi.org/10.1007/s00204-015-1544-3

Stenvang J, Petri A, Lindow M, Obad S, Kauppinen S (2012) Inhibition of microRNA function by antimiR oligonucleotides. Silence 3(1):1. https://doi.org/10.1186/1758-907x-3-1

Suarez-Ulloa V, Fernandez-Tajes J, Aguiar-Pulido V, Prego-Faraldo MV, Florez-Barros F, Sexto-Iglesias A, Mendez J, Eirin-Lopez JM (2015) Unbiased high-throughput characterization of mussel transcriptomic responses to sublethal concentrations of the biotoxin okadaic acid. PeerJ 3:e1429. https://doi.org/10.7717/peerj.1429

Suhett AL, Steinberg CEW, Santangelo JM, Bozelli RL, Farjalla VF (2011) Natural dissolved humic substances increase the lifespan and promote transgenerational resistance to salt stress in the cladoceran *Moina macrocopa*. Environ Sci Pollut Res 18(6):1004–1014. https://doi.org/10.1007/s11356-011-0455-y

Suzuki T, Igarashi T, Ichimi K, Watai M, Suzuki M, Ogiso E, Yasumoto T (2005) Kinetics of diarrhetic shellfish poisoning toxins, okadaic acid, dinophysistoxin-1, pectenotoxin-6 and yessotoxin in scallops *Patinopecten yessoensis*. Fish Sci 71(4):948–955. https://doi.org/10.1111/j.1444-2906.2005.01049.x

Svensson S (2003) Effects, dynamics and management of okadaic acid in blue mussels, *Mytilus edulis*. PhD doctoral thesis, Göteborg University

Svensson S, Förlin L (2004) Analysis of the importance of lipid breakdown for elimination of okadaic acid (diarrhetic shellfish toxin) in mussels, *Mytilus edulis*: results from a field study and a laboratory experiment. Aquat Toxicol 66(4):405–418. https://doi.org/10.1016/j.aquatox.2003.11.002

Svensson S, Särngren A, Förlin L (2003) Mussel blood cells, resistant to the cytotoxic effects of okadaic acid, do not express cell membrane p-glycoprotein activity (multixenobiotic resistance). Aquat Toxicol 65(1):27–37. https://doi.org/10.1016/s0166-445x(03)00097-3

Sykes PF, Huntley ME (1987) Acute physiological reactions of *Calanus pacificus* to selected dinoflagellates: direct observations. Mar Biol 94(1):19–24. https://doi.org/10.1007/BF00392895

Takamatsu Y, Inaba T (1994) Inter-individual variability of human hepatic glutathione *S*-transferase isozymes assessed by inhibitory capacity. Toxicology 88(1–3):191–200. https://doi.org/10.1016/0300-483x(94)90120-1

Tan K, Zheng H (2022) Endogenous LC-PUFA biosynthesis capability in commercially important mollusks. Crit Rev Food Sci Nutr 62(10):2836–2844. https://doi.org/10.1080/10408398.2020.1860896

Teegarden GJ (1999) Copepod grazing selection and particle discrimination on the basis of PSP toxin content. Mar Ecol Prog Ser 181:163–176. https://doi.org/10.3354/meps181163

Teegarden GJ, Cembella AD, Capuano CL, Barron SH, Durbin EG (2003) Phycotoxin accumulation in zooplankton feeding on *Alexandrium fundyense* – vector or sink. J Plankton Res 25(4):429–443. https://doi.org/10.1093/plankt/25.4.429

Tian W-X (2006) Inhibition of fatty acid synthase by polyphenols. Curr Med Chem 13(8):967–977. https://doi.org/10.2174/092986706776361012

Tierbach A, Groh KJ, Schönenberger R, Schirmer K, Suter MJ-F (2018) Glutathione *S*-transferase protein expression in different life stages of zebrafish (*Danio rerio*). Toxicol Sci 162(2):702–712. https://doi.org/10.1093/toxsci/kfx293

Tierno de Figueroa JM, López-Rodríguez MJ (2019) Trophic ecology of Plecoptera (Insecta): a review. Europ Zool J 86(1):79–102. https://doi.org/10.1080/24750263.2019.1592251

Toporowska M, Pawlik-Skowronska B, Kalinowska R (2014) Accumulation and effects of cyanobacterial microcystins and anatoxin-a on benthic larvae of *Chironomus* spp. (Diptera: Chironomidae). Eur J Entomol 111:83–90. https://doi.org/10.14411/eje.2014.010

Tsai WC, Hsu SD, Hsu CS, Lai TC, Chen SJ, Shen R, Huang Y, Chen HC, Lee CH, Tsai TF, Hsu MT, Wu JC, Huang HD, Shiao MS, Hsiao M, Tsou AP (2012) MicroRNA-122 plays a critical role in liver homeostasis and hepatocarcinogenesis. J Clin Invest 122(8):2884–2897. https://doi.org/10.1172/jci63455

Tse AC-K, Li J-W, Chan T-F, Wu RS-S, Lai K-P (2015) Hypoxia induces miR-210, leading to anti-apoptosis in ovarian follicular cells of marine medaka *Oryzias melastigma*. Aquat Toxicol 165:189–196. https://doi.org/10.1016/j.aquatox.2015.06.002

Turnbull A, Malhi N, Seger A, Harwood T, Jolley J, Fitzgibbon Q, Hallegraeff G (2020) Paralytic shellfish toxin uptake, tissue distribution, and depuration in the southern rock lobster *Jasus edwardsii* Hutton. Harmful Algae 95. https://doi.org/10.1016/j.hal.2020.101818

Turner JT, Tester PA (1989) Zooplankton feeding ecology: copepod grazing during an expatriate red tide. In: Cosper EM, Bricelj VM, Carpenter EJ (eds) Costal and estuarine studies, vol 35. Springer Berlin Heidelberg, Berlin/Heidelberg, pp 359–374

Turner E, Hager LJ, Shapiro BM (1988) Ovothiol replaces glutathione peroxidase as a hydrogen peroxide scavenger in sea urchin eggs. Science 242(4880):939–941. https://doi.org/10.1126/science.3187533

Ubero-Pascal N, Aboal M (2024) Cyanobacteria and macroinvertebrate relationships in freshwater benthic communities beyond cytotoxicity. Toxins 16. https://doi.org/10.3390/toxins16040190

Ublacker GA, Johnson JA, Siegel FL, Mulcahy RT (1991) Influence of glutathione *S*-transferases on cellular glutathione determination by flow cytometry using monochlorobimane. Cancer Res 51(7):1783–1788

van Roy F (2014) Beyond E-cadherin: roles of other cadherin superfamily members in cancer. Nat Rev Cancer 14(2):121–134. https://doi.org/10.1038/nrc3647

Van Valen L (1973) A new evolutionary law. Evol Theory 1:1–30

Van Wichelen J, D'Hondt S, Claeys M, Vyverman W, Berney C, Bass D, Vanormelingen P (2016) A hotspot of amoebae diversity: 8 new naked amoebae associated with the planktonic bloom-forming cyanobacterium *Microcystis*. Acta Protozool 55(2):61–87. https://doi.org/10.4467/16890027AP.16.007.4942

Vanderploeg HA, Liebig JR, Carmichael WW, Agy MA, Johengen TH, Fahnenstiel GL, Nalepa TF (2001) Zebra mussel (*Dreissena polymorpha*) selective filtration promoted toxic *Microcystis* blooms in Saginaw Bay (Lake Huron) and Lake Erie. Can J Fish Aquat Sci 58(6):1208–1221. https://doi.org/10.1139/cjfas-58-6-1208

Vehmaa A, Hogfors H, Gorokhova E, Brutemark A, Holmborn T, Engström-Öst J (2013) Projected marine climate change: effects on copepod oxidative status and reproduction. Ecol Evol 3(13):4548–4557. https://doi.org/10.1002/ece3.839

Ventoso P, Pazos AJ, Pérez-Parallé ML, Blanco J, Triviño JC, Sánchez JL (2019) RNA-seq transcriptome profiling of the queen scallop (*Aequipecten opercularis*) digestive gland after exposure to domoic acid-producing *pseudo-nitzschia*. Toxins 11(2). https://doi.org/10.3390/toxins11020097

Ventoso P, Pazos AJ, Blanco J, Pérez-Parallé ML, Triviño JC, Sánchez JL (2021) Transcriptional response in the digestive gland of the king scallop (*Pecten maximus*) after the injection of domoic acid. Toxins 13(5). https://doi.org/10.3390/toxins13050339

Verhoeven KJF, Preite V (2014) Epigenetic variation in asexually reproducting organisms. Evolution 68(3):644–655. https://doi.org/10.1111/evo.12320

Vidt DG, Bravo EL, Fouad FM (1982) Medical intelligence drug therapy: captopril. N Engl J Med 306(4):214–219. https://doi.org/10.1056/nejm198201283060405

Viitasalo M, Koski M, Pellikka K, Johansson S (1995) Seasonal and long-term variations in the body size of planktonic copepods in the northern Baltic Sea. Mar Biol 123(2):241–250. https://doi.org/10.1007/BF00353615

Vinagre TM, Alciati JC, Regoli F, Bocchetti R, Yunes JS, Bianchini A, Monserrat JM (2003) Effect of microcystin on ion regulation and antioxidant system in gills of the estuarine crab *Chasmagnathus granulatus* (Decapoda, Grapsidae). Comparat Biochem Physiol C Toxicol Pharmacol 135(1):67–75. https://doi.org/10.1016/S1532-0456(03)00076-0

Vogt G (2017) Facilitation of environmental adaptation and evolution by epigenetic phenotype variation: insights from clonal, invasive, polyploid, and domesticated animals. Environ Epigenet 3(1). https://doi.org/10.1093/eep/dvx002

Vogt G (2023) Environmental adaptation of genetically uniform organisms with the help of epigenetic mechanisms—an insightful perspective on ecoepigenetics. Epigenomes 7(1). https://doi.org/10.3390/epigenomes7010001

von Elert E, Otte KA (2023) Population-wide differentiation of tolerance to microcystins and other cyanobacterial non-microcystin effects in *daphnia*. Fundam Appl Limnol 196(3–4):267–278. https://doi.org/10.1127/fal/2022/1452

Walters DM, Jardine TD, Cade BS, Kidd KA, Muir DCG, Leipzig-Scott P (2016) Trophic magnification of organic chemicals: a global synthesis. Environ Sci Technol 50(9):4650–4658. https://doi.org/10.1021/acs.est.6b00201

Wang H-P, Shen Z-G (2023) The potential role of epigenetics in aquaculture: insights from different taxa to diverse teleosts. In: Piferrer F, Wang H (eds) Epigenetics in aquaculture. Wiley & Sons, Chichester, pp 1–43

Wang L, Liang X-F, Liao W-Q, Lei L-M, Han B-P (2006) Structural and functional characterization of microcystin detoxification-related liver genes in a phytoplanktivorous fish, Nile tilapia (*Oreochromis niloticus*). Comparat Biochem Physiol Part C Toxicol Pharmacol 144(3):216–227. https://doi.org/10.1016/j.cbpc.2006.08.009

Wang L, Peng Y, Nie X, Pan B, Ku P, Bao S (2016) Gene response of CYP360A, CYP314, and GST and whole-organism changes in *Daphnia magna* exposed to ibuprofen. Comparat Biochem Physiol Part C Toxicol Pharmacol 179:49–56. https://doi.org/10.1016/j.cbpc.2015.08.010

Wang W, Zhong P, Yi JQ, Xu AX, Lin WY, Guo ZC, Wang CG, Sun CB, Chan S (2019) Potential role for microRNA in facilitating physiological adaptation to hypoxia in the Pacific whiteleg shrimp *Litopenaeus vannamei*. Fish Shellf Immunol 84:361–369. https://doi.org/10.1016/j.fsi.2018.09.079

Wang R, Liu X, Wu J, Liu H, Wang W, Chen X, Yuan L, Wang Y, Du X, Ma Y, Losiewicz MD, Zhang X, Zhang H (2020) Role of microRNA-122 in microcystin-leucine arginine-induced dysregulation of hepatic iron homeostasis in mice. Environ Toxicol 35(8):822–830. https://doi.org/10.1002/tox.22918

Wang X, Wang D, Zhang T, Zhang Q, Zhao J (2021) iTRAQ-based quantitative proteomic analysis reveals toxicity mechanisms in *Chlamys farreri* exposed to okadaic acid. Front Mar Sci 8. https://doi.org/10.3389/fmars.2021.792050

Wang R, Liu H, Du X, Ma Y, Tian Z, Zhang S, Shi L, Guo H, Zhang H (2022) MicroRNA-122 overexpression promotes apoptosis and tumor suppressor gene expression induced by microcystin-leucine arginine in mouse liver. Int J Environ Health Res 32(10):2123–2134. https://doi.org/10.1080/09603123.2021.1946489

Wang M, Qiu Y, Zhang Z, Chen Y, Qin W, Guan W, Li G, Yu H, Dai C, Li R, Ma Z (2023) Dual characteristics of *Bellamya aeruginosa* encountering *Microcystis aeruginosa*: algal control and toxin depuration. Ecotoxicol Environ Saf 252:114596. https://doi.org/10.1016/j.ecoenv.2023.114596

Wang Z, Hu J, Mei Z, Zhang Y, Liu Q, Yang D (2024) Mannan-oligosaccharide induces trained immunity activation and alleviates pathological liver injury in turbot (*Scophthalmus maximus*). Aquaculture 578. https://doi.org/10.1016/j.aquaculture.2023.740097

Ward AK, Dahm CN, Cummins KW (1985) *Nostoc* (Cyanophyta) productivity in Oregon stream ecosystems: invertebrate influences and differences between morphologycal types. J Phycol 21(2):223–227. https://doi.org/10.1111/j.0022-3646.1985.00223.x

Watanabe MM, Zhang X, Kaya K (1996) Fate of toxic cyclic heptapeptides, microcystins, in toxic cyanobacteria upon grazing by the mixotrophic flagellate *Poterioochromonas malhamensis* (Ochromonadales, Chrysophyceae). Phycologia 35(Suppl):203–206. https://doi.org/10.2216/i0031-8884-35-6s-203.1

Weger HG, Polasek AK, Wright DM, Damodaran A, Stavrinides J (2024) Grazing preferences of three species of amoebae on cyanobacteria and green algae. J Eukaryot Microbiol 71(2). https://doi.org/10.1111/jeu.13018

Wei XM, Lu MY, Duan GF, Li HY, Liu JS, Yang WD (2019) Responses of CYP450 in the mussel *Perna viridis* after short-term exposure to the DSP toxins-producing dinoflagellate *Prorocentrum lima*. Ecotoxicol Environ Saf 176:178–185. https://doi.org/10.1016/j.ecoenv.2019.03.073

Wei J, Li X, Xu X, Xu W, Chen Y, Zhang L, Yang Z, Huang Y (2022) Elevated temperature mitigates the prolonged effect of high nitrogen on *Microcystis aeruginosa* removal through mixotrophic *Ochromonas gloeopara* grazing. Sci Total Environ 820. https://doi.org/10.1016/j.scitotenv.2022.153267

Weissberger EJ, Glibert PM (2021) Diet of the eastern oyster, *Crassostrea virginica*, growing in a eutrophic tributary of Chesapeake Bay, Maryland, USA. Aquac Rep 20. https://doi.org/10.1016/j.aqrep.2021.100655

Weyrich A, Lenz D, Jeschek M, Chung TH, Rübensam K, Göritz F, Jewgenow K, Fickel J (2016) Paternal intergenerational epigenetic response to heat exposure in male wild Guinea pigs. Mol Ecol 25(8):1729–1740. https://doi.org/10.1111/mec.13494

Whalen KE, Sotka EE, Goldstone JV, Hahn ME (2010) The role of multixenobiotic transporters in predatory marine molluscs as counter-defense mechanisms against dietary allelochemicals. Comparat Biochem Physiol Part C Toxicol Pharmacol 152(3):288–300. https://doi.org/10.1016/j.cbpc.2010.05.003

Whitchcad A, Clark BW, Reid NM, Hahn ME, Nacci D (2017) When evolution is the solution to pollution: key principles, and lessons from rapid repeated adaptation of killifish (*Fundulus heteroclitus*) populations. Evol Appl 10(8):762–783. https://doi.org/10.1111/eva.12470

Wojtal-Frankiewicz A, Bernasińska J, Frankiewicz P, Gwoździński K, Jurczak T (2014) Response of *Daphnia*'s antioxidant system to spatial heterogeneity in cyanobacteria concentrations in a lowland reservoir. PLoS One 9(11):e112597. https://doi.org/10.1371/journal.pone.0112597

Wong LL, Razali SA, Deris ZM, Danish-Daniel M, Tan MP, Nor SAM, Ma H, Min W, Yantao L, Asaduzzaman M, Sung YY, Liu Z, Sorgeloos P, Van de Peer Y, Afiqah-Aleng N (2022) Application of second-generation sequencing (SGS) and third generation sequencing (TGS) in aquaculture breeding program. Aquaculture 548. https://doi.org/10.1016/j.aquaculture.2021.737633

Wood SA, Kelly LT, Bouma-Gregson K, Humbert J-F, Laughinghouse HD IV, Lazorchak J, McAllister TG, McQueen A, Pokrzywinski K, Puddick J, Quiblier C, Reitz LA, Ryan KG, Vadeboncoeur Y, Zastepa A, Davis TW (2020) Toxic benthic freshwater cyanobacterial proliferations: challenges and solutions for enhancing knowledge and improving monitoring and mitigation. Freshw Biol 65(10):1824–1842. https://doi.org/10.1111/fwb.13532

Wu X, Zhang Y (2017) TET-mediated active DNA demethylation: mechanism, function and beyond. Nat Rev Genet 18(9):517–534. https://doi.org/10.1038/nrg.2017.33

Wu JL, Liu WX, Wen CG, Qian GM, Hu BQ, Jian SQ, Yang G, Dong J (2020) Effect of microcystin on the expression of Nrf2 and its downstream antioxidant genes from *Cristaria plicata*. Aquat Toxicol 225. https://doi.org/10.1016/j.aquatox.2020.105526

Wu Q, Gong A, Liu X, Hou J, Liu H, Yang Z, Zhu Y (2024) Probiotics alleviate microcystin-LR-induced developmental toxicity in zebrafish larvae. Toxics 12(7). https://doi.org/10.3390/toxics12070527

Xie L, Xie P, Ozawa K, Honma T, Yokoyama A, Park HD (2004) Dynamics of microcystins-LR and -RR in the phytoplanktivorous silver carp in a sub-chronic toxicity experiment. Environ Pollut 127(3):431–439. https://doi.org/10.1016/j.envpol.2003.08.011

Xinyao L, Miao S, Yonghong L, Yin G, Zhongkai Z, Dong-hui W, Weizhong W, Chencai A (2006) Feeding characteristics of an amoeba (*Lobosea*: *Naegleria*) grazing upon cyanobacteria: food selection, ingestion and digestion progress. Microb Ecol 51:315–325. https://doi.org/10.1007/s00248-006-9031-2

Xu YY, Liang JJ, Yang WD, Wang J, Li HY, Liu JS (2014) Cloning and expression analysis of P-glycoprotein gene in *Crassostrea ariakensis*. Aquaculture 418-419:39–47. https://doi.org/10.1016/j.aquaculture.2013.10.004

Xu L, Li T, Ding W, Cao Y, Ge X, Wang Y (2018a) Combined seven miRNAs for early hepatocellular carcinoma detection with chronic low-dose exposure to microcystin-LR in mice. Sci Total Environ 628–629:271–281. https://doi.org/10.1016/j.scitotenv.2018.02.021

Xu X, Zhang Y-N, Peng S, Wu J, Deng D, Zhou Z (2018b) Effects of *Microcystis aeruginosa* on the expression of nuclear receptor genes in *Daphnia similoides sinensis*. Ecotoxicol Environ Saf 150:344–352. https://doi.org/10.1016/j.ecoenv.2017.12.033

Xu R, Jiang Y, MacIsaac HJ, Chen L, Li J, Xu J, Wang T, Zi Y, Chang X (2019) Blooming cyanobacteria alter water flea reproduction via exudates of estrogen analogues. Sci Total Environ 696. https://doi.org/10.1016/j.scitotenv.2019.133909

Xu W, Li X, Li Y, Sun Y, Zhang L, Huang Y, Yang Z (2021) Rising temperature more strongly promotes low-abundance *Paramecium* to remove *Microcystis* and degrade microcystins. Environ Pollut 291. https://doi.org/10.1016/j.envpol.2021.118143

Xu X, Cao Y, Qi H, Deng D, Zhang YN, Wu J, Peng S, Zhou Z (2022) Effects of toxic *Microcystis aeruginosa* on the expression of Hox genes in *Daphnia similoides sinensis*. Ecol Evol 12(3). https://doi.org/10.1002/ece3.8685

Xue Q, Su X, Steinman AD, Cai Y, Zhao Y, Xie L (2016) Accumulation of microcystins in a dominant chironomid larvae (*Tanypus chinensis*) of a large, shallow and eutrophic Chinese Lake, Lake Taihu. Sci Rep 6(1):31097. https://doi.org/10.1038/srep31097

Xun X, Cheng J, Wang J, Li Y, Li X, Li M, Lou J, Kong Y, Bao Z, Hu X (2020) Solute carriers in scallop genome: gene expansion and expression regulation after exposure to toxic dinoflagellate. Chemosphere 241. https://doi.org/10.1016/j.chemosphere.2019.124968

Yan H, Zhang S, Li C-Z, Chen Y-H, Chen Y-G, Weng S-P, He J-G (2013) Molecular characterization and function of a p38 MAPK gene from *Litopenaeus vannamei*. Fish Shellfish Immunol 34(6):1421–1431. https://doi.org/10.1016/j.fsi.2013.02.030

Yang B, Zhai S, Zhang F, Wang H, Ren L, Li Y, Li Q, Liu S (2022) Genome-wide association study toward efficient selection breeding of resistance to *vibrio alginolyticus* in Pacific oyster, *Crassostrea gigas*. Aquaculture 548. https://doi.org/10.1016/j.aquaculture.2021.737592

Yang X, Zhu J, Hu C, Yang W, Zheng Z (2023a) Integration of transcriptomics and microbiomics reveals the responses of *Bellamya aeruginosa* to toxic cyanobacteria. Toxins 15(2):119. https://doi.org/10.3390/toxins15020119

Yang Y, Wen C, Zheng S, Song F, Liu Y, Yao X, Tang Y, Feng X, Chen J, Yang F (2023b) *Lactobacillus fermentum* alleviates the colorectal inflammation induced by low-dose sub-chronic microcystin-LR exposure. Toxins 15(9). https://doi.org/10.3390/toxins15090579

Yang J, Xu Z, Chen Y, Xu H, Gao Z, Zhang X, Pang M, Zhang S, Liu H (2024) Community changes of gut microbes highlight their importance in the adaptation of copepods to toxic dinoflagellates. Front Mar Sci 11. https://doi.org/10.3389/fmars.2024.1368315

Ye MH, Li DW, Cai QD, Jiao YH, Liu Y, Li HY, Yang WD (2022) Toxic responses of different shellfish species after exposure to *Prorocentrum lima*, a DSP toxins producing dinoflagellate. Toxins 14(7). https://doi.org/10.3390/toxins14070461

Yuan KK, Chen ZM, Liu YX, Li HY, Yang WD (2023) Possible role of docosahexaenoic acid in response to diarrhetic shellfish toxins in the mussel *Perna viridis*. Mar Drugs 21(3). https://doi.org/10.3390/md21030155

Yuan KK, Liu YX, Yang XL, Liu YJ, Li HY, Yang WD (2025) Responses of nuclear receptor HR96 to the toxic dinoflagellate *Prorocentrum lima* in *Crassostrea ariakensis*. Ecotoxicol Environ Saf 292. https://doi.org/10.1016/j.ecoenv.2025.117909

Zhang X, Ji W, Zhang H, Zhang W, Xie P (2011) Studies on the toxic effects of microcystin-LR on the zebrafish (*Danio rerio*) under different temperatures. J Appl Toxicol 31(6):561–567. https://doi.org/10.1002/jat.1597

Zhang G, Fang X, Guo X, Li L, Luo R, Xu F, Yang P, Zhang L, Wang X, Qi H, Xiong Z, Que H, Xie Y, Holland PWH, Paps J, Zhu Y, Wu F, Chen Y, Wang J, Peng C, Meng J, Yang L, Liu J, Wen B, Zhang N, Huang Z, Zhu Q, Feng Y, Mount A, Hedgecock D, Xu Z, Liu Y, Domazet-Loso T, Du Y, Sun X, Zhang S, Liu B, Cheng P, Jiang X, Li J, Fan D, Wang W, Fu W, Wang T, Wang B, Zhang J, Peng Z, Li Y, Li N, Wang J, Chen M, He Y, Tan F, Song X, Zheng Q, Huang R, Yang H, Du X, Chen L, Yang M, Gaffney PM, Wang S, Luo L, She Z, Ming Y, Huang W, Zhang S, Huang B, Zhang Y, Qu T, Ni P, Miao G, Wang J, Wang Q, Steinberg CEW, Wang H, Li N, Qian L, Zhang G, Li Y, Yang H, Liu X, Wang J, Yin Y, Wang J (2012) The oyster genome reveals stress adaptation and complexity of shell formation. Nature 490(7418):49–54. https://doi.org/10.1038/nature11413

Zhang Z, Zhang XX, Wu B, Yin J, Yu Y, Yang L (2016) Comprehensive insights into microcystin-LR effects on hepatic lipid metabolism using cross-omics technologies. J Hazard Mater 315:126–134. https://doi.org/10.1016/j.jhazmat.2016.05.011

Zhang L, Gu L, Hou X, Kong Q, Chen K, Zhu X, Huang Y, Chen Y, Yang Z (2018a) Chlorophytes prolong mixotrophic *Ochromonas* eliminating *Microcystis*: temperature-dependent effect. Sci Total Environ 639:705–713. https://doi.org/10.1016/j.scitotenv.2018.05.196

Zhang L, Lyu K, Wang N, Gu L, Sun Y, Zhu X, Wang J, Huang Y, Yang Z (2018b) Transcriptomic analysis reveals the pathways associated with resisting and degrading microcystin in *Ochromonas*. Environ Sci Technol 52(19):11102–11113. https://doi.org/10.1021/acs.est.8b03106

Zhang L, Wang Z, Wang N, Gu L, Sun Y, Huang Y, Chen Y, Yang Z (2020a) Mixotrophic *Ochromonas* addition improves the harmful *Microcystis*-dominated phytoplankton community in *in situ* microcosms. Environ Sci Technol 54(7):4609–4620. https://doi.org/10.1021/acs.est.9b06438

Zhang Y, Li Z, Kholodkevich S, Sharov A, Feng Y, Ren N, Sun K (2020b) Microcystin-LR-induced changes of hepatopancreatic transcriptome, intestinal microbiota, and histopathology of freshwater crayfish (*Procambarus clarkii*). Sci Total Environ 711. https://doi.org/10.1016/j.scitotenv.2019.134549

Zhang J, Yu M, Gao Y, Zhang M, Dong J, Li M, Li X (2023a) Feeding behavior, microcystin accumulation, biochemical response, and ultramicrostructure changes in edible freshwater bivalve *Corbicula fluminea* exposed to *Microcystis aeruginosa*. Environ Sci Pollut Res 30(5):13560–13570. https://doi.org/10.1007/s11356-022-22833-z

Zhang J, Yu M, Zhang Z, Zhang M, Gao Y, Dong J, Zhou C, Li X (2023b) Integrating regular and transcriptomic analyses reveal resistance mechanisms in *Corbicula fluminea* (Müller, 1774) in response to toxic *Microcystis aeruginosa* exposure. Ecotoxicol Environ Saf 251:114553. https://doi.org/10.1016/j.ecoenv.2023.114553

Zhao M, Xie S, Zhu X, Yang Y, Gan N, Song L (2006) Effect of dietary cyanobacteria on growth and accumulation of microcystins in Nile tilapia (*Oreochromis niloticus*). Aquaculture 261(3):960–966. https://doi.org/10.1016/j.aquaculture.2006.08.019

Zhao J, Tian F, Zhai Q, Yu R, Zhang H, Gu Z, Chen W (2017) Protective effects of a cocktail of lactic acid bacteria on microcystin-LR-induced hepatotoxicity and oxidative damage in BALB/c mice. RSC Adv 7(33):20480–20487. https://doi.org/10.1039/c7ra03035e

Zhao J, Yuan Q, Lei X, Lieke T, Liu Y, Steinberg CEW, Pan B, Xing B (2025) The double-edged environmental effect of dissolved organic matter in global climate change. Environmental and Biogeochemical Processes 1(1). https://doi.org/10.48130/ebp-0025-0009

Zheng Y, Dam HG, Avery DE (2011) Differential responses of populations of the copepod *Acartia hudsonica* to toxic and nutritionally insufficient food algae. Harmful Algae 10(6):723–731. https://doi.org/10.1016/j.hal.2011.06.003

Zhu J, Lu K, Zhang C, Liang J, Hu Z (2011) Biochemical and ultrastructural changes in the hepatopancreas of *Bellamya aeruginosa* (Gastropoda) fed with toxic cyanobacteria. TheScientificWorldJOURNAL 11:2091–2105. https://doi.org/10.1100/2011/402326

Glossary

1.1 a

5-mC 5-methylcytosine

β-oxidation catabolic process by which fatty acid molecules are broken down in the mitochondria in eukaryotes to generate acetyl-CoA, which enters the citric acid cycle, and NADH and $FADH_2$, which are coenzymes used in the electron transport chain (Houten and Wanders 2010)

δ-2-catenin gene involved in development and differentiation (Ruocco et al. 2019)

1.2 a

ACH_{50} alternative complement pathway

ACP acid phosphatase; dephosphorylates compounds; it has an acidic pH optimum

AKT also known as protein kinase B (PKB), is the collective name of a set of three serine/threonine-specific protein kinases that play key roles in multiple cellular processes such as glucose metabolism, apoptosis, cell proliferation, transcription, and cell migration

Akt signaling pathway or PI3K-Akt signaling pathway is a signal transduction pathway that promotes survival and growth in response to extracellular signals. Key proteins involved are PI3K (phosphatidylinositol 3-kinase) and Akt (protein kinase B). Activated Akt mediates downstream responses, including cell survival, growth, proliferation, cell migration, and angiogenesis, by phosphorylating a range of intracellular proteins. The pathway is present in all cells of higher eukaryotes and is highly conserved (Manning and Cantley 2007)

ALP alkaline phosphatase dephosphorylates compounds; it has an alkaline pH optimum

AMP adenosine monophosphate

AMPK adenosine monophosphate–activated protein kinase, an enzyme central in cellular energy homeostasis. The AMPK and TOR (target-of-rapamycin) pathways are interlinked, opposing signaling pathways involved in sensing the availability of nutrients and energy and regulation of cell growth. AMPK (Yin, or

T. Lieke, C. E. W. Steinberg, *The Ecology of Phycotoxins*,
https://doi.org/10.1007/978-3-032-18380-4

the "dark side") is switched on by a lack of energy or nutrients and inhibits cell growth, while TOR (Yang, or the "bright side") is switched on by nutrient availability and promotes cell growth (González et al. 2020).

It has been shown that AMPK can activate silent information regulator 1 (SIRT1), an NAD-dependent deacetylase, which is a key regulator of energy metabolism through the regulation of transcription factors and coregulators via deacetylation (Shi et al. 2018).

ANFs autotrophic nanoflagellates

ASV variants of amplicon sequences; amplicons are DNA fragments produced by PCR and then used for next-generation sequencing

ATP adenosine triphosphate, an organic compound and hydrotrope that provides energy to drive many processes in living cells

Autophagy natural, regulated, destructive mechanism of the cell that disassembles unnecessary or dysfunctional components; key to maintaining cellular homeostasis (Guo et al. 2021)

1.3 b

BMP5–7 gene involved in skeletogenesis (Ruocco et al. 2018)

BW, bw, bwt body weight

1.4 c

C3 complement component 3, a protein of the immune system, central to the activation of the complement system; a connecting link between innate and acquired immunity (Sahu and Lambris 2001)

caspase-3…12 members of the cysteine–aspartic acid protease (caspase) family: Sequential activation of caspases plays a central role in the execution phase of cell apoptosis

CAT catalase: an enzyme that catalyzes the dismutation of hydrogen peroxide (H_2O_2) to water (H_2O) and oxygen (O_2)

Claudins transmembrane proteins that are the most important components of tight junctions

c-myc →*Myc*

Complement system part of the innate immune system that enhances (complements) the ability of antibodies and phagocytic cells to clear microbes and damaged cells from an organism, promotes inflammation, and attacks the pathogen's cell membrane

CpG cytosine phosphate–guanine dinucleotides

crustin encodes an antimicrobial peptide (Vargas-Albores et al. 2004)

C-type lectins a type of lectins with antiviral and antibacterial activity, found in *Penaeus vannamei*

CYP1A Cytochrome P450, family 1, subfamily A, is involved in phase I xenobiotic and drug metabolism
Cytokine cell signaling protein

1.5 d

DEG differentially expressed gene
DHA docosahexaenoic acid, ω3 fatty acid with the chemical formula $C_{22}H_{32}O_2$
Dolastatins especially dolastatin 10; tubulin-targeting anticancer peptides (Nandi et al. 2024), produced by marine and freshwater cyanobacteria (Kallifidas et al. 2024)
Dysbiosis microbial imbalance or maladaptation

1.6 e

EPA eicosapentaenoic acid, ω3 fatty acid with the chemical formula $C_{20}H_{30}O_2$
ER endoplasmic reticulum

1.7 f

FoxOs forkhead box proteins, a family of transcription factors regulating the expression of genes involved in cell growth, proliferation, differentiation, and longevity

1.8 g

GABA γ-aminobutyric acid, a non-proteinogenic amino acid: an inhibitory neurotransmitter
Gluconeogenesis synthesis of glucose from certain non-carbohydrate carbon substrates, such as pyruvate, lactate, or amino acids
Glycogen multibranched polysaccharide of glucose that serves as a form of energy storage in animals
Glycolysis sequence of reactions that metabolizes one molecule of glucose to two molecules of pyruvate with the concomitant net production of two molecules of ATP
GO gene ontology
GOT glutamate oxaloacetate transaminase or aspartate transaminase (→AST)
GPT glutamate pyruvate transaminase (→ALT)
GPx, GPX glutathione peroxidase
GR glutathione reductase
GSH glutathione
GST glutathione transferase

1.9 h

H3K4 histone
HAB harmful algal bloom
HIFs hypoxia-inducible factors: transcription factors that respond to decreases in available oxygen in the cellular environment, or hypoxia (Smith et al. 2008)
HNF heterotrophic nanoflagellate
hpf hours past fertilization
HSPs heat shock proteins: a family of proteins produced by cells in response to exposure to stressful conditions. The number indicates the approximate atomic mass in kDa. Many HSPs perform chaperone functions

1.10 i

IFN-γ interferon-γ, an immunostimulatory cytokine critical for innate and adaptive immunity against viral, some bacterial, and protozoal infections
IgM immunoglobulin M
IL-1 interleukin 1: a family of 11 cytokines, inducing a complex network of pro-inflammatory cytokines
IL-10 interleukin 10: an anti-inflammatory cytokine
IL-17 interleukin 17: a pro-inflammatory cytokine family produced by T-helper cells and induced by IL-23; can be induced in hemocytes after bacterial challenge
IL-1β interleukin 1β: a pro-inflammatory cytokine, involved in cell proliferation, differentiation, and apoptosis
IL-6 interleukin 6 acts as both a pro-inflammatory cytokine and an anti-inflammatory myokine
IL-8 interleukin 8: a pro-inflammatory cytokine

1.11 j

JNK c-Jun N-terminal protein kinases are a mitogen-activated protein kinase family and are responsive to stress stimuli, such as cytokines, ultraviolet irradiation, heat shock, and osmotic shock. For instance, JNK1 is involved in apoptosis, neurodegeneration, cell differentiation and proliferation, inflammatory conditions, and cytokine production mediated by AP-1 (activation protein 1) (Oltmanns et al. 2003)
JUN belongs to a family of basic leucine zipper transcription factors that are evolutionarily conserved; in the sea urchin identified as a component of the gene regulatory network controlling skeletogenesis (Russo et al. 2014)

1.12 k

Keap1-Nrf2 major regulator of cytoprotective responses to oxidative and electrophilic stress

KEGG Kyoto Encyclopedia of Genes and Genomes, a database resource for understanding high-level functions and utilities of biological system

1.13 l

LAB lactic acid bacteria

LC-PUFA long-chain polyunsaturated fatty acid

Lectins carbohydrate-binding proteins; highly specific for sugar moieties, performing recognition on the cellular and molecular levels. Lectins also mediate attachment and binding of bacteria and viruses to their intended targets (Takeuchi and Akira 2010)

LIN often also abbreviated as LA, linoleic acid, a fatty acid: 18:2ω6

LPO lipid peroxidation

LPS Lipopolysaccharides (also lipoglycans, endotoxins): large molecules consisting of lipids and polysaccharides; found in Gram-negative bacteria and cyanobacteria, and elicit strong immune responses in animals

LYZ lysozyme

1.14 m

MAPK or MAP kinase is a type of protein kinase that is specific to the amino acids serine and threonine (i.e., a serine/threonine-specific protein kinase). MAPKs are involved in directing cellular responses to a diverse array of stimuli, such as mitogens, osmotic stress, heat shock, and pro-inflammatory cytokines. They regulate cell functions including proliferation, gene expression, differentiation, mitosis, cell survival, and apoptosis (Pearson et al. 2001)

p38 MAPK p38 mitogen-activated protein kinases are a class of mitogen-activated protein kinases (MAPKs). Persistent activation of the p38 MAPK pathway in muscle satellite cells (muscle stem cells) due to ageing impairs muscle regeneration (Cosgrove et al. 2014)

MC microcystin, a group of heptapeptide toxins from cyanobacteria

MDA Malondialdehyde, a marker for oxidative stress, particularly lipid peroxidation (Davey et al. 2005)

Microbiota resident microbial communities in fishes and invertebrates

Microcystinase a protease that selectively degrades microcystin into a linear peptide, which is 160 times less toxic (Somdee et al. 2013)

microRNA, miRNA, miR noncoding RNAs, about 22 nucleotides long, functioning in RNA silencing and posttranscriptional regulation of gene expression

Mn-SOD →SOD2

MUFA monounsaturated fatty acid containing one ethylenic bond

myd88 encodes the myeloid differentiation factor 88, a pivotal signaling component of the innate immune response, serving as an adaptor for the interleukin 1 receptor and the majority of TLRs (Takeda and Akira 2005)

1.15 n

NAD nicotinamide adenine dinucleotide

Na^+/K^+-ATPase sodium–potassium adenosine triphosphatase central to ion homeostasis. For every ATP molecule that the pump uses, three sodium ions are exported and two potassium ions are imported; involved in controlling perturbations of Na^+ and K^+ homeostasis during apoptosis (Panayiotidis et al. 2006)

Necroptosis cell death resembling apoptosis and necrosis (Nirmala and Lopus 2020)

NF-κB nuclear transcription factor-κB: a protein complex that controls transcription of DNA, cytokine production, and cell survival; found in almost all animal cell types and involved in cellular responses to stimuli such as stress, cytokines, free radicals, ultraviolet irradiation, oxidized low-density lipoprotein, and bacterial or viral antigens

NO nitric oxide

NPY neuropeptide Y: with slight variations, a neurotransmitter in the brain and in the autonomic nervous system of animals; one of the strongest orexigenic signals

Nrf2 nuclear factor erythroid 2-related factor 2: master regulator of the antioxidant response, which is important for the mitigation of oxidative stress (Nguyen et al. 2009)

1.16 o

Occludin integral plasma-membrane protein located at tight junctions

OTA ochratoxin

Oxidative burst →respiratory burst

Oxidative stress states where the balance between generation and elimination of ROS is disturbed in favor of the generation of ROS

1.17 p

p38 encodes p38 mitogen-activated protein kinase; is activated by a variety of cellular stresses (Cuadrado and Nebreda 2010)

p53 encodes tumor protein p53, which prevents cancer formation by inducing apoptosis, and functions as a tumor suppressor (Wu et al. 2019)

PCR polymerase chain reaction

Periphyton aquatic organisms growing on macrophytes or macroalgae

peroxiredoxin a type of peroxidase, causes resistance to oxidative stress and participates in immune responses (Wu et al. 2017)
PIK3R1 Phosphatidylinositol 3-kinase regulatory subunit 1, plays an important role in the metabolic actions of insulin (in humans) (https://www.ncbi.nlm.nih.gov/gene/18708 accessed July 26, 2024)
PI protease inhibitor (=antiprotease)
PI3K/AKT/mTOR signaling Intracellular signaling pathway important in regulating the cell cycle, directly related to cellular quiescence, proliferation, cancer, and longevity. PI3K activation phosphorylates and activates AKT (protein kinase B), localizing it in the plasma membrane (King et al. 2015)
Polyketides class of diverse secondary metabolites imparting survival advantages
PPARs peroxisome proliferator-activated receptors; nuclear receptor proteins functioning as transcription factors. They are central in the regulation of cellular differentiation, development, and metabolism (carbohydrate, lipid, protein)
proPO prophenoloxidase: part of the major innate defense system in invertebrates via melanization of pathogens and damaged tissues. The activated form, phenoloxidase, facilitates the melanization of pathogens and damaged tissues
psu practical salinity units
PUFA polyunsaturated fatty acid containing two or more ethylenic bonds, such as eicosapentaenoic acid (EPA, 20:5n − 3 or 20:5ω − 3)

1.18 q

Quercetin a flavonoid polyphenol

1.19 r

#Rab family proteins small G proteins, which generally possess a GTPase fold. GTPases regulate intracellular membrane trafficking (Homma et al. 2021)
Rapamycin also sirolimus; chemical immunouppressant
RAR Retinoic acid receptor: a type of nuclear receptor that can also act as a transcription factor
RAS recirculating aquaculture system
RB rice bran, consisting of the aleurone, pericarp, subaleurone, and germ, making up approximately 10% of the total weight of raw rice
Respiratory burst rapid production and release of reactive oxygen species
RhoA/ROCK signaling pathway The Rho GTPase/Rho kinase (ROCK) signaling pathway plays an important role in the modulation of cell cytoskeleton by activating its effector molecules, ROCK1 and ROCK2, and regulating diverse signaling effectors and pathophysiological processes (e.g., cell division, cell migration, wound healing, or immune surveillance) (Hauke et al. 2022).
rig-I retinoic *acid-inducible gene I,* an antiviral-related gene

RNA-seq high-throughput transcriptome sequencing, an efficacious first-step methodology to map the most important pathways in a physiological response
ROS reactive oxygen species: chemically reactive molecules containing oxygen, such as peroxides, superoxide, hydroxyl radical, and singlet oxygen (also see electrophilic stress and oxidative stress)
RXR retinoid receptors: nuclear receptors that bind to retinoids; when bound to a retinoid, they act as transcription factors (Rowe 1997)

1.20 s

SAA serum amyloid A, expressed in response to inflammatory stimuli
SBM soybean meal
SCFA short-chain fatty acid, also referred to as volatile fatty acid (VFA): fatty acid with 2–6 C atoms
Septicemia serious bloodstream infection, also known as blood poisoning
SFA also abbreviated SAFA: saturated fatty acid; fatty acids without C=C double bonds
SGR specific growth rate
SLC15 alias H^+-coupled oligopeptide cotransporter family
SMADs structurally similar proteins that are the main signal transducers for receptors of TGF-β superfamily
SOD superoxide dismutase: enzyme that catalyzes the dismutation (= disproportionation) of superoxide ($^{\cdot}O_2^-$) radicals into ordinary molecular oxygen (O_2) and hydrogen peroxide (H_2O_2)
SOD1 Cu-Zn-superoxide dismutase = superoxide dismutase 1
SOD2 Mn-superoxide dismutase = superoxide dismutase 2
spheroplasts cells from which the cell walls have been almost completely removed
Splicing newly made precursor messenger RNA (pre-mRNA) transcripts are transformed into mature messenger RNAs (mRNAs)
srebp1, 2 encodes sterol-regulated element-binding proteins-1, 2 (SREBP-1, 2), a transcriptional activator required for lipid homeostasis; SREBPs are master regulators of lipid homeostasis (Eberlé et al. 2004); its expression is regulated by insulin (Ferré and Foufelle 2010)
STAT1 Signal transducer and activator of transcription 1, can be activated by several ligands such as IFNα, IFNγ, platelet-derived growth factor (PDGF), IL-6, or IL-27 (Baris et al. 2016)

1.21 t

T-2 T-2 toxin
TAK1 transforming growth factor-β-activated kinase 1, a central regulator of cell death and is activated through a diverse set of intra- and extracellular stimuli (Mihaly et al. 2014)

T-cell receptor signaling T-cell receptors (TCR) play a key role in the functioning of T cells and the formation of the immunological synapse. It provides a connection between the T cell and the antigen-presenting cell. The TCR activation promotes a number of signaling cascades that ultimately determine cell fate through regulating cytokine production, cell survival, proliferation, and differentiation (Huse 2009).

T cell T lymphocyte: a type of lymphocyte central to cell-mediated immunity (also see B cells)

T_3 thyroid hormone; more active than T_4 by a factor of 3 to 5

T_4 tyroxine

TCM Traditional Chinese Medicine

TCTP translationally controlled tumor protein, a multifunctional protein that plays essential roles in cell growth, cell cycle progression, and antiapoptotic activity (Lee et al. 2020)

TET enzymes family of ten-eleven translocation (TET) methylcytosine dioxygenases. They are instrumental in DNA demethylation. 5-Methylcytosine is a methylated form of the DNA base cytosine that often regulates gene transcription and has several other functions in the genome (Wu and Zhang 2017)

TF Transferrins are glycoproteins found in vertebrates that bind and consequently mediate the transport of iron (Fe) through blood plasma (Crichton and Charloteaux-Wauters 1987). They are produced in the liver and contain binding sites for two Fe^{3+} ions (Hall et al. 2002).

TG, TAG triglyceride or triacylglycerol: ester derived from glycerol and three fatty acids

TGF-β transforming growth factor β: multifunctional cytokine. The TGF-β superfamily includes endogenous growth-inhibiting proteins with, e.g., anti-inflammatory function

ThQ thymoquinone

Thy thymol

TJ tight junction: an essential component of the differentiated epithelial cell required for polarized transport, intercellular integrity, and signaling. TJs regulate the paracellular pathway for the movement of ions and solutes between cells (Eckert and Fleming 2008)

TLRs toll-like receptors, a class of proteins that play a key role in the innate immune system by recognizing molecules that are broadly shared by pathogens. Once pathogen microbes have breached physical barriers such as the skin or intestinal tract mucosa, they are recognized by TLRs, which activate immune cell responses.

TNF-α tumor necrosis factor α: a pro-inflammatory cytokine, able to induce fever, apoptotic cell death, cachexia, inflammation, and to inhibit tumorigenesis. Moreover, involved in the regulation of a wide spectrum of biological processes, including cell proliferation, differentiation, apoptosis, lipid metabolism, and coagulation

toll encodes members of the toll-like receptor class of proteins, a gene upstream of the NF-κB pathway (Kaleo et al. 2019)

TOR target of rapamycin: highly conserved, nutrient-sensitive protein kinase, a central controller of protein synthesis, cell growth, cell proliferation, cell motility, cell survival, autophagy, transcription, and aging under nutrient availability (see also AMPK) (González et al. 2020)

Trh thyrotropin-releasing hormone: a releasing hormone, produced by the hypothalamus that stimulates the release of thyrotropin

Trp tryptophan: one of the 22 standard amino acids and an essential in diets; distinguishing structural feature is the indole functional group

TYK2/STAT1 pathway Tyrosine kinase 2 is an intracellular enzyme that mediates immune signaling and inflammatory signaling pathways (Min et al. 2015). It contributes to the activation of STAT1, and activated STAT1, in turn, regulates the expression of over 300 IFN-stimulated genes

1.22 u

ubiquitinylation ubiquitin conjugation is a reversible posttranslational modification system that controls the function of key proteins involved in various physiological processes, such as protein degradation, cell cycle, apoptosis, DNA repair, and signal transduction, in all cell types (Hershko and Ciechanover 1998)

1.23 v

VSI viscerosomatic index = (total viscera weight/body weight) × 100

1.24 w

WBC white blood cells

WG weight gain

Wnt signaling pathways a group of signal transduction pathways that begin with proteins that pass signals into a cell through cell surface receptors. Wnt signaling pathways use either nearby cell–cell communication (paracrine) or same-cell communication (autocrine). They are highly evolutionarily conserved in animals, which means they are similar across animal species from fruit flies to humans (Nusse 2005).

1.25 x

xbp1 encodes recombinant X-Box binding protein 1, an important marker reflecting ER stress (Yung et al. 2007)

1.26 z

zo-1, *zo-2* encode zonula occludens-1, -2, tight junction proteins ZO-1, ZO-2: encode proteins located on the cytoplasmic membrane surface of intercellular tight junctions; together with the Claudin group of proteins, they are major components of the tight junctions (Fanning et al. 2012)

References

Baris S, Alroqi F, Kiykim A, Karakoc-Aydiner E, Ogulur I, Ozen A, Charbonnier L-M, Bakır M, Boztug K, Chatila TA, Barlan IB (2016) Severe early-onset combined immunodeficiency due to heterozygous gain-of-function mutations in STAT1. J Clin Immunol 36(7):641–648. https://doi.org/10.1007/s10875-016-0312-3

Cosgrove BD, Gilbert PM, Porpiglia E, Mourkioti F, Lee SP, Corbel SY, Llewellyn ME, Delp SL, Blau HM (2014) Rejuvenation of the muscle stem cell population restores strength to injured aged muscles. Nat Med 20(3):255–264. https://doi.org/10.1038/nm.3464

Crichton RR, Charloteaux-Wauters M (1987) Iron transport and storage. Eur J Biochem 164(3):485–506. https://doi.org/10.1111/j.1432-1033.1987.tb11155.x

Cuadrado A, Nebreda AR (2010) Mechanisms and functions of p38 MAPK signalling. Biochem J 429(3):403–417. https://doi.org/10.1042/bj20100323

Davey MW, Stals E, Panis B, Keulemans J, Swennen RL (2005) High-throughput determination of malondialdehyde in plant tissues. Anal Biochem 347(2):201–207. https://doi.org/10.1016/j.ab.2005.09.041

Eberlé D, Hegarty B, Bossard P, Ferré P, Foufelle F (2004) SREBP transcription factors: master regulators of lipid homeostasis. Biochimie 86(11):839–848. https://doi.org/10.1016/j.biochi.2004.09.018

Eckert JJ, Fleming TP (2008) Tight junction biogenesis during early development. BBA-Biomembranes 1778(3):717–728. https://doi.org/10.1016/j.bbamem.2007.09.031

Fanning AS, Itallie CMV, Anderson JM (2012) Zonula occludens-1 and -2 regulate apical cell structure and the zonula adherens cytoskeleton in polarized epithelia. Mol Biol Cell 23(4):577–590. https://doi.org/10.1091/mbc.e11-09-0791

Ferré P, Foufelle F (2010) Hepatic steatosis: a role for de novo lipogenesis and the transcription factor SREBP-1c. Diabetes Obes Metab 12(s2):83–92. https://doi.org/10.1111/j.1463-1326.2010.01275.x

González A, Hall MN, Lin SC, Hardie DG (2020) AMPK and TOR: the Yin and Yang of cellular nutrient sensing and growth control. Cell Metab 31(3):472–492. https://doi.org/10.1016/j.cmet.2020.01.015

Guo YF, Su T, Yang M, Li CJ, Guo Q, Xiao Y, Huang Y, Liu Y, Luo XH (2021) The role of autophagy in bone homeostasis. J Cell Physiol 236(6):4152–4173. https://doi.org/10.1002/jcp.30111

Hall DR, Hadden JM, Leonard GA, Bailey S, Neu M, Winn M, Lindley PF (2002) The crystal and molecular structures of diferric porcine and rabbit serum transferrins at resolutions of 2.15 and 2.60 a, respectively. Acta Crystallogr D Biol Crystallogr 58 (Pt 1):70-80. https://doi.org/10.1107/s0907444901017309

Hauke M, Eckenstaler R, Ripperger A, Ender A, Braun H, Benndorf RA (2022) Active RhoA exerts an inhibitory effect on the homeostasis and angiogenic capacity of human endothelial cells. J Am Heart Assoc 11(12):e025119. https://doi.org/10.1161/JAHA.121.025119

Hershko A, Ciechanover A (1998) The ubiquitin system. Annu Rev Biochem 67:425–479. https://doi.org/10.1146/annurev.biochem.67.1.425

Homma Y, Hiragi S, Fukuda M (2021) Rab family of small GTPases: an updated view on their regulation and functions. FEBS J 288(1):36–55. https://doi.org/10.1111/febs.15453

Houten SM, Wanders RJA (2010) A general introduction to the biochemistry of mitochondrial fatty acid β-oxidation. J Inherit Metab Dis 33(5):469–477. https://doi.org/10.1007/s10545-010-9061-2

Huse M (2009) The T-cell-receptor signaling network. J Cell Sci 122(9):1269–1273. https://doi.org/10.1242/jcs.042762

Kaleo IV, Gao Q, Liu B, Sun C, Zhou Q, Zhang H, Shan F, Xiong Z, Bo L, Song C (2019) Effects of *Moringa oleifera* leaf extract on growth performance, physiological and immune response, and related immune gene expression of *Macrobrachium rosenbergii* with *Vibrio anguillarum* and ammonia stress. Fish Shellfish Immunol 89:603–613. https://doi.org/10.1016/j.fsi.2019.03.039

Kallifidas D, Dhakal D, Chen M, Chen QY, Kokkaliari S, Colon Rosa NA, Ratnayake R, Bruner SD, Paul VJ, Ding Y, Luesch H (2024) Biosynthesis of dolastatin 10 in marine cyanobacteria, a prototype for multiple approved cancer drugs. Org Lett 26(7):1321–1325. https://doi.org/10.1021/acs.orglett.3c04083

King D, Yeomanson D, Bryant HE (2015) PI3King the lock: targeting the PI3K/Akt/mTOR pathway as a novel therapeutic strategy in neuroblastoma. J Pediatr Hematol Oncol 37(4):245–251. https://doi.org/10.1097/mph.0000000000000329

Lee JS, Jang EH, Woo HA, Lee K (2020) Regulation of autophagy is a novel tumorigenesis-related activity of multifunctional translationally controlled tumor protein. Cells 9(1). https://doi.org/10.3390/cells9010257

Manning BD, Cantley LC (2007) AKT/PKB signaling: navigating downstream. Cell 129(7):1261–1274. https://doi.org/10.1016/j.cell.2007.06.009

Mihaly SR, Ninomiya-Tsuji J, Morioka S (2014) TAK1 control of cell death. Cell Death Differ 21(11):1667–1676. https://doi.org/10.1038/cdd.2014.123

Min X, Ungureanu D, Maxwell S, Hammarén H, Thibault S, Hillert EK, Ayres M, Greenfield B, Eksterowicz J, Gabel C, Walker N, Silvennoinen O, Wang Z (2015) Structural and functional characterization of the JH2 pseudokinase domain of JAK family tyrosine kinase 2 (TYK2). J Biol Chem 290(45):27261–27270. https://doi.org/10.1074/jbc.M115.672048

Nandi S, Khatua S, Nag A, Sen S, Chakraborty N, Naskar A, Acharya K, Mekky RH, Contreras MM, Calina D, Dini I, Sharifi-Rad J (2024) Dolastatins and their analogues present a compelling landscape of potential natural and synthetic anticancer drug candidates. Current Res Biotechnol 7:100167. https://doi.org/10.1016/j.crbiot.2023.100167

Nguyen T, Nioi P, Pickett CB (2009) The Nrf2-antioxidant response element signaling pathway and its activation by oxidative stress. J Biol Chem 284(20):13291–13295. https://doi.org/10.1074/jbc.R900010200

Nirmala JG, Lopus M (2020) Cell death mechanisms in eukaryotes. Cell Biol Toxicol 36(2):145–164. https://doi.org/10.1007/s10565-019-09496-2

Nusse R (2005) Wnt signaling in disease and in development. Cell Res 15(1):28–32. https://doi.org/10.1038/sj.cr.7290260

Oltmanns U, Issa R, Sukkar MB, John M, Chung KF (2003) Role of c-Jun N-terminal kinase in the induced release of GM-CSF, RANTES and IL-8 from human airway smooth muscle cells. Br J Pharmacol 139(6):1228–1234. https://doi.org/10.1038/sj.bjp.0705345

Panayiotidis MI, Bortner CD, Cidlowski JA (2006) On the mechanism of ionic regulation of apoptosis: would the Na+/K+-ATPase please stand up? Acta Physiol 187(1–2):205–215. https://doi.org/10.1111/j.1748-1716.2006.01562.x

Pearson G, Robinson F, Beers Gibson T, Xu BE, Karandikar M, Berman K, Cobb MH (2001) Mitogen-activated protein (MAP) kinase pathways: regulation and physiological functions. Endocr Rev 22(2):153–183. https://doi.org/10.1210/edrv.22.2.0428

Rowe A (1997) Retinoid X receptors. Int J Biochem Cell Biol 29(2):275–278. https://doi.org/10.1016/s1357-2725(96)00101-x

Ruocco N, Costantini S, Zupo V, Lauritano C, Caramiello D, Ianora A, Budillon A, Romano G, Nuzzo G, D'Ippolito G, Fontana A, Costantini M (2018) Toxigenic effects of two benthic diatoms upon grazing activity of the sea urchin: morphological, metabolomic and *de novo* transcriptomic analysis. Sci Rep 8(1):5622. https://doi.org/10.1038/s41598-018-24023-9

Ruocco N, Cavaccini V, Caramiello D, Ianora A, Fontana A, Zupo V, Costantini M (2019) Noxious effects of the benthic diatoms *Cocconeis scutellum* and *Diploneis* sp. on sea urchin development: morphological and de novo transcriptomic analysis. Harmful Algae 86:64–73. https://doi.org/10.1016/j.hal.2019.05.009

Russo R, Pinsino A, Costa C, Bonaventura R, Matranga V, Zito F (2014) The newly characterized *Pl*-Jun is specifically expressed in skeletogenic cells of the *Paracentrotus lividus* sea urchin embryo. FEBS J 281(17):3828–3843. https://doi.org/10.1111/febs.12911

Sahu A, Lambris JD (2001) Structure and biology of complement protein C3, a connecting link between innate and acquired immunity. Immunol Rev 180(1):35–48. https://doi.org/10.1034/j.1600-065X.2001.1800103.x

Shi HJ, Xu C, Liu MY, Wang BK, Liu WB, Chen DH, Zhang L, Xu CY, Li XF (2018) Resveratrol improves the energy sensing and glycolipid metabolism of blunt snout bream *Megalobrama amblycephala* fed high-carbohydrate diets by activating the AMPK-SIRT1-PGC-1α network. Front Physiol 9(Sep):1258. https://doi.org/10.3389/fphys.2018.01258

Smith TG, Robbins PA, Ratcliffe PJ (2008) The human side of hypoxia-inducible factor. Br J Haematol 141(3):325–334. https://doi.org/10.1111/j.1365-2141.2008.07029.x

Somdee T, Thunders M, Ruck J, Lys I, Allison M, Pagwe R (2013) Degradation of [Dha^7]MC-LR by a microcystin degrading bacterium isolated from Lake Rotoiti, New Zealand. Intl Schol Res Not 2013(1):596429. https://doi.org/10.1155/2013/596429

Takeda K, Akira S (2005) Toll-like receptors in innate immunity. Int Immunol 17(1):1–14. https://doi.org/10.1093/intimm/dxh186

Takeuchi O, Akira S (2010) Pattern recognition receptors and inflammation. Cell 140(6):805–820. https://doi.org/10.1016/j.cell.2010.01.022

Vargas-Albores F, Yepiz-Plascencia G, Jiménez-Vega F, Ávila-Villa A (2004) Structural and functional differences of *Litopenaeus vannamei* crustins. Comp Biochem Phys B 138(4):415–422. https://doi.org/10.1016/j.cbpc.2004.05.007

Wu X, Zhang Y (2017) TET-mediated active DNA demethylation: mechanism, function and beyond. Nat Rev Genet 18(9):517–534. https://doi.org/10.1038/nrg.2017.33

Wu L, Zhou Y, Abbas MN, Kausar S, Chen Q, Jiang CX, Dai LS (2017) Molecular structure and functional characterization of the peroxiredoxin 5 in *Procambarus clarkii* following LPS and Poly I:C challenge. Fish Shellfish Immunol 71:28–34. https://doi.org/10.1016/j.fsi.2017.09.072

Wu J, Liu H, Huang H, Yuan L, Liu C, Wang Y, Cheng X, Zhuang D, Xu M, Chen X, Losiewicz MD, Zhang H (2019) *p53*-dependent pathway and the opening of mPTP mediate the apoptosis of co-cultured Sertoli-germ cells induced by microcystin-LR. Environ Toxicol 34(10):1074–1084. https://doi.org/10.1002/tox.22808

Yung HW, Korolchuk S, Tolkovsky AM, Charnock-Jones DS, Burton GJ (2007) Endoplasmic reticulum stress exacerbates ischemia-reperfusion-induced apoptosis through attenuation of Akt protein synthesis in human choriocarcinoma cells. FASEB J 21(3):872–884. https://doi.org/10.1096/fj.06-6054com

The manufacturer's authorised representative in the EU is Springer Nature Customer Service Centre GmbH, Europaplatz 3, 69115 Heidelberg, Germany. If you have any concerns regarding our products, please contact ProductSafety@springernature.com

Printed and bound by CPI Group (UK) Ltd, Croydon, CR0 4YY
17/09/2026
02218457-0002